OVERHEAD CONDUCTOR

MANUAL 2ND EDITION

Southwire Company Overhead Conductor Manual
Second Edition

Edited by:

Ridley Thrash,
Chief Engineer, Energy Division

Amy Murrah,
Product Engineer, Energy Division

Mark Lancaster,
Sr. Product Engineer, Energy Division

Kim Nuckles,
Sr. Product Engineer, Energy Division

Southwire Company
One Southwire Drive
Carrollton, Georgia 30119
800-444-1700
www.southwire.com

Southwire®

SECOND EDITION/FIRST PRINTING

ISBN-13: 978-0-9791427-0-3
ISBN-10: 0-9791427-0-9

Southwire also acknowledges and appreciates the receipt of information from the following organizations. Individual references are contained within the text.

Alcoa Conductor Products Company
ACA Conductor Accessories
The Aluminum Association
American Society of Civil Engineers
American Society of Testing and Materials
Canadian Electrical Association
Electric Power Research Institute
Fargo Manufacturing Company
Hubbell Power Systems, Inc.
Institute of Electrical and Electronics Engineers, Inc.
McGraw-Hill
National Electrical Safety Code®
National Oceanographic and Atmospheric Administration
Power Delivery Consultants, Inc.
Preformed Line Products
United States Department of Agriculture, Rural Utilities Service

INTRODUCTION

This manual was written to assist transmission line engineers and system planners in the selection of overhead conductors and to provide general guidance in transmission line design and/or uprating. It is impossible to fully discuss all conductor alternatives or to provide detailed design, installation and operating data that will cover all applications. Instead, the intent of this manual is to consolidate the engineering reference data most used by line design engineers into a single source with guidance to finding additional, more in-depth information in related reference material. The information presented in the manual should be interpreted as general guidelines only and is intended for users who have an understanding of the engineering fundamentals involved in the design and operation of transmission lines. Further details or assistance are available from the references listed at the end of each chapter or from Southwire Company.

The manual consists of two basic sections – conductor-related information and line design basics. The first three chapters present an overview of conductor characteristics, installation considerations, and thermal rating approaches for overhead conductors; Chapters 4 and 5 discuss the electrical environmental effects of overhead line and the basic design requirements for new lines, while Chapters 6 and 7 discuss the options and design constraints when considering voltage and/or uprating of existing lines.

The design, operational and uprating requirements for overhead transmission lines are constantly changing. It is Southwire's intent to be part of this change. Therefore, we would appreciate hearing from you with your comments and suggestions and how we can improve the manual for future releases.

The overhead products discussed in the manual are only a small sampling of the products manufactured by Southwire Company. For information on new products or any of your wire and cable requirements, please contact your local Southwire representative or call Southwire directly.

ACKNOWLEDGEMENTS

We would like to thank those who have contributed in the preparation of this manual. We want to acknowledge the editorial contribution of Power Delivery Consultants, Inc. (PDC) of Ballston Lake, NY, with whom we contracted for assistance and technical guidance in preparation of the manual. In particular we would like to thank Dale A. Douglass and James R. Stewart who worked with Southwire to develop the basic manuscript for the manual. Their support, patience and knowledge were invaluable. Additionally, we relied heavily on the editing skills of Gerald Hudson. His attention to detail helped immensely in preparing the manuscript for publication.

We remain indebted to the significant efforts extended by the authors of the first edition, Ridley Thrash, Gerald Hudson, Dave Cooper, and Gene Sanders. Their technical proficiency and editorial insight provided the basis for our work, without which we could not have undertaken such a large task. We would also like to thank our colleagues in Southwire's Energy Division for their help and support, as well as their contributions and assistance.

TABLE OF CONTENTS

1

2

3

4

5

6

7

1 INTRODUCTION TO BARE OVERHEAD CONDUCTORS

In the line design process, the most critical decision often involves selecting the phase and ground conductors. At higher transmission voltages, there may be as many as four conductors for each electrical phase of each circuit. To divert lightning strokes to ground, one or more unenergized ground conductors are usually employed.

Conductors usually comprise 20% to 30% of the material and labor costs of a line. The selection of conductors is critical since the wind, ice and tension loads that supporting structures must withstand are dependent on the size and type of conductors used. Structure heights are also largely dependent on the conductors selected since maximum sag of phase conductors under maximum ice or maximum temperature conditions is governed by certain physical, mechanical and dimensional properties of the conductor. Even environmental effects, such as radio noise and audible noise, are heavily influenced by the conductors themselves.

Phase and ground conductors normally consist of multiple strands of aluminum, copper, and/or steel, though the use of copper is uncommon in modern transmission lines. Bare overhead phase conductors are usually classified as homogeneous or non-homogeneous. Homogeneous conductors are those in which the individual strands of wire composing the cable are of the same material. Homogeneous conductors manufactured with relatively pure aluminum are called All-Aluminum Conductors (AAC); those manufactured with an aluminum alloy are called All-Aluminum Alloy Conductors (AAAC).

Non-homogeneous conductors consist of mixtures of different wire materials. The most common type of bare overhead phase conductor is a non-homogeneous cable consisting of aluminum strands covering a steel core, Aluminum Conductor, Steel Reinforced (ACSR). Over 80% of transmission lines around the world have been built with ACSR. Another non-homogeneous conductor is one composed of a mixture of relatively pure aluminum strands and aluminum alloy strands. This conductor is referred to as Aluminum Conductor, Alloy Reinforced (ACAR). Special types of non-homogeneous conductor have been developed which maintain stable physical and electrical properties even when operated continuously at temperatures of 200°C to 250°C. The best known, and most widely used high temperature conductor is Aluminum Conductor Steel Supported (ACSS). It consists of one or more layers of fully annealed aluminum strands helically spiraled around a stranded steel core.

Bare stranded conductors have physical, mechanical and electrical properties that reflect the properties of the individual strands of wire that comprise the conductors. The properties of homogeneous cables quite naturally parallel the properties of their singular-type individual wires. Non-homogeneous conductors, however, possess properties that reflect the individual properties and relative percentages of the different materials forming the composite cable. Due to the stranding-induced helical form of the individual strands, both types of conductors exhibit lower composite tensile strength (4% to 11%), greater weight and higher resistance per unit length of conductor (2% to 4%) than would result from all the component wires being straight.

Selection of the best conductor for a particular transmission or distribution line depends upon many factors, such as power requirements, terrain, ambient conditions, costs of conductor and supporting structures, and governmental and environmental constraints. In addition to the strength and electrical resistance of the conductor, pertinent properties of interest include the stress-strain relationship, thermal characteristics, and inductive and capacitive reactance.

Conductor selection can be complex when an existing line is to be re-conductored in order to increase its electrical power capacity. The replacement conductor must be chosen in order to avoid exceeding the longitudinal tension and transverse ice and wind loads on the existing structures while maintaining electrical clearances and environmental limitations.

Since each overhead line presents a unique set of requirements, an array of conductors ranging from standard AAC, AAAC, ACAR and ACSR to high-temperature types to vibration-resistant designs have been developed to meet the needs of the electrical utility industry.

1.1 TYPES OF CONDUCTOR MATERIALS

The most common types of phase conductors for overhead transmission lines are normally composed of strands of: relatively pure aluminum (1350-H19); an aluminum alloy (6201-T81); 1350-H19 and 6201-T81 combined; or 1350-H19 or 1350-O combined with one or more steel core wires. (These conductors are known as AAC, AAAC, ACAR, and ACSR or ACSS, respectively) In addition to the good corrosion resistance of both 1350 and 6201, each of the aluminum materials possesses certain advantageous characteristics. Wire manufactured from 1350 has high conductivity combined with moderate strength; 6201 exhibits higher strength with good, but less, conductivity. The steel wire typically used as a strength member in overhead conductors possesses high strength and good corrosion resistance provided by a protective coating, but has poor conductivity.

Although used almost exclusively before the turn of the century and commonly up until 1940, bare overhead copper conductors are rarely used today. In addition to their higher cost, copper conductors have fallen into disuse because of their inferior conductivity-to-weight and strength-to-weight ratios as compared to aluminum and aluminum-composite conductors.

1.1.1 Aluminum and Aluminum Alloy Wire

There are two broad classes of aluminum and aluminum alloys: non-heat-treatable and heat-treatable. Non-heat-treatable aluminum and aluminum alloys can only be hardened and strengthened by some form of plastic deformation, such as rolling, drawing, swaging, etc. In general, the greater the deformation, the higher the hardness and strength and the lower the electrical conductivity. The 1350 aluminum used in overhead conductors falls into the non-heat-treatable class. As used in the majority of overhead conductors, the individual strands of 1350 have been greatly deformed, or "work-hardened," in the wire drawing process and are in a full-hard, or H19, temper. To indicate both the material and temper, the wire is referred to as 1350-H19. Since greater deformation during the wire drawing process is required to produce smaller diameter wire, the smaller wire will have been work-hardened to a greater degree and will exhibit a higher tensile strength.

Once hardened, both the non-heat-treatable and heat-treatable materials can only be softened by subjecting them to a thermal treatment. Although a gradation of "softness" can be achieved by controlling the temperature and time of exposure, the only softened temper used in overhead conductors is a full-soft, or O-temper. In this state, the material is said to have been fully annealed. Used only in one of the steel-reinforced composite cables, ACSS, the 1350 wire in this state is designated 1350-O. As compared with 1350-H19, 1350-O wire has a much lower tensile strength, but much greater ductility and slightly increased conductivity.

Heat-treatable aluminum alloys can be strengthened by plastic deformation, thermal treatments, or a combination of the two. For wire, the deformation takes place during the wire drawing process. Two thermal treatments are required: a "solution-heat-treatment" (SHT) and an "aging treatment." The solution-heat-treatment conditions the material to give it the potential to be strengthened. It consists of an elevated temperature-time cycle followed by a quench. The SHT may be accomplished in-line during the manufacture of the redraw rod from which the wire is drawn, or as a separate operation after the manufacture of the redraw rod. After the wire is drawn, it is subjected to an elevated temperature-time aging treatment. The strengthening of the wire that occurs during the aging treatment is added to that achieved during the drawing process. The state of the finished wire is designated as being in the T81 temper.

In applications where higher strength is required and lower conductivity can be tolerated, the heat-treatable aluminum alloy 6201 is used. This is an alloy containing magnesium and silicon that is capable of being processed into wire that is considerably stronger than 1350. In the T81 temper, 6201 wire has a nominal tensile strength range of 44 - 48 ksi as opposed to that of 23 - 28 ksi for 1350-H19 aluminum. However, its nominal conductivity of 52.5% IACS (International Annealed Copper Standard) is somewhat lower than the typical 61 - 62% for 1350-H19.

1.1.2 Coated Steel Core Wire

The steel wire used to form the core in overhead lines is manufactured with one of several types of coatings that impart good corrosion resistance. These coatings include the traditional zinc, zinc-5% aluminum-mischmetal alloy, aluminum coated and aluminum clad.

There are three classes of zinc-coated, or galvanized, steel wire: A, B and C. The distinction among the three is the thickness of coating that progressively increases from Class A to Class C, with Class C offering the highest level of corrosion resistance. Traditionally, as the coating thickness of a given diameter wire increases, the cross-sectional area of the steel must decrease, thereby sacrificing strength for corrosion protection.

Zinc-5% aluminum-mischmetal alloy provides improved thermal stability and corrosion resistance over that of zinc galvanized cores. The use of this coating is indicated by ACSR/MA, MB, and MC designations, with class MC providing the highest level of corrosion protection. The most common of these coatings is Class MA, which offers the same, or slightly higher, corrosion protection as that of zinc coated Class C without any sacrifice in strength.

There are two types of aluminum coatings available: coated or clad. Aluminum-coated steel wire, also known as aluminized, has a relatively thin coating of aluminum, while aluminum-clad steel wire has a much thicker layer of aluminum coating. When used in ACSR, the terminology used to distinguish between the two is ACSR/AZ and ACSR/AW for aluminized and aluminum-clad, respectively. Aluminum-clad steel has the highest level of corrosion protection of any coating type and can also be used in high temperature low sag conductors. At the time of this writing two new strength categories of AW steel core wires are being introduced: High Strength (HS) and Extra High Strength (EHS).

The steel core wire used in most of the overhead conductors is a medium-strength steel containing 0.50 - 0.85 weight percent carbon with a nominal tensile strength of 210 - 230 ksi. For applications requiring conductors with higher strengths, high-strength steel core wire with 0.50 to 0.88% carbon and 0.50 to 1.30% manganese is used. Nominal tensile strength for the high-strength steel is in the 235 - 250 ksi range. The use of high strength steel core wire is indicated by the "/HS" designation (ACSR/HS).

The growing popularity of high temperature conductors using dead soft (annealed) aluminum , 1350-O, has given rise to a new generation of higher strength steel, defined as ultra high-strength steel (UHS) core wire. UHS core wire exhibits similar, or better, ductility and fatigue characteristics as compared with high strength steel core wire, and has an increased tensile strength of 265 - 285 ksi.

Using an UHS core as a direct replacement for a standard strength core can increase the strength of the core by approximately 23%. This gives a 10 to 40% increase in strength of the overall conductor depending on the steel to aluminum ratio. Alternatively, if the UHS steel core is designed to have equivalent strength to a standard strength core, the core diameter will be smaller, thereby allowing more aluminum cross-sectional area within the same conductor diameter.

By incorporating an UHS core, standard ACSS and ACSS/TW constructions can be improved such that their high-temperature low-sag performance characteristics are comparable to those of composite core conductors.

1.1.3 Selected Physical and Mechanical Properties

There are several properties of the various materials used in overhead conductors that are pertinent to their use. Table 1-1 presents a comparison of selected physical and mechanical wire properties for the most common overhead conductor materials.

1.1.4 *Mechanical Properties of 1350 and 6201 Aluminum Alloy Wire*

Aluminum wire used in overhead conductors is drawn from 3/8-inch diameter re-draw rod. This rod can be produced by either a continuous-cast process, or by an older, conventional hot-roll process. Continuous-cast rod combines the manufacturing steps of casting, rolling and solution heat treating, if applicable, into a single tandem operation. The nature of the process allows the continuous production of an unlimited number of rod coils of a size limited only by the capability of material-handling equipment. By contrast, hot-rolled rod is produced by first casting the metal into limited-length billets. These billets are then rolled into rod during a second operation and, if applicable, solution heat-treated in a third operation. Large coils of hot-rolled rod are made possible only by the welding together of smaller coils.

Due to the additional strengthening imparted by greater deformation during the wire drawing process, smaller diameters of 1350-H19 and 6201-T81 wire have higher strengths per cross-sectional unit than larger diameters. This difference is reflected in the requirements of the American Society for Testing and Materials (ASTM) specifications to which this wire must conform. The mechanical property requirements of the pertinent ASTM specification for the 1350-H19 and 6201-T81 wire used in transmission and distribution

TABLE 1-1
NOMINAL OR MINIMUM PROPERTIES OF CONDUCTOR WIRE MATERIALS

PROPERTY	International Annealed Copper Standard	Commercial Hard-Drawn Copper Wire	Standard 1350-H19 Aluminum Wire	Standard 1350-O Aluminum Wire	Standard 6201-T81 Aluminum Wire	Galvanized* Steel Core Wire	Aluminum Clad Steel Core Wire
Conductivity @ 20°C % IACS	100.0	97.0	61.2	61.8	52.5	8.0	20.3
Resistivity @ 20°C Ω - cmil/ft Ω - mm²/m	10.371 0,017241	10.692 0.017774	16.946 0.028172	16.782 0.027899	19.755 0.032841	129.64 0.21552	51.01 0.084815
Ratio of weight for equal dc resistance and length	1.00	1.03	0.50	0.50	0.58	9.1	3.65
Temp. coefficient of resistance per °F @ 68°F per °C @ 20°C	0.00218 0.00393	0.00212 0.00381	0.00224 0.00404	0.00226 0.00408	0.00193 0.00347	0.00182 0.00327	0.00200 0.00360
Density @ 20°C lbs/in³ g/cm³	0.3212 8.89	0.3212 8.89	0.0977 2.705	0.0977 2.705	0.0972 2.69	0.2811 7.78	0.2381 6.59
Coefficient of Linear Expansion 10^{-6} per °F 10^{-6} per °C	9.4 16.9	9.4 16.9	12.8 23.0	12.8 23.0	12.8 23.0	6.4 11.5	7.2 13.0
Modulus of Elasticity 10^6 psi 10^6 Pascal	17 117,000	17 117,000	10 69,000	10 69,000	10 69,000	29 200,000	23.5 162,000
Specific Heat @ 20°C cal/gm-°C	0.0921	0.0921	0.214	0.214	0.214	0.107	0.112
Tensile Strength 10^3 psi 10^6 Pascal	35.0 214	62.0 427	24.0 165	10.0 70	46.0 317	185 1276	175 1206
Minimum Elongation, %	25	1.1	1.5	20	3.0	3.5	1.5

*For Zinc-5% Aluminum-Mischmetal Alloy Coated (Galfan™) steel core wire use values listed for galvanized steel core wire.

conductors are presented in Table 1-2. Fully annealed aluminum, 1350-O, has a tensile strength of 8.5 - 14 ksi and a typical elongation of 28 - 32%. For conductor rated breaking strength (RBS) calculations, typical tensile strengths for 1350-O aluminum, for all wire diameters used in ACSS conductors, is conservatively assumed to be 8.5 ksi.

1.1.5 Mechanical Properties of Steel Wire

For steel, as with aluminum and aluminum alloys, the mechanical properties required by ASTM specifications vary with wire size. The tensile requirements commonly used for Class A steel core wires and aluminum clad steel wires are shown in Table 1-3.

1.2 CONCENTRIC-LAY-STRANDED CONDUCTOR CONSTRUCTION

The term "concentric-lay-stranded conductor" is used to identify a conductor consisting of a straight center core wire surrounded by one or more layers of helically wound wires. The "lay" of the conductor refers to the layers of the individual strands of which the conductor is composed, and is specified in terms of length and direction. The lay length is the axial length of one complete revolution of a helical strand. The lay direction is defined as either right-hand or left-hand, referring to the direction of wrap of the individual strands as viewed axially in a direction away from the observer.

TABLE 1-2
ASTM MECHANICAL PROPERTY REQUIREMENTS FOR
1350-H19 AND 6201-T81 WIRE

Wire Diameter in	1350-H19 Aluminum (ASTM B 230) Minimum Tensile Strength ksi Average for Lot	1350-H19 Aluminum (ASTM B 230) Minimum Tensile Strength ksi Individual Test	1350-H19 Aluminum (ASTM B 230) Elongation in 10" for Individual Tests Min. %	6201-T81 (ASTM B 398) Minimum Tensile Strength ksi Average for Lot	6201-T81 (ASTM B 398) Minimum Tensile Strength ksi Individual Test	6201-T81 (ASTM B 398) Elongation in 10" for Individual Tests Min. %
0.2600 to 0.2101	23.5	22.5	2.2	-	-	-
0.2100 to 0.1801	24.0	23.0	2.0	-	-	-
0.1800 to 0.1501	24.0	23.0	1.9	-	-	-
0.1500 to 0.1401	24.5	23.5	1.8	-	-	-
0.1400 to 0.1201	25.0	23.5	1.7	-	-	-
0.1200 to 0.1101	25.5	24.0	1.6	-	-	-
0.1100 to 0.1001	26.0	24.5	1.5	-	-	-
0.1000 to 0.0901	27.0	25.5	1.5	-	-	-
0.0900 to 0.0801	27.5	26.0	1.5	-	-	-
0.0800 to 0.0701	28.0	26.5	1.4	-	-	-
0.0700 to 0.0601	28.5	27.0	1.3	-	-	-
0.0600 to 0.0501	29.0	27.0	1.2	-	-	-
0.0500 to 0.0105	25.0	23.0	-	-	-	-
0.1878 to 0.1281	-	-	-	46.0	44.0	3.0
0.1280 to 0.0610	-	-	-	48.0	46.0	3.0

The materials of which a conductor is composed may be either homogeneous (all of the same material) or non-homogeneous. The design employed in their construction is one of two individual wire types: conventional-round-wire construction or trapezoidal-shaped-wire (TW) construction. The combinations of strand materials and wire-shape types, coupled with a large choice of conductor sizes, yield an array of conductors with distinct properties from which the line designer can choose for a given application.

1.2.1 Round-Wire Construction

Round-wire, concentric-lay-stranded aluminum conductors for bare overhead power lines are constructed with the outermost layer of wires of the conductor always having right-hand lay, with the inner layers reversing lay direction in succession. For example, consider a conductor composed of 3 layers of aluminum strand over a steel core consisting of one layer of steel strand over a single steel core wire. In this case, the outer aluminum layer is right-hand lay; the middle aluminum layer, left-hand lay;

TABLE 1-3
ASTM STEEL WIRE TENSILE REQUIREMENTS

Nominal Diameter in	**Stress at 1% Extension, Min. ksi**	**Ultimate Tensile Strength, Min. ksi**	**Elongation in 10" Min. %**
Standard Strength Steel Galvanized (Class A Coating) [ASTM B 498] Zn-5% Al-Mischmetal Alloy (Class A Coating) [ASTM B 802]			
0.0500 to 0.0899, incl	190	210	3.0
0.0900 to 0.1199, incl	185	205	3.5
0.1200 to 0.1399, incl	180	205	4.0
0.1400 to 0.1900, incl	170	200	4.0
High Strength Steel Zn-5% Al Mischmetal Alloy (Class A Coating) [ASTM B 803]			
0.0500 to 0.0899, incl	210	235	3.0
0.0900 to 0.1199, incl	205	230	3.0
0.1200 to 0.1399, incl	200	225	3.5
0.1400 to 0.1900, incl	195	220	3.5
Ultra High Strength Steel Zn-5% Al Mischmetal Alloy (Class A Coating)			
0.0500 to 0.0899, incl	-	285	3.0
0.0900 to 0.1199, incl	-	275	3.0
0.1200 to 0.1399, incl	-	270	3.5
0.1400 to 0.1500, incl	-	265	3.5
Aluminum-Clad Steel Wire [ASTM B 502]			
0.0770 to 0.1289, incl	175	195	1.5
0.1290 to 0.1369, incl	170	190	1.5
0.1370 to 0.1443, incl	165	185	1.5
0.1444 to 0.1549, incl	160	180	1.5
0.1550 to 0.1620, incl	160	175	1.5
0.1621 to 0.1729, incl	155	170	1.5
0.1730 to 0.1819, incl	150	165	1.5
0.1820 to 0.1880, incl	145	160	1.5

the inner aluminum layer, right-hand lay; and the outer layer of the steel core, left-hand lay.

As the name implies, a concentric round-wire conductor is composed of individual strands having a circular cross-sectional shape. Homogeneous conventional round-wire, concentric-lay-stranded conductors (AAC and AAAC) and non-homogeneous ACAR conductors are constructed using individual strands of the same diameter. To achieve as much cable compactness as possible using round wire of the same diameter, a simple geometric relationship dictates that the first layer over the core wire will contain six strands, and that each successive layer will contain six strands more than the previous layer. The number of equal-diameter individual strands in concentric round-wire conductors, therefore, is 7 (1+6), 19 (7+12), 37 (19+18), 61 (37+24), etc. It should be noted that in the case of ACAR conductors, the proportions of 1350 aluminum strands and 6201 aluminum alloy strands are very flexible making possible an extensive array of ACAR conductors with unique combinations of strength and conductivity.

Non-homogenous round-wire concentric-lay-stranded ACSR and ACSS conductors are manufactured with either one of two designs. The first design is one in which all the wire strands, both steel and aluminum, are of the same diameter. This is the same design used for the AAC, AAAC and ACAR round-wire concentric-lay-stranded conductors. The "stranding ratio" for ACSR and ACSS is the ratio of the number of aluminum strands to the number of steel strands in the core. Some common stranding ratios corresponding to equal diameter aluminum and steel wires are 6/1, 18/1, 36/1, 12/7, 30/7, 54/7 and 84/7.

The second design consists of using aluminum wires of one diameter and steel wires of a different diameter. With this design, the steel core is composed of a single steel wire or same-diameter steel wires layered as in the first design. The diameter of the individual strands composing the first layer of aluminum wire is that which will yield a compact spacing around the steel core. All additional layers of aluminum strands will contain six more strands than the previous layer. Some of the more common combinations of aluminum/steel strands are 7/1, 8/1, 20/7, 22/7, 24/7, 26/7, 42/7, 45/7, 48/7, 72/7, 16/19, 30/19, 54/19, 76/19 and 84/19.

1.2.2 Trapezoidal-Shaped Wire Construction

Another type of concentric-lay-stranded construction consists of one or more layers of trapezoidal-shaped wire helically applied over a round wire central core. The core may consist of a single round wire for homogenous aluminum conductors, a single wire or concentric-lay-stranded same-size core strands for ACAR, ACSR and ACSS conductors. This type of conductor is routinely referred to as "Trapwire" due to the trapezoidal shaping of the individual wire strands. The Trapezoidal shaped strands "fit" together tighter, thus reducing the size of the interstices (void areas between the strands) found in round-wire constructions.

The cross-sectional area of the trapezoidal-shaped wire, the number of strands per layer, and the number of layers determine the total cross-sectional area of the aluminum in the conductor. Since the individual trapezoidal-shaped wire is not geometrically limited, as in the case of concentric stranded round wire conductor, an almost limitless number of conductor designs are possible. Whereas conventional round wire conductor constructions can be distinguished by the ratio of aluminum/steel stranding, trapezoidal-shaped wire conductors are usually distinguished by a "Type Number." A type number is the ratio of the steel cross-sectional area to the aluminum cross-sectional area, expressed as a percentage rounded to the nearest whole number.

$$\textit{Type Number} = \frac{\textit{Area of Steel}}{\textit{Area of Aluminum}} \times 100$$

A comparison of type number to equivalent strandings is presented in Table 1-4.

Trapwire is available in ACSR, ACSS or AAC constructions. AAC Trapwire constructions are available in either equal kcmil area to a round-wire AAC construction, or in fixed diameter increments. ACSR and ACSS Trapwire constructions are usually either equal in kcmil area or equal in diameter to a standard round-wire construction of the same type. The use of trapezoidal wire in a construction is designated by the /TW suffix applied to the construction and to the codeword used to identify the construction. Examples are ACSR/TW or "Drake/TW."

Trapwire with equal kcmil area to a round-wire construction will have the same codeword. So a 795 kcmil Type 16 "Drake/ACSS/TW" will have the same amount of aluminum as a 795 kcmil-26/7 "Drake/ACSS," but will have a smaller overall diameter due to the more compact construction. A Trapwire conductor that has a diameter equal to a round-wire construction will have approximately 20% more aluminum and steel areas, and thus will have a different codeword associated with it.

TABLE 1-4
COMPARISON OF TYPE NUMBERS WITH
EQUIVALENT STRANDINGS OF ACSR & ACSS

Type Number	Stranding
3	36/1
5	42/7
6	18/1
7	45/7
8	84/19
10	22/7
13	54/7
13	54/19
13	24/7
16	26/7
23	30/7
23	30/19

1.2.3 Stranded Conductor Size

The size of concentric-lay-stranded conductors is usually stated in terms of cross-sectional area or diameter. The most common unit of cross-sectional area measurement used in the United States and Canada is the circular-mil (cmil), where one mil equals 0.001 inch. A cmil equals the cross-sectional area of a circle that is one mil in diameter. The area of any circular shape may be expressed by squaring the diameter expressed in mils. For example, the area of a 1-inch (1000 mil) diameter circle is one million cmils or 1000 kcmils. To reduce the magnitude of the number, the unit of "kcmil" (thousand cmils) is commonly used. To convert the area, A_{kcmil}, from kcmils to square inches, A_{in^2}:

$$A_{in^2} = \left(\frac{\frac{\pi}{4}}{1000} \right) A_{kcmil} = 0.7854 \times 10^{-3} A_{kcmil} \qquad (1\text{-}1)$$

The circular-mil area refers only to the "conducting area" of a conductor, which excludes the steel strands if present. For example, to determine the cross-sectional area of a stranded aluminum or ACSR conductor, measure the diameter, in mils, of one aluminum strand, square it to get the cmil area of the strand, then multiply the cmil area of that single strand by the total number of aluminum strands in the conductor. Because the steel core area is not included, the actual diameter of an ACSR conductor is always larger than that of an all-aluminum conductor of the same cross-sectional area.

The size of concentric-lay-stranded conductors, in terms of overall diameter, will change with conductor stranding and construction. Table 1-5 shows several conductor constructions having the same cross-sectional area of aluminum (795 kcmil in this case) that do not have the same diameter or weight per unit length. As the area of steel increases, so do the strength, weight and overall diameter.

The size of solid wires and small stranded conductors is commonly expressed in terms of the American Wire Gauge (AWG). In this system of measurement, the larger the diameter, the smaller the AWG number. The smallest diameter wire commonly used in bare stranded overhead conductor is 6 AWG. The AWG sizes increase in a logical progression from 6 to 4 to 2 to 1. Beyond this size, the increasing sizes are 0, 00, 000 and 0000, which for ease of reference, are called 1/0, 2/0, 3/0 and 4/0, respectively. An increase of 3 in the AWG size corresponds to a doubling of the aluminum area (e.g., 1/0 vs. 4/0). For sizes larger than 4/0 AWG, the conductor size is commonly stated in terms of the circular mil area.

It should be noted that international practice is to state conductor sizes in terms of cross-sectional areas expressed in square millimeters. In some cases, the cross-sectional area does include the area of the steel strands of ACSR conductors.

1.2.4 ACSR Stranding Ratios

The most common conductor used as a phase conductor in power transmission lines is ACSR. It is manufactured in sizes from 6 AWG to special conductors in excess of 2500 kcmil. In each size, it is usually possible to purchase ACSR in any one of several different stranding ratios. Each stranding ratio corresponds to a particular set of mechanical characteristics, but all have essentially the same electrical characteristics. By choosing the proper stranding ratio, the line designer can ensure that the conductor will have enough steel for mechanical strength, yet avoid the increased cost and weight associated with the higher steel content ACSRs.

Table 1-5 lists some commonly available constructions of 795 kcmil conductors by increasing strength. Since all of these conductors have the same aluminum cross-sectional area, they have nearly the same ac resistance and reactance, and are essentially electrical equivalents. Mechanically, however, these conductors are quite different.

For ACSR and ACSS conductors, the higher the percentage of steel:

- The higher the tension levels that can be tolerated in the conductor during extreme ice and wind loads without breaking.

- The less sag increase that occurs under extreme ice and wind loads and at high temperature.

- The more expensive the conductor. (30/7 "Skimmer" costs about 10% more per foot than all-aluminum "Arbutus" due to the cost of the steel core.)

- The more expensive angle and dead-end structures become as a result of higher tension loads.

1.3 ELECTRICAL CHARACTERISTICS OF CONDUCTORS

The electrical characteristics of a conductor are one of the most important considerations when selecting a conductor. The conductor resistance determines the conductor losses and limits the maximum allowable current carrying capacity of the conductor. The inductive and capacitive reactance losses directly affect system performance.

Metal conductivity, cross-sectional area, construction, temperature, frequency and current density determine the electrical properties of a conductor. In service, these electrical properties are also influenced by the physical location of a conductor in relation to other conductors and to the earth.

General formulas for the calculation of dc and ac resistance, inductance and inductive reactance, and capacitance and capacitive reactance for transmission lines are presented in this section.

1.3.1 DC Resistance

The dc resistance of a conductor is a function of its cross-sectional area, length and volume resistivity and may be expressed as:

$$R = \rho_V \left(\frac{L}{A} \right) \tag{1-2}$$

where:

R = *conductor resistance, Ω/unit length*

ρ_V = *volume resistivity of conductor material*

L = *conductor unit length*

A = *conductor cross-sectional area*

For a concentric-lay-stranded conductor, the dc conductor resistance should be increased by approximately 2% to account for the increase in helical length of the individual strands over the axial length of the conductor. This is commonly referred to as the stranding factor. Exact factors are given in ASTM Standards.

TABLE 1-5
COMPARISON OF MECHANICAL PROPERTIES FOR
DIFFERENT STRANDINGS OF 795 KCMIL CONDUCTORS

Code Name	Aluminum Wires No. x OD in	Steel Wires No. x OD in	Overall Diameter in	Rated Strength lbs	Total Weight lbs/1000 ft	Core Weight lbs/1000 ft
Arbutus	37 x 0.1466	NONE	1.026	13,900	746	-
Coot	36 x 0.1488	1 x 0.1488	1.040	16,800	805	59
Turbit	20 x 0.1994	7 x 0.0886	1.063	21,800	896	150
Puffin	22 x 0.1901	7 x 0.1056	1.077	24,800	958	212
Drake	26 x 0.1749	7 x 0.1360	1.108	31,500	1094	344
Skimmer	30 x 0.1628	7 x 0.1628	1.140	38,300	1246	483

The dc resistance may also be expressed in terms of conductor weight and weight resistivity as follows:

$$R = \rho_w \left(\frac{L^2}{w} \right) \qquad (1\text{-}3)$$

where:

R = *conductor resistance, Ω/unit length*

ρ_w = *weight resistivity of conductor material*

w = *weight of conductor per unit length*

Resistance is universally expressed in ohms; however, conductor length, weight and cross-sectional area may be expressed in many different units. The fundamental quantities of volume resistivity and weight resistivity may also be expressed in many ways. Various conversion factors for volume and weight resistivities are given in Table 1-6[1]. The units used for the resistivity constants define the units to be used for area, length and weight.

The resistance of a material is a function of temperature. Therefore, the dc resistance calculated using the previous equations is at the reference temperature of the selected resistivity values, typically 20°C.

TABLE 1-6
RESISTIVITY AND CONDUCTIVITY CONVERSION FACTORS

Given *N* ⟶	**Volume Resistivity at 20°C**				**Weight Resistivity at 20°C**		**Conductivity at 20°C**	
Perform indicated operation to obtain ↓	Ω•cmils/ft	Ω•mm²/m	μΩ•in	μΩ•cm	Ω•lbs/mile²	Ω•g/m²	% IACS (volume basis)	% IACS (weight basis)
Volume Resistivity at 20°C								
Ω•cmils/ft	—	*N* x 601.52	*N* x 15.279	*N* x 6.0153	*N* x 0.10535 x (1/δ)	*N* x 6.0153 x (1/δ)	(1/*N*) x 1037.1	(1/*N*) x 9220.0 x (1/δ)
Ω•mm²/m	*N* x 0.0016624	—	*N* x 0.025400	*N* x 0.010000	*N* x 0.00017513 x (1/δ)	*N* x (1/δ)	(1/*N*) x 1.7241	(1/*N*) x 15.328 x (1/δ)
μΩ•in	*N* x 0.065450	*N* x 39.370	—	*N* x 0.39370	*N* x 0.0068950 x (1/δ)	*N* x 39.370 x (1/δ)	(1/*N*) x 67.879	(1/*N*) x 603.45 x (1/δ)
μΩ•cm	*N* x 0.16624	*N* x 100.00	*N* x 2.5400	—	*N* x 0.017513 x (1/δ)	*N* x 100.00 x (1/δ)	(1/*N*) x 172.41	(1/*N*) x 1532.8 x (1/δ)
Weight Resistivity at 20°C								
Ω•lbs/mile²	*N* x 9.4924 x δ	*N* x 5710.0 x δ	*N* x 145.03 x δ	*N* x 57.100 x δ	—	*N* x 5710.0	(1/*N*) x 9844.8 xδ	(1/*N*) x 87520
Ω•g/m²	*N* x 0.0016624 xδ	*N* x δ	*N* x 0.025400 xδ	*N* x 0.010000 xδ	*N* x 0.00017513	—	(1/*N*) x 1.7241 xδ	(1/*N*) x 15.328
Conductivity at 20°C								
% IACS (volume basis)	(1/*N*) x 1037.1	(1/*N*) x 1.7241	(1/*N*) x 67.879	(1/*N*) x 172.41	(1/*N*) x 9844.8 xδ	(1/*N*) x 1.7241 xδ	—	*N* x 0.11249 x δ
% IACS (weight basis)	(1/*N*) x 9220.0 x (1/δ)	(1/*N*) x 15.328 x (1/δ)	(1/*N*) x 603.45 x (1/δ)	(1/*N*) x 1532.8 x (1/δ)	(1/*N*) x 87520	(1/*N*) x 15.328	*N* x 8.89 x (1/δ)	—

To determine the dc resistance at other temperatures, the known resistance must be corrected using the temperature coefficient of resistance for the conductor metal. Temperature coefficients of resistance at 20°C for typical conductor materials are given in Tables 1-7[1] and 1-8[2].

TABLE 1-7
TEMPERATURE COEFFICIENTS OF RESISTANCE

Material	Approximate Density, δ @ 20°C, g/cm³	Temperature Coefficient of Resistance, α @ 20°C
Copper, % IACS:		
101.00	8.89	0.00397
100.00	8.89	0.00393
98.40	8.89	0.00387
98.16	8.89	0.00386
97.80	8.89	0.00384
97.66	8.89	0.00384
97.40	8.89	0.00383
97.16	8.89	0.00382
96.66	8.89	0.00380
96.16	8.89	0.00378
94.16	8.89	0.00370
93.15	8.89	0.00366
Aluminum 1350, % IACS:		
64.0	2.705	0.00423
63.0	2.705	0.00416
62.0	2.705	0.00410
61.8	2.705	0.00408
61.5	2.705	0.00406
61.4	2.705	0.00406
61.3	2.705	0.00405
61.2	2.705	0.00404
61.0	2.705	0.00403
Aluminum Alloy, % IACS:		
6201-T81		
52.5	2.69	0.00347
Aluminum Clad Steel, Specification B 415	6.59	0.0036
Copper Clad Steel:		
Grade 30 A, HS, EHS	8.15	0.00378
Grade 40 A, HS, EHS	8.25	0.00378

Over a moderate temperature range, such as 0° to 140°C, resistance and temperature are considered to be linearly related. However, there is some evidence that at conductor temperatures above 140°C this relationship becomes non-linear[3]. Therefore, the resistance at any given temperature may be calculated from a known resistance at a specified temperature by:

$$R_{T_2} = R_{T_{ref}} \left[1 + \alpha_{ref} \left(T_2 - T_{ref} \right) \right] \qquad (1\text{-}4)$$

where:

α_{ref} = *temperature coefficient of resistance at reference temperature* T_{ref}

T_{ref} = *reference temperature, °C*

T_2 = *temperature at which new resistance is desired, °C*

$R_{T_{ref}}$ = *dc resistance at reference temperature* T_{ref}

R_{T_2} = *dc resistance at temperature* T_2

From the previous relationship it can be seen that the resistance of a 61.2% IACS aluminum conductor, having a temperature coefficient of 0.00404 at 20°C, increases approximately 4% for every 10°C increase in temperature.

To use Equation 1-4, the coefficient of resistance is required at the temperature for the known resistance. If the known resistance is at a temperature for which no coefficient of resistance is given, the coefficient of resistance at the new temperature is determined as follows:

$$\alpha_{new} = \frac{1}{\dfrac{1}{\alpha_{ref}} + \left(T_{new} - T_{ref} \right)} \qquad (1\text{-}5)$$

where:

α_{new} = *coefficient of resistance at new temperature* T_{new}

α_{ref} = *coefficient of resistance at temperature* T_{ref}

T_{new} = *temperature at which new coefficient of resistance is desired, °C*

T_{ref} = *reference temperature, °C*

To calculate the dc resistance for a stranded conductor, the conductor is assumed to be a parallel combination of small solid wires twisted together with a predetermined lay length specified by ASTM Standards. The strands of the conductor may or may not be all of the same material. One method of calculating the dc resistance of a non-homogeneous stranded conductor is described by Douglass[4]:

(1) Calculate the cross-sectional area of each of the component wire materials separately.

(2) Multiply each of the component areas by their respective dc resistivities at a reference temperature, T_{ref}.

(3) Multiply each component resistance by the standard increments due to stranding (stranding factor) given in the appropriate ASTM Standard (approximately 2%).

(4) Find the dc resistance at the reference temperature as a parallel combination of the conductor component wires.

(5) Correct the dc resistance to the desired temperature.

1.3.2 AC Resistance

A conductor offers a greater resistance to the flow of alternating current than it does to direct current. The magnitude of this increase is often expressed as an "ac/dc ratio." Factors related to the magnetic fields generated by power currents responsible for this increase are: (1) skin effect, (2) proximity effect, (3) hysteresis and eddy current effect for ACSR and ACSS.

TABLE 1-8

TEMPERATURE - RESISTANCE COEFFICIENTS FOR COPPER AND ALUMINUM

Conductivity % IACS	Temperature, °C					
	0	20	25	50	75	100
	Temperature coefficient of resistance, α, per °C					
Aluminum						
55.0	0.00392	0.00363	0.00357	0.00328	0.00303	0.00282
56.0	0.00400	0.00370	0.00363	0.00333	0.00307	0.00285
57.0	0.00407	0.00377	0.00370	0.00338	0.00312	0.00289
58.0	0.00415	0.00383	0.00376	0.00344	0.00316	0.00239
59.0	0.00423	0.00390	0.00382	0.00349	0.00321	0.00297
60.0	0.00430	0.00396	0.00389	0.00354	0.00325	0.00301
60.6	0.00435	0.00400	0.00392	0.00357	0.00328	0.00303
61.0	0.00438	0.00403	0.00395	0.00360	0.00330	0.00305
61.2	0.00440	0.00404	0.00396	0.00361	0.00331	0.00306
61.3	0.00441	0.00405	0.00397	0.00361	0.00331	0.00306
61.4	0.00442	0.00406	0.00398	0.00362	0.00332	0.00306
61.5	0.00442	0.00406	0.00398	0.00362	0.00332	0.00307
61.8	0.00445	0.00408	0.00400	0.00364	0.00333	0.00308
62.0	0.00446	0.00410	0.00401	0.00365	0.00334	0.00309
63.0	0.00454	0.00416	0.00408	0.00370	0.00339	0.00312
64.0	0.00462	0.00423	0.00414	0.00375	0.00343	0.00316
65.0	0.00470	0.00429	0.00420	0.00380	0.00347	0.00320
6201 Aluminum Alloy 52.5	0.00373	0.00347	0.00341	0.00314	0.00291	0.00272
5005 Aluminum Alloy 53.5	0.00380	0.00353	0.00347	0.00319	0.00296	0.00275
Copper						
95.0	0.00403	0.00373	0.00367	0.00336	0.00310	0.00287
96.0	0.00408	0.00377	0.00370	0.00339	0.00312	0.00290
97.0	0.00413	0.00381	0.00374	0.00342	0.00315	0.00292
97.5	0.00415	0.00383	0.00376	0.00344	0.00316	0.00293
98.0	0.00417	0.00385	0.00378	0.00345	0.00318	0.00294
99.0	0.00422	0.00389	0.00382	0.00348	0.00320	0.00297
100.0	0.00427	0.00393	0.00385	0.00352	0.00323	0.00299
101.0	0.00431	0.00397	0.00389	0.00355	0.00326	0.00301
102.0	0.00436	0.00401	0.00393	0.00358	0.00328	0.00304

Skin Effect

Skin effect, always present in a conductor carrying alternating current, is a function of the internal flux in a conductor. Alternating current tends to flow near the outside of a conductor, yielding a higher current density on the outer layers and increasing the effective resistance. If the conductor is made from, or contains a ferromagnetic material, such as steel, the skin effect is increased.

For a homogeneous, concentric-lay stranded conductor, experimental studies by Kennelly, et al[5] indicate that the skin effect is identical to that of a solid cylindrical conductor having the same dc resistance. Figure 1-1[6], provides a means of estimating skin effect for homogeneous aluminum and copper conductors. This figure can also be used for steel-cored aluminum conductors, treating them as a hollow tube.

It can be seen from Figure 1-1 that skin effect is not directly dependent on the current in the conductor. However, as the current increases, the conductor temperature increases. The increased temperature causes an increase in resistance that results in the ac/dc ratio actually decreasing for all-conductors. To demonstrate the effect of skin effect, consider the following example:

For an AAC Conductor: *Calculate the ac resistance per unit length of 397.5 kcmil-19 strand AAC conductor code name "Canna" with a dc resistance of 0.2296 Ω/mile at 20°C.*

Use Figure 1-1 to estimate the ac/dc ratio. The abscissa value of Figure 1-1 is:

$$\sqrt{\frac{f}{R_{dc}}} = \sqrt{\frac{60}{0.2296}} = 16.2$$

where:

f = *frequency, Hz*

R_{dc} = *dc resistance, Ω/mile*

The ratio $r_o/r_n = 0$ for this conductor because it is a homogeneous all-aluminum conductor. The ac/dc ratio read from the vertical axis of Figure 1-1 is 1.006. Multiplying the conductor dc resistance by the ac/dc ratio gives:

$$R_{ac} = (0.2296)\ (1.006) = 0.2310 \quad \Omega/\text{mile at } 20°C$$

For ACSR Conductor: *Calculate the ac resistance at 75°C, for a 2156 kcmil-84/19 ACSR conductor, code name "Bluebird," with a dc resistance of 0.0423 Ω/mile at 20°C. It has 4 layers of aluminum with a steel core diameter of 0.481 inches and an overall diameter of 1.762 inches.*

Calculate the dc resistance for 75°C using Equation 1-4. The dc resistance accounts for both the steel and aluminum components of the conductor. For temperature correction, the thermal resistance coefficient of aluminum is used.

$$R_{75°C} = 0.0423\ [1 + 0.00404\ (75 - 20)] = 0.0517 \quad \Omega/\text{mile}$$

From Figure 1-1, find the ac/dc ratio. The abscissa value for Figure 1-1 is:

$$\sqrt{\frac{60}{0.0517}} = 34.1$$

The ratio of inner to outer diameter of the aluminum strand layers (r_o/r_n) is 0.27. The ac/dc ratio from Figure 1-1 is approximately 1.075. Therefore, the ac resistance of "Bluebird" at 75°C is:

$$R_{ac} = (0.0517)\ (1.075) = 0.0555 \quad \Omega/\text{mile}$$

Proximity Effect

When two conductors carrying alternating current are spaced relatively close to one another, their mutual induction affects the current distribution in each wire. This distorts the cross-sectional current distribution in the conductors resulting in (1) greater current density on the near sides of the conductor when the current in the conductors is flowing in opposite directions, and (2) at the far sides of the conductor for currents flowing in the same directions. Similar to skin effect, this non-uniform current density, commonly called proximity effect, leads to an increase in resistance and heat loss. Proximity effect is directly proportional to the magnitude of the currents and inversely proportional to the distance between the conductors. However, bare overhead conductors are normally installed with sufficient spacing such that proximity effects may be neglected.

FIGURE 1-1
SKIN EFFECT FACTOR

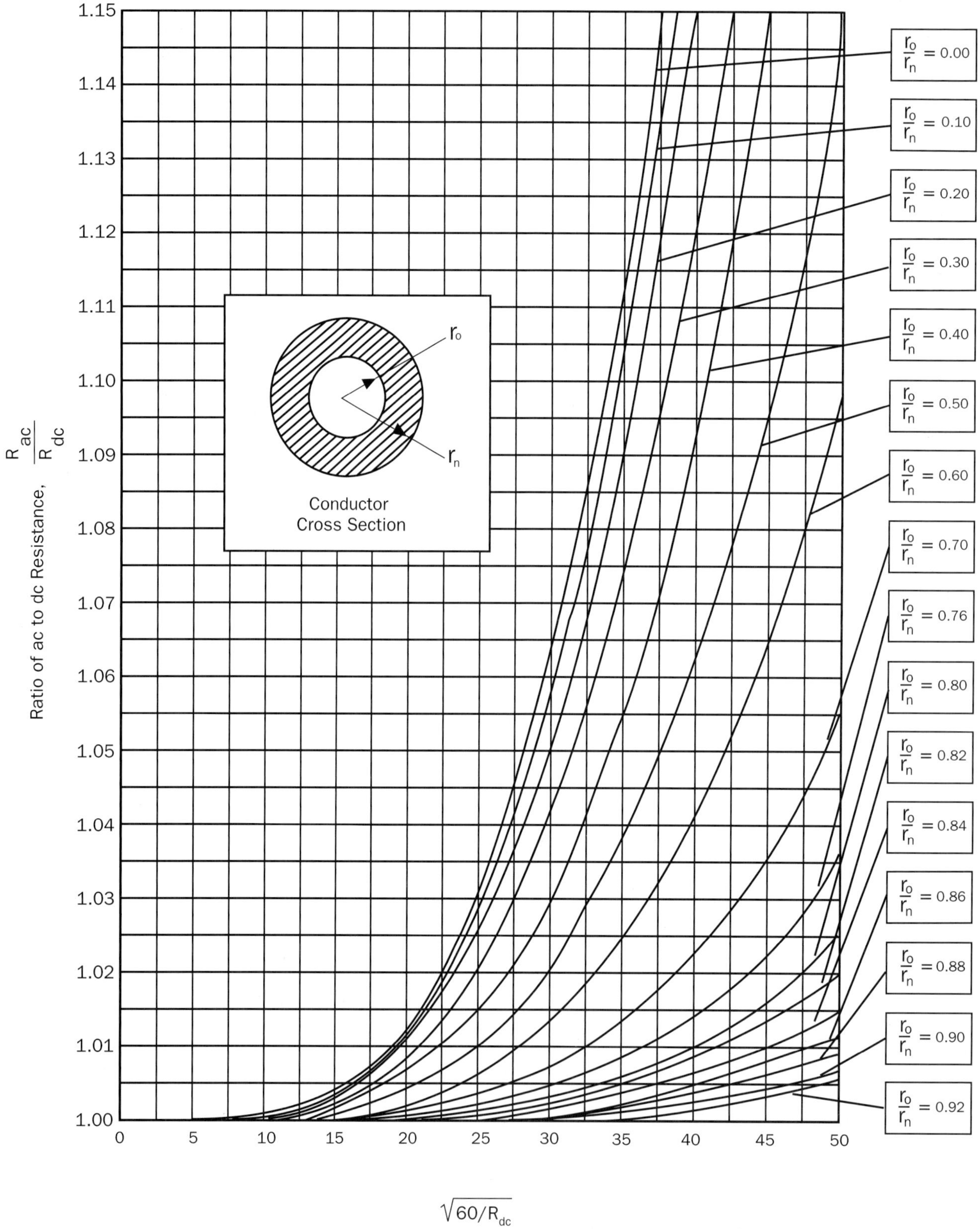

R_{dc} = ohms per mile

Magnetic Hysteresis and Eddy Current Effects

Hysteresis and eddy current losses in conductors and adjacent metal objects add to the effective ac resistance. These losses are important in conductors utilizing steel cores and operating at high current levels or when magnetic materials are present in line hardware. In ACSR conductors, as in all concentric-lay stranded, bare aluminum overhead conductors, the outermost layer of aluminum strands has a right hand lay direction with each layer below the outermost layer alternating in lay direction. The current in the aluminum strands tends to follow the helical path of the strands in each layer. If the number of such layers is even, then the axial magnetic field produced by one layer tends to be canceled by the next. If, however, the number of aluminum strand layers is odd, this cancellation does not take place and the resulting net helical current flow tends to produce an axial magnetic field, which yields magnetically induced losses in the steel core wires. These "core magnetization" losses in the steel core increase the effective conductor resistance per unit length.

Core magnetization is a strong function of current magnitude since the axial magnetic field in the steel core wires is proportional to the total current in the aluminum layers. At very low current levels, core magnetization is negligible. At current levels approaching the conductor's ampacity, however, core magnetization losses are significant.

The effects of core magnetization are in addition to the normal skin and proximity effects that occur with any cylindrical conductor. As noted previously, skin effect actually declines slightly with increasing current as the conductor resistance increases.

With single layer ACSR conductors (6/1, 7/1, 6/7, 12/7) at current levels near conductor ampacity, the core magnetization loss effect can increase the resistance by 20% or more. With three layer ACSR (42/7, 45/7, 48/7, 54/19) the effect of core magnetization at high current levels is much smaller, not exceeding 6%. Table 1-9 lists some examples.

TABLE 1-9
CORE MAGNETIZATION EFFECT ON AC RESISTANCE

Conductor	Number of Aluminum Layers	ac/dc ratio @ 25°C (low current)	ac/dc ratio @ 50°C (high current)
4/0 AWG - 6/1 ACSR "Raven"	1	1.005	1.21
795 kcmil - 26/7 ACSR "Drake"	2	1.00	1.01
1590 kcmil - 54/7 ACSR "Lapwing"	3	1.06	1.10

FIGURE 1-2
AC/DC RATIO FOR CORE MAGNETIZATION FOR THREE LAYER ACSR

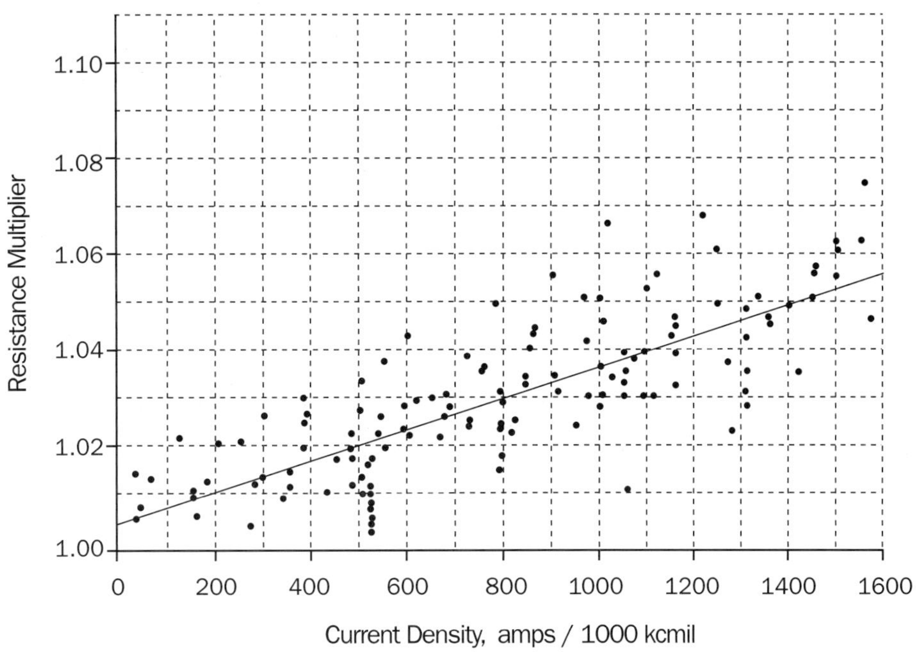

The ac/dc ratios shown in Table 1-9 at low current are due solely to skin effect. The relatively large "Lapwing" conductor has an ac resistance equal to 1.06 times its dc resistance at 25°C. The relatively small "Raven" conductor has an ac resistance which is essentially equal to its dc resistance at 25°C.

The difference in ac/dc ratio between the two columns of the table is due to core magnetization. The ratio is 20% higher for the single layer 6/1 ACSR and 4% higher for the three layer 54/7 ACSR.

Figure 1-2 shows experimental measurements of ac/dc resistance ratio due to core magnetization for three layer ACSR conductors, presented originally in Douglass[4]. Recent research indicates that at current densities above 1400 amps/1000 kcmil the curve may flatten, reducing the effect of core magnetization.

The ac resistance of a three layer ACSR conductor is calculated as follows:

(1) Calculate the dc resistance at 20°C.

(2) Calculate the dc resistance at T_{cdr}.

(3) Calculate skin effect ratio and multiply dc resistance by this ratio.

(4) Multiply by the ac/dc ratio due to core magnetization.

The use of this procedure is demonstrated in the following example.

Calculate the ac resistance for a 1590 kcmil-45/7 ACSR "Lapwing" conductor operated at a temperature of 75°C with a dc resistance of 0.0572 Ω/mile at 20°C.

The dc resistance at 75°C is:

$$R_{75} = 0.0572\ [1 + 0.00404\ (75 - 20)\ ]$$
$$= 0.0699\ \ \Omega/mile$$

The abscissa parameter for Figure 1-1 is:

$$\sqrt{\frac{60}{0.0699}} = 29.3$$

The ratio of the inner to outer aluminum strand layers, r_o/r_n, is 0.25. From Figure 1-1, the ac/dc ratio is 1.048. Therefore, the ac resistance of "Lapwing" at 75°C is:

$$R_{ac} = (1.048)\ (0.0699) = 0.0733\ \ \Omega/mile$$

Since 1590 kcmil "Lapwing" has three layers of aluminum strands, the resistance must be corrected for core magnetization losses. Using the Southwire Ampacity program, the ampacity of "Lapwing" is 1350 amperes for a conductor temperature of 75°C, an air temperature of 25°C and a 2 feet/sec wind in sun. At 1350 amps and an aluminum area of 1590 kcmil, the core magnetization correction factor read from Figure 1-2 is 1.031. The corrected ac resistance is:

$$R_{ac} = (1.031)\ (0.0733)$$
$$= 0.0756\ \ \Omega/mile$$

This is an increase of some 32% over the dc resistance at 20°C.

1.3.3 Inductance and Inductive Reactance

Calculation of the inductance and inductive reactance of a transmission line involves a straight-forward application of the familiar concepts of magnetic fields and magnetic induction. Inductance, L, of an electrical circuit is defined as the ratio of the voltage drop along the conductor, V, to the rate of change of current, i, passing through it:

$$V = L\frac{di}{dt} \tag{1-6}$$

Inductance

The classic inductance formula for a single, round, straight wire in a two-conductor, single-phase circuit is:

$$L = \left[\frac{1}{2} + 2\ ln\left(\frac{D}{r}\right)\right] \times 10^{-7} \quad H/m \tag{1-7}$$

where:

L = *inductance, H/m (Henries/meter)*

D = *distance between the centers of the two conductors*

r = *radius of the conductor*

Note: D and r must be expressed in the same units.

The first term in the preceding equation results from flux linkages inside the conductor and the second from flux linkages outside the conductor. Since inductive reactance is normally calculated to a 1-foot radius from the conductor center, it is convenient to separate the second term as follows:

$$L = \left[\frac{1}{2} + 2\ ln\left(\frac{1}{r}\right) + 2\ ln\left(\frac{D}{1}\right)\right] \times 10^{-7} \quad H/m \tag{1-8}$$

where:

$1/2 \times 10^{-7}$ = *inductance due to the flux inside the conductor*

$2 \times 10^{-7}\ ln(1/r)$ = *inductance due to the flux outside the conductor to a radius of 1 foot*

$2 \times 10^{-7}\ ln(D/1)$ = *inductance due to the flux external to the 1-foot radius of the conductor out to distance D*

Therefore, Equation 1-8 may be written as two equations. The first equation for the inductance out to a 1 foot radius, is:

$$L_1 = \left[\frac{1}{2} + 2\ ln\left(\frac{1}{r}\right)\right] \times 10^{-7} \quad H/m \tag{1-9}$$

The second due to the flux external to the 1-foot radius out to distance D is:

$$L_e = 2 \times 10^{-7}\ ln\left(\frac{D}{1}\right) \quad H/m \tag{1-10}$$

In the case of a two conductor line configuration, D is the distance between conductor centers.

Inductive Reactance

The more general method of calculating inductance and reactance expresses the distance between conductors in terms of geometric mean distance (*GMD*), and conductor radius in terms of geometric mean radius (*GMR*). *GMD* is the geometric mean distance between any number of conductors. *GMR* is the mathematically defined radius for a conductor that describes the inductance of a conductor for both internal and external flux out to a radius of 1 foot.

Substituting and converting to practical units of inductive reactance:

$$X_L = 0.2794\left(\frac{f}{60}\right) log_{10}\left(\frac{1}{GMR}\right) + 0.2794\left(\frac{f}{60}\right) log_{10}\ GMD \quad \Omega/mile \tag{1-11}$$

where:

f = *frequency, Hz*

GMR = *conductor geometric mean radius, ft*

GMD = *conductor separation in terms of geometric mean distance, ft*

Separating terms and letting X_a represent the first term and X_d represent the second term:

$$X_a = 0.2794\left(\frac{f}{60}\right) log_{10}\left(\frac{1}{GMR}\right) \quad \Omega/mile \tag{1-12}$$

$$X_d = 0.2794\left(\frac{f}{60}\right) log_{10}\ GMD \quad \Omega/mile \tag{1-13}$$

FIGURE 1-3

GMR FOR CONDUCTORS AND ANNULAR RINGS

Symmetrical Components

Geometric Mean Radii and Distances

Solid round conductor 0.779r

Full Stranding

7 ....0.726r
19 ....0.758r
37 ....0.768r
61 ....0.772r
91 ....0.774r
127 ....0.776r

Hollow stranded conductors and ACSR (neglecting steel strands)

30-two layer........................0.826r
26-two layer........................0.809r
54-three layer 0.810r
Single layer ACSR........0.35r–0.70r

Point within circle to circle ....r

Point outside circle to circle ....Distance to center of circle

Rectangle section of sides α and β ....0.2235 $(\alpha + \beta)$

Circular tube

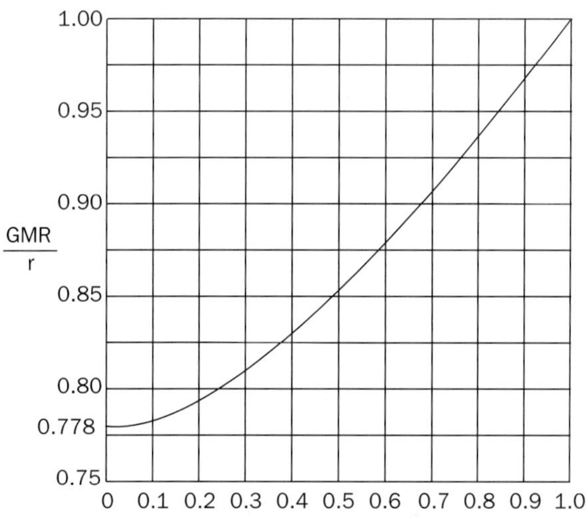

$$\text{Ratio} = \frac{\text{inner radius}}{\text{outer radius}}$$

r = 1/2 outside diameter

FIGURE 1-4

VALUES OF GEOMETRIC MEAN DISTANCE

Conductor Geometry	GMD
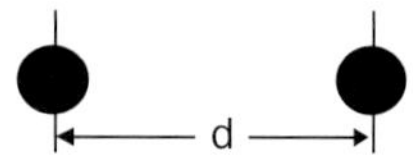	d
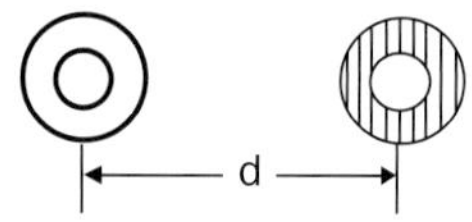	d
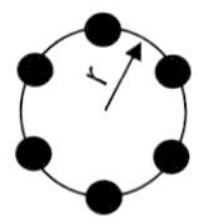 n conductors equally spaced around a circle of radius r	$r\sqrt[n-1]{n}$
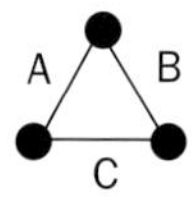 Symmetrical Triangle	A or B or C
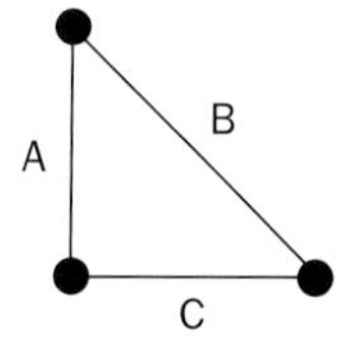 Right Triangle A = C	1.122 A
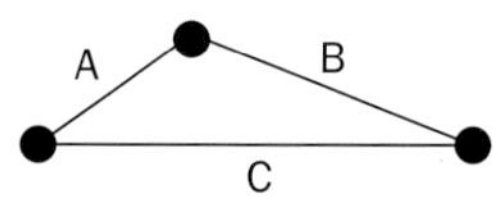 Unequal Triangle	$\sqrt[3]{A \times B \times C}$
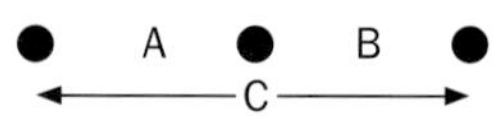 Symmetrical Flat	1.26 A
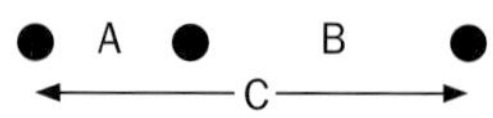 Unsymmetrical Flat	$\sqrt[3]{A \times B \times C}$

therefore:

$$X_L = X_a + X_d \quad \Omega/\text{mile per conductor} \tag{1-14}$$

where:

X_a = *inductive reactance due to the flux internal to the conductor plus the external flux to a radius of 1 foot*

X_d = *inductive reactance due to the flux surrounding the conductor from a 1-foot radius out to the center of the parallel return conductor (see Table 1-10 for calculated values)*

For a two-conductor, single-phase line, the total inductive reactance, phase to neutral, is two times the reactance for one conductor as calculated in Equation 1-14.

$$X_L = 2(X_a + X_d) \quad \Omega/\text{mile of circuit} \tag{1-15}$$

The GMR of a conductor represents the radius of an infinitely thin tube whose inductance under the same current loading is equal to that of the conductor. The basis for calculating GMRs for conductors and common geometrical shapes may be found in [7]. Figure 1-3 may be used to determine the *GMR* for most conductors. Values for the GMR, in feet, of the more common stranded conductors are listed in the conductor tables at the end of this chapter.

The general expression for *GMD* is:

$$GMD = \sqrt[n]{d_1 d_2 d_3 \text{----} d_n} \tag{1-16}$$

where:

d_1, d_2, d_3 ---- d_n *represent the center to center distances for all the conductors of the circuit.*

Figure 1-4 illustrates the geometric mean distance for some common 3-phase conductor configurations[8].

To demonstrate the use of GMR and GMD, consider the following example:

Determine the inductive reactance of a single circuit, balanced three phase 230 kV transmission line with 795 kcmil-26/7 ACSR "Drake" phase conductors arranged horizontally with a phase spacing of 25 feet.

The GMR of the "Drake" conductor, found in the tables at the end of the chapter, is 0.0375 feet (The physical radius of "Drake" is 0.0462 feet). From Equation 1-12 we obtain the inductive reactance to a 1-foot radius:

$$X_a = 0.2794 \log_{10}\left(\frac{1}{0.0375}\right) = 0.3984 \quad \Omega/\text{mile}$$

Referring to Figure 1-4, the GMD for a circuit with flat symmetrical spacing between conductors is 1.26 times the distance from the center conductor to either outside conductor. Therefore, the GMD for the circuit is:

$$GMD = (1.26)(25) = 31.5 \ \text{ft}$$

The inductive reactance related to conductor spacing from Equation 1-13 is:

$$X_d = 0.2794 \log_{10}(31.5) = 0.4186 \quad \Omega/\text{mile}$$

The total inductive reactance for these "Drake" conductors in this three phase circuit is:

$$X_a + X_d = 0.3984 + 0.4186 = 0.8170 \quad \Omega/\text{mile}$$

Since the resistance of "Drake" is only 0.1166 Ω/mile at 25°C, the magnitude of the voltage drop along the conductor is primarily due to inductive reactance, not resistance.

TABLE 1-10

INDUCTIVE REACTANCE SPACING FACTORS (X_d) AT 60 Hz IN Ω/MILE

ft	0.0	0.1	0.2	0.3	0.4	0.5	0.6	0.7	0.8	0.9
0	∞	−0.2794	−0.1953	−0.1461	−0.1112	−0.0841	−0.0620	−0.0433	−0.0271	−0.0128
1	0.0000	0.0116	0.0221	0.0318	0.0408	0.0492	0.0570	0.0644	0.0713	0.0779
2	0.0841	0.0900	0.0957	0.1011	0.1062	0.1112	0.1159	0.1205	0.1249	0.1292
3	0.1333	0.1373	0.1411	0.1449	0.1485	0.1520	0.1554	0.1588	0.1620	0.1651
4	0.1682	0.1712	0.1741	0.1770	0.1798	0.1825	0.1852	0.1878	0.1903	0.1928
5	0.1953	0.1977	0.2001	0.2024	0.2046	0.2069	0.2090	0.2112	0.2133	0.2154
6	0.2174	0.2194	0.2214	0.2233	0.2252	0.2271	0.2290	0.2308	0.2326	0.2344
7	0.2361	0.2378	0.2395	0.2412	0.2429	0.2445	0.2461	0.2477	0.2493	0.2508
8	0.2523	0.2538	0.2553	0.2568	0.2582	0.2597	0.2611	0.2625	0.2639	0.2653
9	0.2666	0.2680	0.2693	0.2706	0.2719	0.2732	0.2744	0.2757	0.2769	0.2782
10	0.2794	0.2806	0.2818	0.2830	0.2842	0.2853	0.2865	0.2876	0.2887	0.2899
11	0.2910	0.2921	0.2932	0.2942	0.2953	0.2964	0.2974	0.2985	0.2995	0.3005
12	0.3015	0.3025	0.3035	0.3045	0.3055	0.3065	0.3074	0.3084	0.3094	0.3103
13	0.3112	0.3122	0.3131	0.3140	0.3149	0.3158	0.3167	0.3176	0.3185	0.3194
14	0.3202	0.3211	0.3219	0.3228	0.3236	0.3245	0.3253	0.3261	0.3270	0.3278
15	0.3286	0.3294	0.3302	0.3310	0.3318	0.3326	0.3334	0.3341	0.3349	0.3357
16	0.3364	0.3372	0.3379	0.3387	0.3394	0.3402	0.3409	0.3416	0.3424	0.3431
17	0.3438	0.3445	0.3452	0.3459	0.3466	0.3473	0.3480	0.3487	0.3494	0.3500
18	0.3507	0.3514	0.3521	0.3527	0.3534	0.3540	0.3547	0.3554	0.3560	0.3566
19	0.3573	0.3579	0.3586	0.3592	0.3598	0.3604	0.3611	0.3617	0.3623	0.3629
20	0.3635	0.3641	0.3647	0.3653	0.3659	0.3665	0.3671	0.3677	0.3683	0.3688
21	0.3694	0.3700	0.3706	0.3711	0.3717	0.3723	0.3728	0.3734	0.3740	0.3745
22	0.3751	0.3756	0.3762	0.3767	0.3773	0.3778	0.3783	0.3789	0.3794	0.3799
23	0.3805	0.3810	0.3815	0.3820	0.3826	0.3831	0.3836	0.3841	0.3846	0.3851
24	0.3856	0.3861	0.3866	0.3871	0.3876	0.3881	0.3886	0.3891	0.3896	0.3901
25	0.3906	0.3911	0.3916	0.3920	0.3925	0.3930	0.3935	0.3939	0.3944	0.3949
26	0.3953	0.3958	0.3963	0.3967	0.3972	0.3977	0.3981	0.3986	0.3990	0.3995
27	0.3999	0.4004	0.4008	0.4013	0.4017	0.4021	0.4026	0.4030	0.4035	0.4039
28	0.4043	0.4048	0.4052	0.4056	0.4061	0.4065	0.4069	0.4073	0.4078	0.4082
29	0.4086	0.4090	0.4094	0.4098	0.4103	0.4107	0.4111	0.4115	0.4119	0.4123
30	0.4127	0.4131	0.4135	0.4139	0.4143	0.4147	0.4151	0.4155	0.4159	0.4163
31	0.4167	0.4171	0.4175	0.4179	0.4182	0.4186	0.4190	0.4194	0.4198	0.4202
32	0.4205	0.4209	0.4213	0.4217	0.4220	0.4224	0.4228	0.4232	0.4235	0.4239
33	0.4243	0.4246	0.4250	0.4254	0.4257	0.4261	0.4265	0.4268	0.4272	0.4275
34	0.4279	0.4283	0.4286	0.4290	0.4293	0.4297	0.4300	0.4304	0.4307	0.4311
35	0.4314	0.4318	0.4321	0.4324	0.4328	0.4331	0.4335	0.4338	0.4342	0.4345
36	0.4348	0.4352	0.4355	0.4358	0.4362	0.4365	0.4368	0.4372	0.4375	0.4378
37	0.4382	0.4385	0.4388	0.4391	0.4395	0.4398	0.4401	0.4404	0.4408	0.4411
38	0.4414	0.4417	0.4420	0.4423	0.4427	0.4430	0.4433	0.4436	0.4439	0.4442
39	0.4445	0.4449	0.4452	0.4455	0.4458	0.4461	0.4464	0.4467	0.4470	0.4473
40	0.4476	0.4479	0.4482	0.4485	0.4488	0.4491	0.4494	0.4497	0.4500	0.4503
41	0.4506	0.4509	0.4512	0.4515	0.4518	0.4521	0.4524	0.4527	0.4530	0.4532
42	0.4535	0.4538	0.4541	0.4544	0.4547	0.4550	0.4553	0.4555	0.4558	0.4561
43	0.4564	0.4567	0.4570	0.4572	0.4575	0.4578	0.4581	0.4584	0.4586	0.4589
44	0.4592	0.4595	0.4597	0.4600	0.4603	0.4606	0.4608	0.4611	0.4614	0.4616
45	0.4619	0.4622	0.4624	0.4627	0.4630	0.4632	0.4635	0.4638	0.4640	0.4643
46	0.4646	0.4648	0.4651	0.4654	0.4656	0.4659	0.4661	0.4664	0.4667	0.4669
47	0.4672	0.4674	0.4677	0.4680	0.4682	0.4685	0.4687	0.4690	0.4692	0.4695
48	0.4697	0.4700	0.4702	0.4705	0.4707	0.4710	0.4712	0.4715	0.4717	0.4720
49	0.4722	0.4725	0.4727	0.4730	0.4732	0.4735	0.4737	0.4740	0.4742	0.4744
50	0.4747	0.4749	0.4752	0.4754	0.4757	0.4759	0.4761	0.4764	0.4766	0.4769

Positive and Negative Sequence Inductive Reactance
The positive and negative sequence inductive reactance per phase for a balanced, equilaterally spaced, three-phase transmission line is the same as that for one conductor of a single-phase, line to neutral, circuit discussed previously. Therefore, for a balanced three-phase system, the positive and negative sequence reactances may be expressed by:

$$X_1 = X_2 = X_L = X_a + X_d \quad \Omega/mile \qquad (1\text{-}17)$$

where:

X_1 = *positive sequence inductive reactance*

X_2 = *negative sequence inductive reactance*

X_L = *inductive reactance for a conductor of geometric mean radius GMR, and an equivalent geometric mean conductor spacing of GMD*

or in terms of Equation 1-11 as follows:

$$X_L = 0.2794\left(\frac{f}{60}\right)\log_{10}\left(\frac{1}{GMR}\right) + 0.2794\left(\frac{f}{60}\right)\log_{10} GMD \quad \Omega/mile \qquad (1\text{-}18)$$

The Equation 1-11 does not hold for an unbalanced three-phase system unless the phases are suitably transposed and it is assumed that balanced loading exists. For a system with asymmetrical spacing, the voltage drop for each conductor is different even if the conductors are carrying equal and balanced currents. To reduce this effect, conductors may be successively transposed in successive line sections[9][10]. Full transposition, called a "barrel transposition," equalizes the total voltage drop for each conductor and reduces the magnetic field external to the conductors.

For a three-phase circuit with non-symmetrical conductor spacing, the *GMD* for the three conductor separations must be substituted for X_d. For this case, the average inductive reactance spacing factor, X_{du}, becomes:

$$X_d = X_{du} = \frac{1}{3}\left(X_{d(12)} + X_{d(23)} + X_{d(31)}\right) \quad \Omega/mile \qquad (1\text{-}19)$$

The *GMD* of the three conductor spacings can also be calculated by:

$$GMD = \sqrt[3]{d_{12}\, d_{23}\, d_{31}} \qquad (1\text{-}20)$$

where:

d_{12}, d_{23}, d_{31} *are the respective distances between phases*

and this *GMD* can be used in the foregoing equations to calculate the inductive reactance spacing factor, X_{du}.

Zero-Sequence Resistance and Inductive Reactance
Zero-sequence currents in three-phase circuits are equal in both phase and magnitude. They flow out through the phase conductors and return through a neutral path consisting of earth or a combination of earth and ground wire paths. Calculation of zero-sequence impedance must be carried out in two steps; the determination of the self-impedance of one conductor with earth return and the determination of the mutual impedance between conductors with common earth return[9][10].

The influence of the earth return path on the calculation of zero sequence reactance is presented in the following equations. The resistance, R_e, and reactance, X_e, of the earth per phase are given by the equations:

$$R_e = 0.2858\left(\frac{f}{60}\right) \quad \Omega/mile \qquad (1\text{-}21)$$

$$X_e = 0.4191\left(\frac{f}{60}\right)\log_{10}\left[77{,}760\left(\frac{60}{f}\right)\rho_e\right] \quad \Omega/mile \qquad (1\text{-}22)$$

where:

f = *frequency, Hz*

ρ_e = *earth's resistivity, Ω-m*

As shown in Table 1-11, the earth's resistivity varies between 1 Ω-m and 10,000 Ω-m. The value of 100 Ω-m is typically used if the resistivity is unknown.

TABLE 1-11
ZERO SEQUENCE RESISTANCE AND REACTANCE OF EARTH AT 60 HZ
FOR ρ_e BETWEEN 1 AND 10,000 Ω-m

Earth Resistivity (ρ_e) Ω-m	**Reactance of Earth** (X_e) Ω/mile per conductor
1	2.050
5	2.343
10	2.469
50	2.762
100	2.888
500	3.181
1,000	3.307
5,000	3.600
10,000	3.726

The zero-sequence reactance, Z_0, is also affected by coupling between the phase and ground conductors. The zero-sequence impedance of a three-phase transmission line without ground conductors excluding capacitance effects is:

$$Z_0 = R_{cdr} + R_e + j(X_e + X_a - 2X_d) \quad \Omega/mile \qquad (1\text{-}23)$$

where:

R_e = *earth resistance, Ω/mile*

R_{cdr} = *conductor ac resistance, Ω/mile*

X_e = *earth reactance, Ω/mile*

X_a = *inductive reactance to 1-foot radius, Ω/mile*

X_d = *inductive reactance spacing factor, Ω/mile*

Calculate the zero-sequence impedance for a 115 kV single-circuit transmission line having one 477 kcmil-26/7 ACSR "Hawk" conductor per phase. The phases are horizontally arranged with 15 feet between conductors and an earth resistivity of 100 Ω-m.

For a frequency of 60 Hz:

$$R_e = 0.2858 \quad \Omega/mile$$

From Equation 1-22, the earth zero sequence reactance is:

$$X_e = 0.4191 \log_{10} \left[(77{,}760) \left(\frac{60}{60} \right) (100) \right]$$
$$= 2.8879 \quad \Omega/mile$$

Conductor ac resistance at 50°C:

$$R_{50°C} = 0.2117 \quad \Omega/mile$$

From Equations 1-12 and 1-13, the inductive reactance out to a 1-foot radius, X_a, and the inductive reactance out from the 1-foot radius to the center of the return conductor, X_d, may be calculated. Therefore:

$$X_a = 0.4296 \quad \Omega/mile$$

and

$$X_d = 0.3286 \quad \Omega/mile$$

From Equation 1-23, the zero sequence reactance is:

$$Z_0 = 0.2117 + 0.2858$$
$$+ j\,[2.8879 + 0.4296 - 2(0.3286)]$$
$$= 0.4975 + j\,2.6604 \quad \Omega/mile$$

Inductive Reactance for Bundled Conductors

For transmission lines operating at voltages above 345 kV, bundles of more than one conductor per phase are often employed[8]. At lower voltages this may be done to increase thermal capacity and reduce voltage drop. At higher voltages, bundling conductors increases the effective electrical radius of the phase, reducing TV and radio interference. Bundled conductors also reduce the reactance of the phase and increase load stability.

Normally, all subconductors making up a conductor bundle are the same size and type. The spacing of the subconductors within the bundle is typically 18 inches and the spacing of the phase bundles far exceeds the bundle diameter. Under these assumptions, it is possible to simplify the calculation of bundle inductive reactance as follows:

(1) Find the X_a for a subconductor.

(2) Calculate the X_d value for each subconductor relative to the other subconductors in the bundle. The average of the X_d's for each subconductor in the bundle is the average inductive reactance spacing factor X_{ds} for the bundle.

(3) Calculate a bundle inductive reactance to 1-foot radius according to the following equations.

For 2- and 3-conductor bundles where n is equal to the number of conductors:

$$X_a^b = \frac{1}{n}[X_a - (n-1)\,X_{ds}] \qquad (1\text{-}24)$$

For a 4-conductor bundle:

$$X_a^b = \frac{1}{n}[X_a - (n-1)\,X_{ds}] + X_d - 0.0105 \qquad (1\text{-}25)$$

(4) Calculate the inductive reactance spacing factor beyond 1 foot $X_{d'}^b$ for the bundles by calculating the $X_{d_{pp}}$ for each phase bundle relative to each other phase bundle, assuming the bundle spacing, S_{pp}, to be the distance between bundle centers. The bundle spacing component of inductive reactance, beyond 1 foot, $X_{d'}^b$ is the average of the $X_{d_{pp}}$ values:

$$X_{d_{pp}} = 0.2794\left(\frac{f}{60}\right)\log_{10} S_{pp} \qquad (1\text{-}26)$$

$$X_d^b = \frac{1}{3}\left(X_{d_{12}} + X_{d_{23}} + X_{d_{13}}\right) \qquad (1\text{-}27)$$

(5) Calculate the positive sequence inductive reactance of a single-bundle using:

$$X^b = X_a^b + X_d^b \quad \Omega/mile \qquad (1\text{-}28)$$

Calculate the positive sequence inductive reactance for a single circuit 345 kV line with horizontal phase arrangement, consisting of two 954 kcmil-45/7 ACSR "Rail" conductors spaced 18 inches apart per phase, and the phases spaced 25 feet apart.

From Equation 1-12,

$$X_a = 0.395 \quad \Omega/mile$$

Calculate X_d for each subconductor relative to the other subconductor in the bundle using Equation 1-13 and a spacing of 18 inches:

$$X_d = 0.0492 \quad \Omega/mile$$

The average X_{ds} for the two subconductors with 18 inches spacing is:

$$X_{ds} = \frac{1}{2}[0.0492 + 0.0492]$$
$$= 0.0492 \quad \Omega/mile$$

The inductive reactance to 1-foot spacing of each bundle is:

$$X_a^b = \frac{1}{2}[0.395 - (2-1)(0.0492)]$$
$$= 0.1729 \quad \Omega/mile$$

The inductive reactance beyond 1 foot due to the spacing of the phase bundles relative to each other is:

$$X_{d_{12}} = 0.2793\,(1)(1.398) = 0.3906 \quad \Omega/mile$$

$$X_{d_{23}} = 0.2793\,(1)(1.398) = 0.3906 \quad \Omega/mile$$

$$X_{d_{13}} = 0.2793\,(1)(1.699) = 0.4747 \quad \Omega/mile$$

The average inductive reactance spacing factor, beyond 1 foot, is:

$$X_d^b = \frac{1}{3}[0.3906 + 0.3906 + 0.4747]$$

$$= 0.4186 \quad \Omega/mile$$

The positive sequence inductive reactance for a single bundle phase of a balanced three phase system is:

$$X^b = 0.1729 + 0.4186 = 0.5915 \quad \Omega/mile$$

1.3.4 Shunt Capacitance and Capacitive Reactance

The capacitance of lower voltage overhead transmission lines is generally negligible under normal operating conditions. However, it becomes a major concern for high voltage transmission lines and must be taken into consideration when determining efficiency, regulation, power factor and voltage distribution along the line. A detailed discussion of the subject can be found in [7][9][10][11]. The following is a general discussion presenting the more commonly seen and used formulas for calculating capacitance and capacitive reactance.

Shunt Capacitance

The capacitance per unit length of a transmission circuit component such as a bare stranded overhead conductor is defined as the ratio of the charging current per unit length, i_{shunt}, to the rate of change of voltage, V, on the conductor:

$$i_{shunt} = c\frac{dV}{dt} \tag{1-29}$$

where:

c = *capacitance per unit length, Farads/mile*

Under alternating current conditions the conductor system charges and discharges during each cycle, producing a capacitive current. This *rms* current is usually termed capacitive charging current and is defined as:

$$I_{charging} = 2\pi\ f\ C\ V \quad amperes \tag{1-30}$$

where:

f = *frequency, Hz*

C = *capacitance, F*

V = *voltage, V*

Consider a single-phase overhead line consisting of two parallel conductors, each with radius r, and separated by a distance D between centers. If the potential difference between the conductors is V and the charge on each conductor is q, the voltage between the two conductors is:

$$V_{12} = 36 \times 10^9 (q)\ ln\left(\frac{D-r}{r}\right) \quad volts \tag{1-31}$$

where:

V_{12} = *potential difference between conductors, volts*

q = *charge on each conductor, coulombs*

r = *conductor radius, meters*

D = *distance between conductor centers, meters*

Equation 1-31 is true only if the charges on the two conductors are equal in magnitude but opposite in sign.

The capacitance between conductors is defined as:

$$C = \frac{q}{V} \quad Farads/meter \tag{1-32}$$

From this:

$$C = \frac{1}{36 \times 10^9\ ln\left[\frac{(D-r)}{r}\right]} \quad Farads/meter \tag{1-33}$$

Simplifying and converting to common logarithms, the line-to-line capacitance is:

$$C = \frac{0.0194}{\log_{10}\left[\frac{(D-r)}{r}\right]} \quad \mu F/mile \qquad (1\text{-}34)$$

where:

r = *conductor radius, ft*

D = *distance between conductor centers, ft*

For overhead lines, the conductor radius r is usually small compared with conductor separation D. For small conductor radii, we can assume that $(D - r) \approx D$. Equation 1-32 may then be written as:

$$C = \frac{0.0194}{\log_{10}\left[\frac{D}{r}\right]} \quad \mu F/mile \qquad (1\text{-}35)$$

The phase-to-neutral capacitance is obtained by dividing the charge, q, on the conductor by one half the voltage. Thus, the capacitance to neutral is twice that given in Equation 1-33:

$$C = \frac{0.0388}{\log_{10}\left[\frac{D}{r}\right]} \quad \mu F/mile \qquad (1\text{-}36)$$

This expression is valid for the line capacitance to neutral for any symmetrically balanced three-phase system with equilateral spacing.

The geometric mean distance, *GMD*, between conductors for any line configuration may be calculated by the methods discussed in Section 1.3.3. Substituting the *GMD*, in feet, for the separation distance D in Equation 1-34 will allow the calculation of the average capacitance between each conductor and neutral for any typical three-phase system:

$$C = \frac{0.0388}{\log_{10}\left[\frac{GMD}{r}\right]} \quad \mu F/mile \qquad (1\text{-}37)$$

Shunt Capacitive Reactance

The capacitive reactance of a conductor system is:

$$X_c = \frac{1}{2\pi f C} \quad M\Omega\text{-}mile \qquad (1\text{-}38)$$

where:

f = *frequency, Hz*

C = *capacitance, μF/mile*

Substituting the expression for C from Equation 1-35 and simplifying gives:

$$X_c = 0.0683\left(\frac{60}{f}\right)\log_{10}\left(\frac{GMD}{r}\right) \quad M\Omega\text{-}mile \qquad (1\text{-}39)$$

In the same manner used for reactance, Equation 1-39 may be divided into two terms; the capacitive reactance to 1-foot spacing and capacitive reactance external to the 1-foot radius. Rewriting Equation 1-39 and separating into components:

$$X_C = \left[0.0683\left(\frac{60}{f}\right)\log_{10}\left(\frac{1}{r}\right)\right] + \left[0.0683\left(\frac{60}{f}\right)\log_{10} GMD\right] \qquad (1\text{-}40)$$

where:

$$X_a' = 0.0683\left(\frac{60}{f}\right)\log_{10}\left(\frac{1}{r}\right) \quad M\Omega\text{-}mile \qquad (1\text{-}41)$$

TABLE 1-12

CAPACITIVE REACTANCE SPACING FACTORS (X'_d) AT 60 HZ IN MΩ-MILE

ft	0.0	0.1	0.2	0.3	0.4	0.5	0.6	0.7	0.8	0.9
0	∞	−0.0683	−0.0477	−0.0357	−0.0272	−0.0206	−0.0152	−0.0106	−0.0066	−0.0031
1	0.0000	0.0028	0.0054	0.0078	0.0100	0.0120	0.0139	0.0157	0.0174	0.0190
2	0.0206	0.0220	0.0234	0.0247	0.0260	0.0272	0.0283	0.0295	0.0305	0.0316
3	0.0326	0.0336	0.0345	0.0354	0.0363	0.0372	0.0380	0.0388	0.0396	0.0404
4	0.0411	0.0419	0.0426	0.0433	0.0440	0.0446	0.0453	0.0459	0.0465	0.0471
5	0.0477	0.0483	0.0489	0.0495	0.0500	0.0506	0.0511	0.0516	0.0521	0.0527
6	0.0532	0.0536	0.0541	0.0546	0.0551	0.0555	0.0560	0.0564	0.0569	0.0573
7	0.0577	0.0581	0.0586	0.0590	0.0594	0.0598	0.0602	0.0606	0.0609	0.0613
8	0.0617	0.0621	0.0624	0.0628	0.0631	0.0635	0.0638	0.0642	0.0645	0.0649
9	0.0652	0.0655	0.0658	0.0662	0.0665	0.0668	0.0671	0.0674	0.0677	0.0680
10	0.0683	0.0686	0.0689	0.0692	0.0695	0.0698	0.0700	0.0703	0.0706	0.0709
11	0.0711	0.0714	0.0717	0.0719	0.0722	0.0725	0.0727	0.0730	0.0732	0.0735
12	0.0737	0.0740	0.0742	0.0745	0.0747	0.0749	0.0752	0.0754	0.0756	0.0759
13	0.0761	0.0763	0.0765	0.0768	0.0770	0.0772	0.0774	0.0776	0.0779	0.0781
14	0.0783	0.0785	0.0787	0.0789	0.0791	0.0793	0.0795	0.0797	0.0799	0.0801
15	0.0803	0.0805	0.0807	0.0809	0.0811	0.0813	0.0815	0.0817	0.0819	0.0821
16	0.0823	0.0824	0.0826	0.0828	0.0830	0.0832	0.0833	0.0835	0.0837	0.0839
17	0.0841	0.0842	0.0844	0.0846	0.0847	0.0849	0.0851	0.0852	0.0854	0.0856
18	0.0857	0.0859	0.0861	0.0862	0.0864	0.0866	0.0867	0.0869	0.0870	0.0872
19	0.0874	0.0875	0.0877	0.0878	0.0880	0.0881	0.0883	0.0884	0.0886	0.0887
20	0.0889	0.0890	0.0892	0.0893	0.0895	0.0896	0.0898	0.0899	0.0900	0.0902
21	0.0903	0.0905	0.0906	0.0907	0.0909	0.0910	0.0912	0.0913	0.0914	0.0916
22	0.0917	0.0918	0.0920	0.0921	0.0922	0.0924	0.0925	0.0926	0.0928	0.0929
23	0.0930	0.0931	0.0933	0.0934	0.0935	0.0937	0.0938	0.0939	0.0940	0.0942
24	0.0943	0.0944	0.0945	0.0947	0.0948	0.0949	0.0950	0.0951	0.0953	0.0954
25	0.0955	0.0956	0.0957	0.0958	0.0960	0.0961	0.0962	0.0963	0.0964	0.0965
26	0.0967	0.0968	0.0969	0.0970	0.0971	0.0972	0.0973	0.0974	0.0976	0.0977
27	0.0978	0.0979	0.0980	0.0981	0.0982	0.0983	0.0984	0.0985	0.0986	0.0987
28	0.0989	0.0990	0.0991	0.0992	0.0993	0.0994	0.0995	0.0996	0.0997	0.0998
29	0.0999	0.1000	0.1001	0.1002	0.1003	0.1004	0.1005	0.1006	0.1007	0.1008
30	0.1009	0.1010	0.1011	0.1012	0.1013	0.1014	0.1015	0.1016	0.1017	0.1018
31	0.1019	0.1020	0.1021	0.1022	0.1023	0.1023	0.1024	0.1025	0.1026	0.1027
32	0.1028	0.1029	0.1030	0.1031	0.1032	0.1033	0.1034	0.1035	0.1035	0.1036
33	0.1037	0.1038	0.1039	0.1040	0.1041	0.1042	0.1043	0.1044	0.1044	0.1045
34	0.1046	0.1047	0.1048	0.1049	0.1050	0.1050	0.1051	0.1052	0.1053	0.1054
35	0.1055	0.1056	0.1056	0.1057	0.1058	0.1059	0.1060	0.1061	0.1061	0.1062
36	0.1063	0.1064	0.1065	0.1066	0.1066	0.1067	0.1068	0.1069	0.1070	0.1070
37	0.1071	0.1072	0.1073	0.1074	0.1074	0.1075	0.1076	0.1077	0.1078	0.1078
38	0.1079	0.1080	0.1081	0.1081	0.1082	0.1083	0.1084	0.1085	0.1085	0.1086
39	0.1087	0.1088	0.1088	0.1089	0.1090	0.1091	0.1091	0.1092	0.1093	0.1094
40	0.1094	0.1095	0.1096	0.1097	0.1097	0.1098	0.1099	0.1100	0.1100	0.1101
41	0.1102	0.1102	0.1103	0.1104	0.1105	0.1105	0.1106	0.1107	0.1107	0.1108
42	0.1109	0.1110	0.1110	0.1111	0.1112	0.1112	0.1113	0.1114	0.1114	0.1115
43	0.1116	0.1117	0.1117	0.1118	0.1119	0.1119	0.1120	0.1121	0.1121	0.1122
44	0.1123	0.1123	0.1124	0.1125	0.1125	0.1126	0.1127	0.1127	0.1128	0.1129
45	0.1129	0.1130	0.1131	0.1131	0.1132	0.1133	0.1133	0.1134	0.1135	0.1135
46	0.1136	0.1136	0.1137	0.1138	0.1138	0.1139	0.1140	0.1140	0.1141	0.1142
47	0.1142	0.1143	0.1143	0.1144	0.1145	0.1145	0.1146	0.1147	0.1147	0.1148
48	0.1148	0.1149	0.1150	0.1150	0.1151	0.1152	0.1152	0.1153	0.1153	0.1154
49	0.1155	0.1155	0.1156	0.1156	0.1157	0.1158	0.1158	0.1159	0.1159	0.1160
50	0.1161	0.1161	0.1162	0.1162	0.1163	0.1164	0.1164	0.1165	0.1165	0.1166

and

$$X'_d = 0.0683\left(\frac{60}{f}\right)\log_{10} GMD \qquad M\Omega\text{-}mile \tag{1-42}$$

therefore:

$$X_c = X'_a + X'_d \qquad M\Omega\text{-}mile \tag{1-43}$$

X'_a = *electrostatic flux within a 1-foot radius*

X'_d = *electrostatic flux external to the 1-foot radius out to distance D, feet (see Table 1-12 for calculated values)*

The previous equations do not consider the effect of the earth on capacitance or capacitive reactance. Even under balanced line conditions the presence of earth near a transmission line causes a small increase in the capacitance to neutral. Determining the effect of the earth on capacitance involves evaluating the variation of charge density along the earth's surface with the distance to each conductor and is beyond the scope of this book. The effects of earth on transmission line constants are discussed in [7][9][10][11].

Positive and Negative Sequence Capacitive Reactance

As with inductive reactance, the equations developed for the capacitive reactance for a single-phase line to neutral circuit may be used to calculate the positive and negative sequence capacitive reactance for a balanced, equilaterally spaced (symmetrical), three-phase line. Writing the equations in terms of their component factors:

$$X_{c1} = X_{c2} = X_c = X'_a + X'_d \qquad M\Omega\text{-}mile \tag{1-44}$$

or from Equation 1-40,

$$X_c = \left[0.0683\left(\frac{60}{f}\right)\log_{10}\left(\frac{1}{r}\right)\right] + \left[0.0683\left(\frac{60}{f}\right)\log_{10} GMD\right] \qquad M\Omega\text{-}mile \tag{1-45}$$

Equation 1-45 may be used for a three-phase circuit with asymmetrical conductor spacing if the *GMD* for the three conductor separations is substituted for X'_d. For this case the average spacing factor, X'_{du}, between all conductors becomes:

$$X'_{du} = \frac{1}{3}\left(X'_{d(12)} + X'_{d(23)} + X'_{d(31)}\right) \qquad M\Omega\text{-}mile \tag{1-46}$$

or the *GMD* of the three conductor spacings may be calculated by:

$$GMD = \sqrt[3]{d_{12}d_{23}d_{31}} \tag{1-47}$$

This new *GMD* should then be substituted in Equation 1-46 to calculate the spacing factor, X'_{du}.

Zero-Sequence Capacitive Reactance

The zero-sequence capacitive reactance, X_{co}, of a 3-phase circuit with earth return but without ground conductors is:

$$X_{co} = X'_a + X_{ce} - 2X'_d \qquad M\Omega\text{-}mile \tag{1-48}$$

The term X_{ce} accounts for the electrostatic flux outside the 1-foot radius to the earth return and is determined by:

$$X_{ce} = 0.205\left(\frac{60}{f}\right)\log_{10}(2h) \qquad M\Omega\text{-}mile \tag{1-49}$$

where:

f = *frequency, Hz*

h = *conductor height above ground, ft*

Table 1-13 gives values for X_{ce} assuming a frequency of 60 Hz and conductor heights above ground ranging from 10 to 100 feet.

TABLE 1-13
ZERO SEQUENCE CAPACITIVE REACTANCE OF EARTH AT 60 HZ

Conductor Height Above Ground (*h*) ft	X_{ce} MΩ-Mile Per Conductor
10	0.267
15	0.303
20	0.328
25	0.348
30	0.364
40	0.390
50	0.410
60	0.426
70	0.440
80	0.452
90	0.462
100	0.472

Methods and equations for calculating the zero sequence capacitive reactance for circuits having ground wires as well as earth return, or consisting of multiple circuits, may be found in [7][9][10][11].

Shunt Capacitive Reactance of Bundled Conductors

The capacitive reactance of bundled conductors can be found by use of the same equations as were used for inductive reactance except the appropriate capacitive reactance terms must be substituted for the inductive reactance terms.

The capacitive reactance of 2-, 3- or 4-conductor bundles is calculated using equations similar to those for calculating inductive reactance.

For 2- or 3-conductor bundles:

$$X'^{b}_{a} = \frac{1}{n}\left[X'_a - (n-1)\,X'_{ds}\right] \qquad (1\text{-}50)$$

For a 4-conductor bundle:

$$X'^{b}_{a} = \frac{1}{n}\left[X'_a - (n-1)\,X'_{ds}\right] + X'_d - 0.0105 \qquad (1\text{-}51)$$

where:

X'_{ds} = *the average capacitive reactance spacing factor for the conductor bundle, and is the average of the* X'_d *values for each subconductor in the bundle*

1.4 MECHANICAL CHARACTERISTICS OF STRANDED CONDUCTORS

In addition to electrical and thermal properties, the mechanical characteristics of a conductor are an important consideration in conductor selection. Two of the most important mechanical properties of a conductor are the rated breaking strength and the stress-strain relationship. Both of these properties are a function of the mechanical properties of the individual strands of which the conductor is composed and the design used to incorporate the strands into the conductor.

The method of calculating rated breaking strength and the general stress-strain characteristics of stranded conductor are presented in the following sections. A more complete discussion of the effects of stress-strain behavior on conductor sag and tension is presented in Chapter 2.

1.4.1 Rated Strength of Concentric-Lay-Stranded Conductors

The method of calculating the rated breaking strength of a stranded conductor is specified by ASTM Standards based on conductor material, type and stranding. It is a function of the minimum average tensile strength of the component wires and rating factors that are based on the number of strand layers. For composite conductors, the rated breaking strength is the sum of the calculated rated breaking strengths for each material.

The values of rated strength given in the conductor tables at the end of the chapter have been calculated in accordance with the various ASTM specifications for the specified conductor construction. Calculation of the rated strength for an AAC and ACSR conductor may be performed as demonstrated in the following examples.

Rated Strength of an AAC Conductor

The rated strength for an all-aluminum conductor (AAC) is calculated in accordance with ASTM Standards B 230 and B 231. ASTM Standard B 230 defines the mechanical characteristics for the individual aluminum strands of 1350-H19 aluminum. ASTM Standard B 231 gives the requirements for concentric-lay stranded AAC conductors and defines the rated strength of the conductor as being the sum of the minimum average tensile strengths of the individual wires for the specified wire diameter multiplied by the stranding factors given in Table 1-14[1].

TABLE 1-14
AAC RATING FACTORS

Stranding		Rating Factor %
Number of Wires in Conductor	Number of Layers	
7	1	96
19	2	93
37	3	91
61	4	90
91 and above	5 and above	89

Calculate the rated breaking strength of a 795 kcmil-37 strand AAC "Arbutus" conductor.

From Table 1-20, the diameter of the individual strands is 0.1466 inches. From Table 1-2, the minimum average tensile strength for a 0.1466 inch diameter 1350-H19 wire is 24.5 ksi (24,500 psi). The breaking strength of a single strand is:

$$\text{Wire Strength} = \frac{\pi}{4}\,(0.1446)^2\,(24{,}500)$$
$$= 413.5 \text{ lbs}$$

Multiplying the rated breaking strength of a single strand by the number of strands in the cable and the specified rating factor of 91% from Table 1-14, the rated breaking strength of the conductor is:

$$\text{Rated Breaking Strength} = (37)(413.5)(0.91)$$
$$= 13{,}924 \text{ lbs}$$

Rounding the calculated breaking strength to three significant places gives:

$$\text{Rated Breaking Strength} = 13{,}900 \text{ lbs}$$

Rated Strength of an ACSR Conductor

The calculation of the rated strength for an ACSR conductor is similar to that of an AAC conductor except the strength of ACSR is the sum of the strengths of two different materials multiplied by the appropriate stranding factors specified in ASTM Standards.

ACSR conductors, with galvanized core wire, are manufactured in accordance with ASTM Standard B 232. The 1350-H19 aluminum wires and the galvanized steel core wires must meet the requirements of ASTM Standard B 230 and ASTM Standard B 498, respectively. ASTM Standard B 232 defines the rated strength of ACSR conductors as being the aggregate sum of the strengths of the individual aluminum and steel component wires. The tensile strength of the individual aluminum strands is the minimum average tensile strengths for the specified wire diameter. Since the elongation of 1350-H19 wire is approximately 1%, the tensile strength of the steel wire at 1% elongation is used in calculating the composite rated breaking strength. The rating factors for the aluminum and steel components are listed in Table 1-15[1].

TABLE 1-15
ACSR RATING FACTORS

Stranding				Rating Factor %	
Number of Wires		Number of Layers			
Aluminum	Steel	Aluminum	Steel	Aluminum	Steel
6	1	1	1	96	96
7	1	1	1	96	96
8	1	1	1	96	96
18	1	2	1	93	96
36	1	3	1	91	96
12	7	1	1	96	96
24	7	2	1	93	96
26	7	2	1	93	96
30	7	2	1	93	96
42	7	3	1	91	96
45	7	3	1	91	96
48	7	3	1	91	96
54	7	3	1	91	96
72	7	4	1	90	96
16	19	1	2	96	93
30	19	2	2	93	93
54	19	3	2	91	93
76	19	4	2	90	93
84	19	4	2	90	93

Calculate the rated breaking strength of 795 kcmil-26/7 ACSR "Drake" conductor (26 × 0.1749 inch 1350-H19 strands and 7 × 0.1360 inch steel strands).

From Table 1-2, the minimum average tensile strength for a 0.1749 inch diameter 1350-H19 wire is 24,000 ksi. The breaking strength of a single strand is:

$$\text{Wire Strength} = \frac{\pi}{4}(0.1749)^2\ (24{,}000)$$

$$= 576.6 \text{ lbs}$$

From Table 1-3, the minimum average tensile stress at 1% elongation for a 0.1360 inch diameter Class A galvanized steel wire is 180 ksi. The rated breaking strength of a single steel strand is:

$$\text{St. Wire Strength} = \frac{\pi}{4}(0.1360)^2\ (180{,}000)$$

$$= 2615 \text{ lbs}$$

FIGURE 1-5

TYPICAL STRESS-STRAIN TEST PLOT

THE ONE HOUR MODULUS CURVE, THE INITIAL MODULUS AND FINAL MODULUS CURVES ARE INDICATED

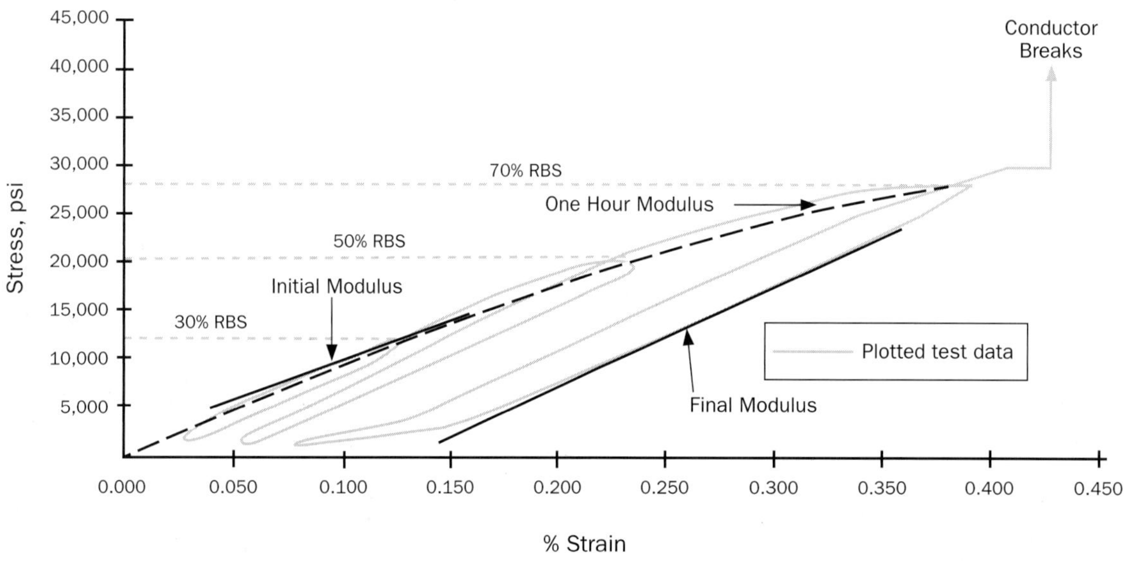

Note:

1. The conductor length increases due to non-elastic elongation during each hold period and the final modulus (slope) after each hold period is approximately the same.
2. The curve drawn through the end of the hold periods is called the ONE HOUR MODULUS curve of the initial composite stress-strain curve.
3. The slope of the curves during each unloading, decreasing tension level, for each hold period is essentially parallel.
4. Since the INITIAL composite curve is non-linear, it is normally expressed as a polynomial equation. The FINAL curve is linear and is normally expressed as a simple linear modulus.

Taking the rating factors for 26/7 ACSR from Table 1-15, the rated breaking strength of the conductor is:

Rated Breaking Strength = (26) (576.6) (0.93) + (7) (2615) (0.96) = 31,515 lbs

and rounding the rated breaking strength to three significant places,

Rated Breaking Strength = 31,500 lbs

1.4.2 Stress-Strain Characteristics of Stranded Conductors

The elastic and non-elastic (plastic) elongation of conductors under tensile stress (tension) must be determined to predict the behavior of conductors in service. Single strands of conductor materials initially elongate elastically and linearly when subjected to sub-yield point tensile stresses for short periods of time. The application of such stresses is used to develop the tensile properties shown in Table 1-1. Some of these properties, particularly the elastic modulus, are not valid for the more complex behavior of stranded conductors that remain under mechanical stress for very long periods of time. Under such conditions, the conductors exhibit both elastic and plastic elongation even though the applied stress is below the yield point of the conductor materials. The plastic elongation is due to a metallurgical phenomenon called "creep."

To predict the elongation of conductors under tension, two (2) types of test are performed. The first is a stress-strain test to measure the short time behavior of a conductor under varying tensile loads, and the second is a series of creep tests to measure the long-term non-elastic behavior of a conductor under tension. Details of these test methods are provided by The Aluminum Association[12][13].

To perform a stress-strain test, a sample of new

FIGURE 1-6
CREEP PLOT

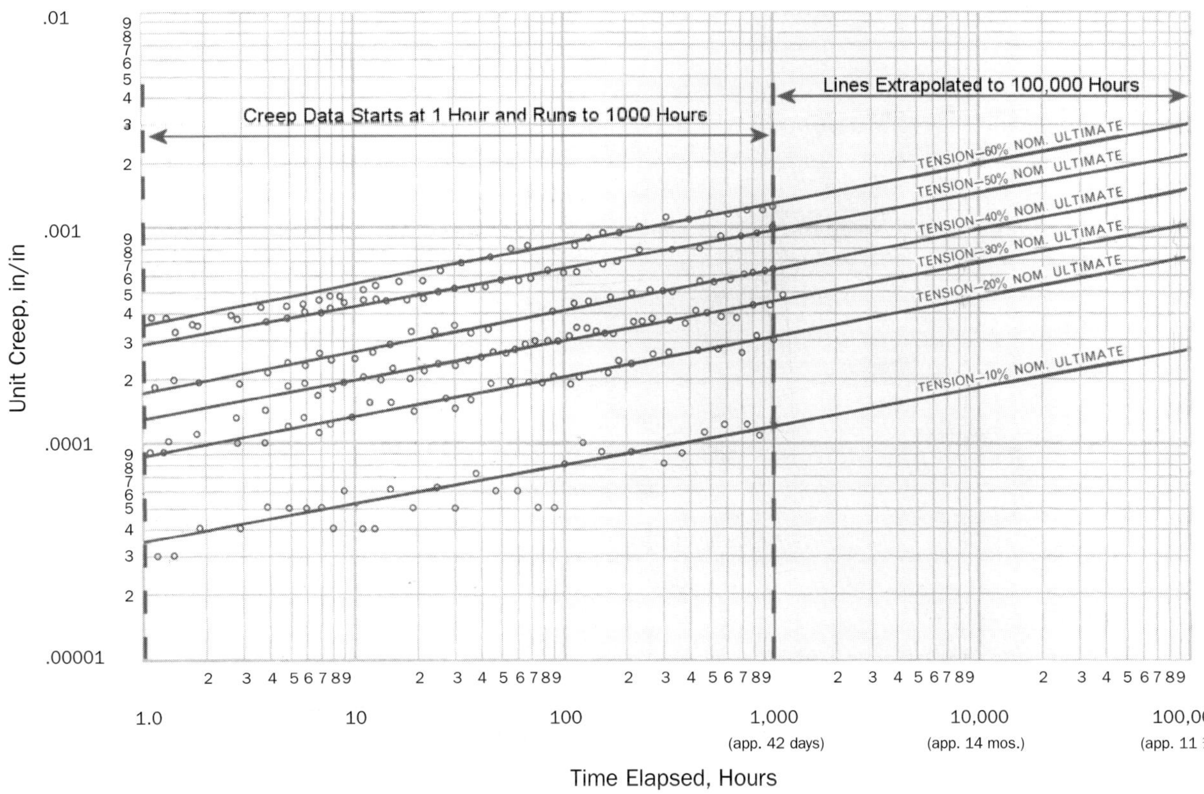

Courtesy of Alcoa Conductor Products Company.

conductor with an active gauge length of approximately 500 inches (15 meters) is placed in a long bed tensile test machine with the conductor ends gripped in such a manner as to assure positive holding of the conductor strands during the test. An initial bias tension load of 1000 lbs or 8% of the conductor rated breaking strength (RBS), whichever is smaller, is applied to the conductor to remove any looseness in the conductor. Leveling jack supports are placed under the cable to keep the conductor straight. Tension is then applied three consecutive times, at increasing tension levels of 30%, 50% and 70% of the conductor rated breaking strength, RBS, measuring the elongation of the conductor during each loading series. The tension is held for 30 minutes at the 30% tension level, and one hour at both the 50% and 70% tension levels. At the end of each hold period, the tension is reduced to the starting bias tension. The conductor is loaded a fourth, and final time, to 75% of its calculated RBS. The extensometers, elongation measuring devices, are removed from the sample and the conductor is pulled to tensile failure. A fitted curve is drawn through the endpoints of each hold period. This curve is typically referred to as the one hour modulus of the composite conductor. A straight line is

FIGURE 1-7
STRESS-STRAIN CURVE FOR 37-STRAND AAC CONDUCTOR

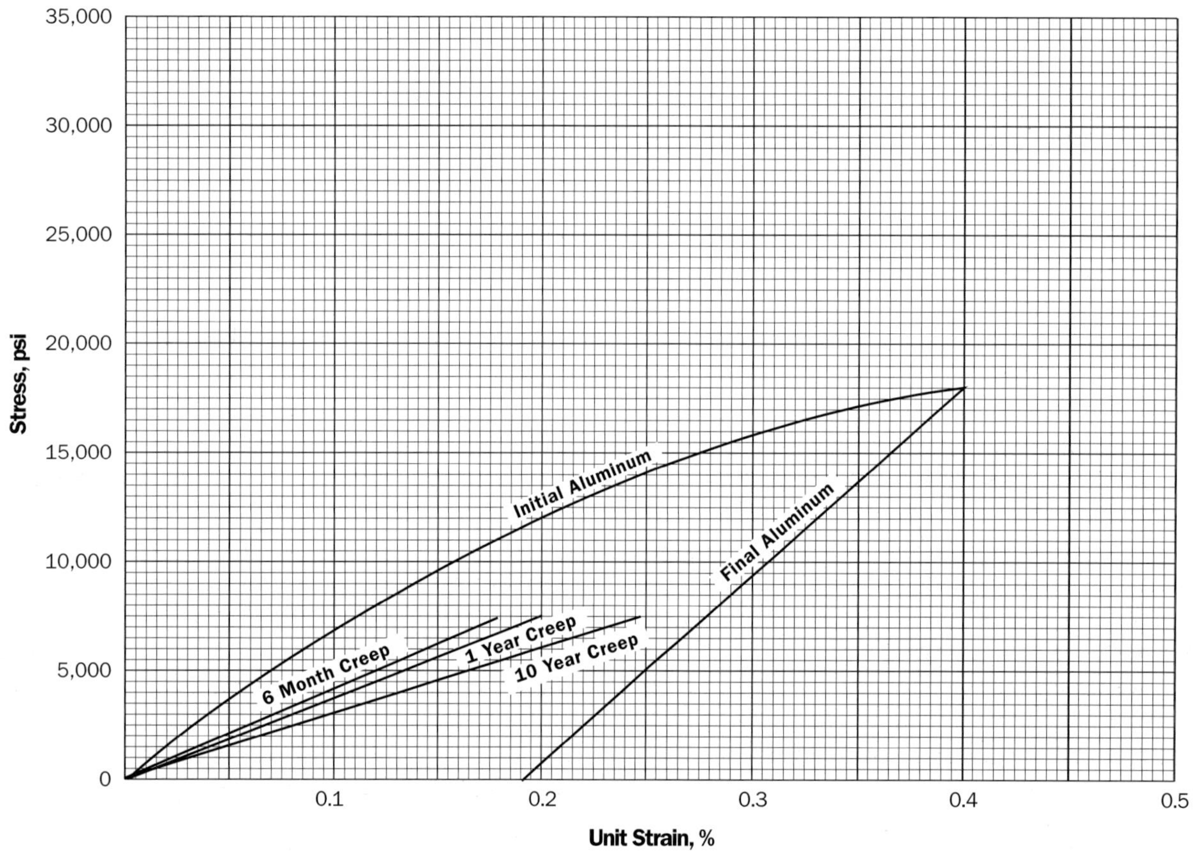

Equations for Curves (X = unit strain in %: Y = stress in psi):

Initial Aluminum: $X = -5.31 \times 10^{-3} + (1.74 \times 10^{-5})\,Y - (6.17 \times 10^{-10})\,Y^2 + (5.05 \times 10^{-14})\,Y^3$

$Y = 136 + (7.46 \times 10^4)\,X - (8.51 \times 10^4)\,X^2 + (2.33 \times 10^4)\,X^3$

Final Aluminum: $Y = (85.20\,X - 16.14) \times 10^3$

6 Month Creep: $Y = (42.30 \times 10^3)\,X$

1 Year Creep: $Y = (38.20 \times 10^3)\,X$

10 Year Creep: $Y = (30.60 \times 10^3)\,X$

Test Temperature 70°F to 75°F

drawn tangent to the one hour modulus curve between a stress of approximately 8% and 35% of the conductor RBS. The slope of this line is referred to as the initial modulus. The slopes of the stress-strain curves during unloading, at each tension level, are essentially the same. This slope is called the final modulus of the composite conductors. Typical results of such a test are shown in Figure 1-5.

If the conductor being tested is made entirely of aluminum, only one stress-strain test is required. If the conductor is a non-homogenous conductor, such as an ACSR, having a core with different tensile and elongation characteristics, a second stress-strain test is required to predict the stress-strain behavior of the core. The stress-strain characteristic of the core is then subtracted form the composite stress-strain curve to determine the initial and final stress-strain characteristic of the aluminum.

Next a series of creep tests are performed to determine the non-elastic elongation of a conductor under constant tension. These tests consist of loading a sample of new conductor in a long-bed tensile test machine and gripping the ends to prevent slippage of the conductor strands relative to each other. A pre-determined tension, typically between 10% and 50% of the conductor RBS, is applied to the conductor, and the elongation of the conductor is monitored as a function of time. A creep test cycle lasts for 1000 hours, approximately 42 days. The room temperature must be held constant during the test to prevent thermally induced elongation. Upon completion of the test, the percent strain (percent elongation of original sample length) between one hour and 1000 hours is recorded on a log-log plot, and the curve is extrapolated to 100,000 hours. Multiple creep tests are typically performed at increasing tension levels to predict the creep characteristic of a conductor. A representative creep plot is shown in Figure 1-6.

The creep strain is added to the stress-strain curve to indicate the total stress-strain-creep characteristic of a conductor. Figure 1-7 shows a typical stress-strain curve for an AAC conductor published by The Aluminum Association[13]. Stress, the conductor tension divided by the total area of the conductor strands, expressed in psi, is plotted on the vertical axis. Conductor strain, expressed as percent of conductor length, is plotted on the horizontal axis.

Stress-stain characteristics are typically grouped by stranding (i.e. 7, 19, 37, 61, etc) for all-aluminum conductors, and by steel to aluminum ratio for ACSR and ACSS strandings, such as 26/7, 54/9, etc. Stress-strain curves are available for all stranded conductor. Their application in the sag-tension calculation process is discussed in considerable detail in Chapter 2.

1.5 THERMAL RATING OF BARE OVERHEAD CONDUCTORS

The maximum continuous current carrying capacity of a conductor is often referred to as the conductor "ampacity" or the "normal" rating. This is one of the major factors considered when selecting a conductor for an overhead transmission line. The thermal rating is a function of local weather conditions, conductor characteristics, and the designed maximum operating temperature of the line.

If the line current produces a conductor temperature that exceeds the line's thermal rating, problems may occur. Temperature in excess of the intended maximum operating temperature of the line can cause excessive sag, which may result in violations of minimum electrical clearance requirements. Another very serious problem which occurs in conductors that depend on hardened aluminum strands for a portion of their strength, such as ACSR, AAC, AAAC and ACAR, is the permanent loss of strength due to annealing of the aluminum strands at temperatures in excess of 94°C. The degree of harm in both cases is dependent on the magnitude and duration of the over-temperature event.

The maximum allowable continuous current that yields the maximum allowable conductor temperature under worst-case weather conditions is usually calculated by assuming that the conductor is in thermal equilibrium at the maximum allowable conductor temperature. If values of ambient conditions and conductor variables are measured, calculated, or estimated, the ampacity can be calculated using the steady-state heat balance equation. The calculation of thermal ratings is discussed in considerable detail in Chapter 3.

In determining steady-state ratings, a continuous current is calculated that would yield the maximum allowable conductor temperature for certain weather assumptions and conductor parameters. Conservative weather conditions of low wind speed (2 to 3 feet/sec perpendicular to the conductor) and high ambient temperature (35°C to 40°C) are normally used to calculate heat loss rates.

The "IEEE Standard for Calculating the Current-Temperature Relationship of Bare Overhead Conductors" (IEEE Standard 738)[14] describes a heat balance calculation method which is widely used in North America and is presented in detail in Chapter 3. This standard is based largely on the work of House and Tuttle, published in 1958[15], as modified by a group sponsored by the East Central Area Reliability Coordination Agreement (ECAR)[16]. It is also the basis for the SWRATE thermal rating computer program supplied with this book. The use of this program allows the evaluation of a variety of conductor thermal rating conditions in determining the rating most suitable for a given system.

1.6 CONDUCTOR SELECTION

Since the conductor is one of the major cost components of a line design, it is essential that the most appropriate conductor type and size be selected for optimum operating efficiency[17][18]. A systematic consideration of factors such as tension loads, ice and wind loads, the current loading of the line, voltage stability, environmental effects, electrical losses, ambient conditions and many others must be considered in the conductor selection process. The goal is to select a conductor that exemplifies the best conductivity-to-weight ratio and/or strength-to-weight ratio at a minimal cost for the application. The electrical and mechanical properties, thermal properties, and stress-strain relationship of the conductor are key factors in the choice of conductor type and size for a given line design problem.

The choice of conductor can be quite different when designing a new line as opposed to reconductoring an existing line. In a new line, structure and foundations costs are certainly affected by the conductor tension and wind/ice loads, but moderate changes in structure and foundation load capacity can be accommodated easily. In reconductoring an existing line, the choice of replacement conductor is more limited due to the constraints caused by limitations of the existing structures and foundations.

In either design application, there are distinct advantages and disadvantages to using a certain conductor type under given conditions. Some conductors, such as "self-damping" or "vibration resistant" conductors, are designed to better withstand the effects of wind-induced vibrations. These conductors can be installed with less sag in new lines, thereby, reducing required structure height. Other conductors, such as trapezoidal wire or steel supported conductors, may be a better choice in reconductoring applications. The lower electrical resistance and reduced high-temperature sag of these conductors can allow large increases in the line thermal rating without rebuilding structures and foundations.

Whether the conductor offers the advantage of increased conductivity, increased tensile strength, or superior corrosion resistance, its characteristics should be fully evaluated for the intended application in the line design process. Regardless of the conductor type, there are certain desirable characteristics for any conductor:

- **Lower electrical resistance**. Lower electrical resistance yields reduced electrical losses over the life of a line and allows increased emergency loadings.

- **Lower thermal elongation**. For a conductor having a lower thermal elongation, the same maximum allowable conductor temperature will result in less sag.

- **Higher annealing temperatures**. Conductors made with 1350-H19 and 6201-T81 aluminum strands cannot be operated continuously above 94°C without significant loss of strength over time. Steel-supported conductors made using dead soft aluminum, 1350-O, and high temperature aluminum alloys, such as most aluminum-zirconium (Al-Zr) alloys, can operate continuously at temperatures of 200°C or greater.

- **Reduced susceptibility to aeolian vibration and galloping.** Conductors designed to reduce aeolian vibration and galloping can be strung to a higher unloaded tension, thereby allowing longer spans and/or closer phase spacings.

The choice of conductor type and size has a major impact on transmission line design. Some of the physical and economic considerations that affect the choice of conductor in the design process are:

- Conductor diameter directly affects the ice and wind loading on the conductor, which in turn affects the required conductor strength and tower loading. In new lines, as the size of the conductors increase, the cost of the conductor and structure/ foundations increase.

- Compared to standard round wire constructions, TW (Trapezoidal Wire) type conductors with an equal conducting area have a smaller diameter. Used in a new installation, this diameter reduction results in lower ice and wind loading, as well as lower structure loading.

- Emergency and contingency operation of transmission lines is increasingly becoming an issue. The use of high temperature, low sag conductors can mitigate, and sometimes eliminate, risk associated with operating a transmission line above its normal rating.

- Operating cost of a line is affected by the electrical properties of the conductor. Conductor material and construction should be selected to balance strength and operating loss requirements for a line.

- The required tension, and thus structure strength, is affected by the selection of the conductor size and type. Tension is also a consideration in the potential for aeolian vibration and increased transverse tension loads on angle structures.

- Magnetic fields, which are unaffected by the type and size of conductor, depend primarily on phase spacing, configuration and distance to edge of right-of-way. Since magnetic fields are calculated for the maximum power flow allowed, calculated magnetic fields will increase as line rating increases.

- The radial electric field around the phase conductors and shield wire increases as the conductor size decreases, yielding higher levels of corona-induced noise.

Southwire supplies many sizes and types of conductors that address these considerations. Some representative types have been selected for discussion in the following sections. Each of these conductor types has unique characteristics and properties that make them especially suited for specific applications. Tables 1-17 through 1-42 provide a more complete listing of these conductors and their electrical and mechanical properties.

FIGURE 1-8
STRANDINGS OF ACAR

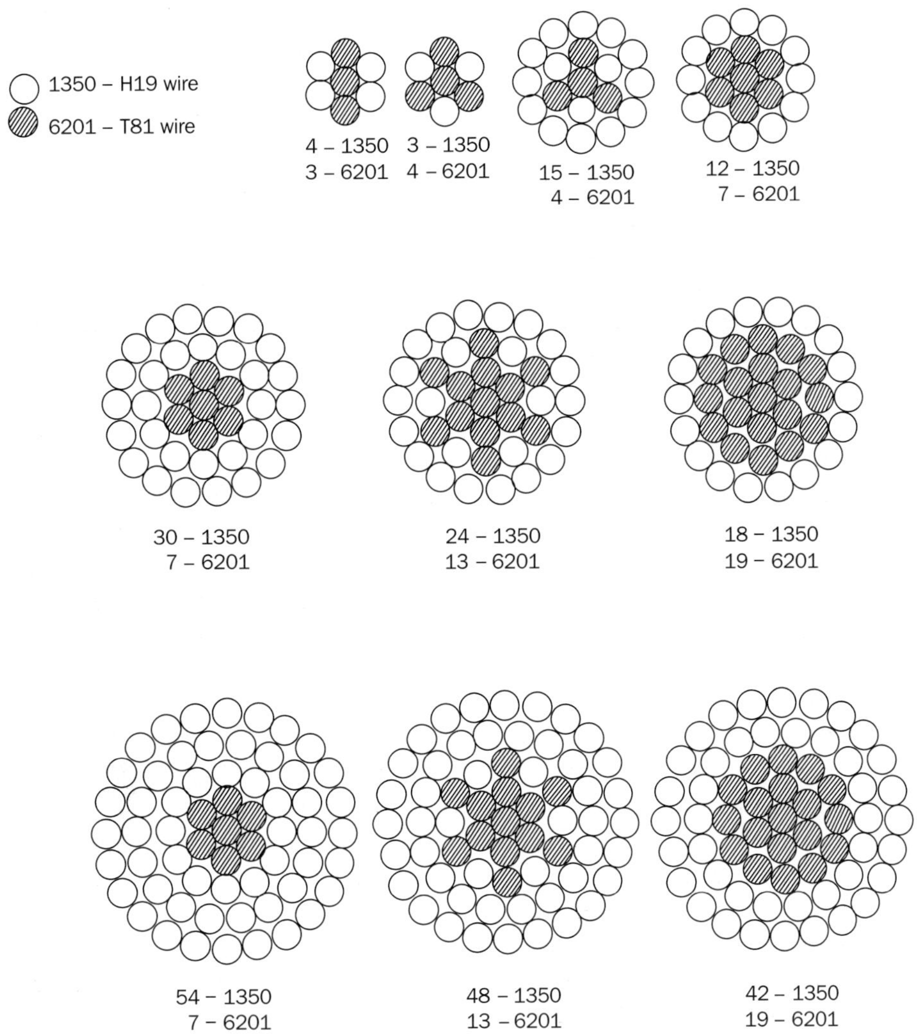

1.6.1 Conventional Round Wire Conductors

The five most commonly used conductor types in the utility industry are: all-aluminum (AAC); all-aluminum alloy (AAAC); aluminum conductor, alloy reinforced (ACAR); aluminum conductor, steel reinforced (ACSR); and aluminum conductor, steel supported (ACSS). All of these conductor types have long, proven records of performance under a variety of conditions. ACSS conductors will be discussed and described under Section 1.6.4 on High Temperature Conductors.

FIGURE 1-9
STRANDINGS OF ACSR

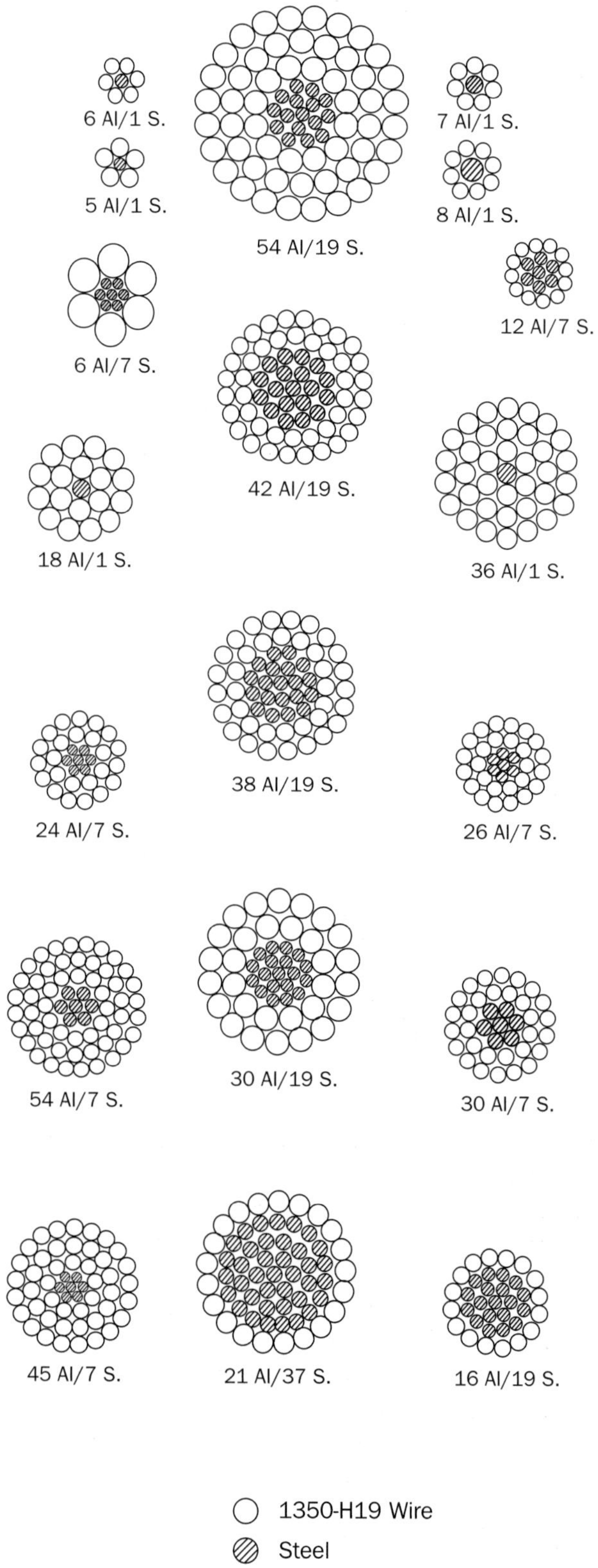

All-Aluminum Conductor (AAC)

AAC is composed entirely of 1350-H19 aluminum wires. It is a low cost conductor with good corrosion resistance, moderate strength and a conductivity of 61.2% IACS, minimum. AAC has the highest conductivity-to-weight ratio of all the overhead conductors, making it ideal for installations in urban areas limited in space where short spans with maximum current transfer are required.

All-Aluminum Alloy Conductor (AAAC)

AAAC was originally developed as a replacement for 6/1 and 26/7 ACSR conductors when steel was in short supply. AAAC is made of high strength 6201-T81 aluminum alloy which has a minimum conductivity of 52.5% IACS, giving it comparable and, in some instances, improved qualities over both ACSR and AAC conductors. Although available in standard sizes, this conductor is typically used to replace 6/1 and 26/7 ACSR diameter equivalent constructions. AAAC conductors have comparable thermal ratings, improved strength-to-weight ratio, lower electrical losses and superior corrosion resistance to these ACSR conductors. These factors make AAAC conductors the preferred choice for distribution installations on the seacoast, farm and industrial areas severely impacted by corrosion problems.

Aluminum Conductor, Aluminum-Alloy Reinforced (ACAR)

ACAR, a mix of 6201-T81 and 1350-H19 strands typically of the same diameter, has an excellent balance between mechanical and electrical properties. ACAR is often designed as diameter equivalent to ACSR, and as with ACSR, the 1350-H19 aluminum strands are usually located in the outside layers of the conductor. The 6201-T81 aluminum alloy wires are primarily used in the center of the conductor in lieu of steel to provide the necessary strength. Unlike ACSR, the alloy wires not only provide the required strength, but also add conductive cross-sectional area to the complete conductor. As a result, an ACAR conductor of the same diameter as an equivalent ACSR can have less electrical resistance and higher current carrying capacity. Because the 6201 and 1350 strands are equal in diameter, they can be interchanged to optimize the properties of ACAR, creating an excellent balance between mechanical and electrical properties. An example of this flexibility is the increase of the relative number of 6201 strands to achieve a strength-to-weight ratio matching that of a wide range of ACSR conductors. These conductors exhibit excellent corrosion resistance and utilize simple termination hardware, making them excellent choices for many transmission line applications. Figure 1-8 shows some of the standard strandings for ACAR[8].

Aluminum Conductor, Steel-Reinforced (ACSR)

The utility industry's predominate choice of conductor in transmission and rural distribution circuits for many years has been ACSR. It is used extensively on long spans as both ground and phase conductors because of its high mechanical strength-to-weight ratio and good current-carrying capacity. ACSR consists of a solid or stranded galvanized steel core surrounded by one or more layers of 1350-H19 aluminum. The use of 1350 aluminum in ACSR gives it equivalent, or higher, thermal ratings in comparison with equivalent sizes of AAC. The circular mil area of ACSR is specified according to the cross-sectional area of aluminum to be contained in the construction.

The steel content of ACSR typically ranges from 11% to 18% by weight for larger-than-AWG sizes, and is available in 18/1, 45/7, 72/7 and 84/19 strandings. The content can vary up to 40%, depending on the desired tensile strength. When used as ground wires or extra long spans, such as river crossings, strandings of 8/1, 12/7 or 16/19 are often used to provide higher tensile strength. Figure 1-9 shows some standard strandings of ACSR[8].

The high tensile strength combined with the good conductivity of ACSR gives several advantages:

(1) Because of the presence of the steel core, lines designed with ACSR elongate less than other conductors, yielding less sag at a given tension. Therefore, the maximum allowable conductor temperature can be increased to allow a higher thermal rating when replacing other standard conductors with ACSR.

(2) The high tensile strength of ACSR allows it to be installed in areas subject to extreme ice and wind loading.

(3) ACSR is less likely to be broken by falling tree limbs.

1.6.2 Trapezoidal-Shaped Wire Conductors

Trapezoidal wire (TW), called Trapwire, differs from the conventional round wire conductor construction by having strands shaped in the form of a trapezoid. For homogenous Trapwire constructions, all strands are trapezoidal shaped with the exception of the round center core wire For non-homogeneous conductors, the multi-strand core is usually composed of round strands of one material surrounded by the trapezoidal strands of another. Further research is leading to the development of many combinations of layers of trapezoidal strands and round wire.

Traditional round wire constructions have voids, called interstices, between the strands and layers. By shaping the wires into trapezoids, these interstices can be practically eliminated, creating a more compact conductor. TW constructions are usually designed in one of two ways:

- **Area equivalent constructions.** An area equivalent construction is one wherein the same cross-sectional area of aluminum is the same as that of an equivalent round wire construction. By reducing the interstices, greater compaction and a smaller conductor diameter are produced with a TW design. The TW design yields a conductor that is as much as 10% smaller in diameter than its equivalent round wire conductor, thus it is subject to considerably less ice and wind load.

- **Diameter equivalent constructions.** Trapezoidal conductors with diameters equivalent to concentric round wire constructions pack more aluminum into the same physical area. These diameter-equivalent constructions usually contain approximately 20% more aluminum, thereby reducing the conductor resistance and increasing the current carrying capacity by 8% to 10% over that of a concentric conductor of equivalent diameter. Thus, a diameter equivalent TW construction delivers more ampacity for the same ice and wind loading.

There are three types of Trapwire constructions that have been in use for many years in the U.S.: AAC/TW, ACSR/TW and ACSS/TW. ACSS/TW will be discussed later in Section 1.6.4, High Temperature Conductors.

All-Aluminum Conductor/Trapwire (AAC/TW)

AAC/TW is mechanically and electrically similar to AAC conductor except all the strands but the center strand are shaped into a trapezoidal cross-section prior to stranding. AAC/TW is manufactured having either the same kcmil area (resistance) as a conventional AAC round wire conductor, or in a series of diameters in 0.1 inch increments. AAC/TW, therefore, is useful for reducing the conductor diameter for a given resistance, or decreasing the resistance for a given diameter. Table 1-16 shows two possible AAC/TW alternatives to a standard 795 kcmil AAC "Arbutus."

TABLE 1-16
COMPARISON OF AAC WITH AAC/TW ALTERNATIVES

Conductor Description	Type	Diameter in	dc Resistance @ 20°C Ω/1000 ft
"Arbutus" AAC	Round Wire	1.026 (100%)	0.0217 (100%)
"Arbutus/TW"	Area Equivalent	0.919 (89.5%)	0.0217 (100%)
"Rainier/TW"	Diameter Equiv.	1.00 (97.4%)	0.0188 (86.6%)

Aluminum Conductor, Steel-Reinforced/Trapwire (ACSR/TW)

ACSR/TW differs from conventional round wire ACSR in that its aluminum strands are trapezoidal shaped. ACSR/TW is available in both area and diameter equivalent constructions, offering a reduced overall diameter or increased current carrying capacity, as shown in Figure 1-10. Both constructions have the same steel-to-aluminum area ratios as the equivalent round wire ACSR. ACSR/TW constructions are grouped by "Type Numbers" that reflect the steel-to-aluminum area ratio, expressed in percentages, typically ranging from 3 to 23.

FIGURE 1-11
VR™ CONDUCTOR

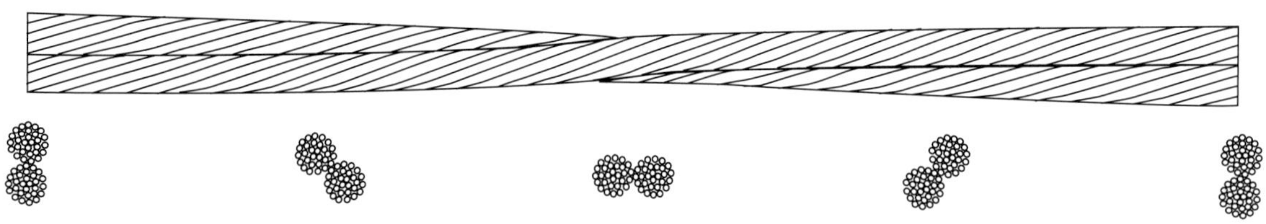

$$D = d(1 + \sin\theta)$$

(D varies by a factor of 2)

where:

D = projected diameter at any cross section

θ = angle between the wind vector and a line through the conductor centers

d = diameter of each subconductor

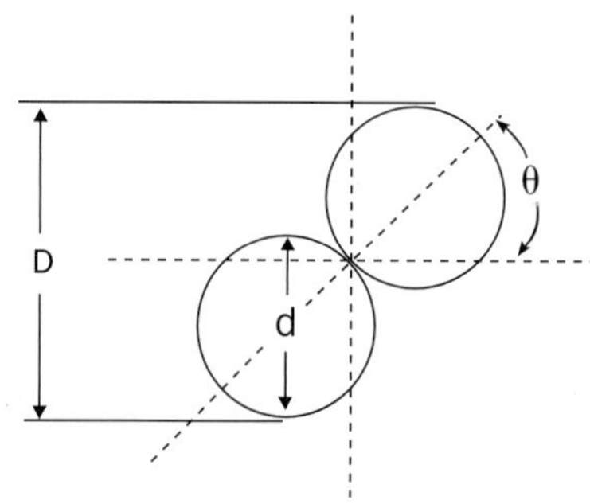

FIGURE 1-10
ACSR TRAPEZOIDAL SHAPED CONDUCTOR

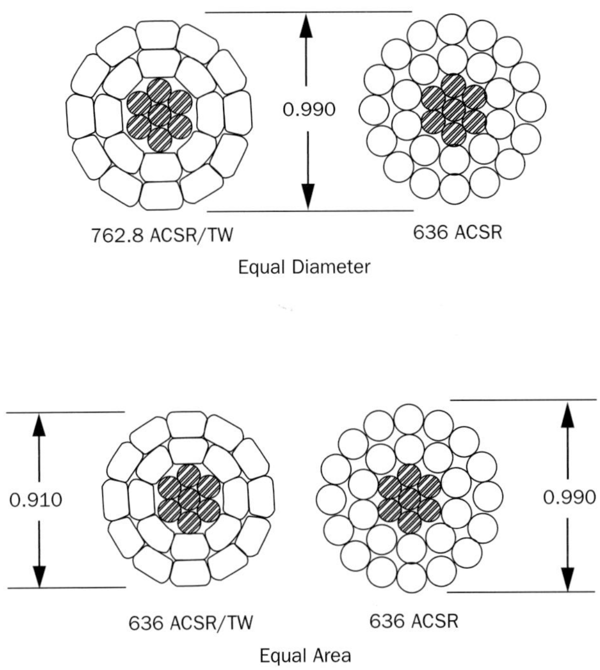

1.6.3 *Wind-Motion Resistant Conductors*

Overhead lines are subject to two types of wind-induced motions: aeolian vibration and ice galloping. Aeolian vibration occurs at relatively low wind speeds and can cause conductor fatigue failure. Ice galloping occurs when ice-covered conductors are exposed to relatively high wind speeds. The resulting large undulations can cause flashover between phase conductors, or between phase and ground conductors. By using conductors that are resistant to wind-induced motion, lines with reduced tower height and reduced phase spacing can be designed. The reduced visual impact of such lines often leads to an increased likelihood of regulatory approval for new lines.

Vibration Resistant Conductors (VR™)

Vibration Resistant conductors, as defined in ASTM B 911/B 911M[1], are commonly known by the trade names VR™ (Vibration Resistant) and T2™ conductor. These conductors are designed for use in overhead lines normally subject to aeolian vibration and galloping [19]. They consist of two conductors twisted together with a left-hand lay length of approximately 9 feet. The conductor cross-section forms a rotating "figure-8" as shown in Figure 1-11.

The two subconductors used to make VR™ conductor are selected based on the thermal and mechanical strength requirements of the line. The subconductors may be any of the conventional conductors - AAC, AAAC, ACAR or ACSR - or even the non-standard ACSR/TW and AAC/TW conductors. Some of the major advantages of VR™ conductor are:

- The amplitude and frequency of galloping due to ice-shedding and high winds are reduced or eliminated due to the non-uniform shaped cross-section.
- The amplitude and frequency of aeolian vibration and the accompanying fatigue inducing strains near clamps are reduced due to the varying profile along the length of the conductor. As a result, VR™ conductor can be installed to higher tension levels without the need for additional dampers.

Self-Damping Conductor (SDC)

SDC is an ACSR construction designed to limit aeolian vibration by internal damping of the strands[20][21]. The layers of aluminum consist of trapezoidal strands whose dimensions and lay lengths deliberately leave gaps between the two innermost layers of aluminum and the innermost layer of aluminum and the steel core. These gaps remain even when the conductor is under full tension. The onset of motion due to aeolian vibration causes impact between the separated layers, imparting the self-damping characteristics of the conductor. This impact-induced damping allows the elimination of tension limit requirements normally used to control aeolian vibration. Even with the small inter-layer gaps, the use of trapezoidal aluminum strands still results in a reduced conductor diameter for a given ac resistance per mile. Figure 1-12 is an illustration of a self-damping conductor.

FIGURE 1-12
SELF-DAMPING CONDUCTOR

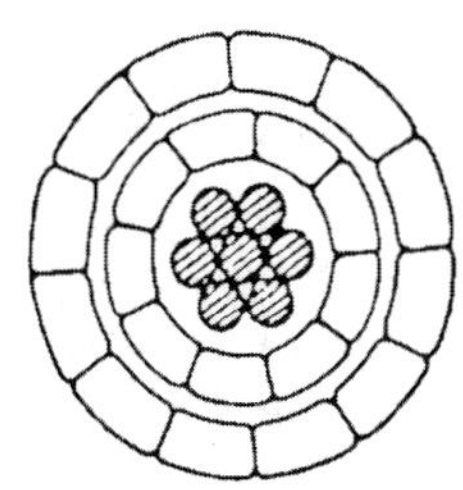

Some of the major advantages that SDC offers are:

- Its high impact damping ability allows an increase of unloaded tension levels, resulting in reduced sag and possibly reduced structure costs.

- The reduced conductor diameter for a given ac resistance yields reduced wind and ice loading, thereby reducing structural loading.

1.6.4 High Temperature Conductors

The need to move larger quantities of power through existing right-of-ways (ROW) has led to increased interest in high temperature conductors. These conductors fall into a category known as High Temperature, Low Sag (HTLS) conductors. In addition to ACSS, which has extensive usage, several new types of conductors are being marketed. With the exception of ACSS and ACSS/TW, all HTLS conductors require special installation procedures and hardware.

Aluminum Conductor, Steel-Supported (ACSS)

ACSS conductor has been in use since the early 1970's. Originally known as SSAC[22], ACSS was designed to operate at higher temperatures than ACSR, increasing current carrying capacity at reduced sags. Since construction costs were low and ROW access easily obtainable, ACSS did not have a large impact on the design of transmission lines when it was originally introduced. As the transmission line environment has changed, the advantages of ACSS can now be fully utilized in transmission line design.

ACSS consists of a core composed of coated steel wires surrounded by one or more layers of fully annealed, 1350-O aluminum round wires. Core wire coatings used include galvanized zinc, aluminum coated, aluminum clad and Zinc-5% Aluminum-Mischmetal Alloy. There are three strength ratings available for the steel core wire: standard, high and ultra high (Southwire HS285). Unlike an ACSR conductor, which receives a significant portion of its strength rating from its 1350-H19 aluminum wires, only a very small portion of an ACSS conductor's strength comes from its 1350-O aluminum wires.

This has two significant implications for ACSS: an ACSS conductor looses only minimal strength when operated at temperatures above 94°C (the onset temperature for annealing of aluminum wires); and the sag response for an ACSS conductor is determined primarily by the sag response of the steel core. This means that the rate of change in sag with temperature for an ACSS conductor is much less than that of an equivalent area ACSR conductor.

ACSS is available in three constructions: round wire, area equivalent Trapwire and diameter equivalent Trapwire. Some advantages of ACSS are:

- ACSS has a conductivity of 63% IACS or better due to the aluminum strands being "dead-soft" (fully annealed). This can translate to higher ratings and lower operating costs.

- Since the aluminum strands are dead-soft, the conductor may be operated at temperatures in excess of 250°C without loss of strength.

- After the conductor has been stressed, the tension in the aluminum strands is low, providing excellent self-damping. This allows the conductor to be installed at high unloaded tension levels without the need for dampers.

- When operated above 180°C, testing has shown that the internal temperature of the conductor can cause standard zinc coatings to degrade. Therefore, it is recommended that either an aluminum clad or Zinc-5% Aluminum-Mischmetal Alloy coated core be used.

- Reconductoring of existing lines with ACSS and ACSS/TW conductors can yield a substantial increase in current carrying capacity while maintaining existing clearance limits.

- Construction of new lines with ACSS has less environmental impact by reducing the number and/or height of required structures.

- Because of its lower thermal elongation, ACSS exhibits less sag with increasing temperature than ACSR.

As discussed in more detail in Chapter 7, these characteristics have made ACSS an attractive choice, both for reconductoring existing transmission and distribution lines and for use in new lines. ACSS is especially suitable for new line and uprating applications where high load currents are the result of occasional emergency and contingency situations.

Aluminum-Zirconium Alloy Conductor, Steel-Reinforced (TACSR)

This family of conductors is constructed using thermal-resistant aluminum-zirconium (Al-Zr) alloy wires stranded over a reinforcing steel core. There are several Al-Zr alloys available, each having a different temperature rating as summarized in Table 1-17. All types of these conductors have traditional concentric round wire strandings similar to ACSR.

TABLE 1-17
TEMPERATURE RATINGS OF
ALUMINUM-ZIRCONIUM ALLOYS

Symbol	Max. Temperature		Conductivity
	Continuous °C	Emergency °C	% IACS
TACSR	150	180	60
UTACSR	200	230	57
ZTACSR	210	240	60
XTACSR	230	310	58

Thermal-Resistant Aluminum Conductor Invar Reinforced (TACIR)

TACIR conductors are similar to the TACSR family of conductors except the standard steel core is replaced with a special iron-nickel alloy ("INVAR") steel core. Invar is typically aluminum clad to provide corrosion resistance at elevated operating temperatures. This core material has a smaller coefficient of thermal expansion and lower tensile strength than galvanized steel. Conventional galvanized steel core wire has a tensile strength of 170 to 190 ksi. Invar steel wires have a 15-20% lower tensile strength but also have a much lower coefficient of thermal expansion. The thermal expansion coefficient of conventional steel is 11.5×10^{-6} /°C whereas the thermal coefficient of Invar steel is only 2.8×10^{-6} /°C.

These conductor are similarly designated TACIR, UTACIR, etc. Both TACSR and TACIR type conductors have been used mainly in Asia to date.

Gap-Type Thermal-Resistant ACSR (GTACSR)

Gap-type ACSR conductors are constructed with a gap between the galvanized steel core and innermost shaped aluminum layer, as shown in Figure 1-13. The gap is filled with a heat resistant grease to reduce friction between the steel core and aluminum and to prevent water penetration. This allows the conductor to be tensioned by gripping the galvanized steel core only. The conductor is then able to take full advantage of the lower thermal elongation properties of steel.

FIGURE 1-13
GAP-TYPE ACSR (GTACSR)

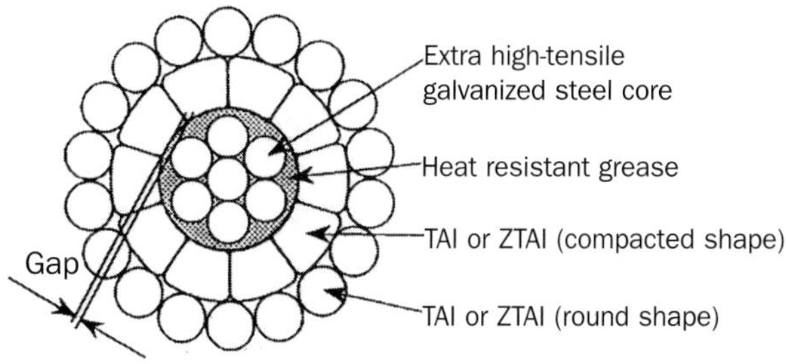

Gap-type conductors employ the same thermal-resistant aluminum-zirconium (Al-Zr) alloy wires used in TACIR conductors and are designated GTACSR, G(Z)TACSR, etc. Gap-type conductors have seen only limited usage in the United States to date.

Aluminum Conductor Composite Reinforced (ACCR)

ACCR conductors are constructed with Al-Zr alloy wires (round or trapezoidal) over a reinforcing core consisting of stranded ceramic filaments. Each filament consists of ceramic fibers encased in an aluminum matrix. The composite core has a lower thermal elongation and equal or greater strength than galvanized steel. This conductor is typically rated 210°C to 240°C. Currently, 3M is the only manufacturer of this type of composite core. The 3M ACCR conductors have had extensive laboratory testing, but limited service application to date.

FIGURE 1-14
ALUMINUM CONDUCTOR COMPOSITE REINFORCED
(ACCR)

Courtesy of 3M.

Aluminum Conductor Composite Core (ACCC)
ACCC conductors have a core consisting of polymer-bound carbon-fibers encased in a fiberglass tube. This core may be described better as a "rod" than a single strand core. ACCC is typically constructed using trapezoidal shaped, fully annealed 1350-O aluminum wires over a single "strand" (rod) composite core, as seen in Figure 1-15, and is typically rated at 200°C.

ACCC was developed by Composite Technology Corporation (CTC) and is relatively new to the utility market. It has seen limited service application to date.

FIGURE 1-15
ALUMINUM CONDUCTOR
COMPOSITE CORE (ACCC)

Courtesy of CTC.

1.7 REFERENCES

[1] *2006 Annual Book of ASTM Standards*, Section 2 - Nonferrous Metal Products, Volume 02.03, Electrical Conductors, American Society for Testing and Materials (ASTM).

[2] Fink, D.G. and Beaty, H.W., *Standard Handbook for Electrical Engineers*, Fourteenth Edition, Section 4, McGraw-Hill, Inc., 1999.

[3] Chen, S.L., Black, W.Z., and Loard, H.W., "High-Temperature Ampacity Model for Overhead Conductors," *IEEE Transactions on Power Delivery*, Vol. 17, No. 4, October 2002.

[4] Douglass, D.A., "AC Resistance of ACSR – Magnetic and Temperature Effects," *IEEE Transactions on Power Apparatus and Systems*, Vol. PAS-104, No. 6, June 1985.

[5] Kennelly, A.E., Laws, F.A., and Pierce, P.H., "Experimental Researches of Skin Effect in Conductors," AIEE Transactions, Vol. 34, Part 2, pp. 1953-2021, 1915.

[6] Dwight, H.B., "Skin Effect in Tubular and Flat Conductors," AIEE Transactions, p. 1379, 1918.

[7] Woodruff, L. F., "Principles of Electric Power Transmission," Second Edition, John Wiley & Sons, Inc., New York.

[8] *Aluminum Electrical Conductor Handbook*, Third Edition, The Aluminum Association, 1989.

[9] Wagner, C. F., and Evans, R. D., "Symmetrical Components," McGraw-Hill Book Company, 1933, Copyright renewed 1961 by Ruth L. Evans.

[10] *Electrical Transmission and Distribution Reference Book*, Fourth Edition Westinghouse Electric Corporation, 1964.

[11] Kimbark, Edward W., *Electrical Transmission of Power and Signals*, John Wiley and Sons, Inc., 1949.

[12] The Aluminum Association, "A Method of Stress-Strain Testing of Aluminum Conductor and ACSR" and "A Test Method for Determining the Long Time Tensile Creep of Aluminum Conductors in Overhead Lines," January, 1999.

[13] Stress-Strain-Creep Curves for Aluminum Electrical Conductors," A Technical Report of the Aluminum Association's Technical Committee on Electrical Conductors, The Aluminum Association, 1969.

[14] "IEEE Standard for Calculating the Current-Temperature Relationship of Bare Overhead Conductors," IEEE Standard 738, 2006.

[15] House, H.E. and Tuttle, P. D., "Current Carrying Capacity of ACSR," *IEEE Transactions Power Apparatus and Systems*, pp.1169-1178, February 1958.

[16] "Transmission Conductors Thermal Ratings," 68-TAP-28. Report by Transmission Advisory Panel, East Central Area Reliability Coordination Agreement.

[17] Douglass, D.A., "Economic Measures of Bare Overhead Conductor Characteristics," *IEEE Transactions on Power Delivery*, Vol. 3, No. 2, pp. 754-761, April 1988.

[18] Kennon, Richard E., and Douglass, Dale A., "EHV Transmission Line Design Opportunities for Cost Reduction," IEEE Paper 89 TD 434-2 PWRD.

[19] Douglass, D.A., and Roche, J.B., "T2 Wind Motion Resistant Conductor," *IEEE Transactions on Power Apparatus and Systems*, Vol. PAS-104, No. 10, pp. 2879-2887, October 1985.

[20] Edwards, A.T., and Livingston, A.E., "Self-Damping Conductors for the Control of Vibration and Galloping of Transmission Lines," IEEE Paper 68 C 59 PWR.

[21] Kirkpatrick, L.A., McCulloch, A.R., and Pue-Gilchrist, A.C., "Ten Years of Progress with Self-Damping Conductor," IEEE Paper F 79 736-0, Presented at the IEEE PES Summer Meeting, 1979.

[22] Adams, H.W., "Steel Supported Aluminum Conductors(SSAC) for Overhead Transmission Lines," IEEE Paper T 74 054-3, Presented at the IEEE PES Winter Power Meeting, 1974.

CHAPTER 1
APPENDIX

APPENDIX 1.1 - CONDUCTOR TABLES

TABLE 1-18
BARE SOLID COPPER WIRES

Size AWG	Nominal Diameter in	Cross-Sectional Area		Nominal Weight	Hard Drawn		Medium Hard Drawn			Annealed	
		Circular Mils	Sq. in	lbs/ 1000 ft	Minimum Breaking Strength lbs	Max. dc Resistance @ 20°C Ω/1000 ft	Minimum Breaking Strength lbs	Maximum Breaking Strength lbs	Max. dc Resistance @ 20°C Ω/1000 ft	Maximum Breaking Strength lbs	Max. dc Resistance @ 20°C Ω/1000 ft
50	0.0010	1.00	7.854E-07	0.00303	......	......	......	......	......	......	......
49	0.0011	1.21	9.503E-07	0.00366	......	......	......	......	......	......	......
48	0.0012	1.44	1.131E-06	0.00436	......	......	......	......	......	......	......
47	0.0014	1.96	1.539E-06	0.00593	......	......	......	......	......	......	......
46	0.0016	2.56	2.011E-06	0.00775	......	......	......	......	......	......	......
45	0.0018	3.24	2.545E-06	0.00981	......	......	......	......	......	......	......
44	0.0020	4.00	3.142E-06	0.0121	......	2700	......	......	2680	......	2590
43	0.0022	4.84	3.801E-06	0.0147	......	2230	......	......	2220	......	2140
42	0.0025	6.25	4.909E-06	0.0189	......	1720	......	......	1720	......	1660
41	0.0028	7.84	6.158E-06	0.0237	......	1380	......	......	1370	......	1320
40	0.0031	9.61	7.548E-06	0.0291	......	1120	......	......	1120	......	1080
39	0.0035	12.2	9.621E-06	0.0371	......	880	......	......	876	......	847
38	0.0040	16.0	1.257E-05	0.0484	......	674	......	......	671	......	648
37	0.0045	20.2	1.590E-05	0.0613	......	533	......	......	530	......	512
36	0.0050	25.0	1.963E-05	0.0757	......	431	......	......	429	......	415
35	0.0056	31.4	2.463E-05	0.0949	......	344	......	......	342	......	331
34	0.0063	39.7	3.117E-05	0.120	......	272	......	......	270	......	261
33	0.0071	50.4	3.959E-05	0.153	......	214	......	......	213	......	206
32	0.0080	64.0	5.027E-05	0.194	......	168	......	......	168	......	162
31	0.0089	79.2	6.221E-05	0.240	......	136	......	......	135	......	131
30	0.0100	100	7.854E-05	0.303	......	108	......	......	107	......	104
29	0.0113	128	1.003E-04	0.387	......	84.5	......	......	84.0	......	81.2
28	0.0126	159	1.247E-04	0.481	......	67.9	......	......	67.6	......	65.3
27	0.0142	202	1.584E-04	0.610	......	53.5	......	......	53.2	......	51.4
26	0.0159	253	1.986E-04	0.765	......	42.7	......	......	42.4	......	41.0
25	0.0179	320	2.516E-04	0.970	......	33.7	......	......	33.5	......	32.4
24	0.0201	404	3.173E-04	1.22	......	26.7	......	......	26.6	......	25.7
23	0.0226	511	4.011E-04	1.55	......	21.1	......	......	21.0	15.4	20.3
22	0.0253	640	5.027E-04	1.94	......	16.9	......	......	16.8	19.4	16.2
21	0.0285	812	6.379E-04	2.46	......	13.3	......	......	13.2	24.6	12.8
20	0.0320	1,020	8.042E-04	3.10	......	10.5	......	......	10.5	31.0	10.1
19	0.0359	1,290	0.00101	3.90	......	8.37	......	......	8.32	39.0	8.05
18	0.0403	1,620	0.00128	4.92	85.5	6.64	67.6	76.5	6.61	49.1	6.39
17	0.0453	2,050	0.00161	6.21	108	5.26	84.9	96.2	5.23	62.0	5.05
16	0.0508	2,580	0.00203	7.81	135	4.18	106	120	4.16	73.0	4.02
15	0.0571	3,260	0.00256	9.87	170	3.31	133	151	3.29	98.6	3.18
14	0.0641	4,110	0.00323	12.4	214	2.63	167	189	2.61	124	2.52
13	0.0720	5,180	0.00407	15.7	268	2.09	209	237	2.07	157	2.00
12	0.0808	6,530	0.00513	19.8	337	1.65	262	297	1.64	197	1.59
11	0.0907	8,230	0.00646	24.9	423	1.31	327	372	1.30	249	1.26
10	0.1019	10,380	0.00816	31.4	529	1.0390	410	467	1.03	314	0.9988
9	0.1144	13,090	0.01028	39.6	661	0.8241	514	586	0.8199	380	0.7925
8	0.1285	16,510	0.01297	50.0	826	0.6532	644	735	0.6498	480	0.6281
7	0.1443	20,820	0.01635	63.0	1030	0.5180	807	921	0.5153	605	0.4981
6	0.1620	26,240	0.02061	79.4	1280	0.4110	1010	1154	0.4088	763	0.3952
5	0.1819	33,090	0.02599	100.2	1590	0.3260	1265	1446	0.3243	962	0.3134
4	0.2043	41,740	0.03278	126.3	1970	0.2584	1584	1814	0.2571	1213	0.2485
3	0.2294	52,620	0.04133	159.3	2439	0.2050	1984	2273	0.2039	1529	0.1971
2	0.2576	66,360	0.05212	200.9	3002	0.1625	2450	2814	0.1617	1928	0.1563
1	0.2893	83,690	0.06573	253.3	3688	0.1289	3024	3484	0.1282	2432	0.1239
1/0	0.3249	105,600	0.08291	319.5	4518	0.1022	3731	4311	0.1016	2985	0.09825
2/0	0.3648	133,100	0.1045	402.8	5519	0.08021	4599	5330	0.0798	3763	0.07793
3/0	0.4096	167,800	0.1318	507.8	6720	0.06362	5666	6588	0.0633	4744	0.06182
4/0	0.4600	211,600	0.1662	640.5	8143	0.05045	6980	8143	0.0502	5983	0.04901

TABLE 1-19

COPPER CONDUCTORS

Conductor Size AWG or Kcmil	Circular Mils	Stranding		Diameter		Cross-Sectional Area Sq. in	Weight lbs/ 1000 ft	Hard Drawn Min. Ultimate Strength lbs	Resistance				GMR ft	Reactance @ 1 ft Spacing 60 Hz		Ampacity @ 75°C Amps
		No. of Wires	Class	Individual Wire in	Complete Cable in				dc @ 20°C Ω/mile	ac-60 Hz @ 25°C Ω/mile	ac-60 Hz @ 50°C Ω/mile	ac-60 Hz @75°C Ω/mile		Inductive Xa Ω/mile	Capacitive X'a MΩ-mile	
SOLID																
8	16,510	1	---	0.1285	---	0.0130	50	826	3.4487	3.5141	3.8409	4.1676	0.00417	0.665	0.1552	96
7	20,820	1	---	0.1443	---	0.0164	63	1,030	2.7348	2.7868	3.0459	3.3050	0.00468	0.651	0.1517	111
6	26,250	1	---	0.1620	---	0.0206	79	1,280	2.1702	2.2114	2.4170	2.6226	0.00526	0.637	0.1483	128
5	33,100	1	---	0.1819	---	0.0260	100	1,590	1.7204	1.7532	1.9162	2.0791	0.00590	0.623	0.1449	147
4	41,740	1	---	0.2043	---	0.0328	126	1,970	1.3643	1.3903	1.5196	1.6488	0.00663	0.609	0.1414	170
3	52,620	1	---	0.2294	---	0.0413	159	2,440	1.0822	1.1030	1.2055	1.3080	0.00745	0.595	0.1380	197
2	66,360	1	---	0.2576	---	0.0521	201	3,000	0.8581	0.8747	0.9560	1.0373	0.00836	0.581	0.1345	228
1	83,690	1	---	0.2893	---	0.0657	253	3,690	0.6804	0.6938	0.7582	0.8226	0.00939	0.566	0.1311	263
1/0	105,600	1	---	0.3249	---	0.0829	320	4,520	0.5392	0.5500	0.6011	0.6521	0.01054	0.553	0.1276	306
2/0	133,100	1	---	0.3648	---	0.1045	403	5,520	0.4241	0.4329	0.4733	0.5136	0.01184	0.538	0.1242	354
3/0	167,800	1	---	0.4096	---	0.1318	508	6,720	0.3364	0.3437	0.3757	0.4077	0.01329	0.524	0.1208	409
4/0	211,600	1	---	0.4600	---	0.1662	641	8,140	0.2668	0.2730	0.2983	0.3237	0.01493	0.510	0.1173	473
CONCENTRIC																
4	41,740	3	AA	0.1180	0.2543	0.0328	127.6	1,880	1.3779	1.4042	1.5348	1.6653	0.00717	0.599	0.1349	179
4	41,740	7	A,B	0.0772	0.2316	0.0328	128.9	1,940	1.3916	1.4181	1.5500	1.6818	0.00700	0.602	0.1377	174
3	52,620	3	AA	0.1325	0.2855	0.0414	160.9	2,360	1.0930	1.1140	1.2175	1.3211	0.00805	0.585	0.1315	207
3	52,620	7	A,B	0.0867	0.2601	0.0413	162.5	2,430	1.1036	1.1248	1.2294	1.3339	0.00786	0.588	0.1342	201
2	66,360	3	AA	0.1487	0.3204	0.0521	202.9	2,910	0.8667	0.8835	0.9656	1.0477	0.00903	0.571	0.1281	239
2	66,360	7	A,B	0.0974	0.2922	0.0522	204.9	3,050	0.8752	0.8921	0.9750	1.0579	0.00883	0.574	0.1308	233
1	83,690	3	AA	0.1670	0.3599	0.0657	255.9	3,620	0.6872	0.7007	0.7658	0.8309	0.01016	0.557	0.1246	277
1	83,690	7	A	0.1093	0.3279	0.0657	258.4	3,800	0.6940	0.7076	0.7733	0.8391	0.00991	0.560	0.1273	269
1/0	105,600	7	A,AA	0.1228	0.3684	0.0829	325.9	4,750	0.5506	0.5615	0.6137	0.6658	0.01114	0.546	0.1239	311
1/0	105,600	19	B	0.0745	0.3725	0.0828	325.6	4,900	0.5506	0.5615	0.6137	0.6658	0.01175	0.539	0.1236	312
2/0	133,100	7	A,AA	0.1379	0.4137	0.1045	411.0	5,930	0.4364	0.4454	0.4866	0.5279	0.01251	0.532	0.1204	360
2/0	133,100	19	B	0.0837	0.4185	0.1045	411.0	6,150	0.4364	0.4454	0.4866	0.5279	0.01320	0.525	0.1201	361
3/0	167,800	7	A,AA	0.1548	0.4644	0.1317	517.9	7,360	0.3461	0.3536	0.3863	0.4190	0.01404	0.518	0.1170	417
3/0	167,800	12	A	0.1183	0.4915	0.1319	518.5	7,560	0.3461	0.3536	0.3863	0.4190	0.01559	0.505	0.1153	423
3/0	167,800	19	B	0.0940	0.4700	0.1319	518.3	7,180	0.3461	0.3536	0.3863	0.4190	0.01483	0.511	0.1167	414
4/0	211,600	7	A,AA	0.1739	0.5217	0.1663	653.6	9,160	0.2745	0.2808	0.3067	0.3326	0.01577	0.504	0.1136	482
4/0	211,600	12	A	0.1328	0.5518	0.1662	653.4	9,480	0.2745	0.2808	0.3067	0.3326	0.01750	0.491	0.1119	489
4/0	211,600	19	B	0.1055	0.5275	0.1661	652.9	9,610	0.2745	0.2808	0.3067	0.3326	0.01664	0.497	0.1132	483
250	---	12	AA	0.1443	0.5996	0.1962	771.5	11,100	0.2323	0.2380	0.2599	0.2819	0.01902	0.481	0.1094	542
250	---	19	A	0.1147	0.5735	0.1963	771.8	11,400	0.2323	0.2380	0.2599	0.2819	0.01810	0.487	0.1108	537
250	---	37	B	0.0822	0.5754	0.1964	771.9	11,600	0.2323	0.2380	0.2599	0.2819	0.01840	0.485	0.1107	537
300	---	12	AA	0.1581	0.6569	0.2356	926.1	13,200	0.1936	0.1988	0.2170	0.2353	0.02080	0.470	0.1068	608
300	---	19	A	0.1257	0.6285	0.2358	926.9	13,500	0.1936	0.1988	0.2170	0.2353	0.01983	0.476	0.1080	601
300	---	37	B	0.0900	0.6300	0.2354	925.3	13,900	0.1936	0.1988	0.2170	0.2353	0.02015	0.474	0.1080	602
350	---	12	AA	0.1708	0.7097	0.2749	1080.9	15,100	0.1660	0.1709	0.1865	0.2021	0.02250	0.460	0.1044	669
350	---	19	A	0.1357	0.6785	0.2748	1080.3	15,600	0.1660	0.1709	0.1865	0.2021	0.02141	0.466	0.1058	662
350	---	37	B	0.0973	0.6811	0.2751	1081.5	16,100	0.1660	0.1709	0.1865	0.2021	0.02178	0.464	0.1057	662
400	---	19	A,AA	0.1451	0.7255	0.3142	1235.1	17,800	0.1452	0.1500	0.1636	0.1772	0.02289	0.458	0.1038	719
400	---	37	B	0.1040	0.7280	0.3143	1235.6	18,300	0.1452	0.1500	0.1636	0.1772	0.02328	0.456	0.1037	720
450	---	19	AA	0.1539	0.7695	0.3534	1389.5	19,800	0.1291	0.1338	0.1459	0.1579	0.02428	0.451	0.1020	774
450	---	37	A,B	0.1103	0.7721	0.3535	1389.8	20,500	0.1291	0.1338	0.1459	0.1579	0.02469	0.449	0.1019	774
500	---	19	AA	0.1622	0.8110	0.3926	1543.4	21,900	0.1162	0.1209	0.1317	0.1425	0.02559	0.445	0.1005	825
500	---	37	A,B	0.1162	0.8134	0.3924	1542.5	22,500	0.1162	0.1209	0.1317	0.1425	0.02601	0.443	0.1004	826
600	---	37	A,AA	0.1273	0.8911	0.4709	1851.3	27,000	0.0968	0.1017	0.1106	0.1196	0.02850	0.432	0.0977	923
600	---	61	B	0.0992	0.8928	0.4715	1853.4	27,500	0.0968	0.1017	0.1106	0.1196	0.02872	0.431	0.0976	924
700	---	37	AA	0.1375	0.9625	0.5494	2159.8	31,200	0.0830	0.0881	0.0956	0.1033	0.03078	0.422	0.0954	1014
700	---	61	A,B	0.1071	0.9639	0.5495	2160.3	31,800	0.0830	0.0881	0.0956	0.1033	0.03100	0.422	0.0954	1015
750	---	37	AA	0.1424	0.9968	0.5893	2316.5	33,400	0.0774	0.0827	0.0897	0.0968	0.03188	0.418	0.0944	1057
750	---	61	A,B	0.1109	0.9981	0.5892	2316.4	34,100	0.0774	0.0827	0.0897	0.0968	0.03210	0.417	0.0943	1058
800	---	37	AA	0.1470	1.0290	0.6280	2468.6	35,100	0.0726	0.0780	0.0845	0.0911	0.03291	0.414	0.0934	1099
800	---	61	A,B	0.1145	1.0305	0.6281	2469.2	36,300	0.0726	0.0780	0.0845	0.0911	0.03315	0.413	0.0934	1099
900	---	37	AA	0.1560	1.0920	0.7072	2780.1	39,500	0.0645	0.0702	0.0760	0.0818	0.03492	0.407	0.0917	1178
900	---	61	A,B	0.1215	1.0935	0.7072	2780.3	40,500	0.0645	0.0702	0.0760	0.0818	0.03517	0.406	0.0916	1179
1000	---	37	AA	0.1644	1.1508	0.7854	3087.6	43,800	0.0581	0.0640	0.0692	0.0744	0.03681	0.401	0.0901	1252
1000	---	61	A,B	0.1280	1.1520	0.7849	3085.8	45,000	0.0581	0.0640	0.0692	0.0744	0.03705	0.400	0.0901	1253

Notes: (1) Data based on a nominal cable manufactured in accordance with ASTM B 8.

(2) Resistance and ampacity based on a copper conductivity of 96.16% IACS at 20°C for individual wire sizes smaller than 0.325 inches and 97.16% IACS for sizes 0.325 inches and larger.

(3) Ampacity based on a 75°C conductor temperature 25°C ambient temperature 2 ft/sec wind in sun with an emissivity of 0.5 and a coefficient of solar absorption of 0.5 at sea level.

TABLE 1-20
SOLID ROUND 1350 ALUMINUM WIRES

Wire Size AWG	Wire Diameter in	Cross-Sectional Area		1350 Aluminum Wire					GMR ft
						Minimum Breaking Strength, lbs			
		Circular Mils	Sq. in	Weight lbs/1000 ft	Resistance dc @ 20°C Ω/Mile	Full Hard 1350-H19	Three-quarter Hard 1350-H16 1350-H26	One-half Hard 1350-H14 1350-H24	
29	0.0113	128	0.000100	0.118	700.7	2.510	1.700	1.500	0.00037
28	0.0126	159	0.000125	0.146	563.6	3.120	2.125	1.875	0.00041
27	0.0142	202	0.000158	0.186	443.7	3.960	2.686	2.370	0.00046
26	0.0159	253	0.000199	0.233	353.9	4.960	3.383	2.985	0.00052
25	0.0179	320	0.000252	0.295	279.3	6.290	4.284	3.780	0.00058
24	0.0201	404	0.000317	0.372	221.5	7.930	5.389	4.755	0.00065
23	0.0226	511	0.000401	0.470	175.2	10.03	6.817	6.015	0.00073
22	0.0253	640	0.000503	0.589	139.8	12.57	8.551	7.545	0.00082
21	0.0285	812	0.000638	0.748	110.2	15.95	10.85	9.570	0.00092
20	0.0320	1,020	0.000804	0.942	87.38	20.11	13.67	12.06	0.00104
19	0.0359	1,290	0.00101	1.186	69.42	25.31	17.17	15.15	0.00116
18	0.0403	1,620	0.00128	1.495	55.09	31.89	21.76	19.20	0.00131
17	0.0453	2,050	0.00161	1.890	43.60	40.29	27.37	24.15	0.00147
16	0.0508	2,580	0.00203	2.38	34.67	58.78	34.51	30.45	0.00165
15	0.0571	3,260	0.00256	3.00	27.44	74.26	43.52	38.40	0.00185
14	0.0641	4,110	0.00323	3.78	21.78	91.97	54.91	48.45	0.00208
13	0.0720	5,180	0.00407	4.77	17.26	114.0	69.19	61.05	0.00234
12	0.0808	6,530	0.00513	6.01	13.70	141.0	87.21	76.95	0.00262
11	0.0907	8,230	0.00646	7.57	10.876	174.4	109.8	96.90	0.00294
10	0.1019	10,380	0.00816	9.56	8.617	212.0	138.6	122.3	0.00331
9	0.1144	13,090	0.01028	12.04	6.837	262.1	174.8	154.2	0.00371
8	0.1285	16,510	0.01297	15.20	5.419	324.2	220.5	194.6	0.00417
7	0.1443	20,820	0.01635	19.16	4.297	400.7	278.0	245.3	0.00468
6	0.1620	26,240	0.02061	24.15	3.409	494.7	350.4	309.2	0.00526
5	0.1819	33,090	0.02599	30.45	2.704	623.7	441.8	389.9	0.00590
4	0.2043	41,740	0.03278	38.41	2.144	786.8	557.3	491.7	0.00663
3	0.2294	52,620	0.04133	48.43	1.700	971.3	702.6	620.0	0.00744
2	0.2576	66,360	0.05212	61.07	1.348	1225	886.0	781.8	0.00836
1	0.2893	83,690	0.06573	77.03	1.0691	1512	1117	986.0	0.00939
0	0.3249	105,600	0.08291	97.15	0.8476	1907	1409	1244	0.0105
00	0.3648	133,100	0.1045	122.5	0.6723	2404	1777	1568	0.0118
000	0.4096	167,800	0.1318	154.4	0.5333	3031	2241	1977	0.0133
0000	0.4600	211,600	0.1662	194.7	0.4228	3822	2825	2493	0.0149

(1) Breaking strength value of 1350-H19 are based on minimum average tensile strength of ASTM B 230 and ASTM B 609.
(2) Electrical resistance is based on the electrical conductivity of 61.2% IACS at 20°C.

TABLE 1-21
ALL-ALUMINUM CONDUCTOR (AAC)
Stranding Classes AA and A

Code Word	Conductor Size AWG or kcmil	Stranding		Diameter		Cross-Sectional Area Sq. in	Weight lbs/ 1000 ft	Rated Strength lbs	Resistance				GMR ft	Reactance @ 1 ft Spacing 60Hz		Ampacity @ 75°C Amps
		No. of Wires	Class	Individual Wire in	Complete Cable in				dc @ 20°C Ω/mile	ac-60 Hz @ 25°C Ω/mile	ac-60 Hz @ 50°C Ω/mile	ac-60 Hz @ 75°C Ω/mile		Inductive Xa Ω/mile	Capacitive X'a MΩ-mile	
Peachbell	6	7	A	0.0612	0.184	0.0206	24.6	563	3.4781	3.5484	3.8997	4.2510	0.00555	0.630	0.1445	103
Rose	4	7	A	0.0772	0.232	0.0328	39.1	881	2.1865	2.2308	2.4516	2.6725	0.00700	0.602	0.1377	138
Iris	2	7	AA,A	0.0974	0.292	0.0522	62.2	1,350	1.3751	1.4031	1.5419	1.6808	0.00883	0.574	0.1308	185
Pansy	1	7	AA,A	0.1093	0.328	0.0657	78.4	1,640	1.0905	1.1128	1.2229	1.3330	0.00991	0.560	0.1273	214
Poppy	1/0	7	AA,A	0.1228	0.368	0.0829	98.9	1,990	0.8642	0.8821	0.9693	1.0566	0.0111	0.546	0.1239	247
Aster	2/0	7	AA,A	0.1379	0.414	0.1045	124.8	2,510	0.6857	0.7000	0.7692	0.8384	0.0125	0.532	0.1204	286
Phlox	3/0	7	AA,A	0.1548	0.464	0.1317	157.2	3,040	0.5439	0.5543	0.6103	0.6652	0.0140	0.518	0.1170	331
Oxlip	4/0	7	AA,A	0.1739	0.522	0.1663	198.4	3,830	0.4313	0.4407	0.4842	0.5277	0.0158	0.504	0.1136	383
Sneezewort	250	7	AA	0.1890	0.567	0.1964	234.4	4,520	0.3651	0.3733	0.4101	0.4469	0.0171	0.493	0.1111	425
Valerian	250	19	A	0.1147	0.574	0.1963	234.3	4,660	0.3651	0.3733	0.4101	0.4469	0.0181	0.487	0.1108	426
Daisy	266.8	7	AA	0.1953	0.586	0.2097	250.2	4,830	0.3421	0.3499	0.3843	0.4188	0.0177	0.489	0.1101	443
Laurel	266.8	19	A	0.1185	0.593	0.2095	250.1	4,970	0.3421	0.3499	0.3843	0.4188	0.0187	0.483	0.1098	444
Peony	300	19	A	0.1257	0.629	0.2358	281.4	5,480	0.3042	0.3113	0.3420	0.3726	0.0198	0.476	0.1080	478
Tulip	336.4	19	A	0.1331	0.666	0.2644	315.5	6,150	0.2713	0.2779	0.3052	0.3325	0.0210	0.469	0.1063	513
Daffodil	350	19	A	0.1357	0.679	0.2748	327.9	6,390	0.2608	0.2672	0.2934	0.3197	0.0214	0.466	0.1058	526
Canna	397.5	19	AA,A	0.1447	0.724	0.3125	372.9	7,110	0.2296	0.2355	0.2586	0.2817	0.0228	0.459	0.1039	570
Goldentuft	450	19	AA	0.1539	0.769	0.3534	421.8	7,890	0.2028	0.2084	0.2287	0.2491	0.0243	0.451	0.1020	616
Cosmos	477	19	AA	0.1584	0.793	0.3744	446.8	8,360	0.1913	0.1968	0.2159	0.2352	0.0250	0.448	0.1012	639
Syringa	477	37	A	0.1135	0.795	0.3744	446.8	8,690	0.1913	0.1968	0.2159	0.2352	0.0254	0.446	0.1011	639
Zinnia	500	19	AA	0.1622	0.811	0.3926	468.5	8,760	0.1825	0.1879	0.2061	0.2245	0.0256	0.445	0.1005	658
Hyacinth	500	37	A	0.1162	0.813	0.3924	468.3	9,110	0.1825	0.1879	0.2061	0.2245	0.0260	0.443	0.1004	658
Dahlia	556.5	19	AA	0.1711	0.856	0.4369	521.4	9,750	0.1640	0.1691	0.1855	0.2020	0.0270	0.438	0.0989	703
Mistletoe	556.5	37	A	0.1226	0.858	0.4368	521.3	9,940	0.1640	0.1691	0.1855	0.2020	0.0275	0.436	0.0988	704
Meadowsweet	600	37	AA,A	0.1273	0.891	0.4709	562.0	10,700	0.1521	0.1571	0.1723	0.1875	0.0285	0.432	0.0977	738
Orchid	636	37	AA,A	0.1311	0.918	0.4995	596.1	11,400	0.1435	0.1485	0.1628	0.1771	0.0294	0.428	0.0968	765
Heuchera	650	37	AA	0.1326	0.928	0.5110	609.8	11,600	0.1404	0.1454	0.1594	0.1734	0.0297	0.427	0.0965	775
Verbena	700	37	AA	0.1375	0.963	0.5494	655.7	12,500	0.1304	0.1353	0.1483	0.1613	0.0308	0.422	0.0954	812
Flag	700	61	A	0.1071	0.964	0.5495	655.8	12,900	0.1304	0.1353	0.1483	0.1613	0.0310	0.421	0.0954	812
Violet	715.5	37	AA	0.1391	0.974	0.5623	671.0	12,800	0.1276	0.1325	0.1451	0.1578	0.0312	0.421	0.0951	823
Nasturtium	715.5	61	A	0.1083	0.975	0.5619	670.6	13,100	0.1276	0.1325	0.1451	0.1578	0.0314	0.420	0.0950	823
Petunia	750	37	AA	0.1424	0.997	0.5893	703.2	13,100	0.1217	0.1266	0.1387	0.1508	0.0319	0.418	0.0944	847
Cattail	750	61	A	0.1109	0.998	0.5892	703.2	13,500	0.1217	0.1266	0.1387	0.1508	0.0321	0.417	0.0943	847
Arbutus	795	37	AA	0.1466	1.026	0.6245	745.3	13,900	0.1148	0.1197	0.1311	0.1425	0.0328	0.415	0.0935	878
Lilac	795	61	A	0.1142	1.028	0.6248	745.7	14,300	0.1148	0.1197	0.1311	0.1425	0.0331	0.414	0.0935	879
Cockscomb	900	37	AA	0.1560	1.092	0.7072	844.0	16,400	0.1014	0.1064	0.1163	0.1264	0.0349	0.407	0.0917	948
Snapdragon	900	61	A	0.1215	1.094	0.7072	844.0	15,900	0.1014	0.1064	0.1163	0.1264	0.0352	0.406	0.0916	948
Magnolia	954	37	AA	0.1606	1.124	0.7495	894.5	16,400	0.0957	0.1007	0.1101	0.1195	0.0360	0.404	0.0908	982
Goldenrod	954	61	A	0.1251	1.126	0.7498	894.8	16,900	0.0957	0.1007	0.1101	0.1195	0.0362	0.403	0.0908	983
Hawkweed	1000	37	AA	0.1644	1.150	0.7854	937.3	17,200	0.0913	0.0963	0.1053	0.1142	0.0368	0.401	0.0901	1010
Camellia	1000	61	A	0.1280	1.152	0.7849	936.8	17,700	0.0913	0.0963	0.1053	0.1142	0.0371	0.400	0.0901	1011
Bluebell	1033.5	37	AA	0.1671	1.170	0.8114	968.3	17,700	0.0883	0.0934	0.1020	0.1107	0.0374	0.399	0.0896	1031
Larkspur	1033.5	61	A	0.1302	1.172	0.8122	969.2	18,300	0.0883	0.0934	0.1020	0.1107	0.0377	0.398	0.0896	1032
Marigold	1113	61	AA,A	0.1351	1.216	0.8744	1044	19,700	0.0820	0.0872	0.0952	0.1032	0.0391	0.393	0.0885	1079
Hawthorn	1192.5	61	AA,A	0.1398	1.258	0.9363	1117	21,100	0.0765	0.0819	0.0893	0.0967	0.0405	0.389	0.0875	1124
Narcissus	1272	61	AA,A	0.1444	1.300	0.9990	1192	22,000	0.0717	0.0772	0.0841	0.0912	0.0418	0.385	0.0865	1169
Columbine	1351.5	61	AA,A	0.1488	1.340	1.0608	1266	23,400	0.0675	0.0731	0.0796	0.0861	0.0431	0.381	0.0856	1212
Carnation	1431	61	AA,A	0.1532	1.379	1.1244	1342	24,300	0.0638	0.0695	0.0756	0.0817	0.0444	0.378	0.0847	1253
Gladiolas	1510.5	61	AA,A	0.1574	1.417	1.1869	1417	25,600	0.0604	0.0663	0.0720	0.0778	0.0456	0.375	0.0839	1294
Coreopsis	1590	61	AA	0.1614	1.454	1.2480	1489	27,000	0.0574	0.0635	0.0689	0.0743	0.0468	0.372	0.0832	1333
Jessamine	1750	61	AA	0.1694	1.525	1.3748	1641	29,700	0.0522	0.0585	0.0634	0.0683	0.0490	0.366	0.0818	1408
Cowslip	2000	91	A	0.1482	1.630	1.5697	1873	34,200	0.0456	0.0525	0.0567	0.0609	0.0526	0.357	0.0798	1518
Sagebrush	2250	91	A	0.1572	1.729	1.7662	2128	37,700	0.0410	0.0482	0.0519	0.0556	0.0558	0.350	0.0780	1612
Lupine	2500	91	A	0.1657	1.823	1.9623	2365	41,900	0.0369	0.0446	0.0478	0.0511	0.0588	0.344	0.0765	1706
Bitterroot	2750	91	A	0.1738	1.913	2.1589	2602	46,100	0.0335	0.0416	0.0445	0.0475	0.0617	0.338	0.0750	1793
Trillium	3000	127	A	0.1537	1.996	2.3564	2840	50,300	0.0307	0.0392	0.0419	0.0445	0.0646	0.332	0.0737	1874
Bluebonnet	3500	127	A	0.1660	2.158	2.7486	3345	58,700	0.0263	0.0355	0.0377	0.0399	0.0697	0.323	0.0715	2014

Notes: (1) Data based on a nominal cable manufactured in accordance with ASTM B 231.
(2) Resistance and ampacity based on an aluminum conductivity of 61.2% IACS at 20°C.
(3) Ampacity based on a 75°C conductor temperature, 25°C ambient temperature, 2 ft/sec wind, in sun, with an emissivity of 0.5 and a coefficient of solar absorption of 0.5, at sea level.

TABLE 1-22

SHAPED WIRE COMPACT CONCENTRIC-LAY ALUMINUM CONDUCTOR (AAC/TW)

Area Equal to Standard AAC Sizes

Code Word	Conductor Size kcmil	No. of Wires	No. of Layers	Cross-Sectional Area Sq. in	Conductor Diameter in	Weight lbs/1000 ft	Rated Strength lbs	Resistance: dc @ 20°C Ω/mile	Resistance: ac-60 Hz @ 25°C Ω/mile	Resistance: ac-60 Hz @ 50°C Ω/mile	Resistance: ac-60 Hz @ 75°C Ω/mile	GMR ft	Reactance @ 1 ft Spacing 60 Hz: Inductive Xa Ω/mile	Reactance @ 1 ft Spacing 60 Hz: Capacitive X'a MΩ-mile	Ampacity @ 75° C Amps
Tulip/TW	336.4	17	2	0.2642	0.612	315.3	6,020	0.2713	0.2779	0.3052	0.3325	0.0192	0.480	0.1088	502
Canna/TW	397.5	17	2	0.3122	0.661	372.6	6,960	0.2296	0.2355	0.2586	0.2817	0.0208	0.470	0.1066	556
Cosmos/TW	477	17	2	0.3746	0.720	447.1	8,360	0.1913	0.1968	0.2159	0.2352	0.0226	0.460	0.1040	623
Zinnia/TW	500	17	2	0.3927	0.736	468.7	8,760	0.1825	0.1879	0.2061	0.2245	0.0232	0.457	0.1033	641
Mistletoe/TW	556.5	17	2	0.4371	0.775	521.6	9,750	0.1640	0.1691	0.1855	0.2020	0.0244	0.451	0.1018	685
Meadowsweet/TW	600	17	2	0.4712	0.803	562.4	10,520	0.1521	0.1571	0.1723	0.1875	0.0253	0.446	0.1008	717
Orchid/TW	636	17	2	0.4995	0.825	596.1	11,100	0.1435	0.1485	0.1628	0.1771	0.0260	0.443	0.0999	743
Verbena/TW	700	17	2	0.5498	0.864	656.1	12,300	0.1304	0.1353	0.1483	0.1613	0.0272	0.437	0.0986	789
Nasturtium/TW	750	17	2	0.5890	0.893	702.1	13,100	0.1217	0.1266	0.1387	0.1508	0.0281	0.433	0.0976	823
Arbutus/TW	795	17	2	0.6244	0.919	745.1	13,600	0.1148	0.1197	0.1311	0.1425	0.0290	0.430	0.0968	853
Cockscomb/TW	900	31	3	0.7069	0.990	843.6	15,400	0.1014	0.1064	0.1163	0.1264	0.0316	0.419	0.0946	923
Magnolia/TW	954	31	3	0.7493	1.018	894.2	16,400	0.0957	0.1007	0.1101	0.1195	0.0325	0.416	0.0938	956
Hawkweed/TW	1000	31	3	0.7854	1.041	937.3	17,100	0.0913	0.0963	0.1053	0.1142	0.0333	0.413	0.0931	984
Bluebell/TW	1033.5	31	3	0.8117	1.057	968.7	17,700	0.0883	0.0934	0.1020	0.1107	0.0338	0.411	0.0927	1003
Marigold/TW	1113	31	3	0.8741	1.095	1043.2	19,100	0.0820	0.0872	0.0952	0.1032	0.0350	0.407	0.0916	1049
Hawthorn/TW	1192.5	31	3	0.9366	1.132	1117.7	20,400	0.0765	0.0819	0.0893	0.0967	0.0362	0.403	0.0906	1093
Narcissus/TW	1272	31	3	0.9990	1.168	1192.2	21,800	0.0717	0.0772	0.0841	0.0911	0.0374	0.399	0.0896	1136
Columbine/TW	1351.5	31	3	1.0615	1.202	1266.3	23,200	0.0675	0.0731	0.0796	0.0861	0.0386	0.395	0.0888	1177
Carnation/TW	1431	31	3	1.1239	1.236	1341.3	24,000	0.0638	0.0695	0.0756	0.0817	0.0397	0.391	0.0880	1217
Coreopsis/TW	1590	49	4	1.2488	1.315	1490.3	27,000	0.0574	0.0635	0.0689	0.0743	0.0425	0.383	0.0861	1297
Jessamine/TW	1750	49	4	1.3744	1.377	1640.3	29,700	0.0522	0.0585	0.0634	0.0683	0.0446	0.377	0.0848	1370
Cowslip/TW	2000	49	4	1.5708	1.468	1893.0	33,900	0.0456	0.0525	0.0567	0.0609	0.0478	0.369	0.0829	1476
Lupine/TW	2500	71	5	1.9635	1.648	2366.2	41,900	0.0369	0.0446	0.0478	0.0511	0.0543	0.353	0.0794	1660
Trillium/TW	3000	71	5	2.3562	1.799	2839.5	50,300	0.0307	0.0392	0.0419	0.0445	0.0597	0.342	0.0768	1821

Notes: (1) Data based on a nominal cable manufactured in accordance with ASTM B 778.

(2) Resistance and ampacity based on an aluminum conductivity of 61.2% IACS at 20°C.

(3) Ampacity based on a 75°C conductor temperature, 25°C ambient temperature, 2 ft/sec wind, in sun, with an emissivity of 0.5 and a coefficient of solar absorption of 0.5, at sea level.

(4) Rated strengths are based on ASTM B 778.

(5) The final design of a shaped wire compact conductor is contingent upon several factors such as: layer diameter, wire width and wire thickness. The actual configuration of a given size may vary between manufacturers. This may result in a slight variation in the number of wires, number of layers and dimensions of individual wires from that shown in the chart.

TABLE 1-23
SHAPED WIRE COMPACT CONCENTRIC-LAY ALUMINUM CONDUCTOR (AAC/TW)
Fixed Diameter Increments

Code Word	Conductor Size kcmil	No. of Wires	No. of Layers	Cross-Sectional Area Sq. in	Conductor Diameter in	Weight lbs/1000 ft	Rated Strength lbs	Resistance				GMR ft	Reactance @ 1 ft Spacing 60 Hz		Ampacity @ 75°C Amps
								dc @ 20°C Ω/mile	ac-60 Hz @ 25°C Ω/mile	ac-60 Hz @ 50°C Ω/mile	ac-60 Hz @ 75°C Ω/mile		Inductive Xa Ω/mile	Capacitive X'a MΩ-mile	
Logan/TW	322.5	17	2	0.2533	0.60	302.3	5,880	0.2830	0.2898	0.3183	0.3468	0.0188	0.482	0.1094	489
Wheeler/TW	449.4	17	2	0.3530	0.70	421.2	7,880	0.2031	0.2087	0.2290	0.2494	0.0220	0.463	0.1048	600
Robson/TW	595.8	17	2	0.4679	0.80	558.4	10,440	0.1532	0.1582	0.1735	0.1888	0.0252	0.447	0.1009	714
McKinley/TW	761.5	17	2	0.5981	0.90	713.7	13,070	0.1198	0.1247	0.1366	0.1486	0.0284	0.432	0.0974	830
Rainier/TW	918.8	31	3	0.7216	1.00	861.2	15,760	0.0993	0.1043	0.1141	0.1239	0.0319	0.418	0.0912	935
Helens/TW	1123.1	31	3	0.8821	1.10	1052.7	19,260	0.0813	0.0865	0.0944	0.1023	0.0352	0.406	0.0915	1055
Baker/TW	1346.8	31	3	1.0578	1.20	1262.3	23,100	0.0678	0.0734	0.0799	0.0864	0.0385	0.395	0.0889	1174
Hood/TW	1583.2	34	3	1.2434	1.30	1483.9	26,590	0.0576	0.0637	0.0691	0.0746	0.0419	0.385	0.0865	1291
Whitney/TW	1812.7	49	4	1.4237	1.40	1699.0	30,740	0.0503	0.0568	0.0615	0.0662	0.0454	0.375	0.0843	1397
Powell/TW	2093.6	49	4	1.6443	1.50	1981.6	35,510	0.0440	0.0510	0.0550	0.0591	0.0489	0.366	0.0822	1507
Jefferson/TW	2388.1	52	4	1.8756	1.60	2260.3	39,670	0.0386	0.0461	0.0495	0.0530	0.0524	0.357	0.0803	1617
Shasta/TW	2667.2	71	5	2.0948	1.70	2524.5	44,740	0.0346	0.0425	0.0456	0.0486	0.0561	0.349	0.0785	1717
Adams/TW	3006.2	71	5	2.3611	1.80	2873.0	50,430	0.0310	0.0394	0.0421	0.0448	0.0578	0.342	0.0768	1816

Notes:
(1) Data based on a nominal cable manufactured in accordance with ASTM B 778.
(2) Resistance and ampacity based on an aluminum conductivity of 61.2% IACS at 20°C.
(3) Ampacity based on a 75°C conductor temperature, 25°C ambient temperature, 2 ft/sec wind, in sun, with an emissivity of 0.5 and a coefficient of solar absorption of 0.5, at sea level.
(4) Rated strengths are based on ASTM B 778.
(5) The final design of a shaped wire compact conductor is contingent upon several factors such as: layer diameter, wire width and wire thickness. The actual configuration of a given size may vary between manufacturers. This may result in a slight variation in the number of wires, number of layers and dimensions of individual wires from that shown in the chart.

TABLE 1-24
ALL-ALUMINUM ALLOY CONCENTRIC-LAY-STRANDED CONDUCTOR (AAAC)
ACSR Diameter Equivalent

Code Word	Conductor Size kcmil	No. of Wires	Size of ACSR of Equal Diameter AWG or kcmil and Stranding	Diameter		Cross-Sectional Area Sq. in	Weight lbs/ 1000 ft	Rated Strength lbs	Resistance				GMR ft	Reactance @ 1 ft Spacing 60 Hz		Ampacity @ 75°C Amps
				Individual Wire in	Complete Cable in				dc @ 20°C Ω/mile	ac-60 Hz @ 25°C Ω/mile	ac-60 Hz @ 50°C Ω/mile	ac-60 Hz @ 75°C Ω/mile		Inductive Xa Ω/mile	Capacitive X'a MΩ-mile	
Akron	30.58	7	6-6/1	0.0661	0.198	0.0240	28.5	1,110	3.4790	3.5394	3.8412	4.1430	0.00599	0.6209	0.1423	107
Alton	48.69	7	4-6/1	0.0834	0.250	0.0382	45.4	1,760	2.1854	2.2235	2.4126	2.6026	0.00756	0.5927	0.1354	143
Ames	77.47	7	2-6/1	0.1052	0.316	0.0608	72.2	2,800	1.3733	1.3973	1.5164	1.6355	0.00954	0.5645	0.1285	191
Azusa	123.3	7	1/0-6/1	0.1327	0.398	0.0968	114.9	4,270	0.8628	0.8782	0.9530	1.0278	0.01204	0.5363	0.1216	256
Anaheim	155.4	7	2/0-6/1	0.1490	0.447	0.1221	144.9	5,390	0.6846	0.6969	0.7563	0.8156	0.01315	0.5223	0.1182	296
Amherst	195.7	7	3/0-6/1	0.1672	0.502	0.1537	182.5	6,790	0.5436	0.5536	0.6007	0.6478	0.01516	0.5080	0.1147	342
Alliance	246.9	7	4/0-6/1	0.1878	0.563	0.1939	230.2	8,560	0.4309	0.4391	0.4764	0.5137	0.01703	0.4942	0.1113	395
Butte	312.8	19	266.8-26/7	0.1283	0.642	0.2456	291.6	10,500	0.3401	0.3469	0.3763	0.4058	0.02024	0.4732	0.1074	460
Canton	394.5	19	366.4-26/7	0.1441	0.721	0.3098	367.9	13,300	0.2697	0.2755	0.2988	0.3221	0.02273	0.4591	0.1040	533
Cairo	465.4	19	397.5-26/7	0.1565	0.783	0.3655	433.9	15,600	0.2286	0.2339	0.2536	0.2733	0.02469	0.4491	0.1015	590
Darien	559.5	19	477.0-26/7	0.1716	0.858	0.4394	521.7	18,800	0.1901	0.1950	0.2114	0.2278	0.02707	0.4379	0.0988	663
Elgin	652.4	19	556.5-26/7	0.1853	0.927	0.5124	608.3	21,900	0.1631	0.1677	0.1817	0.1958	0.02923	0.4286	0.0965	729
Flint	740.8	37	636.0-26/7	0.1415	0.991	0.5818	690.8	24,400	0.1436	0.1482	0.1605	0.1728	0.03168	0.4189	0.0946	790
Greeley	927.2	37	795.0-26/7	0.1583	1.108	0.7282	864.6	30,500	0.1147	0.1193	0.1291	0.1389	0.03544	0.4053	0.0912	908

Notes: (1) Data based on a nominal cable manufactured in accordance with ASTM B 399.
(2) Resistance and ampacity based on an aluminum alloy (AAAC/6201) conductivity of 52.5% IACS at 20°C.
(3) Ampacity based on a 75°C conductor temperature, 25°C ambient temperature, 2 ft/sec wind, in sun, with an emissivity of 0.5 and a coefficient of solar absorption of 0.5, at sea level.

TABLE 1-25

ALL-ALUMINUM ALLOY CONCENTRIC-LAY-STRANDED CONDUCTOR (AAAC)

Even AWG and kcmil Sizes

Conductor Size AWG or kcmil	No. of Wires	Diameter		Cross-Sectional Area Sq. in	Weight lbs/1000 ft	Rated Strength lbs	Resistance				GMR ft	Reactance @ 1 ft Spacing 60 Hz		Ampacity @ 75°C Amps
		Individual Wire in	Complete Cable in				dc @ 20°C Ω/mile	ac-60 Hz @ 25°C Ω/mile	ac-60 Hz @ 50°C Ω/mile	ac-60 Hz @ 75°C Ω/mile		Inductive Xa Ω/mile	Capacitive X'a MΩ-mile	
6	7	0.0612	0.184	0.0206	24.49	949	4.0544	4.1248	4.4765	4.8282	0.00555	0.6302	0.1445	97
4	7	0.0772	0.232	0.0328	38.90	1,510	2.5488	2.5931	2.8142	3.0353	0.00700	0.6020	0.1377	130
2	7	0.0974	0.292	0.0521	61.92	2,400	1.6029	1.6309	1.7700	1.9090	0.00883	0.5738	0.1308	173
1/0	7	0.1228	0.368	0.0829	98.43	3,820	1.0075	1.0252	1.1126	1.2000	0.01114	0.5457	0.1239	232
2/0	7	0.1379	0.414	0.1045	124.1	4,620	0.7993	0.8135	0.8829	0.9522	0.01251	0.5316	0.1204	268
3/0	7	0.1548	0.464	0.1317	156.4	5,820	0.6340	0.6455	0.7005	0.7554	0.01404	0.5176	0.1170	310
4/0	7	0.1739	0.522	0.1663	197.4	7,340	0.5028	0.5121	0.5557	0.5992	0.01577	0.5035	0.1136	359
250	19	0.1147	0.574	0.1963	233.1	8,760	0.4255	0.4336	0.4705	0.5074	0.01810	0.4868	0.1108	400
300	19	0.1257	0.629	0.2358	280.0	10,500	0.3546	0.3616	0.3923	0.4230	0.01983	0.4575	0.1080	448
350	19	0.1357	0.679	0.2748	326.3	11,800	0.3040	0.3102	0.3365	0.3628	0.02141	0.4664	0.1058	494
400	19	0.1451	0.726	0.3142	373.0	13,400	0.2660	0.2717	0.2947	0.3177	0.02289	0.4583	0.1038	537
450	19	0.1539	0.770	0.3534	419.6	15,100	0.2364	0.2418	0.2622	0.2826	0.02428	0.4512	0.1020	578
500	19	0.1622	0.811	0.3926	466.1	16,800	0.2128	0.2179	0.2362	0.2546	0.02559	0.4448	0.1005	618
550	37	0.1219	0.853	0.4318	512.7	18,900	0.1934	0.1983	0.2150	0.2317	0.02729	0.4370	0.0990	656
600	37	0.1273	0.891	0.4709	559.1	20,600	0.1774	0.1821	0.1973	0.2126	0.02850	0.4317	0.0977	693
650	37	0.1325	0.928	0.5102	605.7	21,400	0.1637	0.1683	0.1824	0.1964	0.02966	0.4269	0.0965	728
700	37	0.1375	0.963	0.5494	652.3	23,000	0.1520	0.1566	0.1696	0.1827	0.03078	0.4224	0.0954	763
750	37	0.1424	0.997	0.5893	699.6	24,700	0.1419	0.1464	0.1586	0.1707	0.03188	0.4181	0.0944	796
800	37	0.1470	1.029	0.6280	745.6	26,300	0.1330	0.1375	0.1489	0.1603	0.03291	0.4143	0.0934	828
900	37	0.1560	1.092	0.7072	839.7	29,600	0.1182	0.1228	0.1328	0.1429	0.03492	0.4070	0.0917	891
1000	37	0.1644	1.151	0.7854	932.5	32,900	0.1064	0.1110	0.1200	0.1291	0.03681	0.4007	0.0901	951

Notes: (1) Data based on a nominal cable manufactured in accordance with ASTM B 399.
(2) Resistance and ampacity based on an aluminum alloy (AAAC/6201) conductivity of 52.5% IACS at 20°C.
(3) Ampacity based on a 75°C conductor temperature, 25°C ambient temperature, 2 ft/sec wind, in sun, with an emissivity of 0.5 and a coefficient of solar absorption of 0.5, at sea level.

TABLE 1-26
ALUMINUM ALLOY CONDUCTOR STEEL REINFORCED (AACSR)

		Cross-Sectional Area Sq. in			Diameter				Weight / 1000 ft		
					Individual Wires						
Conductor Size kcmil	Stranding (Al/St)	Aluminum Alloy	Total	Layers of Al Alloy	Aluminum Alloy in	Steel in	Steel Core in	Complete Cable in	Aluminum Alloy lbs	Steel lbs	Total lbs
266.8	26/7	0.2095	0.2436	2	0.1013	0.0788	0.2364	0.642	250.0	115.6	365.6
266.8	30/7	0.2095	0.2584	2	0.0943	0.0943	0.2829	0.660	250.6	165.5	416.1
300	26/7	0.2355	0.2738	2	0.1074	0.0835	0.2505	0.680	281.0	129.8	410.8
336.4	26/7	0.2640	0.3070	2	0.1137	0.0884	0.2652	0.720	315.0	145.4	460.4
336.4	30/7	0.2642	0.3259	2	0.1059	0.1059	0.3177	0.741	316.0	208.7	524.7
397.5	26/7	0.3120	0.3628	2	0.1236	0.0961	0.2883	0.783	372.2	171.9	544.1
397.5	30/7	0.3121	0.3849	2	0.1151	0.1151	0.3453	0.806	373.3	246.5	619.9
477	26/7	0.3744	0.4354	2	0.1354	0.1053	0.3159	0.858	446.7	206.4	653.0
477	30/7	0.3747	0.4621	2	0.1261	0.1261	0.3783	0.883	448.1	295.9	744.0
556.5	26/7	0.4371	0.5083	2	0.1463	0.1138	0.3414	0.927	521.5	241.0	762.5
556.5	30/7	0.4371	0.5391	2	0.1362	0.1362	0.4086	0.953	522.8	345.2	868.0
605	26/7	0.4749	0.5522	2	0.1525	0.1186	0.3558	0.966	566.6	261.8	828.4
605	30/7	0.4751	0.5860	2	0.1420	0.1420	0.4260	0.994	568.2	375.3	943.5
605	30/19	0.4751	0.5834	2	0.1420	0.0852	0.4260	0.994	568.2	367.4	935.6
636	26/7	0.4995	0.5808	2	0.1564	0.1216	0.3648	0.990	596.0	275.2	871.1
636	30/7	0.4995	0.6160	2	0.1456	0.1456	0.4368	1.019	597.4	394.5	991.9
636	30/19	0.4995	0.6135	2	0.1456	0.0874	0.4370	1.019	597.4	386.6	984.0
666.6	26/7	0.5234	0.6086	2	0.1601	0.1245	0.3735	1.014	624.5	288.5	912.9
715.5	26/7	0.5620	0.6535	2	0.1659	0.1290	0.3870	1.051	670.6	309.7	980.2
715.5	30/19	0.5617	0.6897	2	0.1544	0.0926	0.4630	1.081	671.8	434.0	1105.8
795	26/7	0.6247	0.7264	2	0.1749	0.1360	0.4080	1.108	745.3	344.2	1089.5
795	30/7	0.6245	0.7702	2	0.1628	0.1628	0.4884	1.140	746.9	493.2	1240.1
795	45/7	0.6242	0.6674	3	0.1329	0.0886	0.2658	1.063	744.8	146.1	890.9
795	54/7	0.6240	0.7049	3	0.1213	0.1213	0.3639	1.092	744.5	273.8	1018.4
795	30/19	0.6245	0.7669	2	0.1628	0.0977	0.4885	1.140	746.9	483.1	1230.0
900	45/7	0.7066	0.7555	3	0.1414	0.0943	0.2829	1.131	843.1	165.5	1008.6
900	54/7	0.7069	0.7985	3	0.1291	0.1291	0.3873	1.162	843.4	310.2	1153.5
954	45/7	0.7492	0.8010	3	0.1456	0.0971	0.2913	1.165	893.9	175.5	1069.4
954	54/7	0.7491	0.8462	3	0.1329	0.1329	0.3987	1.196	893.7	328.7	1222.4
954	30/19	0.7491	0.9199	2	0.1783	0.1070	0.5350	1.248	895.9	579.5	1475.4
1033.5	45/7	0.8112	0.8673	3	0.1515	0.1010	0.3030	1.212	967.8	189.8	1157.7
1033.5	54/7	0.8112	0.9164	3	0.1383	0.1383	0.4149	1.246	967.8	356.0	1323.8
1113	45/7	0.8745	0.9350	3	0.1573	0.1049	0.3147	1.259	1043.4	204.8	1248.1
1113	54/19	0.8746	0.9855	3	0.1436	0.0862	0.4310	1.293	1048.5	376.1	1424.6
1192.5	45/7	0.9367	1.0014	3	0.1628	0.1085	0.3255	1.302	1117.6	219.1	1336.7
1192.5	54/19	0.9365	1.0552	3	0.1486	0.0892	0.4460	1.338	1122.8	402.7	1525.5
1272	45/7	0.9987	1.0678	3	0.1681	0.1121	0.3363	1.345	1191.6	233.9	1425.4
1272	54/19	0.9993	1.1259	3	0.1535	0.0921	0.4605	1.382	1198.1	429.3	1627.4
1351.5	45/7	1.0614	1.1347	3	0.1733	0.1155	0.3465	1.386	1266.4	248.3	1514.7
1351.5	54/19	1.0614	1.1958	3	0.1582	0.0949	0.4745	1.424	1272.6	455.8	1728.4
1431	45/7	1.1236	1.2013	3	0.1783	0.1189	0.3567	1.427	1340.5	263.1	1603.6
1431	54/19	1.1241	1.2665	3	0.1628	0.0977	0.4885	1.465	1347.7	483.1	1830.8
1510.5	45/7	1.1862	1.2682	3	0.1832	0.1221	0.3663	1.466	1415.2	277.4	1692.7
1510.5	54/19	1.1856	1.3357	3	0.1672	0.1003	0.5015	1.505	1421.5	509.2	1930.7
1590	45/7	1.2492	1.3355	3	0.1880	0.1253	0.3759	1.504	1490.4	292.2	1782.5
1590	54/19	1.2489	1.4072	3	0.1716	0.1030	0.5150	1.545	1497.3	537.0	2034.3
1780	84/19	1.3986	1.5126	4	0.1456	0.0874	0.4370	1.602	1676.8	386.6	2063.4
2034.5	72/7	1.5979	1.6671	4	0.1681	0.1122	0.3366	1.681	1915.8	234.3	2150.1
2057	76/19	1.6152	1.7032	4	0.1645	0.0768	0.3840	1.700	1936.5	298.5	2235.1
2156	84/19	1.6931	1.8309	4	0.1602	0.0961	0.4805	1.762	2029.9	467.4	2497.4
2167	72/7	1.7022	1.7758	4	0.1735	0.1157	0.3471	1.735	2040.8	249.1	2290.0
2312	76/19	1.8155	1.9144	4	0.1744	0.0814	0.4070	1.802	2176.6	335.4	2512.0
2515	76/19	1.9750	2.0826	4	0.1819	0.0849	0.4245	1.880	2367.9	364.8	2732.7

Notes: (1) Data based on a nominal cable manufactured in accordance with ASTM B 711.
(2) Resistance and ampacity based on an aluminum alloy (6201) conductivity of 52.5% IACS at 20°C and a steel conductivity of 8% IACS at 20°C.
(3) Ampacity based on a 75°C conductor temperature, 25°C ambient temperature, 2 ft/sec wind, in sun, with an emissivity of 0.5 and a coefficient of solar absorption of 0.5, at sea level.
(4) Rated strengths based on Class A galvanized steel core wire in accordance with ASTM B 498.

TABLE 1-26
ALUMINUM ALLOY CONDUCTOR STEEL REINFORCED (AACSR)

Metal Content % Weight			Resistance					Reactance @ 1 ft Spacing 60 Hz			
				ac-60 Hz							
Aluminum Alloy %	Steel %	Rated Strength lbs	dc @ 20°C Ω/mile	@ 25°C Ω/mile	@ 50°C Ω/mile	@ 75°C Ω/mile	GMR ft	Inductive Xa Ω/mile	Capacitive X'a MΩ-mile	Ampacity @ 75°C Amps	Conductor Size kcmil
68.39	31.61	16,200	0.3908	0.3980	0.4318	0.4656	0.0217	0.465	0.1073	430	266.8
60.23	39.77	19,000	0.3876	0.3946	0.4281	0.4616	0.0227	0.459	0.1066	435	266.8
68.41	31.59	18,200	0.3476	0.3540	0.3841	0.4142	0.0230	0.458	0.1056	463	300
68.41	31.59	20,400	0.3100	0.3158	0.3426	0.3694	0.0243	0.451	0.1039	497	336.4
60.23	39.77	23,900	0.3074	0.3131	0.3397	0.3663	0.0255	0.445	0.1030	503	336.4
68.41	31.59	23,900	0.2623	0.2675	0.2901	0.3128	0.0265	0.441	0.1014	552	397.5
60.23	39.77	28,300	0.2601	0.2651	0.2876	0.3101	0.0277	0.435	0.1006	559	397.5
68.40	31.60	28,000	0.2186	0.2231	0.2420	0.2608	0.0290	0.430	0.0987	619	477
60.23	39.77	33,900	0.2168	0.2211	0.2398	0.2586	0.0304	0.424	0.0979	627	477
68.39	31.61	32,700	0.1874	0.1915	0.2076	0.2238	0.0313	0.420	0.0964	682	556.5
60.23	39.77	38,800	0.1858	0.1897	0.2058	0.2218	0.0328	0.415	0.0956	690	556.5
68.40	31.60	35,500	0.1723	0.1763	0.1911	0.2060	0.0327	0.415	0.0952	719	605
60.23	39.77	41,600	0.1709	0.1746	0.1894	0.2041	0.0342	0.410	0.0943	728	605
60.73	39.27	41,500	0.1711	0.1748	0.1895	0.2043	0.0342	0.410	0.0943	727	605
68.41	31.59	37,400	0.1639	0.1678	0.1819	0.1960	0.0335	0.412	0.0946	742	636
60.23	39.77	43,700	0.1626	0.1662	0.1802	0.1942	0.0351	0.406	0.0936	751	636
60.71	39.29	43,600	0.1627	0.1663	0.1804	0.1944	0.0351	0.406	0.0936	750	636
68.40	31.60	39,200	0.1564	0.1601	0.1736	0.1871	0.0343	0.409	0.0939	764	666.6
68.41	31.59	42,000	0.1457	0.1493	0.1619	0.1744	0.0355	0.405	0.0927	799	715.5
60.75	39.25	48,400	0.1447	0.1481	0.1605	0.1730	0.0372	0.399	0.0919	808	715.5
68.41	31.59	46,700	0.1312	0.1346	0.1459	0.1572	0.0375	0.399	0.0911	853	795
60.23	39.77	54,700	0.1301	0.1333	0.1445	0.1557	0.0392	0.393	0.0904	864	795
83.60	16.40	34,800	0.1330	0.1370	0.1491	0.1609	0.0352	0.406	0.0925	834	795
73.11	26.89	43,200	0.1318	0.1354	0.1501	0.1625	0.0368	0.401	0.0916	836	795
60.72	39.28	53,900	0.1302	0.1334	0.1446	0.1558	0.0392	0.393	0.0903	863	795
83.59	16.41	39,200	0.1175	0.1214	0.1320	0.1424	0.0374	0.399	0.0905	901	900
73.11	26.89	48,900	0.1164	0.1200	0.1328	0.1437	0.0392	0.393	0.0897	904	900
83.59	16.41	41,600	0.1109	0.1148	0.1248	0.1345	0.0385	0.395	0.0896	934	954
73.11	26.89	50,500	0.1099	0.1133	0.1254	0.1357	0.0404	0.389	0.0889	937	954
60.72	39.28	64,600	0.1085	0.1116	0.1208	0.1302	0.0430	0.382	0.0877	967	954
83.60	16.40	45,000	0.1023	0.1062	0.1154	0.1244	0.0401	0.390	0.0885	982	1033.5
73.11	26.89	54,700	0.1014	0.1049	0.1159	0.1254	0.0420	0.385	0.0877	985	1033.5
83.59	16.41	48,500	0.0950	0.0990	0.1075	0.1158	0.0416	0.386	0.0873	1028	1113
73.60	26.40	58,300	0.0947	0.0981	0.1084	0.1172	0.0436	0.380	0.0866	1029	1113
83.61	16.39	51,900	0.0887	0.0927	0.1006	0.1083	0.0431	0.382	0.0863	1072	1192.5
73.60	26.40	62,400	0.0883	0.0918	0.1014	0.1096	0.0451	0.376	0.0855	1074	1192.5
83.59	16.41	55,400	0.0832	0.0872	0.0946	0.1018	0.0445	0.378	0.0854	1116	1272
73.62	26.38	66,000	0.0828	0.0864	0.0952	0.1029	0.0466	0.372	0.0846	1118	1272
83.61	16.39	58,900	0.0783	0.0824	0.0893	0.0961	0.0459	0.374	0.0845	1158	1351.5
73.63	26.37	70,100	0.0780	0.0815	0.0898	0.0970	0.0480	0.368	0.0837	1161	1351.5
83.59	16.41	62,300	0.0739	0.0781	0.0846	0.0910	0.0472	0.371	0.0836	1200	1431
73.61	26.39	74,200	0.0736	0.0772	0.0851	0.0918	0.0494	0.365	0.0828	1202	1431
83.61	16.39	65,800	0.0700	0.0743	0.0804	0.0864	0.0485	0.367	0.0828	1238	1510.5
73.63	26.37	78,300	0.0697	0.0734	0.0808	0.0872	0.0508	0.362	0.0821	1242	1510.5
83.61	16.39	69,300	0.0665	0.0709	0.0767	0.0824	0.0497	0.364	0.0821	1277	1590
73.60	26.4	82,500	0.0663	0.0700	0.0770	0.0830	0.0521	0.358	0.0813	1282	1590
81.26	18.74	80,200	0.0596	0.0640	0.0690	0.0739	0.0534	0.355	0.0802	1372	1780
89.10	10.90	79,800	0.0525	0.0577	0.0620	0.0663	0.0553	0.351	0.0788	1467	2034.5
86.64	13.36	84,100	0.0518	0.0569	0.0611	0.0653	0.0562	0.345	0.0785	1482	2057
81.28	18.72	96,400	0.0492	0.0541	0.0580	0.0621	0.0588	0.344	0.0774	1535	2156
89.12	10.88	85,000	0.0492	0.0547	0.0587	0.0627	0.0570	0.348	0.0778	1521	2167
86.65	13.35	94,500	0.0461	0.0516	0.0552	0.0589	0.0595	0.342	0.0767	1585	2312
86.65	13.35	102,800	0.0424	0.0481	0.0515	0.0548	0.0621	0.337	0.0755	1662	2515

TABLE 1-27
ALUMINUM CONDUCTOR ALUMINUM ALLOY REINFORCED (ACAR)
Diameter Equivalent to ACSR

Conductor Size kcmil	Stranding	ACSR with Equivalent Diameter	Cross-Sectional Area Sq. in	Diameter: Individual Wires in	Diameter: Complete Cable in	Weight lbs/1000 ft	Rated Strength lbs	Resistance: dc @ 20°C Ω/mile	Resistance ac-60 Hz: @ 25°C Ω/mile	Resistance ac-60 Hz: @ 50°C Ω/mile	Resistance ac-60 Hz: @ 75°C Ω/mile	GMR ft	Reactance @ 1 ft Spacing 60 Hz: Inductive Xa Ω/mile	Reactance @ 1 ft Spacing 60 Hz: Capacitive X'a MΩ-mile	Ampacity @ 75°C Amps
355	15/4	336.4 - 18/1	0.2788	0.1367	0.684	332.4	7,580	0.2650	0.2714	0.2973	0.3233	0.0216	0.465	0.1054	524
355	12/7	336.4 - 18/1	0.2788	0.1367	0.684	332.2	8,540	0.2713	0.2776	0.3036	0.3296	0.0216	0.465	0.1054	519
465.9	12/7	397.5 - 26/7	0.3659	0.1566	0.783	435.9	11,000	0.2067	0.2121	0.2319	0.2517	0.0247	0.449	0.1015	616
503.6	15/4	477 - 18/1	0.3955	0.1628	0.814	471.5	10,500	0.1868	0.1921	0.2103	0.2286	0.0257	0.444	0.1003	652
503.6	12/7	477 - 18/1	0.3955	0.1628	0.814	471.1	11,900	0.1912	0.1965	0.2147	0.2330	0.0257	0.444	0.1003	646
653.1	15/4	556.5 - 26/7	0.5129	0.1854	0.927	611.5	13,600	0.1441	0.1490	0.1629	0.1769	0.0293	0.429	0.0964	767
653.1	12/7	556.5 - 26/7	0.5129	0.1854	0.927	611.0	15,400	0.1475	0.1523	0.1663	0.1804	0.0293	0.429	0.0964	759
739.8	33/4	636 - 26/7	0.5810	0.1414	0.990	693.0	14,200	0.1253	0.1302	0.1424	0.1547	0.0317	0.419	0.0945	834
739.8	30/7	636 - 26/7	0.5810	0.1414	0.990	692.7	15,300	0.1268	0.1316	0.1439	0.1562	0.0317	0.419	0.0945	830
739.8	24/13	636 - 26/7	0.5810	0.1414	0.990	692.1	16,900	0.1298	0.1346	0.1470	0.1593	0.0317	0.419	0.0945	822
739.8	18/19	636 - 26/7	0.5810	0.1414	0.990	691.6	18,800	0.1331	0.1378	0.1502	0.1625	0.0317	0.419	0.0945	814
853.7	30/7	795 - 45/7	0.6705	0.1519	1.063	799.4	17,500	0.1099	0.1147	0.1253	0.1359	0.0340	0.410	0.0923	907
853.7	24/13	795 - 45/7	0.6705	0.1519	1.063	798.7	19,300	0.1125	0.1173	0.1280	0.1386	0.0340	0.410	0.0923	898
853.7	18/19	795 - 45/7	0.6705	0.1519	1.063	798.1	21,500	0.1153	0.1201	0.1307	0.1414	0.0340	0.410	0.0923	889
927.2	30/7	795 - 26/7	0.7282	0.1583	1.108	868.2	19,000	0.1012	0.1061	0.1158	0.1255	0.0354	0.405	0.0911	954
927.2	24/13	795 - 26/7	0.7282	0.1583	1.108	867.5	20,900	0.1036	0.1084	0.1182	0.1280	0.0354	0.405	0.0911	945
927.2	18/19	795 - 26/7	0.7282	0.1583	1.108	866.7	23,400	0.1062	0.1110	0.1207	0.1305	0.0354	0.405	0.0911	936
1024.5	30/7	954 - 45/7	0.8046	0.1664	1.165	959.3	20,900	0.0916	0.0966	0.1054	0.1142	0.0373	0.399	0.0896	1014
1024.5	24/13	954 - 45/7	0.8046	0.1664	1.165	958.5	23,100	0.0938	0.0987	0.1075	0.1164	0.0373	0.399	0.0896	1004
1024.5	18/19	954 - 45/7	0.8046	0.1664	1.165	957.7	25,800	0.0961	0.1010	0.1098	0.1187	0.0373	0.399	0.0896	994
1081	30/7	954 - 54/7	0.8490	0.1709	1.196	1011.9	22,100	0.0868	0.0918	0.1001	0.1084	0.0383	0.396	0.0888	1047
1081	24/13	954 - 54/7	0.8490	0.1709	1.196	1011.1	24,400	0.0889	0.0939	0.1022	0.1105	0.0383	0.396	0.0888	1037
1081	18/19	954 - 54/7	0.8490	0.1709	1.196	1010.2	27,200	0.0911	0.0960	0.1043	0.1127	0.0383	0.396	0.0888	1027
1109	30/7	1033.5 - 45/7	0.8710	0.1731	1.212	1038.1	22,700	0.0846	0.0897	0.0977	0.1058	0.0388	0.394	0.0885	1064
1109	24/13	1033.5 - 45/7	0.8710	0.1731	1.212	1037.2	25,000	0.0866	0.0917	0.0997	0.1078	0.0388	0.394	0.0885	1054
1109	18/19	1033.5 - 45/7	0.8710	0.1731	1.212	1036.4	27,900	0.0888	0.0937	0.1018	0.1100	0.0388	0.394	0.0885	1044
1172	33/4	1033.5 - 54/7	0.9205	0.1780	1.246	1098.2	22,100	0.0791	0.0843	0.0919	0.0995	0.0399	0.391	0.0876	1105
1172	30/7	1033.5 - 54/7	0.9205	0.1780	1.246	1097.7	24,000	0.0800	0.0852	0.0928	0.1004	0.0399	0.391	0.0876	1100
1172	24/13	1033.5 - 54/7	0.9205	0.1780	1.246	1096.8	26,500	0.0820	0.0871	0.0947	0.1024	0.0399	0.391	0.0876	1090
1172	18/19	1033.5 - 54/7	0.9205	0.1780	1.246	1095.9	29,500	0.0840	0.0891	0.0967	0.1044	0.0399	0.391	0.0876	1079
1197	30/7	1113 - 45/7	0.9401	0.1799	1.259	1121.3	24,500	0.0784	0.0836	0.0910	0.0985	0.0403	0.390	0.0873	1114
1197	24/13	1113 - 45/7	0.9401	0.1799	1.259	1120.3	27,000	0.0803	0.0854	0.0929	0.1003	0.0403	0.390	0.0873	1104
1197	18/19	1113 - 45/7	0.9401	0.1799	1.259	1119.4	30,200	0.0823	0.0873	0.0948	0.1023	0.0403	0.390	0.0873	1093
1280	30/7	1192.5 - 45/7	1.0053	0.1860	1.302	1198.6	26,200	0.0733	0.0787	0.0856	0.0925	0.0417	0.386	0.0863	1160
1280	24/13	1192.5 - 45/7	1.0053	0.1860	1.302	1197.6	28,900	0.0750	0.0803	0.0873	0.0942	0.0417	0.386	0.0863	1149
1280	18/19	1192.5 - 45/7	1.0053	0.1860	1.302	1196.6	32,200	0.0769	0.0821	0.0891	0.0961	0.0417	0.386	0.0863	1138
1362	54/7	1272 - 45/7	1.0697	0.1494	1.345	1275.4	26,300	0.0682	0.0737	0.0803	0.0866	0.0433	0.381	0.0854	1209
1362	42/19	1272 - 45/7	1.0697	0.1494	1.345	1274.1	30,300	0.0702	0.0756	0.0821	0.0886	0.0433	0.381	0.0854	1195
1362	24/37	1272 - 45/7	1.0697	0.1494	1.345	1272.2	35,400	0.0734	0.0787	0.0852	0.0917	0.0433	0.381	0.0854	1174
1527	24/37	1351.5 - 54/19	1.1993	0.1582	1.424	1426.5	39,500	0.0654	0.0710	0.0767	0.0825	0.0458	0.374	0.0838	1257
1527	42/19	1351.5 - 54/19	1.1993	0.1582	1.424	1428.6	33,600	0.0625	0.0683	0.0740	0.0798	0.0458	0.374	0.0838	1278
1703	54/7	1590 - 45/7	1.3375	0.1671	1.504	1595.5	32,300	0.0545	0.0607	0.0660	0.0708	0.0484	0.368	0.0821	1377
1703	48/13	1590 - 45/7	1.3375	0.1671	1.504	1594.7	34,600	0.0553	0.0614	0.0665	0.0715	0.0484	0.368	0.0821	1369
1703	42/19	1590 - 45/7	1.3375	0.1671	1.504	1593.9	37,500	0.0561	0.0622	0.0672	0.0723	0.0484	0.368	0.0821	1362
1703	24/37	1590 - 45/7	1.3375	0.1671	1.504	1591.5	44,000	0.0586	0.0645	0.0696	0.0748	0.0484	0.368	0.0821	1340
1933	54/7	1780 - 84/19	1.5182	0.1780	1.602	1810.4	36,700	0.0480	0.0546	0.0591	0.0634	0.0515	0.360	0.0802	1479
1933	48/13	1780 - 84/19	1.5182	0.1780	1.602	1809.5	39,200	0.0487	0.0553	0.0596	0.0641	0.0515	0.360	0.0802	1472
1933	42/19	1780 - 84/19	1.5182	0.1780	1.602	1808.6	42,500	0.0494	0.0559	0.0603	0.0647	0.0515	0.360	0.0802	1464
1933	24/37	1780 - 84/19	1.5182	0.1780	1.602	1805.9	49,900	0.0517	0.0580	0.0624	0.0669	0.0515	0.360	0.0802	1441
2267	54/7	2167 - 72/7	1.7805	0.1928	1.735	2144.9	43,000	0.0413	0.0486	0.0523	0.0560	0.0558	0.350	0.0778	1608
2267	48/13	2167 - 72/7	1.7805	0.1928	1.735	2143.8	46,000	0.0419	0.0491	0.0528	0.0565	0.0558	0.350	0.0778	1601
2267	42/19	2167 - 72/7	1.7805	0.1928	1.735	2142.7	49,900	0.0425	0.0496	0.0533	0.0571	0.0558	0.350	0.0778	1593
2267	24/37	2167 - 72/7	1.7805	0.1928	1.735	2139.5	58,600	0.0445	0.0514	0.0551	0.0589	0.0558	0.350	0.0778	1568
2338	54/7	2156 - 84/19	1.8363	0.1958	1.762	2212.1	44,400	0.0401	0.0474	0.0512	0.0549	0.0567	0.348	0.0774	1635
2338	48/13	2156 - 84/19	1.8363	0.1958	1.762	2211.0	47,500	0.0406	0.0479	0.0515	0.0551	0.0567	0.348	0.0774	1628
2338	42/19	2156 - 84/19	1.8363	0.1958	1.762	2209.9	51,500	0.0412	0.0484	0.0520	0.0556	0.0567	0.348	0.0774	1620
2338	24/37	2156 - 84/19	1.8363	0.1958	1.762	2206.6	60,400	0.0431	0.0501	0.0537	0.0573	0.0567	0.348	0.0774	1595
2442	72/19	2312 - 76/19	1.9176	0.1638	1.802	2308.4	50,400	0.0389	0.0464	0.0498	0.0532	0.0581	0.345	0.0768	1667
2442	63/28	2312 - 76/19	1.9176	0.1638	1.802	2307.3	53,600	0.0395	0.0469	0.0503	0.0537	0.0581	0.345	0.0768	1659
2442	54/37	2312 - 76/19	1.9176	0.1638	1.802	2306.1	57,600	0.0401	0.0474	0.0508	0.0543	0.0581	0.345	0.0768	1650

Notes: (1) Data based on a nominal cable manufactured in accordance with ASTM B 524.
(2) Resistance and ampacity based on 1350 aluminum conductivity of 61.2% IACS at 20°C and an aluminum alloy (6201) conductivity of 52.5% IACS at 20°C.
(3) Ampacity based on a 75°C conductor temperature, 25°C ambient temperature, 2 ft/sec wind, in sun, with an emissivity of 0.5 and a coefficient of solar absorption of 0.5, at sea level.
(4) Weights do not match those in ASTM B 524/B 524M.

TABLE 1-28

SINGLE LAYER ALUMINUM CONDUCTOR STEEL REINFORCED (ACSR)

Code Word	Conductor Size AWG or kcmil	Stranding (Al/St)	Cross-Sectional Area Sq. in		Diameter				Weight/1000 ft			Metal Content % Weight		Rated Strength lbs	Approximate Resistance				Reactance @ 1 ft Spacing 60 Hz		Ampacity @ 75°C Amps
					Individual Wires											ac-60 Hz					
			Aluminum	Total	Aluminum in	Steel in	Steel Core in	Complete Cable in	Aluminum lbs	Steel lbs	Total lbs	Aluminum %	Steel %		dc @ 20°C Ω/mile	@ 25°C Ω/mile	@ 50°C Ω/mile	@ 75°C Ω/mile	Inductive Xa Ω/mile	Capacitive X'a MΩ-mile	
Turkey	6	6/1	0.0206	0.0240	0.0661	0.0661	0.0661	0.198	24.5	11.6	36.1	67.9	32.1	1,190	3.3862	3.4544	3.9360	4.3013	0.7574	0.1423	105
Swan	4	6/1	0.0328	0.0383	0.0834	0.0834	0.0834	0.250	38.9	18.4	57.3	67.9	32.1	1,860	2.1287	2.1716	2.5191	2.7482	0.7117	0.1354	139
Swanate	4	7/1	0.0328	0.0411	0.0772	0.1029	0.1029	0.257	38.9	28.0	66.9	58.1	41.9	2,360	2.1049	2.1473	2.5201	2.7705	0.7291	0.1346	139
Sparrow	2	6/1	0.0522	0.0609	0.1052	0.1052	0.1052	0.316	61.9	29.3	91.2	67.9	32.1	2,850	1.3389	1.3660	1.6178	1.7661	0.6624	0.1285	184
Sparate	2	7/1	0.0522	0.0655	0.0974	0.1299	0.1299	0.325	61.9	44.7	106.6	58.1	41.9	3,640	1.3239	1.3506	1.6190	1.7845	0.6707	0.1276	184
Robin	1	6/1	0.0657	0.0767	0.1181	0.1182	0.1182	0.355	78.1	37.0	115.1	67.9	32.1	3,550	1.0616	1.0832	1.2975	1.4184	0.6375	0.1250	211
Raven	1/0	6/1	0.0830	0.0968	0.1327	0.1327	0.1327	0.398	98.5	46.6	145.1	67.9	32.1	4,380	0.8414	0.8585	1.0412	1.1400	0.6134	0.1216	243
Quail	2/0	6/1	0.1045	0.1219	0.1489	0.1489	0.1489	0.447	124.1	58.7	182.8	67.9	32.1	5,310	0.6676	0.6813	0.8402	0.9203	0.5909	0.1182	278
Pigeon	3/0	6/1	0.1317	0.1537	0.1672	0.1672	0.1672	0.502	156.4	74.0	230.4	67.9	32.1	6,620	0.5295	0.5405	0.6795	0.7433	0.5686	0.1147	319
Penguin	4/0	6/1	0.1662	0.1939	0.1878	0.1878	0.1878	0.563	197.4	93.4	290.8	67.9	32.1	8,350	0.4199	0.4288	0.5510	0.5999	0.5468	0.1113	366
Owl	266.8	6/7	0.2096	0.2368	0.2109	0.0703	0.2109	0.633	251.4	92.0	343.4	73.2	26.8	9,645	0.3379	0.3452	0.4679	0.4834	0.5680	0.1078	417
High Mechanical Strength																					
Grouse	80	8/1	0.0628	0.0847	0.1000	0.1670	0.1670	0.367	75.0	73.9	148.9	50.4	49.6	5,200	1.0901	1.1121	1.3843	1.5252	0.6658	0.1240	206
Petrel	101.8	12/7	0.0799	0.1265	0.0921	0.0921	0.2763	0.461	95.9	157.9	253.8	37.8	62.2	10,400	0.8358	0.8526	1.1698	1.3168	0.7375	0.1172	234
Minorca	110.8	12/7	0.0870	0.1378	0.0961	0.0961	0.2883	0.481	104.4	171.9	276.3	37.8	62.2	11,300	0.7679	0.7833	1.0894	1.2304	0.7260	0.1160	245
Leghorn	134.6	12/7	0.1057	0.1674	0.1059	0.1059	0.3177	0.530	126.8	208.7	335.5	37.8	62.2	13,600	0.6322	0.6449	0.9263	1.0559	0.6987	0.1131	271
Guinea	159	12/7	0.1249	0.1977	0.1151	0.1151	0.3453	0.576	149.7	246.6	396.3	37.8	62.2	16,000	0.5351	0.5459	0.8075	0.9281	0.6753	0.1106	296
Dotterel	176.9	12/7	0.1389	0.2199	0.1214	0.1214	0.3642	0.607	166.6	274.3	440.9	37.8	62.2	17,300	0.4810	0.4907	0.7395	0.8542	0.6596	0.1091	313
Dorking	190.8	12/7	0.1499	0.2373	0.1261	0.1261	0.3783	0.631	179.7	296.0	475.7	37.8	62.2	18,700	0.4460	0.4550	0.6953	0.8059	0.6486	0.1079	326
Brahma	203.2	16/19	0.1596	0.3020	0.1127	0.0977	0.4885	0.714	191.4	483.2	674.6	28.4	71.6	28,400	0.4034	0.4115	0.6419	0.8203	0.5920	0.1043	341
Cochin	211.3	12/7	0.1660	0.2628	0.1327	0.1327	0.3981	0.664	199.0	327.8	526.8	37.8	62.2	20,700	0.4027	0.4109	0.6396	0.7447	0.6335	0.1064	344

Notes:
(1) Data based on a nominal cable manufactured in accordance with ASTM B 232.
(2) Resistance and ampacity based on an aluminum conductivity of 61.2% IACS at 20°C and a steel conductivity of 8% IACS at 20°C.
(3) Ampacity based on a 75°C conductor temperature, 25°C ambient temperature, 2 ft/sec wind, in sun, with an emissivity of 0.5 and a coefficient of solar absorption of 0.5, at sea level.
(4) Rated strengths based on Class A galvanized steel core wire in accordance with ASTM B 498.
(5) The ac resistance and reactance for single layer ACSR conductors is affected by the magnetic flux in the core and varies with load current. The values given are approximate and are based on the current to produce a 75°C conductor temperature with a 25°C ambient. The coefficient of resistance of the steel is assumed to be 0.00327 per degree C.

TABLE 1-29
ALUMINUM CONDUCTOR STEEL REINFORCED (ACSR)

Code Word	Conductor Size kcmil	Stranding (Al/St)	Cross-Sectional Area Sq. in		Layers of Al	Diameter				Weight/1000 ft		
						Individual Wires		Steel Core in	Complete Cable in			
			Aluminum	Total		Aluminum in	Steel in			Aluminum lbs	Steel lbs	Total lbs
Waxwing	266.8	18/1	0.2094	0.2210	2	0.1217	0.1217	0.1217	0.609	249.9	39.2	289.1
Partridge	266.8	26/7	0.2095	0.2436	2	0.1013	0.0788	0.2364	0.642	251.3	115.6	366.9
Junco	266.8	30/7	0.2095	0.2584	2	0.0943	0.0943	0.2829	0.660	251.9	165.5	417.4
Ostrich	300	26/7	0.2355	0.2738	2	0.1074	0.0835	0.2505	0.680	282.5	129.7	412.2
Merlin	336.4	18/1	0.2642	0.2789	2	0.1367	0.1367	0.1367	0.684	315.3	49.5	364.8
Linnet	336.4	26/7	0.2640	0.3070	2	0.1137	0.0884	0.2652	0.720	316.6	145.4	462.0
Oriole	336.4	30/7	0.2642	0.3259	2	0.1059	0.1059	0.3177	0.741	317.7	208.7	526.4
Chickadee	397.5	18/1	0.3122	0.3295	2	0.1486	0.1486	0.1486	0.743	372.5	58.5	431.0
Ptarmigan	397.5	20/7	0.3123	0.3339	2	0.1410	0.0627	0.1881	0.752	374.5	73.2	447.7
Brant	397.5	24/7	0.3122	0.3527	2	0.1287	0.0858	0.2574	0.772	374.4	137.0	511.4
Ibis	397.5	26/7	0.3120	0.3628	2	0.1236	0.0961	0.2883	0.783	374.1	171.9	546.0
Lark	397.5	30/7	0.3121	0.3849	2	0.1151	0.1151	0.3453	0.806	375.3	246.5	621.8
Pelican	477	18/1	0.3747	0.3955	2	0.1628	0.1628	0.1628	0.814	447.1	70.2	517.3
Tailorbird	477	20/7	0.3745	0.4004	2	0.1544	0.0686	0.2058	0.824	449.1	87.6	536.7
Flicker	477	24/7	0.3747	0.4233	2	0.1410	0.0940	0.2820	0.846	449.4	164.5	613.9
Hawk	477	26/7	0.3744	0.4354	2	0.1354	0.1053	0.3159	0.858	449.0	206.4	655.3
Hen	477	30/7	0.3747	0.4621	2	0.1261	0.1261	0.3783	0.883	450.4	296.0	746.4
Osprey	556.5	18/1	0.4369	0.4612	2	0.1758	0.1758	0.1758	0.879	521.4	81.8	603.3
Sapsucker	556.5	22/7	0.4368	0.4797	2	0.1590	0.0883	0.2649	0.901	523.9	145.1	669.0
Parakeet	556.5	24/7	0.4372	0.4938	2	0.1523	0.1015	0.3045	0.914	524.3	191.8	716.1
Dove	556.5	26/7	0.4371	0.5083	2	0.1463	0.1138	0.3414	0.927	524.2	241.0	765.2
Eagle	556.5	30/7	0.4371	0.5391	2	0.1362	0.1362	0.4086	0.953	525.4	345.2	870.7
Peacock	605	24/7	0.4753	0.5370	2	0.1588	0.1059	0.3177	0.953	570.1	208.7	778.8
Squab	605	26/7	0.4749	0.5522	2	0.1525	0.1186	0.3558	0.966	569.5	261.8	831.3
Wood Duck	605	30/7	0.4751	0.5860	2	0.1420	0.1420	0.4260	0.994	571.2	375.3	946.5
Teal	605	30/19	0.4751	0.5834	2	0.1420	0.0852	0.4260	0.994	571.2	367.4	938.6
Kingbird	636	18/1	0.4997	0.5275	2	0.1880	0.1880	0.1880	0.940	596.3	93.6	689.9
Swift	636	36/1	0.4994	0.5133	3	0.1329	0.1329	0.1329	0.930	596.0	46.8	642.8
Goldfinch	636	22/7	0.4994	0.5484	2	0.1700	0.0944	0.2832	0.963	598.8	165.8	764.7
Rook	636	24/7	0.4996	0.5643	2	0.1628	0.1085	0.3255	0.977	599.1	219.1	818.2
Grosbeak	636	26/7	0.4995	0.5808	2	0.1564	0.1216	0.3648	0.990	599.0	275.2	874.2
Scoter	636	30/7	0.4995	0.6160	2	0.1456	0.1456	0.4368	1.019	600.5	394.6	995.1
Egret	636	30/19	0.4995	0.6135	2	0.1456	0.0874	0.4370	1.019	600.5	386.7	987.2
Flamingo	666.6	24/7	0.5238	0.5917	2	0.1667	0.1111	0.3333	1.000	628.2	229.7	857.9
Gannet	666.6	26/7	0.5234	0.6086	2	0.1601	0.1245	0.3735	1.014	627.7	288.5	916.2
Stilt	715.5	24/7	0.5622	0.6350	2	0.1727	0.1151	0.3453	1.036	674.2	246.6	920.8
Starling	715.5	26/7	0.5620	0.6535	2	0.1659	0.1290	0.3870	1.051	674.0	309.7	983.7
Crow	715.5	54/7	0.5619	0.6347	3	0.1151	0.1151	0.3450	1.040	673.8	246.5	920.4
Redwing	715.5	30/19	0.5617	0.6897	2	0.1544	0.0926	0.4630	1.081	675.3	434.0	1109.3
Coot	795	36/1	0.6244	0.6417	3	0.1486	0.1486	0.1486	1.040	745.1	58.5	803.6
Cuckoo	795	24/7	0.6244	0.7053	2	0.1820	0.1213	0.3639	1.092	748.8	273.9	1022.6
Drake	795	26/7	0.6247	0.7264	2	0.1749	0.1360	0.4080	1.108	749.1	344.3	1093.4
Skimmer	795	30/7	0.6245	0.7702	2	0.1628	0.1628	0.4884	1.140	750.7	493.3	1244.0
Macaw	795	42/7	0.6246	0.6567	3	0.1376	0.0764	0.2292	1.055	749.0	108.6	857.6
Tern	795	45/7	0.6242	0.6674	3	0.1329	0.0886	0.2658	1.063	748.6	146.1	894.7
Condor	795	54/7	0.6240	0.7049	3	0.1213	0.1213	0.3639	1.092	748.4	273.9	1022.2
Mallard	795	30/19	0.6245	0.7669	2	0.1628	0.0977	0.4885	1.140	750.7	483.2	1233.9
Ruddy	900	45/7	0.7066	0.7555	3	0.1414	0.0943	0.2829	1.131	847.4	165.5	1013.0
Canary	900	54/7	0.7069	0.7985	3	0.1291	0.1291	0.3873	1.162	847.7	310.2	1157.9
Corncrake	954	20/7	0.7492	0.8010	2	0.2184	0.0971	0.2910	1.165	898.5	175.5	1074.0
Redbird	954	24/7	0.7495	0.8466	2	0.1994	0.1329	0.3990	1.196	898.8	328.7	1227.5
Rail	954	45/7	0.7492	0.8010	3	0.1456	0.0971	0.2913	1.165	898.5	175.5	1074.0
Cardinal	954	54/7	0.7491	0.8462	3	0.1329	0.1329	0.3987	1.196	898.3	328.7	1227.1
Canvasback	954	30/19	0.7491	0.9199	2	0.1783	0.1070	0.5350	1.248	900.5	579.6	1480.1
Snowbird	1033.5	42/7	0.8121	0.8539	3	0.1569	0.0872	0.2616	1.203	973.9	141.5	1115.4
Ortolan	1033.5	45/7	0.8112	0.8673	3	0.1515	0.1010	0.3030	1.212	972.8	189.9	1162.7
Curlew	1033.5	54/7	0.8112	0.9164	3	0.1383	0.1383	0.4149	1.245	972.8	356.0	1328.8
Bluejay	1113	45/7	0.8745	0.9350	3	0.1573	0.1049	0.3147	1.259	1048.7	204.8	1253.6
Finch	1113	54/19	0.8746	0.9855	3	0.1436	0.0862	0.4310	1.293	1053.9	376.1	1430.1
Bunting	1192.5	45/7	0.9367	1.0014	3	0.1628	0.1085	0.3255	1.302	1123.4	219.1	1342.5
Grackle	1192.5	54/19	0.9365	1.0552	3	0.1486	0.0892	0.4460	1.338	1128.6	402.8	1531.4
Bittern	1272	45/7	0.9987	1.0678	3	0.1681	0.1121	0.3363	1.345	1197.7	233.9	1431.6
Pheasant	1272	54/19	0.9993	1.1259	3	0.1535	0.0921	0.4605	1.382	1204.3	429.4	1633.7
Dipper	1351.5	45/7	1.0614	1.1347	3	0.1733	0.1155	0.3465	1.386	1272.9	248.3	1521.2
Martin	1351.5	54/19	1.0614	1.1958	3	0.1582	0.0949	0.4745	1.424	1279.1	455.9	1735.0
Bobolink	1431	45/7	1.1236	1.2013	3	0.1783	0.1189	0.3567	1.427	1347.5	263.1	1610.6
Plover	1431	54/19	1.1241	1.2665	3	0.1628	0.0977	0.4885	1.465	1354.6	483.2	1837.8
Nuthatch	1510.5	45/7	1.1862	1.2682	3	0.1832	0.1221	0.3663	1.466	1422.5	277.5	1700.0
Parrot	1510.5	54/19	1.1856	1.3357	3	0.1672	0.1003	0.5015	1.505	1428.8	509.2	1938.1
Lapwing	1590	45/7	1.2492	1.3355	3	0.1880	0.1253	0.3759	1.504	1498.1	292.2	1790.3
Falcon	1590	54/19	1.2489	1.4072	3	0.1716	0.1030	0.5150	1.545	1505.0	537.0	2042.1
Chukar	1780	84/19	1.3986	1.5126	4	0.1456	0.0874	0.4370	1.602	1685.4	386.7	2072.1
Mockingbird	2034.5	72/7	1.5979	1.6671	4	0.1681	0.1122	0.3360	1.681	1925.7	234.3	2160.0
Roadrunner	2057	76/19	1.6152	1.7032	4	0.1645	0.0768	0.3840	1.700	1946.5	298.6	2245.1
Bluebird	2156	84/19	1.6931	1.8309	4	0.1602	0.0961	0.4805	1.762	2040.4	467.5	2507.9
Kiwi	2167	72/7	1.7022	1.7758	4	0.1735	0.1157	0.3471	1.735	2051.4	249.2	2300.5
Thrasher	2312	76/19	1.8155	1.9144	4	0.1744	0.0814	0.4070	1.802	2187.9	335.4	2523.3
Joree	2515	76/19	1.9750	2.0826	4	0.1819	0.0849	0.4245	1.880	2380.1	364.9	2745.0

Notes: (1) Data based on a nominal cable manufactured in accordance with ASTM B 232.

(2) Resistance and ampacity based on an aluminum conductivity of 61.2% IACS at 20°C and a steel conductivity of 8% IACS at 20°C.

(3) Ampacity based on a 75°C conductor temperature, 25°C ambient temperature, 2 ft/sec wind, in sun, with an emissivity of 0.5 and a coefficient of solar absorption of 0.5, at sea level.

(4) Rated strengths based on Class A galvanized steel core wire in accordance with ASTM B 498.

TABLE 1-29

ALUMINUM CONDUCTOR STEEL REINFORCED (ACSR)

Metal Content % Weight			Resistance					Reactance @ 1 ft Spacing 60 Hz			
				ac-60 Hz							
Aluminum %	Steel %	Rated Strength lbs	dc @ 20°C Ω/mile	@ 25°C Ω/mile	@ 50°C Ω/mile	@ 75°C Ω/mile	GMR ft	Inductive Xa Ω/mile	Capacitive X'a MΩ-mile	Ampacity @ 75°C Amps	Code Word
86.4	13.6	6,880	0.3396	0.3471	0.3813	0.4155	0.0197	0.477	0.1090	449	Waxwing
68.5	31.5	11,300	0.3364	0.3437	0.3775	0.4113	0.0217	0.465	0.1074	457	Partridge
60.3	39.7	13,900	0.3342	0.3413	0.3748	0.4083	0.0227	0.459	0.1066	462	Junco
68.5	31.5	12,700	0.2992	0.3058	0.3358	0.3659	0.0230	0.458	0.1057	492	Ostrich
86.4	13.6	8,680	0.2693	0.2756	0.3027	0.3298	0.0221	0.463	0.1056	519	Merlin
68.5	31.5	14,100	0.2668	0.2728	0.2996	0.3264	0.0243	0.451	0.1040	529	Linnet
60.3	39.7	17,300	0.2650	0.2708	0.2974	0.3240	0.0255	0.445	0.1031	535	Oriole
86.4	13.6	9,940	0.2279	0.2336	0.2565	0.2794	0.0240	0.452	0.1031	576	Chickadee
83.7	16.3	11,100	0.2286	0.2342	0.2571	0.2801	0.0246	0.450	0.1027	577	Ptarmigan
73.2	26.8	14,600	0.2268	0.2321	0.2549	0.2777	0.0259	0.444	0.1019	584	Brant
68.5	31.5	16,300	0.2258	0.2311	0.2537	0.2764	0.0265	0.441	0.1015	587	Ibis
60.3	39.7	20,300	0.2243	0.2294	0.2518	0.2743	0.0277	0.435	0.1007	594	Lark
86.4	13.6	11,800	0.1899	0.1950	0.2141	0.2331	0.0263	0.441	0.1004	646	Pelican
83.7	16.3	13,100	0.1905	0.1955	0.2146	0.2337	0.0270	0.439	0.1000	647	Tailorbird
73.2	26.8	17,200	0.1890	0.1937	0.2127	0.2316	0.0283	0.433	0.0992	655	Flicker
68.5	31.5	19,500	0.1882	0.1928	0.2117	0.2305	0.0290	0.430	0.0988	659	Hawk
60.3	39.7	23,800	0.1869	0.1913	0.2100	0.2288	0.0304	0.424	0.0980	666	Hen
86.4	13.6	13,700	0.1628	0.1676	0.1838	0.2002	0.0284	0.432	0.0981	711	Osprey
78.3	21.7	17,600	0.1627	0.1672	0.1834	0.1997	0.0298	0.426	0.0974	717	Sapsucker
73.2	26.8	19,800	0.1620	0.1663	0.1825	0.1988	0.0306	0.423	0.0969	721	Parakeet
68.5	31.5	22,600	0.1613	0.1655	0.1817	0.1978	0.0313	0.420	0.0965	726	Dove
60.3	39.7	27,800	0.1602	0.1642	0.1802	0.1963	0.0328	0.415	0.0957	734	Eagle
73.2	26.8	21,600	0.1490	0.1532	0.1681	0.1830	0.0319	0.418	0.0957	760	Peacock
68.5	31.5	24,300	0.1484	0.1524	0.1672	0.1821	0.0327	0.415	0.0953	765	Squab
60.3	39.7	28,900	0.1474	0.1512	0.1659	0.1806	0.0342	0.410	0.0944	773	Wood Duck
60.9	39.1	30,000	0.1475	0.1513	0.1660	0.1808	0.0342	0.410	0.0944	773	Teal
86.4	13.6	15,700	0.1424	0.1470	0.1612	0.1755	0.0304	0.424	0.0961	773	Kingbird
92.7	7.30	13,800	0.1430	0.1477	0.1620	0.1763	0.0301	0.425	0.0964	769	Swift
78.3	21.7	19,800	0.1423	0.1466	0.1608	0.1750	0.0319	0.418	0.0954	779	Goldfinch
73.2	26.8	22,600	0.1418	0.1458	0.1600	0.1742	0.0327	0.415	0.0950	784	Rook
68.5	31.5	25,200	0.1411	0.1451	0.1592	0.1733	0.0335	0.412	0.0946	789	Grosbeak
60.3	39.7	30,400	0.1402	0.1439	0.1579	0.1719	0.0351	0.407	0.0937	798	Scoter
60.8	39.2	31,500	0.1403	0.1440	0.1581	0.1720	0.0351	0.407	0.0937	798	Egret
73.2	26.8	23,700	0.1352	0.1393	0.1528	0.1663	0.0335	0.412	0.0943	807	Flamingo
68.5	31.5	26,400	0.1347	0.1385	0.1520	0.1654	0.0343	0.409	0.0939	812	Gannet
73.2	26.8	25,500	0.1260	0.1299	0.1425	0.1551	0.0347	0.408	0.0932	844	Stilt
68.5	31.5	28,400	0.1255	0.1292	0.1417	0.1543	0.0355	0.405	0.0928	849	Starling
73.2	26.8	26,000	0.1260	0.1299	0.1460	0.1596	0.0350	0.407	0.0932	833	Crow
60.9	39.1	34,600	0.1247	0.1282	0.1407	0.1531	0.0372	0.399	0.0920	859	Redwing
92.7	7.30	16,800	0.1144	0.1190	0.1303	0.1417	0.0337	0.411	0.0931	884	Coot
73.2	26.8	27,900	0.1134	0.1172	0.1285	0.1398	0.0365	0.402	0.0917	901	Cuckoo
68.5	31.5	31,500	0.1129	0.1166	0.1278	0.1390	0.0375	0.399	0.0912	907	Drake
60.3	39.7	38,300	0.1121	0.1156	0.1267	0.1379	0.0392	0.393	0.0904	918	Skimmer
87.3	12.7	20,100	0.1146	0.1189	0.1310	0.1427	0.0346	0.408	0.0927	887	Macaw
83.7	16.3	22,100	0.1143	0.1185	0.1305	0.1422	0.0352	0.406	0.0925	887	Tern
73.2	26.8	28,200	0.1134	0.1172	0.1315	0.1438	0.0368	0.401	0.0917	889	Condor
60.8	39.2	38,400	0.1122	0.1156	0.1268	0.1380	0.0392	0.393	0.0904	917	Mallard
83.7	16.3	24,400	0.1010	0.1051	0.1157	0.1260	0.0374	0.399	0.0906	958	Ruddy
73.2	26.8	31,900	0.1002	0.1039	0.1165	0.1273	0.0392	0.393	0.0898	960	Canary
83.7	16.3	25,600	0.0953	0.0994	0.1088	0.1183	0.0381	0.396	0.0897	996	Corncrake
73.2	26.8	33,500	0.0945	0.0982	0.1076	0.1170	0.0400	0.391	0.0890	1009	Redbird
83.7	16.3	25,900	0.0953	0.0994	0.1094	0.1191	0.0385	0.395	0.0897	993	Rail
73.2	26.8	33,800	0.0945	0.0982	0.1100	0.1202	0.0404	0.390	0.0890	996	Cardinal
60.8	39.2	46,100	0.0935	0.0968	0.1060	0.1153	0.0430	0.382	0.0877	1028	Canvasback
87.3	12.7	25,400	0.0881	0.0925	0.1012	0.1099	0.0395	0.392	0.0888	1043	Snowbird
83.7	16.3	27,700	0.0879	0.0921	0.1013	0.1102	0.0401	0.390	0.0886	1043	Ortolan
73.2	26.8	36,600	0.0872	0.0910	0.1018	0.1112	0.0420	0.385	0.0878	1047	Curlew
83.7	16.3	29,800	0.0816	0.0859	0.0943	0.1026	0.0416	0.386	0.0874	1092	Bluejay
73.7	26.3	39,100	0.0814	0.0852	0.0952	0.1039	0.0436	0.380	0.0867	1093	Finch
83.7	16.3	32,000	0.0762	0.0805	0.0884	0.0960	0.0431	0.382	0.0864	1139	Bunting
73.7	26.3	41,900	0.0760	0.0798	0.0891	0.0972	0.0451	0.376	0.0856	1141	Grackle
83.7	16.3	34,100	0.0714	0.0759	0.0832	0.0903	0.0445	0.378	0.0855	1184	Bittern
73.7	26.3	43,600	0.0712	0.0751	0.0838	0.0914	0.0466	0.372	0.0847	1187	Pheasant
83.7	16.3	36,200	0.0672	0.0718	0.0786	0.0853	0.0459	0.374	0.0846	1228	Dipper
73.7	26.3	46,300	0.0671	0.0710	0.0791	0.0862	0.0480	0.368	0.0838	1231	Martin
83.7	16.3	38,300	0.0635	0.0681	0.0745	0.0808	0.0472	0.371	0.0837	1272	Bobolink
73.7	26.3	49,100	0.0633	0.0673	0.0749	0.0816	0.0494	0.365	0.0829	1275	Plover
83.7	16.3	40,000	0.0602	0.0649	0.0709	0.0769	0.0485	0.367	0.0829	1313	Nuthatch
73.7	26.3	51,700	0.0600	0.0641	0.0712	0.0776	0.0508	0.362	0.0821	1317	Parrot
83.7	16.3	42,200	0.0572	0.0620	0.0677	0.0733	0.0497	0.364	0.0822	1354	Lapwing
73.7	26.3	54,500	0.0570	0.0611	0.0679	0.0739	0.0521	0.359	0.0814	1359	Falcon
81.3	18.7	51,000	0.0512	0.0561	0.0609	0.0658	0.0534	0.355	0.0803	1453	Chukar
89.2	10.8	46,800	0.0450	0.0509	0.0551	0.0593	0.0553	0.351	0.0789	1551	Mockingbird
86.7	13.3	50,400	0.0445	0.0502	0.0543	0.0585	0.0562	0.349	0.0785	1567	Roadrunner
81.4	18.6	60,300	0.0423	0.0477	0.0516	0.0555	0.0588	0.344	0.0775	1622	Bluebird
89.2	10.8	49,800	0.0423	0.0484	0.0523	0.0562	0.0570	0.348	0.0779	1607	Kiwi
86.7	13.3	56,700	0.0396	0.0457	0.0493	0.0529	0.0595	0.342	0.0768	1672	Thrasher
86.7	13.3	61,700	0.0364	0.0428	0.0460	0.0493	0.0621	0.337	0.0755	1751	Joree

TABLE 1-30

SHAPED WIRE CONCENTRIC-LAY COMPACT ALUMINUM CONDUCTOR STEEL REINFORCED (ACSR/TW)

Diameters Equal to Standard ACSR Conductors

Code Word	Conductor Size kcmil	Type No.	Size & Stranding of ACSR with Equal Diameter		Cross-Sectional Area Sq. in		Layers of Al	Stranding		Diameter	
			kcmil	Stranding	Aluminum	Total		No. of Aluminum Wires	No. & Dia. Individual Steel Wire	Steel Core in	Complete Conductor in
Monongahela/TW	405.1	6	336.4	18/1	0.3181	0.3362	2	14	1 x 0.1520	0.1520	0.680
Mohawk/TW	571.7	13	477	24/7	0.4490	0.5074	2	18	7 x 0.1030	0.3090	0.846
Calumet/TW	565.3	16	477	26/7	0.4439	0.5162	2	20	7 x 0.1146	0.3441	0.858
Mystic/TW	666.6	13	556.5	24/7	0.5236	0.5914	2	20	7 x 0.1111	0.3333	0.913
Oswego/TW	664.8	16	556.5	26/7	0.5221	0.6072	2	20	7 x 0.1244	0.3732	0.927
Nechako/TW	768.9	3	636	36/1	0.6039	0.6220	3	27	1 x 0.1520	0.1520	0.930
Maumee/TW	768.2	13	636	24/7	0.6034	0.6819	2	20	7 x 0.1195	0.3585	0.977
Wabash/TW	762.8	16	636	26/7	0.5992	0.6966	2	20	7 x 0.1331	0.3343	0.990
Kettle/TW	957.2	7	795	45/7	0.7518	0.8038	3	32	7 x 0.0973	0.2919	1.060
Fraser/TW	946.7	10	795	22/7	0.7436	0.8168	3	35	7 x 0.1154	0.3462	1.077
Columbia/TW	966.2	13	795	54/7	0.7589	0.8573	2	21	7 x 0.1338	0.4014	1.092
Suwannee/TW	959.6	16	795	26/7	0.7537	0.8762	2	22	7 x 0.1493	0.4479	1.108
Cheyenne/TW	1168.1	5	954	42/7	0.9175	0.9646	3	30	7 x 0.0926	0.2778	1.155
Genesee/TW	1158	7	954	45/7	0.9095	0.9733	3	33	7 x 0.1078	0.3234	1.165
Hudson/TW	1158.4	13	954	54/7	0.9098	1.0281	2	25	7 x 0.1467	0.4401	1.196
Catawba/TW	1272	5	1033.5	42/7	0.9991	1.0505	3	30	7 x 0.0967	0.2901	1.203
Nelson/TW	1257.1	7	1033.5	45/7	0.9874	1.0557	3	35	7 x 0.1115	0.3345	1.213
Yukon/TW	1233.6	13	1033.5	54/7	0.9689	1.0925	3	38	19 x 0.0910	0.4550	1.245
Truckee/TW	1372.5	5	1113	42/7	1.0780	1.1334	3	30	7 x 0.1004	0.3012	1.248
Mackenzie/TW	1359.7	7	1113	45/7	1.0679	1.1418	3	36	7 x 0.1159	0.3477	1.259
Thames/TW	1334.6	13	1113	54/19	1.0480	1.1809	3	39	19 x 0.0944	0.4720	1.290
St. Croix/TW	1467.8	5	1192.5	42/7	1.1529	1.2124	3	33	7 x 0.1041	0.3123	1.292
Miramichi/TW	1455.3	7	1192.5	45/7	1.1430	1.2222	3	36	7 x 0.1200	0.3600	1.302
Merrimack/TW	1433.6	13	1192.5	54/19	1.1250	1.2677	3	39	19 x 0.0978	0.4890	1.340
Platte/TW	1569	5	1272	42/7	1.2323	1.2957	3	33	7 x 0.1074	0.3222	1.334
Potomac/TW	1557.4	7	1272	45/7	1.2232	1.3079	3	36	7 x 0.1241	0.3723	1.345
Rio Grande/TW	1533.3	13	1272	54/19	1.2043	1.3571	3	39	19 x 0.1012	0.5060	1.382
Schuylkill/TW	1657.4	7	1351.5	45/7	1.3020	1.3920	3	36	7 x 0.1280	0.3840	1.386
Pecos/TW	1622	13	1351.5	54/19	1.2739	1.4429	3	39	19 x 0.1064	0.5320	1.424
Pee Dee/TW	1758.6	7	1431	45/7	1.3810	1.4770	3	37	7 x 0.1319	0.3957	1.427
James/TW	1730.6	13	1431	54/19	1.3590	1.5314	3	39	19 x 0.1075	0.5375	1.470
Athabaska/TW	1949.6	7	1590	45/7	1.5312	1.6377	3	42	7 x 0.1392	0.4176	1.504
Cumberland/TW	1926.9	13	1590	54/19	1.5134	1.7049	3	42	19 x 0.1133	0.5665	1.545
Powder/TW	2153.8	8	1780	84/19	1.6912	1.8290	4	64	19 x 0.0961	0.4805	1.602
Santee/TW	2627.3	8	2156	84/19	2.0630	2.2268	4	64	19 x 0.1062	0.5310	1.762

Notes: (1) Data based on a nominal cable manufactured in accordance with ASTM B 779.

(2) Resistance and ampacity based on an aluminum conductivity of 61.2% IACS at 20°C and a steel conductivity of 8% IACS at 20°C.

(3) Ampacity based on a 75°C conductor temperature, 25°C ambient temperature, 2 ft/sec wind, in sun, with an emissivity of 0.5 and a coefficient of solar absorption of 0.5, at sea level.

(4) Rated strengths based on Class A galvanized steel core wire in accordance with ASTM B 498 or ASTM B 606 for high strength steel core wire.

(5) The final design of a shaped wire compact conductor is contingent upon several factors such as: layer diameter, wire width and wire thickness. The actual configuration of a given size may vary between manufacturers. This may result in a slight variation in the number of wires, number of layers and dimensions of individual wires from that shown in the chart.

TABLE 1-30

SHAPED WIRE CONCENTRIC-LAY COMPACT ALUMINUM CONDUCTOR STEEL REINFORCED (ACSR/TW)

Diameters Equal to Standard ACSR Conductors

Weight/1000 ft				Resistance					Reactance @ 1 ft Spacing 60 Hz			
					ac-60 Hz							
Aluminum lbs	Steel lbs	Total lbs	Rated Strength lbs	dc @ 20°C Ω/mile	@ 25°C Ω/mile	@ 50°C Ω/mile	@ 75°C Ω/mile	GMR ft	Inductive Xa Ω/mile	Capacitive X'a MΩ-mile	Ampacity @ 75°C Amps	Code Word
379.8	61.2	441.0	10,200	0.2236	0.2291	0.2516	0.2740	0.0218	0.464	0.1097	568	Monongahela/TW
537.3	197.4	734.7	20,600	0.1572	0.1613	0.1770	0.1928	0.0281	0.433	0.0991	717	Mohawk/TW
531.0	244.8	775.8	22,900	0.1584	0.1625	0.1783	0.1941	0.0288	0.430	0.0988	718	Calumet/TW
626.6	229.7	856.3	24,000	0.1348	0.1387	0.1521	0.1656	0.0304	0.424	0.0970	790	Mystic/TW
625.4	288.0	913.4	26,600	0.1347	0.1385	0.1519	0.1653	0.0310	0.421	0.0964	794	Oswego/TW
720.7	61.2	781.9	16,400	0.1187	0.1232	0.1381	0.1510	0.0300	0.425	0.0965	832	Nechako/TW
722.1	265.7	987.8	27,700	0.1170	0.1207	0.1323	0.1440	0.0325	0.416	0.0949	862	Maumee/TW
717.0	330.0	1047	30,500	0.1174	0.1212	0.1329	0.1446	0.0330	0.413	0.0946	863	Wabash/TW
902.8	176.2	1079	26,000	0.0949	0.0990	0.1108	0.1210	0.0350	0.407	0.0925	962	Kettle/TW
894.0	248.0	1142	29,600	0.0957	0.0995	0.1114	0.1217	0.0358	0.404	0.0919	963	Fraser/TW
908.0	333.0	1241	34,000	0.0930	0.0966	0.1058	0.1150	0.0364	0.402	0.0917	993	Columbia/TW
903.0	415.0	1318	37,000	0.0933	0.0967	0.1059	0.1152	0.0373	0.399	0.0913	996	Suwannee/TW
1100.4	159.6	1260	28,900	0.0779	0.0823	0.0918	0.1001	0.0378	0.397	0.0901	1082	Cheyenne/TW
1092.0	216.0	1308	31,600	0.0785	0.0826	0.0922	0.1005	0.0384	0.395	0.0897	1082	Genesee/TW
1089.0	400.0	1489	39,600	0.0776	0.0811	0.0888	0.0964	0.0400	0.391	0.0889	1111	Hudson/TW
1198.0	174.0	1372	31,400	0.0715	0.0760	0.0847	0.0922	0.0394	0.392	0.0889	1139	Catawba/TW
1185.7	231.3	1417	34,200	0.0723	0.0765	0.0853	0.0930	0.0400	0.390	0.0886	1137	Nelson/TW
1166.5	419.5	1586	42,900	0.0733	0.0769	0.0858	0.0937	0.0420	0.385	0.0877	1140	Yukon/TW
1293.4	187.6	1481	33,400	0.0663	0.0709	0.0788	0.0858	0.0409	0.388	0.0877	1192	Truckee/TW
1280.0	250.0	1530	36,900	0.0668	0.0712	0.0792	0.0863	0.0420	0.386	0.0874	1192	Mackenzie/TW
1261.6	451.4	1713	46,300	0.0677	0.0714	0.0796	0.0868	0.0436	0.380	0.0866	1196	Thames/TW
1383.0	202.0	1585	35,800	0.0620	0.0667	0.0741	0.0806	0.0424	0.840	0.0867	1242	St. Croix/TW
1372.0	268.0	1640	39,200	0.0624	0.0669	0.0743	0.0809	0.0431	0.382	0.0867	1242	Miramichi/TW
1354.0	484.2	1838	49,700	0.0631	0.0668	0.0744	0.0811	0.0450	0.376	0.0856	1250	Merrimack/TW
1478.0	215.0	1693	38,200	0.0580	0.0629	0.0697	0.0758	0.0439	0.379	0.0858	1291	Platte/TW
1468.0	287.0	1755	41,900	0.0584	0.0629	0.0698	0.0759	0.0445	0.378	0.0853	1293	Potomac/TW
1449.0	519.0	1968	53,200	0.0590	0.0628	0.0698	0.0760	0.0466	0.372	0.0847	1301	Rio Grande/TW
1563.0	305.0	1868	44,000	0.0548	0.0595	0.0660	0.0717	0.0459	0.374	0.0845	1341	Schuylkill/TW
1533.0	574.0	2107	57,500	0.0557	0.0595	0.0662	0.0720	0.0481	0.368	0.0839	1348	Pecos/TW
1658.0	324.0	1982	46,700	0.0517	0.0565	0.0625	0.0679	0.0473	0.370	0.0837	1389	Pee Dee/TW
1636.0	585.0	2221	59,400	0.0522	0.0562	0.0624	0.0679	0.0494	0.365	0.0829	1400	James/TW
1838.0	361.0	2199	51,900	0.0466	0.0517	0.0562	0.0608	0.0500	0.363	0.0822	1488	Athabaska/TW
1821.0	650.0	2471	65,300	0.0469	0.0511	0.0566	0.0614	0.0523	0.358	0.0815	1491	Cumberland/TW
2030.0	468.0	2498	61,100	0.0424	0.0475	0.0515	0.0554	0.0538	0.355	0.0803	1583	Powder/TW
2477.0	571.0	3048	74,500	0.0348	0.0405	0.0436	0.0468	0.0594	0.343	0.0775	1767	Santee/TW

TABLE 1-31

SHAPED WIRE CONCENTRIC-LAY COMPACT ALUMINUM CONDUCTOR STEEL REINFORCED (ACSR/TW)

Area Equal to Standard ACSR Sizes

Code Word	Conductor Size kcmil	Type No.	Cross-Sectional Area Sq. in		Layers of Al	Stranding		Diameter	
			Aluminum	Total		No. of Aluminum Wires	No. & Dia. Individual Steel Wires	Steel Core in	Complete Conductor in
Merlin/TW	336.4	6	0.2642	0.2788	2	14	1 x 0.1367	0.1367	0.630
Flicker/TW	477	13	0.3747	0.4233	2	18	7 x 0.0940	0.2820	0.776
Hawk/TW	477	16	0.3746	0.4356	2	18	7 x 0.1053	0.3159	0.789
Parakeet/TW	556.5	13	0.4371	0.4937	2	18	7 x 0.1015	0.3045	0.835
Dove/TW	556.5	16	0.4371	0.5083	2	20	7 x 0.1138	0.3414	0.852
Swift/TW	636	3	0.4995	0.5133	3	27	1 x 0.1329	0.1329	0.850
Rook/TW	636	13	0.4995	0.5643	2	18	7 x 0.1085	0.3255	0.890
Grosbeak/TW	636	16	0.4995	0.5808	2	20	7 x 0.1216	0.3648	0.908
Tern/TW	795	7	0.6244	0.6675	2	17	7 x 0.0886	0.2658	0.960
Puffin/TW	795	10	0.6244	0.6919	2	18	7 x 0.1108	0.3324	0.980
Condor/TW	795	13	0.6244	0.7053	2	20	7 x 0.1213	0.3639	0.993
Drake/TW	795	16	0.6244	0.7261	2	20	7 x 0.1360	0.4080	1.010
Phoenix/TW	954	5	0.7493	0.7876	3	30	7 x 0.0837	0.2511	1.044
Rail/TW	954	7	0.7493	0.8011	3	32	7 x 0.0971	0.2913	1.061
Cardinal/TW	954	13	0.7493	0.8464	2	20	7 x 0.1329	0.3987	1.084
Snowbird/TW	1033.5	5	0.8117	0.8534	3	30	7 x 0.0871	0.2613	1.089
Ortolan/TW	1033.5	7	0.8117	0.8678	3	32	7 x 0.1010	0.3030	1.102
Curlew/TW	1033.5	13	0.8117	0.9169	2	21	7 x 0.1363	0.4149	1.129
Avocet/TW	1113	5	0.8742	0.9191	3	30	7 x 0.0904	0.2712	1.129
Bluejay/TW	1113	7	0.8742	0.9347	3	33	7 x 0.1049	0.3147	1.143
Finch/TW	1113	13	0.8742	0.9851	3	38	19 x 0.0862	0.4310	1.185
Oxbird/TW	1192.5	5	0.9366	0.9848	3	30	7 x 0.0936	0.2808	1.167
Bunting/TW	1192.5	7	0.9366	1.0013	3	33	7 x 0.1085	0.3255	1.181
Grackle/TW	1192.5	13	0.9366	1.0554	3	38	19 x 0.0892	0.4460	1.225
Scissortail/TW	1272	5	0.9991	1.0505	3	30	7 x 0.0967	0.2901	1.203
Bittern/TW	1272	7	0.9990	1.0681	3	35	7 x 0.1121	0.3363	1.220
Pheasant/TW	1272	13	0.9990	1.1256	3	39	19 x 0.0921	0.4605	1.264
Dipper/TW	1351.5	7	1.0615	1.1348	3	35	7 x 0.1155	0.3465	1.256
Martin/TW	1351.5	13	1.0615	1.1959	3	39	19 x 0.0949	0.4745	1.300
Bobolink/TW	1431	7	1.1236	1.2017	3	36	7 x 0.1189	0.3567	1.291
Plover/TW	1431	13	1.1239	1.2664	3	39	19 x 0.0977	0.4885	1.337
Lapwing/TW	1590	7	1.2488	1.3351	3	36	7 x 0.1253	0.3759	1.358
Falcon/TW	1590	13	1.2488	1.4071	3	42	19 x 0.1030	0.5150	1.408
Chukar/TW	1780	8	1.3986	1.5126	3	37	19 x 0.0874	0.4370	1.445
Bluebird/TW	2156	8	1.6931	1.8309	4	64	19 x 0.0961	0.4805	1.608

Notes: (1) Data based on a nominal cable manufactured in accordance with ASTM B 779.

(2) Resistance and ampacity based on an aluminum conductivity of 61.2% IACS at 20°C and a steel conductivity of 8% IACS at 20°C.

(3) Ampacity based on a 75°C conductor temperature, 25°C ambient temperature, 2 ft/sec wind, in sun, with an emissivity of 0.5 and a coefficient of solar absorption of 0.5, at sea level.

(4) Rated strengths based on Class A galvanized steel core wire in accordance with ASTM B 498.

(5) The final design of a shaped wire compact conductor is contingent upon several factors such as: layer diameter, wire width and wire thickness. The actual configuration of a given size may vary between manufacturers. This may result in a slight variation in the number of wires, number of layers and dimensions of individual wires from that shown in the chart.

TABLE 1-31

SHAPED WIRE CONCENTRIC-LAY COMPACT ALUMINUM CONDUCTOR STEEL REINFORCED (ACSR/TW)

Area Equal to Standard ACSR Sizes

Weight/1000 ft			Rated Strength	Resistance				GMR	Reactance @ 1 ft Spacing 60 Hz		Ampacity	Code Word
					ac-60 Hz							
Aluminum lbs	Steel lbs	Total lbs	lbs	dc @ 20°C Ω/mile	@ 25°C Ω/mile	@ 50°C Ω/mile	@ 75°C Ω/mile	ft	Inductive Xa Ω/mile	Capacitive X'a MΩ-mile	@ 75°C Amps	
315.5	49.5	365.0	8,560	0.2693	0.2756	0.3027	0.3298	0.0200	0.457	0.1079	508	Merlin/TW
448.4	164.4	612.8	17,200	0.1884	0.1930	0.2119	0.2307	0.0257	0.444	0.1017	641	Flicker/TW
448.7	206.3	655.0	19,400	0.1877	0.1923	0.2110	0.2298	0.0264	0.441	0.1013	645	Hawk/TW
523.2	191.7	714.9	20,000	0.1615	0.1657	0.1818	0.1980	0.0277	0.435	0.0994	706	Parakeet/TW
523.5	241.0	764.5	22,600	0.1609	0.1650	0.1811	0.1972	0.0286	0.431	0.0991	711	Dove/TW
599.2	46.8	646.0	13,500	0.1435	0.1482	0.1665	0.1821	0.0273	0.437	0.0991	740	Swift/TW
597.0	219.1	816.0	22,900	0.1413	0.1452	0.1593	0.1735	0.0296	0.427	0.0978	766	Rook/TW
598.4	275.1	873.5	25,400	0.1408	0.1446	0.1587	0.1727	0.0305	0.423	0.0971	772	Grosbeak/TW
745.9	146.1	892.0	21,800	0.1138	0.1178	0.1291	0.1405	0.0312	0.421	0.0955	869	Tern/TW
746.9	228.4	975.3	25,900	0.1133	0.1171	0.1283	0.1396	0.0323	0.417	0.0949	876	Puffin/TW
747.2	273.8	1021	28,200	0.1130	0.1167	0.1279	0.1392	0.0331	0.414	0.0945	880	Condor/TW
747.8	344.2	1092	31,800	0.1126	0.1161	0.1273	0.1386	0.0339	0.411	0.0940	887	Drake/TW
897.7	130.4	1028	23,700	0.0954	0.0996	0.1114	0.1217	0.0343	0.409	0.0928	955	Phoenix/TW
900.0	175.0	1075	25,900	0.0953	0.0993	0.1111	0.1214	0.0349	0.407	0.0925	960	Rail/TW
897.3	328.7	1226	33,900	0.0942	0.0977	0.1071	0.1164	0.0362	0.403	0.0919	985	Cardinal/TW
973.8	141.2	1115	25,700	0.0881	0.0923	0.1032	0.1126	0.0356	0.405	0.0917	1004	Snowbird/TW
975.2	189.8	1165	28,100	0.0879	0.0920	0.1028	0.1123	0.0363	0.402	0.0914	1009	Ortolan/TW
971.1	355.9	1327	36,300	0.0869	0.0905	0.0991	0.1077	0.0377	0.398	0.0906	1036	Curlew/TW
1048.9	152.1	1201	27,500	0.0818	0.0861	0.0961	0.1048	0.0369	0.400	0.0906	1051	Avocet/TW
1048.0	204.8	1253	30,300	0.0816	0.0858	0.0958	0.1045	0.0376	0.398	0.0903	1056	Bluejay/TW
1052.6	376.4	1429	39,100	0.0812	0.0848	0.0948	0.1035	0.0399	0.391	0.0891	1071	Finch/TW
1123.0	163.0	1286	29,500	0.0763	0.0807	0.0900	0.0981	0.0382	0.396	0.0896	1095	Oxbird/TW
1124.0	219.0	1343	32,400	0.0762	0.0804	0.0897	0.0978	0.0390	0.394	0.0893	1101	Bunting/TW
1127.0	403.0	1530	41,900	0.0758	0.0794	0.0887	0.0968	0.0412	0.387	0.0883	1117	Grackle/TW
1198.0	174.0	1372	31,400	0.0715	0.0760	0.0847	0.0922	0.0394	0.392	0.0888	1139	Scissortail/TW
1199.0	234.0	1433	34,600	0.0714	0.0757	0.0843	0.0919	0.0403	0.390	0.0884	1145	Bittern/TW
1202.0	430.0	1632	44,100	0.0711	0.0747	0.0834	0.0909	0.0426	0.383	0.0874	1162	Pheasant/TW
1274.0	248.0	1522	36,700	0.0672	0.0716	0.0797	0.0868	0.0415	0.386	0.0874	1188	Dipper/TW
1278.0	456.0	1734	46,800	0.0669	0.0706	0.0787	0.0858	0.0438	0.377	0.0865	1205	Martin/TW
1350.0	263.0	1613	38,900	0.0635	0.0679	0.0755	0.0822	0.0427	0.383	0.0867	1229	Bobolink/TW
1353.0	483.0	1836	49,600	0.0632	0.0669	0.0745	0.0812	0.0451	0.376	0.0860	1248	Plover/TW
1498.0	292.2	1790	42,200	0.0572	0.0618	0.0685	0.0745	0.0449	0.377	0.0851	1309	Lapwing/TW
1503.0	537.0	2040	55,100	0.0569	0.0607	0.0675	0.0735	0.0476	0.370	0.0841	1330	Falcon/TW
1676.0	387.0	2063	50,700	0.0509	0.0555	0.0615	0.0667	0.0482	0.368	0.0832	1405	Chukar/TW
2047.0	468.0	2515	61,100	0.0424	0.0475	0.0514	0.0554	0.0538	0.355	0.0801	1586	Bluebird/TW

TABLE 1-32

COMPACT ROUND ALUMINUM CONDUCTOR STEEL REINFORCED (ACSR/COMP)

Conductor Size AWG or kcmil	Stranding Al/St	Cross-Sectional Area Sq. in		Diameter		Weight/1000 ft			Rated Strength lbs	Resistance				Reactance @ 1 ft Spacing 60 Hz			Ampacity @ 75°C Amps
										dc	ac-60 Hz			GMR ft			
		Aluminum	Total	Steel Core in	Complete Compact Cable in	Aluminum lbs	Steel lbs	Total lbs		@ 20°C Ω/mile	@ 25°C Ω/mile	@ 50°C Ω/mile	@ 75°C Ω/mile		Inductive Xa Ω/mile	Capacitive X'a MΩ-mile	
6	6/1	0.0206	0.0240	0.0661	0.182	24.4	11.6	36.0	1,190	3.3862	3.4544	3.9513	4.3115	0.00589	0.6231	0.1448	102
4	6/1	0.0328	0.0383	0.0834	0.229	38.9	18.4	57.3	1,860	2.1287	2.1716	2.5326	2.7590	0.00741	0.5952	0.1380	135
4	7/1	0.0328	0.0411	0.1029	0.236	39.0	28.0	67.0	2,360	2.1049	2.1472	2.5380	2.7870	0.01543	0.5061	0.1371	136
2	6/1	0.0521	0.0608	0.1052	0.290	61.9	29.3	91.2	2,850	1.3389	1.3660	1.6289	1.7560	0.00938	0.5666	0.1310	179
2	7/1	0.0521	0.0654	0.1299	0.298	61.9	44.7	106.6	3,640	1.3239	1.3506	1.6326	1.7983	0.01885	0.4819	0.1302	179
1	6/1	0.0657	0.0767	0.1181	0.326	78.1	36.9	115.0	3,550	1.0617	1.0832	1.3077	1.4271	0.01054	0.5524	0.1275	206
1/0	6/1	0.0829	0.0967	0.1327	0.365	98.6	46.6	145.2	4,380	0.8414	0.8585	1.0507	1.1489	0.01187	0.5386	0.1242	236
2/0	6/1	0.1045	0.1219	0.1489	0.410	124.0	58.8	182.8	5,300	0.6676	0.6812	0.8488	0.9289	0.01326	0.5245	0.1207	271
3/0	6/1	0.1318	0.1538	0.1672	0.461	156.4	74.1	230.5	6,620	0.5295	0.5405	0.6871	0.7511	0.01419	0.5103	0.1172	310
4/0	6/1	0.1662	0.1939	0.1878	0.517	197.4	93.4	290.8	8,350	0.4199	0.4287	0.5581	0.6073	0.01672	0.4964	0.1138	356
266.8	18/1	0.2095	0.2211	0.1217	0.559	249.9	39.2	289.1	6,880	0.3396	0.3471	0.3813	0.4155	0.03126	0.4205	0.1115	439
300	18/1	0.2356	0.2487	0.1291	0.593	282.0	44.0	326.0	7,740	0.3020	0.3088	0.3392	0.3696	0.03310	0.4133	0.1098	473
336.4	18/1	0.2642	0.2789	0.1367	0.628	315.3	49.5	364.8	8,680	0.2693	0.2756	0.3027	0.3298	0.03512	0.4064	0.1081	508
477	18/1	0.3746	0.3954	0.1628	0.742	447.8	70.2	518.0	11,800	0.1899	0.1950	0.2140	0.2331	0.04147	0.3862	0.1031	631

Notes: (1) Data based on a nominal cable manufactured in accordance with ASTM B 401.
(2) Resistance and ampacity based on an aluminum conductivity of 61.2% IACS at 20°C and a steel conductivity of 8% IACS at 20°C.
(3) Ampacity based on a 75°C conductor temperature, 25°C ambient temperature, 2 ft/sec wind, in sun, with an emissivity of 0.5 and a coefficient of solar absorption of 0.5, at sea level.
(4) Reactance values for the equivalent non-compacted conductor may be used with reasonable accuracy for practical purposes.
(5) Rated strengths based on Class A galvanized steel core wire in accordance with ASTM B 498.

TABLE 1-33
SINGLE LAYER ALUMINUM CONDUCTOR ALUMINUM-CLAD STEEL REINFORCED (ACSR/AW)

Code Word	Conductor Size AWG or kcmil	Stranding (Al/AW)	Cross-Sectional Area Sq. in		Diameter				Weight / 1000 ft			Metal Content % Weight		Rated Strength lbs	Approximate Resistance				Reactance @ 1 ft Spacing 60 Hz		Ampacity @ 75°C Amps
					Individual Wires											ac-60 Hz					
			Aluminum	Total	Aluminum in	AW in	AW Core in	Complete Cable in	Aluminum lbs	AW lbs	Total lbs	Aluminum %	AW %		dc @ 20°C Ω/mile	@ 25°C Ω/mile	@ 50°C Ω/mile	@ 75°C Ω/mile	Inductive Xa Ω/mile	Capacitive X'a MΩ-mile	
Swan/AW	4	6/1	0.0328	0.0383	0.0834	0.0834	0.0834	0.250	38.9	15.6	54.5	71.4	28.6	1,780	2.0602	2.1017	2.4439	2.6618	0.5858	0.1354	141
Swanate/AW	4	7/1	0.0328	0.0411	0.0772	0.1029	0.1029	0.257	38.9	23.8	62.7	62.0	38.0	2,280	2.0046	2.0448	2.4100	2.6437	0.5789	0.1346	142
Sparrow/AW	2	6/1	0.0522	0.0609	0.1052	0.1052	0.1052	0.316	61.9	24.8	86.7	71.4	28.6	2,760	1.2958	1.3220	1.5703	1.7114	0.5576	0.1284	187
Sparate/AW	2	7/1	0.0522	0.0655	0.0974	0.1299	0.1299	0.325	61.9	37.9	99.8	62.0	38.0	3,510	1.2606	1.2860	1.5500	1.7010	0.5507	0.1276	188
Robin/AW	1	6/1	0.0658	0.0768	0.1182	0.1182	0.1182	0.355	78.2	31.4	109.6	71.4	28.6	3,450	1.0274	1.0482	1.2601	1.3748	0.5435	0.1250	214
Raven/AW	1/0	6/1	0.0830	0.0968	0.1327	0.1327	0.1327	0.398	98.5	39.5	138.0	71.4	28.6	4,250	0.8143	0.8309	1.0115	1.1057	0.5294	0.1216	246
Quail/AW	2/0	6/1	0.1045	0.1219	0.1489	0.1489	0.1489	0.447	124.1	49.8	173.9	71.4	28.6	5,130	0.6461	0.6593	0.8165	0.8932	0.5154	0.1182	282
Pigeon/AW	3/0	6/1	0.1317	0.1537	0.1672	0.1672	0.1672	0.502	156.4	62.7	219.1	71.4	28.6	6,300	0.5125	0.5231	0.6607	0.7219	0.5014	0.1147	324
Penguin/AW	4/0	6/1	0.1662	0.1939	0.1878	0.1878	0.1878	0.563	197.4	79.1	276.5	71.4	28.6	7,690	0.4064	0.4150	0.5359	0.5830	0.4873	0.1113	371
High Mechanical Strength																					
Grouse/AW	80.0	8/1	0.0628	0.0847	0.1000	0.1670	0.1670	0.367	75.0	62.6	137.6	54.5	45.5	4,890	1.0204	1.0409	1.3082	1.4387	0.5330	0.1240	212
Petrel/AW	101.8	12/7	0.0799	0.1265	0.0921	0.0921	0.2763	0.461	95.9	133.8	229.7	41.8	58.2	9,910	0.7523	0.7673	1.0848	1.2085	0.4982	0.1172	244
Minorca/AW	110.8	12/7	0.0870	0.1378	0.0961	0.0961	0.2883	0.481	104.4	145.6	250.0	41.8	58.2	10,800	0.6912	0.7050	1.0115	1.1313	0.4931	0.1160	255
Leghorn/AW	134.6	12/7	0.1057	0.1674	0.1059	0.1059	0.3177	0.530	126.8	176.9	303.7	41.8	58.2	13,000	0.5690	0.5804	0.8635	0.9752	0.4813	0.1131	282
Guinea/AW	159.0	12/7	0.1249	0.1977	0.1151	0.1151	0.3453	0.576	149.7	208.9	358.6	41.7	58.3	15,300	0.4817	0.4914	0.7553	0.8608	0.4712	0.1106	307
Dotterel/AW	176.9	12/7	0.1389	0.2199	0.1214	0.1214	0.3642	0.607	166.6	232.4	399.0	41.8	58.2	16,900	0.4329	0.4417	0.6926	0.7945	0.4647	0.1091	324
Dorking/AW	190.8	12/7	0.1499	0.2373	0.1261	0.1261	0.3783	0.631	179.7	250.8	430.5	41.7	58.3	18,300	0.4014	0.4095	0.6518	0.7510	0.4601	0.1079	337
Brahma/AW	203.2	16/19	0.1596	0.3020	0.1127	0.0977	0.4885	0.714	191.4	409.4	600.8	31.9	68.1	27,100	0.3467	0.3537	0.5991	0.7232	0.4409	0.1043	355
Cochin/AW	211.3	12/7	0.1660	0.2628	0.1327	0.1327	0.3981	0.664	199.0	277.7	476.7	41.7	58.3	19,800	0.3624	0.3698	0.6002	0.6958	0.4539	0.1064	355

Notes: (1) Data based on a nominal cable manufactured in accordance with ASTM B 549.
(2) Resistance and ampacity based on an aluminum conductivity of 61.2% IACS at 20°C and an aluminum-clad steel conductivity of 20.3% IACS at 20°C.
(3) Ampacity based on a 75°C conductor temperature, 25°C ambient temperature, 2 ft/sec wind, in sun, with an emissivity of 0.5 and a coefficient of solar absorption of 0.5, at sea level.
(4) Rated strengths based on Class A galvanized steel core wire in accordance with ASTM B 502.
(5) The ac resistance and reactance for single layer ACSR conductors is affected by the magnetic flux in the core and varies with load current. The values given are approximate and are based on the current to produce a 75°C conductor temperature with a 25°C ambient. The coefficient of resistance of the aluminum-clad steel is assumed to be 0.0036 per degree C.
(6) The Inductive reactance can be assumed to be approximately equal to that for single layer ACSR with reasonable accuracy for practical purposes.

TABLE 1-34
ALUMINUM CONDUCTOR ALUMINUM-CLAD STEEL REINFORCED (ACSR/AW)

Code Word	Conductor Size kcmil	Stranding (Al/AW)	Cross-Sectional Area Sq. in		Layers of Al	Diameter				Weight/1000 ft		
						Individual Wires		AW Core in	Complete Cable in			
			Aluminum	Total		Aluminum in	AW in			Aluminum lbs	AW lbs	Total lbs
Waxwing/AW	266.8	18/1	0.2094	0.2210	2	0.1217	0.1217	0.1217	0.609	249.9	33.2	283.1
Partridge/AW	266.8	26/7	0.2095	0.2436	2	0.1013	0.0788	0.2364	0.642	251.3	97.9	349.2
Junco/AW	266.8	30/7	0.2095	0.2584	2	0.0943	0.0943	0.2829	0.660	251.9	140.3	392.2
Ostrich/AW	300	26/7	0.2355	0.2738	2	0.1074	0.0835	0.2505	0.680	282.5	110.0	392.5
Merlin/AW	336.4	18/1	0.2642	0.2789	2	0.1367	0.1367	0.1367	0.684	315.3	41.9	357.2
Linnet/AW	336.4	26/7	0.2640	0.3070	2	0.1137	0.0884	0.2652	0.720	316.6	123.3	439.9
Oriole/AW	336.4	30/7	0.2642	0.3259	2	0.1059	0.1059	0.3177	0.741	317.7	176.9	494.6
Chickadee/AW	397.5	18/1	0.3122	0.3295	2	0.1486	0.1486	0.1486	0.743	372.6	49.6	422.2
Ptarmigan/AW	397.5	20/7	0.3123	0.3339	2	0.1410	0.0627	0.1881	0.752	374.5	62.0	436.5
Brant/AW	397.5	24/7	0.3122	0.3527	2	0.1287	0.0858	0.2574	0.772	374.4	116.1	490.5
Ibis/AW	397.5	26/7	0.3120	0.3628	2	0.1236	0.0961	0.2883	0.783	374.1	145.7	519.8
Lark/AW	397.5	30/7	0.3121	0.3849	2	0.1151	0.1151	0.3453	0.806	375.3	209.0	584.3
Pelican/AW	477	18/1	0.3747	0.3955	2	0.1628	0.1628	0.1628	0.814	447.2	59.5	506.7
Tailorbird/AW	477	20/7	0.3745	0.4004	2	0.1544	0.0686	0.2058	0.824	449.1	74.2	523.3
Flicker/AW	477	24/7	0.3747	0.4233	2	0.1410	0.0940	0.2820	0.846	449.4	139.4	588.8
Hawk/AW	477	26/7	0.3744	0.4354	2	0.1354	0.1053	0.3159	0.858	449.0	174.9	623.9
Hen/AW	477	30/7	0.3747	0.4621	2	0.1261	0.1261	0.3783	0.883	450.4	250.8	701.2
Osprey/AW	556.5	18/1	0.4369	0.4612	2	0.1758	0.1758	0.1758	0.879	521.4	69.4	590.8
Sapsucker/AW	556.5	22/7	0.4368	0.4797	2	0.1590	0.0883	0.2649	0.901	523.9	123.0	646.9
Parakeet/AW	556.5	24/7	0.4372	0.4938	2	0.1523	0.1015	0.3045	0.914	524.3	162.5	686.8
Dove/AW	556.5	26/7	0.4371	0.5083	2	0.1463	0.1138	0.3414	0.927	524.2	204.3	728.5
Eagle/AW	556.5	30/7	0.4371	0.5391	2	0.1362	0.1362	0.4086	0.953	525.5	292.6	818.1
Peacock/AW	605	24/7	0.4753	0.5370	2	0.1588	0.1059	0.3177	0.953	570.0	176.9	746.9
Squab/AW	605	26/7	0.4749	0.5522	2	0.1525	0.1186	0.3558	0.966	569.5	221.9	791.4
Wood Duck/AW	605	30/7	0.4751	0.5860	2	0.1420	0.1420	0.4260	0.994	571.2	318.1	889.3
Teal/AW	605	30/19	0.4751	0.5834	2	0.1420	0.0852	0.4260	0.994	571.2	311.4	882.6
Kingbird/AW	636	18/1	0.4997	0.5275	2	0.1880	0.1880	0.1880	0.940	596.3	79.3	675.6
Swift/AW	636	36/1	0.4994	0.5133	3	0.1329	0.1329	0.1329	0.930	596.0	39.6	635.6
Goldfinch/AW	636	22/7	0.4994	0.5484	2	0.1700	0.0944	0.2832	0.963	598.9	140.6	739.5
Rook/AW	636	24/7	0.4996	0.5643	2	0.1628	0.1085	0.3255	0.977	599.1	185.7	784.8
Grosbeak/AW	636	26/7	0.4995	0.5808	2	0.1564	0.1216	0.3648	0.990	599.0	233.2	832.2
Scoter/AW	636	30/7	0.4995	0.6160	2	0.1456	0.1456	0.4368	1.019	600.5	334.4	934.9
Egret/AW	636	30/19	0.4995	0.6135	2	0.1456	0.0874	0.4370	1.019	600.5	327.7	928.2
Flamingo/AW	666.6	24/7	0.5238	0.5917	2	0.1667	0.1111	0.3333	1.000	628.2	194.7	822.9
Gannet/AW	666.6	26/7	0.5234	0.6086	2	0.1601	0.1245	0.3735	1.014	627.7	244.5	872.2
Stilt/AW	715.5	24/7	0.5622	0.6350	2	0.1727	0.1151	0.3453	1.036	674.2	209.0	883.2
Starling/AW	715.5	26/7	0.5620	0.6535	2	0.1659	0.1290	0.3870	1.051	674.0	262.5	936.5
Crow/AW	715.5	54/7	0.5619	0.6347	3	0.1151	0.1151	0.3450	1.040	673.8	209.0	882.8
Redwing/AW	715.5	30/19	0.5617	0.6897	2	0.1544	0.0926	0.4630	1.081	675.3	367.9	1043.2
Coot/AW	795	36/1	0.6244	0.6417	3	0.1486	0.1486	0.1486	1.040	745.1	49.6	794.7
Cuckoo/AW	795	24/7	0.6244	0.7053	2	0.1820	0.1213	0.3639	1.092	748.8	232.1	980.9
Drake/AW	795	26/7	0.6247	0.7264	2	0.1749	0.1360	0.4080	1.108	749.1	291.7	1040.8
Skimmer/AW	795	30/7	0.6245	0.7702	2	0.1628	0.1628	0.4884	1.140	750.7	418.1	1168.8
Macaw/AW	795	42/7	0.6246	0.6567	3	0.1376	0.0764	0.2292	1.055	749.0	92.1	841.1
Tern/AW	795	45/7	0.6242	0.6674	3	0.1329	0.0886	0.2658	1.063	748.6	123.8	872.4
Condor/AW	795	54/7	0.6240	0.7049	3	0.1213	0.1213	0.3639	1.092	748.4	232.1	980.5
Mallard/AW	795	30/19	0.6245	0.7669	2	0.1628	0.0977	0.4885	1.140	750.7	409.5	1160.2
Ruddy/AW	900	45/7	0.7066	0.7555	3	0.1414	0.0943	0.2829	1.131	847.4	140.3	987.7
Canary/AW	900	54/7	0.7069	0.7985	3	0.1291	0.1291	0.3873	1.162	847.7	262.9	1110.6
Corncrake/AW	954	20/7	0.7492	0.8010	2	0.2184	0.0971	0.2910	1.165	898.5	148.7	1047.2
Redbird/AW	954	24/4	0.7495	0.8466	2	0.1994	0.1329	0.3990	1.196	898.8	278.6	1177.4
Rail/AW	954	45/7	0.7492	0.8010	3	0.1456	0.0971	0.2913	1.165	898.5	148.7	1047.2
Cardinal/AW	954	54/7	0.7491	0.8462	3	0.1329	0.1329	0.3987	1.196	898.3	278.6	1176.9
Canvasback/AW	954	30/19	0.7491	0.9199	2	0.1783	0.1070	0.5350	1.248	900.5	491.2	1391.7
Snowbird/AW	1033.5	42/7	0.8121	0.8539	3	0.1569	0.0872	0.2616	1.203	973.9	119.9	1093.8
Ortolan/AW	1033.5	45/7	0.8112	0.8673	3	0.1515	0.1010	0.3030	1.212	972.8	160.9	1133.7
Curlew/AW	1033.5	54/7	0.8112	0.9164	3	0.1383	0.1383	0.4149	1.245	972.8	301.7	1274.5
Bluejay/AW	1113	45/7	0.8745	0.9350	3	0.1573	0.1049	0.3147	1.259	1048.7	173.6	1222.3
Finch/AW	1113	54/19	0.8746	0.9855	3	0.1436	0.0862	0.4310	1.293	1053.9	318.8	1372.7
Bunting/AW	1192.5	45/7	0.9367	1.0014	3	0.1628	0.1085	0.3255	1.302	1123.4	185.7	1309.1
Grackle/AW	1192.5	54/19	0.9365	1.0552	3	0.1486	0.0892	0.4460	1.333	1128.6	341.3	1469.9
Bittern/AW	1272	45/7	0.9987	1.0678	3	0.1681	0.1121	0.3363	1.345	1197.7	198.2	1395.9
Pheasant/AW	1272	54/19	0.9993	1.1259	3	0.1535	0.0921	0.4605	1.382	1204.3	363.9	1568.2
Dipper/AW	1351.5	45/7	1.0614	1.1347	3	0.1733	0.1155	0.3465	1.386	1272.9	210.4	1483.3
Martin/AW	1351.5	54/19	1.0614	1.1958	3	0.1582	0.0949	0.4745	1.424	1279.1	386.4	1665.5
Bobolink/AW	1431	45/7	1.1236	1.2013	3	0.1783	0.1189	0.3567	1.427	1347.5	223.0	1570.5
Plover/AW	1431	54/19	1.1241	1.2665	3	0.1628	0.0977	0.4885	1.465	1354.6	409.5	1764.1
Nuthatch/AW	1510.5	45/7	1.1862	1.2682	3	0.1832	0.1221	0.3663	1.466	1422.5	235.2	1657.7
Parrot/AW	1510.5	54/19	1.1856	1.3357	3	0.1672	0.1003	0.5015	1.505	1428.8	431.6	1860.4
Lapwing/AW	1590	45/7	1.2492	1.3355	3	0.1880	0.1253	0.3759	1.504	1498.1	247.6	1745.7
Falcon/AW	1590	54/19	1.2489	1.4072	3	0.1716	0.1030	0.5150	1.545	1505.0	455.1	1960.1
Chukar/AW	1780	84/19	1.3986	1.5126	4	0.1456	0.0874	0.4370	1.602	1685.4	327.7	2013.1
Mockingbird/AW	2034.5	72/7	1.5979	1.6671	4	0.1681	0.1122	0.3360	1.681	1925.7	198.6	2124.3
Roadrunner/AW	2057	76/19	1.6152	1.7032	4	0.1645	0.0768	0.3840	1.700	1946.5	253.0	2199.5
Bluebird/AW	2156	84/19	1.6931	1.8309	4	0.1602	0.0961	0.4805	1.762	2040.4	396.2	2436.6
Kiwi/AW	2167	72/7	1.7022	1.7758	4	0.1735	0.1157	0.3471	1.735	2051.4	211.2	2262.6
Thrasher/AW	2312	76/19	1.8155	1.9144	4	0.1744	0.0814	0.4070	1.802	2187.9	284.2	2472.1
Joree/AW	2515	76/19	1.9750	2.0826	4	0.1819	0.0849	0.4245	1.880	2380.1	309.2	2689.3

Notes: (1) Data based on a nominal cable manufactured in accordance with ASTM B 549.
(2) Resistance and ampacity based on an aluminum conductivity of 61.2% IACS at 20°C and an aluminum-clad steel conductivity of 20.3% IACS at 20°C.
(3) Ampacity based on a 75°C conductor temperature, 25°C ambient temperature, 2 ft/sec wind, in sun, with an emissivity of 0.5 and a coefficient of solar absorption of 0.5, at sea level.
(4) Rated strengths based on aluminum-clad steel core wire in accordance with ASTM B 502.

TABLE 1-34
ALUMINUM CONDUCTOR ALUMINUM-CLAD STEEL REINFORCED (ACSR/AW)

Metal Content % Weight		Rated Strength	Resistance				GMR	Reactance @ 1 ft Spacing 60 Hz		Ampacity	Code Word
			dc	ac-60 Hz				Inductive	Capacitive		
Aluminum %	AW %	lbs	@ 20°C Ω/mile	@ 25°C Ω/mile	@ 50°C Ω/mile	@ 75°C Ω/mile	ft	Xa Ω/mile	X'a MΩ-mile	@ 75°C Amps	
88.3	11.7	6,820	0.3358	0.3433	0.3770	0.4108	0.0197	0.477	0.1090	451	Waxwing/AW
72.0	28.0	10,800	0.3258	0.3328	0.3655	0.3981	0.0217	0.465	0.1074	465	Partridge/AW
64.2	35.8	13,500	0.3193	0.3261	0.3581	0.3900	0.0227	0.459	0.1066	473	Junco/AW
72.0	28.0	12,100	0.2897	0.2961	0.3252	0.3542	0.0230	0.458	0.1057	500	Ostrich/AW
88.3	11.7	8,540	0.2663	0.2726	0.2993	0.3261	0.0221	0.463	0.1056	522	Merlin/AW
72.0	28.0	13,500	0.2584	0.2642	0.2901	0.3160	0.0243	0.451	0.1040	537	Linnet/AW
64.2	35.8	16,700	0.2532	0.2588	0.2841	0.3095	0.0255	0.445	0.1031	547	Oriole/AW
88.3	11.7	9,780	0.2254	0.2310	0.2536	0.2762	0.0240	0.452	0.1031	580	Chickadee/AW
85.8	14.2	10,700	0.2254	0.2309	0.2536	0.2762	0.0246	0.450	0.1027	581	Ptarmigan/AW
76.3	23.7	14,100	0.2210	0.2263	0.2484	0.2706	0.0259	0.444	0.1019	591	Brant/AW
72.0	28.0	15,800	0.2187	0.2238	0.2457	0.2676	0.0265	0.441	0.1015	597	Ibis/AW
64.2	35.8	19,600	0.2143	0.2192	0.2406	0.2621	0.0277	0.435	0.1007	608	Lark/AW
88.3	11.7	11,500	0.1878	0.1929	0.2117	0.2305	0.0263	0.441	0.1004	650	Pelican/AW
85.8	14.2	12,700	0.1879	0.1928	0.2116	0.2305	0.0270	0.439	0.1000	652	Tailorbird/AW
76.3	23.7	16,700	0.1842	0.1888	0.2073	0.2257	0.0283	0.433	0.0992	663	Flicker/AW
72.0	28.0	18,900	0.1822	0.1868	0.2050	0.2232	0.0290	0.430	0.0988	669	Hawk/AW
64.2	35.8	23,400	0.1786	0.1829	0.2007	0.2186	0.0304	0.424	0.0980	682	Hen/AW
88.3	11.7	13,200	0.1610	0.1657	0.1818	0.1979	0.0284	0.432	0.0981	715	Osprey/AW
81.0	19.0	17,000	0.1595	0.1639	0.1799	0.1958	0.0298	0.426	0.0974	724	Sapsucker/AW
76.3	23.7	19,300	0.1579	0.1622	0.1779	0.1937	0.0306	0.423	0.0969	731	Parakeet/AW
72.0	28.0	21,900	0.1562	0.1603	0.1759	0.1915	0.0313	0.420	0.0965	737	Dove/AW
64.2	35.8	26,800	0.1531	0.1570	0.1723	0.1875	0.0328	0.415	0.0957	751	Eagle/AW
76.3	23.7	21,000	0.1452	0.1493	0.1638	0.1783	0.0319	0.418	0.0957	770	Peacock/AW
72.0	28.0	23,600	0.1437	0.1477	0.1620	0.1763	0.0327	0.415	0.0953	777	Squab/AW
64.2	35.8	28,400	0.1408	0.1445	0.1586	0.1726	0.0342	0.410	0.0944	791	Wood Duck/AW
64.7	35.3	28,500	0.1411	0.1448	0.1589	0.1729	0.0342	0.410	0.0944	790	Teal/AW
88.3	11.7	15,000	0.1408	0.1454	0.1594	0.1735	0.0304	0.424	0.0961	778	Kingbird/AW
93.8	6.2	13,600	0.1422	0.1469	0.1611	0.1753	0.0302	0.425	0.0964	771	Swift/AW
81.0	19.0	19,400	0.1396	0.1438	0.1577	0.1716	0.0319	0.418	0.0954	787	Goldfinch/AW
76.3	23.7	22,000	0.1381	0.1422	0.1560	0.1698	0.0327	0.415	0.0950	794	Rook/AW
72.0	28.0	24,800	0.1367	0.1406	0.1542	0.1678	0.0335	0.412	0.0946	801	Grosbeak/AW
64.2	35.8	29,300	0.1340	0.1376	0.1509	0.1643	0.0351	0.407	0.0937	816	Scoter/AW
64.7	35.3	29,900	0.1405	0.1443	0.1584	0.1724	0.0351	0.407	0.0937	797	Egret/AW
76.3	23.7	23,100	0.1318	0.1358	0.1489	0.1621	0.0335	0.412	0.0943	818	Flamingo/AW
72.0	28.0	26,000	0.1304	0.1342	0.1472	0.1602	0.0343	0.409	0.0939	825	Gannet/AW
76.3	23.7	24,800	0.1228	0.1267	0.1389	0.1511	0.0347	0.408	0.0932	855	Stilt/AW
72.0	28.0	27,500	0.1215	0.1252	0.1373	0.1494	0.0355	0.405	0.0928	863	Starling/AW
76.3	23.7	25,300	0.1228	0.1267	0.1423	0.1556	0.0350	0.407	0.0932	844	Crow/AW
64.7	35.3	33,400	0.1193	0.1228	0.1346	0.1465	0.0372	0.399	0.0920	878	Redwing/AW
93.8	6.2	16,600	0.1137	0.1184	0.1296	0.1409	0.0337	0.411	0.0931	886	Coot/AW
76.3	23.7	27,500	0.1105	0.1143	0.1253	0.1363	0.0366	0.402	0.0917	913	Cuckoo/AW
72.0	28.0	30,500	0.1093	0.1130	0.1238	0.1347	0.0375	0.399	0.0912	922	Drake/AW
64.2	35.8	35,000	0.1072	0.1105	0.1211	0.1318	0.0392	0.393	0.0904	939	Skimmer/AW
89.1	10.9	19,400	0.1134	0.1177	0.1296	0.1412	0.0346	0.408	0.0927	889	Macaw/AW
85.8	14.2	21,500	0.1127	0.1169	0.1288	0.1403	0.0352	0.406	0.0925	893	Tern/AW
76.3	23.7	27,800	0.1105	0.1143	0.1283	0.1402	0.0368	0.401	0.0917	900	Condor/AW
64.7	35.3	37,100	0.1073	0.1107	0.1213	0.1320	0.0392	0.393	0.0904	938	Mallard/AW
85.8	14.2	24,000	0.0996	0.1037	0.1141	0.1243	0.0374	0.399	0.0906	965	Ruddy/AW
76.3	23.7	31,000	0.0976	0.1014	0.1136	0.1241	0.0392	0.393	0.0898	973	Canary/AW
85.8	14.2	25,100	0.0939	0.0981	0.1074	0.1167	0.0381	0.396	0.0897	1003	Corncrake/AW
76.3	23.7	15,800	0.0921	0.0958	0.1049	0.1140	0.0400	0.391	0.0890	1022	Redbird/AW
85.8	14.2	25,400	0.0939	0.0981	0.1079	0.1174	0.0385	0.395	0.0897	1000	Rail/AW
76.3	23.7	32,900	0.0921	0.0958	0.1073	0.1172	0.0404	0.390	0.0890	1008	Cardinal/AW
64.7	35.3	44,500	0.0895	0.0927	0.1015	0.1104	0.0430	0.382	0.0877	1050	Canvasback/AW
89.0	11.0	24,700	0.0872	0.0916	0.1006	0.1095	0.0395	0.392	0.0888	1045	Snowbird/AW
85.8	14.2	27,100	0.0867	0.0909	0.0999	0.1087	0.0401	0.390	0.0886	1050	Ortolan/AW
76.3	23.7	35,600	0.0850	0.0888	0.0993	0.1084	0.0420	0.385	0.0878	1060	Curlew/AW
85.8	14.2	29,300	0.0805	0.0848	0.0931	0.1012	0.0416	0.386	0.0874	1099	Bluejay/AW
76.8	23.2	37,500	0.0794	0.0832	0.0929	0.1014	0.0436	0.380	0.0867	1107	Finch/AW
85.8	14.2	31,300	0.0751	0.0795	0.0872	0.0948	0.0431	0.382	0.0864	1146	Bunting/AW
76.8	23.2	40,200	0.0741	0.0779	0.0870	0.0949	0.0451	0.376	0.0856	1155	Grackle/AW
85.8	14.2	33,400	0.0705	0.0749	0.0821	0.0891	0.0445	0.378	0.0855	1192	Bittern/AW
76.8	23.2	42,400	0.0695	0.0733	0.0818	0.0892	0.0466	0.372	0.0847	1201	Pheasant/AW
85.8	14.2	35,500	0.0663	0.0708	0.0776	0.0842	0.0459	0.374	0.0846	1237	Dipper/AW
76.8	23.2	45,100	0.0654	0.0693	0.0772	0.0842	0.0480	0.368	0.0838	1246	Martin/AW
85.8	14.2	37,600	0.0626	0.0673	0.0736	0.0798	0.0472	0.371	0.0837	1280	Bobolink/AW
76.8	23.2	47,700	0.0617	0.0658	0.0732	0.0797	0.0494	0.365	0.0829	1290	Plover/AW
85.8	14.2	39,700	0.0593	0.0641	0.0700	0.0759	0.0485	0.367	0.0829	1322	Nuthatch/AW
76.8	23.2	50,300	0.0585	0.0626	0.0696	0.0758	0.0508	0.362	0.0821	1333	Parrot/AW
85.8	14.2	41,800	0.0564	0.0612	0.0668	0.0724	0.0497	0.364	0.0822	1363	Lapwing/AW
76.8	23.2	53,000	0.0556	0.0597	0.0664	0.0722	0.0521	0.359	0.0814	1375	Falcon/AW
83.7	16.3	49,400	0.0504	0.0553	0.0601	0.0649	0.0534	0.355	0.0803	1464	Chukar/AW
90.7	9.3	46,100	0.0446	0.0506	0.0547	0.0588	0.0553	0.351	0.0789	1557	Mockingbird/AW
88.5	11.5	49,200	0.0440	0.0497	0.0538	0.0579	0.0562	0.346	0.0785	1574	Roadrunner/AW
83.7	16.3	59,000	0.0416	0.0471	0.0509	0.0548	0.0588	0.344	0.0775	1634	Bluebird/AW
90.7	9.3	49,100	0.0419	0.0481	0.0519	0.0558	0.0570	0.348	0.0779	1613	Kiwi/AW
88.5	11.5	55,300	0.0391	0.0453	0.0488	0.0524	0.0595	0.342	0.0768	1680	Thrasher/AW
88.5	11.5	60,200	0.0360	0.0424	0.0456	0.0489	0.0621	0.337	0.0755	1759	Joree/AW

TABLE 1-35
ALUMINUM CONDUCTOR STEEL SUPPORTED (ACSS)

Code Word	Conductor Size kcmil	Stranding (Al/St)	Cross-Sectional Area Sq. in Aluminum	Cross-Sectional Area Sq. in Total	Layers of Al	Diameter Individual Wires Aluminum in	Diameter Individual Wires Steel in	Diameter Steel Core in	Diameter Complete Cable in	Weight / 1000 ft Aluminum lbs	Weight / 1000 ft Steel lbs	Weight / 1000 ft Total lbs	Metal Content % Weight Aluminum %	Metal Content % Weight Steel %
Partridge/ACSS	266.8	26/7	0.2095	0.2436	2	0.1013	0.0788	0.2364	0.642	251.3	115.6	366.9	68.5	31.5
Junco/ACSS	266.8	30/7	0.2095	0.2584	2	0.0943	0.0943	0.2829	0.660	251.3	165.5	416.8	60.3	39.7
Ostrich/ACSS	300	26/7	0.2355	0.2738	2	0.1074	0.0835	0.2505	0.680	282.5	129.8	412.3	68.5	31.5
Woodcock/ACSS	336.4	22/7	0.2644	0.2903	2	0.1237	0.0687	0.2061	0.701	317.1	87.8	404.9	78.3	21.7
Linnet/ACSS	336.4	26/7	0.2640	0.3070	2	0.1137	0.0884	0.2652	0.720	316.6	145.5	462.1	68.5	31.5
Oriole/ACSS	336.4	30/7	0.2642	0.3259	2	0.1059	0.1059	0.3177	0.741	317.7	208.7	526.4	60.4	39.6
Ptarmigan/ACSS	397.5	20/7	0.3123	0.3339	2	0.1410	0.0627	0.1881	0.752	374.5	73.2	447.7	83.6	16.4
Brant/ACSS	397.5	24/7	0.3122	0.3527	2	0.1287	0.0858	0.2574	0.772	374.4	137.0	511.4	73.2	26.8
Ibis/ACSS	397.5	26/7	0.3120	0.3628	2	0.1236	0.0961	0.2883	0.783	374.1	171.9	546.0	68.5	31.5
Lark/ACSS	397.5	30/7	0.3121	0.3849	2	0.1151	0.1151	0.3453	0.806	375.3	246.6	621.9	60.3	39.7
Tailorbird/ACSS	477	20/7	0.3745	0.4004	2	0.1544	0.0686	0.2058	0.824	449.1	87.6	536.7	83.7	16.3
Flicker/ACSS	477	24/7	0.3747	0.4233	2	0.1410	0.0940	0.2820	0.846	449.4	164.5	613.9	73.2	26.8
Hawk/ACSS	477	26/7	0.3744	0.4354	2	0.1354	0.1053	0.3159	0.858	449.0	206.4	655.4	68.5	31.5
Hen/ACSS	477	30/7	0.3747	0.4621	2	0.1261	0.1261	0.3783	0.883	450.4	296.0	746.4	60.3	39.7
Sapsucker/ACSS	556.5	22/7	0.4368	0.4797	2	0.1590	0.0883	0.2649	0.901	523.9	145.1	669.0	78.3	21.7
Parakeet/ACSS	556.5	24/7	0.4372	0.4938	2	0.1523	0.1015	0.3045	0.914	524.3	191.8	716.1	73.2	26.8
Dove/ACSS	556.5	26/7	0.4371	0.5083	2	0.1463	0.1138	0.3414	0.927	524.2	241.0	765.2	68.5	31.5
Eagle/ACSS	556.5	30/7	0.4371	0.5391	2	0.1362	0.1362	0.4086	0.953	525.5	345.3	870.8	60.3	39.7
Peacock/ACSS	605	24/7	0.4753	0.5370	2	0.1588	0.1059	0.3177	0.953	570.0	208.7	778.7	73.2	26.8
Squab/ACSS	605	26/7	0.4749	0.5522	2	0.1525	0.1186	0.3558	0.966	569.5	261.8	831.3	68.5	31.5
Wood Duck/ACSS	605	30/7	0.4751	0.5860	2	0.1420	0.1420	0.4260	0.994	571.2	375.3	946.5	60.3	39.7
Teal/ACSS	605	30/19	0.4751	0.5834	2	0.1420	0.0852	0.4260	0.994	571.2	367.5	938.7	60.9	39.1
Goldfinch/ACSS	636	22/7	0.4994	0.5484	2	0.1700	0.0944	0.2832	0.963	598.9	165.9	764.8	78.3	21.7
Rook/ACSS	636	24/7	0.4996	0.5643	2	0.1628	0.1085	0.3255	0.977	599.1	219.1	818.2	73.2	26.8
Grosbeak/ACSS	636	26/7	0.4995	0.5808	2	0.1564	0.1216	0.3648	0.991	599.0	275.2	874.2	68.5	31.5
Scoter/ACSS	636	30/7	0.4995	0.6160	2	0.1456	0.1456	0.4368	1.019	600.5	394.6	995.1	60.3	39.7
Egret/ACSS	636	30/19	0.4995	0.6135	2	0.1456	0.0874	0.4370	1.019	600.5	386.7	987.2	60.8	39.2
Flamingo/ACSS	666.6	24/7	0.5238	0.5917	2	0.1667	0.1111	0.3333	1.000	628.2	229.7	857.9	73.2	26.8
Gannet/ACSS	666.6	26/7	0.5234	0.6086	2	0.1601	0.1245	0.3735	1.014	627.7	288.5	916.2	68.5	31.5
Stilt/ACSS	715.5	24/7	0.5622	0.6350	2	0.1727	0.1151	0.3453	1.036	674.2	246.6	920.8	73.2	26.8
Starling/ACSS	715.5	26/7	0.5620	0.6535	2	0.1659	0.1290	0.3870	1.051	674.0	309.7	983.7	68.5	31.5
Redwing/ACSS	715.5	30/19	0.5617	0.6897	2	0.1544	0.0926	0.4630	1.081	675.3	434.1	1109.4	60.9	39.1
Puffin/ACSS	795	22/7	0.6244	0.6857	2	0.1901	0.1056	0.3168	1.077	748.8	207.6	956.4	78.3	21.7
Cuckoo/ACSS	795	24/7	0.6244	0.7053	2	0.1820	0.1213	0.3639	1.092	748.8	273.9	1022.7	73.2	26.8
Drake/ACSS	795	26/7	0.6247	0.7264	2	0.1749	0.1360	0.4080	1.108	749.1	344.3	1093.4	68.5	31.5
Macaw/ACSS	795	42/7	0.6246	0.6567	3	0.1376	0.0764	0.2292	1.055	749.0	108.6	857.6	87.3	12.7
Tern/ACSS	795	45/7	0.6242	0.6674	3	0.1329	0.0886	0.2658	1.063	748.6	146.1	894.7	83.7	16.3
Condor/ACSS	795	54/7	0.6240	0.7049	3	0.1213	0.1213	0.3639	1.092	748.4	273.9	1022.3	73.2	26.8
Mallard/ACSS	795	30/19	0.6245	0.7669	2	0.1628	0.0977	0.4885	1.140	750.7	483.2	1233.9	60.8	39.2
Ruddy/ACSS	900	45/7	0.7066	0.7555	3	0.1414	0.0943	0.2829	1.131	847.4	165.5	1012.9	83.7	16.3
Canary/ACSS	900	54/7	0.7069	0.7985	3	0.1291	0.1291	0.3873	1.162	847.7	310.2	1157.9	73.2	26.8
Corncrake/ACSS	954	20/7	0.7492	0.8010	2	0.2184	0.0971	0.2913	1.165	898.5	175.5	1074.0	83.7	16.3
Redbird/ACSS	954	24/7	0.7495	0.8466	2	0.1994	0.1329	0.3987	1.196	898.8	328.7	1227.5	73.2	26.8
Rail/ACSS	954	45/7	0.7492	0.8010	3	0.1456	0.0971	0.2913	1.165	898.5	175.5	1074.0	83.7	16.3
Towhee/ACSS	954	48/7	0.7495	0.8157	3	0.1410	0.1097	0.3291	1.175	898.8	224.0	1122.8	80.0	20.0
Cardinal/ACSS	954	54/7	0.7491	0.8462	3	0.1329	0.1329	0.3987	1.196	898.3	328.7	1227.0	73.2	26.8
Canvasback/ACSS	954	30/19	0.7491	0.9199	2	0.1783	0.1070	0.5350	1.248	900.5	579.6	1480.1	60.8	39.2
Snowbird/ACSS	1033.5	42/7	0.8121	0.8539	3	0.1569	0.0872	0.2616	1.203	973.9	141.5	1115.4	87.3	12.7
Ortolan/ACSS	1033.5	45/7	0.8112	0.8673	3	0.1515	0.1010	0.3030	1.212	972.8	189.9	1162.7	83.7	16.3
Curlew/ACSS	1033.5	54/7	0.8112	0.9164	3	0.1383	0.1383	0.4149	1.245	972.8	356.0	1328.8	73.2	26.8
Moose/ACSS	1043.3	54/7	0.8194	0.9256	3	0.1390	0.1390	0.4170	1.251	982.7	359.6	1342.3	73.2	26.8
Bluejay/ACSS	1113	45/7	0.8745	0.9350	3	0.1573	0.1049	0.3147	1.258	1048.7	204.8	1253.5	83.7	16.3
Finch/ACSS	1113	54/19	0.8746	0.9855	3	0.1436	0.0862	0.4310	1.292	1053.9	376.1	1430.0	73.7	26.3
Bunting/ACSS	1192.5	45/7	0.9367	1.0014	3	0.1628	0.1085	0.3255	1.302	1123.4	219.1	1342.5	83.7	16.3
Grackle/ACSS	1192.5	54/19	0.9365	1.0552	3	0.1486	0.0892	0.4460	1.337	1128.6	402.8	1531.4	73.7	26.3
Bittern/ACSS	1272	45/7	0.9987	1.0678	3	0.1681	0.1121	0.3363	1.345	1197.7	233.9	1431.6	83.7	16.3
Diver/ACSS	1272	48/7	0.9992	1.0873	3	0.1628	0.1266	0.3798	1.357	1198.3	298.3	1496.6	80.1	19.9
Pheasant/ACSS	1272	54/19	0.9993	1.1259	3	0.1535	0.0921	0.4605	1.381	1204.3	429.4	1633.7	73.7	26.3
Dipper/ACSS	1351.5	45/7	1.0614	1.1347	3	0.1733	0.1155	0.3465	1.386	1272.9	248.3	1521.2	83.7	16.3
Martin/ACSS	1351.5	54/19	1.0614	1.1958	3	0.1582	0.0949	0.4745	1.424	1279.1	455.9	1735.0	73.7	26.3
Bobolink/ACSS	1431	45/7	1.1236	1.2013	3	0.1783	0.1189	0.3567	1.427	1347.5	263.1	1610.6	83.7	16.3
Plover/ACSS	1431	54/19	1.1241	1.2665	3	0.1628	0.0977	0.4885	1.465	1354.6	483.2	1837.8	73.7	26.3
Nuthatch/ACSS	1510.5	45/7	1.1862	1.2682	3	0.1832	0.1221	0.3663	1.466	1422.5	277.5	1700.0	83.7	16.3
Parrot/ACSS	1510.5	54/19	1.1856	1.3357	3	0.1672	0.1003	0.5015	1.505	1428.8	509.2	1938.0	73.7	26.3
Ratite/ACSS	1590	42/7	1.2492	1.3134	3	0.1946	0.1081	0.3243	1.492	1498.1	217.5	1715.6	87.3	12.7
Lapwing/ACSS	1590	45/7	1.2492	1.3355	3	0.1880	0.1253	0.3759	1.504	1498.1	292.2	1790.3	83.7	16.3
Falcon/ACSS	1590	54/19	1.2489	1.4072	3	0.1716	0.1030	0.5150	1.544	1505.0	537.0	2042.0	73.7	26.3
Chukar/ACSS	1780	84/19	1.3986	1.5126	4	0.1456	0.0874	0.4370	1.601	1685.4	386.7	2072.1	81.3	18.7
Mockingbird/ACSS	2034.5	72/7	1.5979	1.6670	4	0.1681	0.1121	0.3363	1.681	1925.7	233.9	2159.6	89.2	10.8
Roadrunner/ACSS	2057	76/19	1.6152	1.7032	4	0.1645	0.0768	0.3840	1.700	1946.5	298.6	2245.1	86.7	13.3
Bluebird/ACSS	2156	84/19	1.6931	1.8309	4	0.1602	0.0961	0.4805	1.762	2040.4	467.5	2507.9	81.4	18.6
Kiwi/ACSS	2167	72/7	1.7022	1.7758	4	0.1735	0.1157	0.3471	1.735	2051.4	249.2	2300.6	89.2	10.8
Thrasher/ACSS	2312	76/19	1.8155	1.9144	4	0.1744	0.0814	0.4070	1.802	2187.9	335.4	2523.3	86.7	13.3
Joree/ACSS	2515	76/19	1.9750	2.0826	4	0.1819	0.0849	0.4245	1.880	2380.1	364.9	2745.0	86.7	13.3

Notes: (1) Data based on a nominal cable manufactured in accordance with ASTM B 856.

(2) Resistance and ampacity based on an aluminum conductivity of 63% IACS at 20°C and a steel conductivity of 8% IACS at 20°C.

(3) Ampacity based on reference conductor temperature , 25°C ambient temperature, 2 ft/sec wind, in sun, with an emissivity of 0.5 and a coefficient of solar absorption of 0.5, at sea level.

(4) Rated strengths for standard strength core based on Class A Galfan coated steel core wire in accordance with ASTM B 802.

(5) Rated Strength for high strength core based on Class A Galfan coated high strength steel core wire in accordance with ASTM B 803.

TABLE 1-35
ALUMINUM CONDUCTOR STEEL SUPPORTED (ACSS)

Rated Strength			Resistance					Reactance @ 1 ft Spacing 60 Hz		Ampacity					
				ac-60 Hz											
Standard Strength lbs	High Strength lbs	HS285 Strength lbs	dc @ 20°C Ω/mile	@ 25°C Ω/mile	@ 50°C Ω/mile	@ 75°C Ω/mile	GMR ft	Inductive Xa Ω/mile	Capacitive X'a MΩ-mile	@ 75°C Amps	@ 100°C Amps	@ 150°C Amps	@ 200°C Amps	@ 250°C Amps	Code Word
8,880	9,730	11,400	0.3270	0.3343	0.3681	0.4019	0.0217	0.465	0.1074	463	564	707	813	901	Partridge/ACSS
11,700	13,000	15,200	0.3249	0.3320	0.3655	0.3991	0.0227	0.459	0.1066	468	571	715	823	913	Junco/ACSS
10,000	10,900	12,800	0.2908	0.2974	0.3275	0.3575	0.0230	0.458	0.1057	498	608	763	877	974	Ostrich/ACSS
7,610	8,260	9,560	0.2615	0.2677	0.2947	0.3218	0.0232	0.457	0.1048	529	646	811	933	1036	WoodCock/ACSS
11,200	12,300	14,400	0.2594	0.2654	0.2922	0.3190	0.0243	0.451	0.1040	535	654	821	946	1050	Linnet/ACSS
14,800	16,300	19,100	0.2577	0.2635	0.2901	0.3167	0.0255	0.445	0.1031	541	662	831	958	1064	Oriole/ACSS
7,090	7,630	8,710	0.2221	0.2277	0.2507	0.2737	0.0246	0.450	0.1027	584	715	898	1035	1150	Ptarmigan/ACSS
11,000	12,100	14,100	0.2204	0.2258	0.2486	0.2714	0.0259	0.444	0.1019	591	723	909	1048	1165	Brant/ACSS
13,000	14,200	16,500	0.2195	0.2248	0.2474	0.2701	0.0265	0.441	0.1015	594	727	915	1055	1173	Ibis/ACSS
17,500	19,300	22,600	0.2181	0.2232	0.2457	0.2681	0.0277	0.435	0.1007	601	736	926	1069	1188	Lark/ACSS
8,490	9,140	10,400	0.1851	0.1901	0.2093	0.2284	0.0270	0.439	0.1000	655	803	1011	1166	1297	Tailorbird/ACSS
13,000	14,200	16,400	0.1837	0.1885	0.2074	0.2264	0.0283	0.433	0.0992	663	812	1023	1181	1314	Flicker/ACSS
15,600	17,100	19,800	0.1829	0.1876	0.2064	0.2253	0.0290	0.430	0.0988	667	817	1030	1189	1323	Hawk/ACSS
21,000	22,700	26,700	0.1817	0.1862	0.2049	0.2236	0.0304	0.424	0.0980	674	827	1043	1204	1341	Hen/ACSS
12,600	13,600	15,800	0.1581	0.1626	0.1789	0.1952	0.0298	0.426	0.0974	725	890	1123	1298	1445	Sapsucker/ACSS
15,200	16,600	19,200	0.1574	0.1618	0.1780	0.1943	0.0306	0.423	0.0969	730	896	1131	1307	1456	Parakeet/ACSS
18,200	19,900	23,200	0.1568	0.1610	0.1772	0.1933	0.0313	0.420	0.0965	734	902	1138	1316	1466	Dove/ACSS
24,500	26,500	31,100	0.1558	0.1598	0.1758	0.1919	0.0328	0.415	0.0957	743	912	1152	1332	1485	Eagle/ACSS
16,500	18,100	20,800	0.1448	0.1490	0.1639	0.1789	0.0319	0.418	0.0957	769	945	1194	1380	1539	Peacock/ACSS
19,700	21,700	25,200	0.1442	0.1483	0.1631	0.1780	0.0327	0.415	0.0953	774	951	1201	1390	1549	Squab/ACSS
26,000	28,300	33,300	0.1433	0.1471	0.1619	0.1766	0.0342	0.410	0.0944	783	962	1217	1408	1570	Wood Duck/ACSS
26,600	29,300	34,800	0.1434	0.1472	0.1620	0.1767	0.0342	0.410	0.0945	782	962	1216	1407	1570	Teal/ACSS
14,100	15,300	17,500	0.1383	0.1426	0.1568	0.1711	0.0319	0.418	0.0954	789	969	1225	1416	1579	Goldfinch/ACSS
17,300	19,000	21,900	0.1378	0.1419	0.1561	0.1702	0.0327	0.415	0.0950	793	975	1233	1426	1590	Rook/ACSS
20,700	22,400	26,000	0.1372	0.1412	0.1553	0.1694	0.0335	0.412	0.0946	798	982	1241	1436	1602	Grosbeak/ACSS
27,400	29,700	35,000	0.1363	0.1401	0.1541	0.1681	0.0351	0.407	0.0937	807	993	1257	1455	1623	Scoter/ACSS
28,000	30,900	36,600	0.1364	0.1402	0.1542	0.1682	0.0351	0.407	0.0937	807	993	1256	1454	1623	Egret/ACSS
18,200	19,900	22,900	0.1314	0.1355	0.1490	0.1625	0.0335	0.412	0.0943	817	1005	1271	1471	1641	Flamingo/ACSS
21,700	23,400	27,300	0.1309	0.1348	0.1482	0.1617	0.0343	0.409	0.0939	822	1011	1279	1481	1652	Gannet/ACSS
19,500	21,300	24,600	0.1225	0.1264	0.1390	0.1516	0.0347	0.408	0.0932	854	1051	1330	1541	1719	Stilt/ACSS
23,300	25,200	29,800	0.1219	0.1258	0.1383	0.1508	0.0355	0.405	0.0928	859	1058	1340	1552	1732	Starling/ACSS
30,800	34,000	39,800	0.1212	0.1248	0.1372	0.1497	0.0372	0.399	0.0920	869	1070	1356	1571	1755	Redwing/ACSS
17,700	19,200	22,000	0.1107	0.1147	0.1260	0.1374	0.0357	0.405	0.0921	906	1116	1415	1640	1831	Puffin/ACSS
21,700	23,300	26,900	0.1102	0.1141	0.1254	0.1367	0.0365	0.402	0.0917	912	1124	1424	1651	1844	Cuckoo/ACSS
25,900	28,000	32,600	0.1097	0.1135	0.1247	0.1359	0.0375	0.399	0.0912	918	1131	1434	1663	1858	Drake/ACSS
11,800	12,600	14,300	0.1113	0.1157	0.1271	0.1395	0.0346	0.408	0.0927	895	1101	1394	1614	1800	Macaw/ACSS
14,200	15,200	17,400	0.1111	0.1153	0.1273	0.1390	0.0352	0.406	0.0925	898	1105	1399	1620	1808	Tern/ACSS
21,700	23,300	26,900	0.1102	0.1141	0.1284	0.1406	0.0368	0.401	0.0917	900	1107	1400	1620	1807	Condor/ACSS
34,300	37,900	44,300	0.1091	0.1125	0.1237	0.1349	0.0392	0.393	0.0904	928	1145	1452	1684	1882	Mallard/ACSS
15,800	17,000	19,200	0.0981	0.1023	0.1129	0.1232	0.0374	0.399	0.0906	970	1196	1516	1757	1963	Ruddy/ACSS
24,600	26,400	30,500	0.0974	0.1012	0.1137	0.1245	0.0392	0.393	0.0898	972	1198	1518	1759	1964	Canary/ACSS
16,700	18,000	20,400	0.0926	0.0968	0.1062	0.1156	0.0381	0.396	0.0897	1008	1245	1581	1835	2052	Corncrake/ACSS
26,000	28,000	32,300	0.0918	0.0956	0.1050	0.1144	0.0400	0.391	0.0890	1021	1261	1602	1800	2081	Redbird/ACSS
16,700	18,000	20,400	0.0926	0.0968	0.1067	0.1164	0.0385	0.395	0.0897	1006	1241	1574	1826	2041	Rail/ACSS
19,700	21,300	24,300	0.0923	0.0964	0.1058	0.1152	0.0391	0.393	0.0895	1010	1246	1581	1834	2050	Towhee/ACSS
26,000	28,000	32,300	0.0918	0.0956	0.1074	0.1176	0.0404	0.390	0.0890	1008	1243	1576	1827	2041	Cardinal/ACSS
41,100	45,400	53,100	0.0909	0.0942	0.1035	0.1128	0.0430	0.382	0.0877	1040	1285	1634	1898	2125	Canvasback/ACSS
15,400	16,500	18,500	0.0856	0.0901	0.0988	0.1074	0.0395	0.392	0.0888	1052	1299	1651	1916	2143	Snowbird/ACSS
18,100	19,500	22,000	0.0854	0.0897	0.0988	0.1077	0.0401	0.390	0.0886	1056	1304	1657	1924	2152	Ortolan/ACSS
28,200	30,300	35,000	0.0848	0.0886	0.0994	0.1087	0.0420	0.385	0.0878	1059	1308	1660	1926	2153	Curlew/ACSS
28,500	30,600	35,400	0.0840	0.0878	0.0985	0.1077	0.0422	0.384	0.0876	1066	1315	1670	1938	2167	Moose/ACSS
19,500	21,100	23,800	0.0793	0.0837	0.0921	0.1003	0.0416	0.386	0.0874	1105	1367	1738	2019	2260	Bluejay/ACSS
30,400	33,200	38,700	0.0791	0.0830	0.0930	0.1017	0.0436	0.380	0.0867	1106	1367	1737	2017	2257	Finch/ACSS
21,400	23,500	25,400	0.0740	0.0785	0.0863	0.0939	0.0431	0.382	0.0864	1153	1427	1817	2112	2366	Bunting/ACSS
32,600	35,500	41,500	0.0739	0.0777	0.0870	0.0951	0.0451	0.376	0.0856	1154	1427	1816	2111	2363	Grackle/ACSS
22,300	24,000	27,200	0.0694	0.0739	0.0812	0.0884	0.0448	0.378	0.0855	1199	1485	1893	2203	2469	Bittern/ACSS
26,200	28,000	31,900	0.0692	0.0736	0.0808	0.0880	0.0452	0.376	0.0852	1204	1492	1902	2214	2481	Diver/ACSS
34,100	37,300	43,000	0.0692	0.0732	0.0818	0.0894	0.0466	0.372	0.0847	1201	1487	1893	2202	2466	Pheasant/ACSS
23,700	25,500	28,800	0.0653	0.0699	0.0768	0.0835	0.0459	0.374	0.0846	1243	1542	1968	2291	2569	Dipper/ACSS
36,200	39,600	45,600	0.0652	0.0692	0.0773	0.0844	0.0480	0.368	0.0838	1246	1544	1969	2291	2568	Martin/ACSS
25,100	27,000	30,500	0.0617	0.0664	0.0728	0.0791	0.0472	0.371	0.0837	1287	1597	2041	2378	2669	Bobolink/ACSS
38,400	41,900	48,300	0.0615	0.0656	0.0732	0.0799	0.0494	0.365	0.0829	1290	1600	2042	2378	2667	Plover/ACSS
26,500	28,100	31,800	0.0585	0.0633	0.0693	0.0752	0.0485	0.367	0.0829	1329	1651	2112	2463	2765	Nuthatch/ACSS
40,400	44,200	51,000	0.0583	0.0624	0.0696	0.0759	0.0508	0.362	0.0821	1333	1655	2114	2464	2764	Parrot/ACSS
23,400	25,000	27,900	0.0557	0.0608	0.0663	0.0718	0.0490	0.366	0.0824	1364	1696	2171	2534	2847	Ratite/ACSS
27,900	29,600	33,500	0.0555	0.0605	0.0662	0.0718	0.0497	0.364	0.0822	1370	1704	2181	2546	2860	Lapwing/ACSS
42,600	46,600	53,700	0.0554	0.0596	0.0664	0.0724	0.0521	0.359	0.0814	1375	1708	2184	2547	2860	Falcon/ACSS
35,400	38,200	43,900	0.0498	0.0547	0.0596	0.0645	0.0534	0.355	0.0803	1469	1832	2353	2752	3097	Chukar/ACSS
27,200	28,900	32,000	0.0438	0.0498	0.0539	0.0581	0.0553	0.351	0.0789	1568	1960	2526	2962	3340	Mockingbird/ACSS
31,700	33,900	38,300	0.0432	0.0490	0.0531	0.0573	0.0562	0.349	0.0785	1584	1980	2553	2994	3376	Roadrunner/ACSS
42,100	45,500	51,700	0.0411	0.0466	0.0505	0.0544	0.0588	0.344	0.0775	1640	2053	2649	3108	3507	Bluebird/ACSS
29,000	30,800	34,100	0.0411	0.0473	0.0512	0.0551	0.0570	0.348	0.0779	1624	2033	2625	3082	3478	Kiwi/ACSS
35,600	38,100	43,000	0.0385	0.0447	0.0483	0.0519	0.0595	0.342	0.0768	1690	2119	2740	3221	3638	Thrasher/ACSS
38,700	41,400	46,800	0.0353	0.0419	0.0451	0.0484	0.0621	0.337	0.0755	1770	2223	2882	3393	3838	Joree/ACSS

TABLE 1-36
SHAPED WIRE CONCENTRIC-LAY COMPACT ALUMINUM CONDUCTOR STEEL SUPPORTED (ACSS/TW)
Diameters Equal to Standard ACSS Conductors

Code Word	Conductor Size kcmil	Type No.	Size & Stranding of ACSR with Equal Diameter		Cross-Sectional Area Sq. in		Layers of Al	Stranding		Diameter		Weight/1000 feet		
			kcmil	Stranding	Aluminum	Total		No. of Aluminum Wires	No. & Dia. Individual Steel Wire	Steel Core in	Complete Conductor in	Aluminum lbs	Steel lbs	Total lbs
———	403.4	16	336.4	26/7	0.3168	0.3681	2	18	7 x 0.0966	0.2898	0.720	378.7	173.7	552.4
Mohawk/ACSS/TW	571.7	13	477.0	24/7	0.4490	0.5074	2	18	7 x 0.1030	0.3090	0.846	536.6	197.5	734.1
Calumet/ACSS/TW	565.3	16	477.0	26/7	0.4439	0.5162	2	20	7 x 0.1146	0.3441	0.858	531.2	244.4	775.6
Mystic/ACSS/TW	666.6	13	556.5	24/7	0.5236	0.5914	2	20	7 x 0.1111	0.3333	0.913	625.7	229.7	855.4
Oswego/ACSS/TW	664.8	16	556.5	26/7	0.5221	0.6072	2	20	7 x 0.1244	0.3732	0.927	624.6	288.0	912.6
Maumee/ACSS/TW	768.2	13	636.0	24/7	0.6034	0.6819	2	20	7 x 0.1195	0.3585	0.977	721.1	265.8	986.9
Wabash/ACSS/TW	762.8	16	636.0	26/7	0.5992	0.6966	2	20	7 x 0.1331	0.3993	0.990	716.7	329.7	1046
Kettle/ACSS/TW	957.2	7	795.0	45/7	0.7518	0.8038	3	32	7 x 0.0973	0.2919	1.060	901.6	176.2	1078
Fraser/ACSS/TW	946.7	10	795.0	22/7	0.7436	0.8168	3	35	7 x 0.1154	0.3462	1.077	892.6	247.9	1141
Columbia/ACSS/TW	966.2	13	795.0	54/7	0.7589	0.8573	2	21	7 x 0.1338	0.4014	1.092	906.9	333.2	1240
Suwannee/ACSS/TW	959.6	16	795.0	26/7	0.7537	0.8762	2	22	7 x 0.1493	0.4479	1.108	901.6	414.9	1317
Cheyenne/ACSS/TW	1168.1	5	954.0	42/7	0.9175	0.9646	3	30	7 x 0.0926	0.2778	1.155	1099	159.6	1259
Genesee/ACSS/TW	1158.0	7	954.0	45/7	0.9095	0.9733	3	33	7 x 0.1078	0.3234	1.165	1091	216.3	1307
Hudson/ACSS/TW	1158.4	13	954.0	54/7	0.9098	1.0281	2	25	7 x 0.1467	0.4401	1.196	1087	400.6	1488
Catawba/ACSS/TW	1272.0	5	1033.5	42/7	0.9991	1.0505	3	30	7 x 0.0967	0.2901	1.203	1197	174.0	1371
Nelson/ACSS/TW	1257.1	7	1033.5	45/7	0.9874	1.0557	3	35	7 x 0.1115	0.3345	1.213	1184	231.4	1415
Yukon/ACSS/TW	1233.6	13	1033.5	54/7	0.9689	1.0925	3	38	19 x 0.0910	0.4550	1.245	1165	419.2	1584
Truckee/ACSS/TW	1372.5	5	1113.0	42/7	1.0780	1.1334	3	30	7 x 0.1004	0.3012	1.248	1292	187.6	1479
Mackenzie/ACSS/TW	1359.7	7	1113.0	45/7	1.0679	1.1418	3	36	7 x 0.1159	0.3477	1.259	1281	250.0	1531
Thames/ACSS/TW	1334.6	13	1113.0	54/19	1.0480	1.1809	3	39	19 x 0.0944	0.4720	1.290	1260	451.1	1711
St. Croix/ACSS/TW	1467.8	5	1192.5	42/7	1.1529	1.2124	3	33	7 x 0.1041	0.3123	1.292	1381	201.7	1583
Miramichi/ACSS/TW	1455.3	7	1192.5	45/7	1.1430	1.2222	3	36	7 x 0.1200	0.3600	1.302	1371	268.0	1639
Merrimack/ACSS/TW	1433.6	13	1192.5	54/19	1.1250	1.2677	3	39	19 x 0.0978	0.4890	1.340	1354	484.2	1838
Platte/ACSS/TW	1569.0	5	1272.0	42/7	1.2323	1.2957	3	33	7 x 0.1074	0.3222	1.334	1476	214.7	1691
Potomac/ACSS/TW	1557.4	7	1272.0	45/7	1.2232	1.3079	3	36	7 x 0.1241	0.3723	1.345	1467	286.7	1754
Rio Grande/ACSS/TW	1533.3	13	1272.0	54/19	1.2043	1.3571	3	39	19 x 0.1012	0.5060	1.382	1448	518.4	1966
Schuylkill/ACSS/TW	1657.4	7	1351.5	45/7	1.3020	1.3920	3	36	7 x 0.1280	0.3840	1.386	1561	305.0	1866
Pecos/ACSS/TW	1622.0	13	1351.5	54/19	1.2739	1.4429	3	39	19 x 0.1064	0.5320	1.424	1532	573.1	2105
Pee Dee/ACSS/TW	1758.6	7	1431.0	45/7	1.3810	1.4770	3	37	7 x 0.1319	0.3957	1.427	1656	323.8	1980
James/ACSS/TW	1730.6	13	1431.0	54/19	1.3590	1.5314	3	39	19 x 0.1075	0.5375	1.470	1634	585.0	2219
Athabaska/ACSS/TW	1949.6	7	1590.0	45/7	1.5312	1.6377	3	42	7 x 0.1392	0.4176	1.504	1836	360.7	2197
Cumberland/ACSS/TW	1926.9	13	1590.0	54/19	1.5134	1.7049	3	42	19 x 0.1133	0.5665	1.545	1819	649.8	2469
Powder/ACSS/TW	2153.8	8	1780.0	84/19	1.6912	1.8290	4	64	19 x 0.0961	0.4805	1.602	2043	467.5	2510
Santee/ACSS/TW	2627.3	8	2156.0	84/19	2.0630	2.2268	4	64	19 x 0.1062	0.5310	1.762	2492	570.9	3062

Notes: (1) Data based on a nominal cable manufactured in accordance with ASTM B 857.

(2) Resistance and ampacity based on an aluminum conductivity of 63% IACS at 20°C and a steel conductivity of 8% IACS at 20°C.

(3) Ampacity based on reference conductor temperature, 25°C ambient temperature, 2 ft/sec wind, in sun, with an emissivity of 0.5 and a coefficient of solar absorption of 0.5, at sea level.

(4) Rated strengths for standard core based on Class A Galfan coated steel core wire in accordance with ASTM B 802.

(5) Rated Strength for high strength core based on Class A Galfan coated high strength steel core wire in accordance with ASTM B 803.

(6) The final design of a shaped wire compact conductor is contingent upon several factors such as: layer diameter, wire width and wire thickness. The actual configuration of a given size may vary between manufacturers. This may result in a slight variation in the number of wires, number of layers and dimensions of individual wires from that shown in the chart.

TABLE 1-36

SHAPED WIRE CONCENTRIC-LAY COMPACT ALUMINUM CONDUCTOR STEEL SUPPORTED (ACSS/TW)

Diameters Equal to Standard ACSS Conductors

Rated Breaking Strength			Resistance				GMR	Reactance @ 1 ft Spacing 60 Hz		Ampacity					Code Word
				ac-60 Hz											
Standard Strength lbs	High Strength lbs	HS285 Strength lbs	dc @ 20°C Ω/mile	@ 25°C Ω/mile	@ 50°C Ω/mile	@ 75°C Ω/mile	ft	Inductive Xa Ω/mile	Capacitive X'a MΩ-mile	@ 75°C amps	@ 100°C amps	@ 150°C amps	@ 200°C amps	@ 250°C amps	
13,100	14,400	16,700	0.2158	0.2209	0.2432	0.2655	0.0241	0.452	0.1040	586	716	900	1036	1151	————
15,600	17,100	19,700	0.1527	0.1570	0.1727	0.1884	0.0281	0.433	0.0991	725	889	1121	1294	1441	Mohawk/ACSS/TW
18,400	20,200	23,500	0.1540	0.1581	0.1739	0.1898	0.0288	0.430	0.0988	725	890	1122	1295	1442	Calumet/ACSS/TW
18,200	19,900	22,900	0.1310	0.1349	0.1484	0.1619	0.0304	0.424	0.0970	798	980	1238	1431	1595	Mystic/ACSS/TW
21,700	23,400	27,200	0.1309	0.1347	0.1482	0.1617	0.0310	0.421	0.0964	802	985	1244	1439	1604	Oswego/ACSS/TW
21,000	23,000	26,500	0.1137	0.1174	0.1291	0.1407	0.0325	0.416	0.0949	872	1072	1356	1569	1750	Maumee/ACSS/TW
24,900	26,800	31,200	0.1141	0.1177	0.1294	0.1411	0.0330	0.413	0.0946	873	1074	1359	1573	1755	Wabash/ACSS/TW
16,800	18,100	20,400	0.0923	0.0964	0.1081	0.1183	0.0350	0.407	0.0925	973	1197	1514	1753	1955	Kettle/ACSS/TW
21,100	22,900	26,200	0.0930	0.0969	0.1087	0.1190	0.0358	0.404	0.0919	974	1199	1517	1756	1959	Fraser/ACSS/TW
26,400	28,300	32,800	0.0904	0.0940	0.1032	0.1124	0.0364	0.402	0.0917	1005	1239	1571	1822	2035	Columbia/ACSS/TW
30,700	33,100	38,600	0.0907	0.0941	0.1034	0.1127	0.0373	0.399	0.0913	1008	1243	1576	1828	2042	Suwannee/ACSS/TW
17,200	18,300	20,500	0.0757	0.0801	0.0896	0.0979	0.0378	0.397	0.0901	1095	1350	1712	1986	2219	Cheyenne/ACSS/TW
20,500	22,100	25,000	0.0762	0.0805	0.0900	0.0983	0.0384	0.395	0.0897	1094	1350	1712	1985	2218	Genesee/ACSS/TW
31,100	33,500	38,800	0.0754	0.0790	0.0866	0.0943	0.0400	0.391	0.0889	1124	1389	1766	2051	2295	Hudson/ACSS/TW
18,700	20,000	22,300	0.0695	0.0741	0.0827	0.0902	0.0394	0.392	0.0889	1152	1423	1807	2098	2346	Catawba/ACSS/TW
22,100	23,800	26,900	0.0702	0.0745	0.0833	0.0909	0.0400	0.390	0.0886	1150	1420	1804	2094	2342	Nelson/ACSS/TW
33,200	36,300	41,900	0.0712	0.0749	0.0838	0.0916	0.0420	0.385	0.0877	1154	1425	1810	2101	2350	Yukon/ACSS/TW
20,200	21,500	24,000	0.0644	0.0691	0.0770	0.0840	0.0409	0.388	0.0877	1206	1491	1896	2203	2466	Truckee/ACSS/TW
23,900	25,700	29,000	0.0649	0.0694	0.0774	0.0844	0.0420	0.386	0.0874	1206	1490	1895	2202	2465	Mackenzie/ACSS/TW
35,800	39,100	45,100	0.0658	0.0696	0.0777	0.0849	0.0436	0.380	0.0866	1210	1495	1902	2209	2472	Thames/ACSS/TW
21,600	23,100	25,800	0.0602	0.0650	0.0724	0.0789	0.0424	0.840	0.0867	1256	1554	1979	2302	2578	St. Croix/ACSS/TW
25,600	27,100	30,700	0.0607	0.0652	0.0726	0.0792	0.0431	0.382	0.0867	1256	1554	1979	2301	2577	Miramichi/ACSS/TW
38,400	42,000	48,400	0.0613	0.0651	0.0727	0.0793	0.0450	0.376	0.0856	1265	1565	1992	2317	2595	Merrimack/ACSS/TW
23,100	24,600	27,500	0.0564	0.0613	0.0681	0.0742	0.0439	0.379	0.0858	1306	1618	2063	2402	2692	Platte/ACSS/TW
27,300	29,000	32,800	0.0567	0.0613	0.0682	0.0743	0.0445	0.378	0.0853	1308	1620	2065	2403	2694	Potomac/ACSS/TW
41,200	45,000	51,900	0.0573	0.0612	0.0682	0.0744	0.0466	0.372	0.0847	1316	1630	2078	2418	2710	Rio Grande/ACSS/TW
29,100	30,900	34,900	0.0533	0.0581	0.0645	0.0702	0.0459	0.374	0.0845	1357	1682	2147	2501	2805	Schuylkill/ACSS/TW
45,000	49,300	56,900	0.0541	0.0580	0.0646	0.0705	0.0481	0.368	0.0839	1363	1690	2157	2511	2816	Pecos/ACSS/TW
30,900	32,800	37,100	0.0502	0.0551	0.0611	0.0665	0.0473	0.370	0.0837	1405	1743	2228	2598	2916	Pee Dee/ACSS/TW
46,400	50,800	58,500	0.0508	0.0549	0.0610	0.0664	0.0494	0.365	0.0829	1416	1758	2245	2617	2937	James/ACSS/TW
34,300	36,500	41,300	0.0453	0.0505	0.0550	0.0595	0.0500	0.363	0.0822	1505	1873	2403	2808	3157	Athabaska/ACSS/TW
51,600	56,400	65,000	0.0456	0.0499	0.0553	0.0602	0.0523	0.358	0.0815	1508	1875	2400	2802	3148	Cumberland/ACSS/TW
42,100	45,500	51,700	0.0412	0.0464	0.0504	0.0543	0.0538	0.355	0.0803	1599	1996	2569	3009	3391	Powder/ACSS/TW
51,300	55,600	63,100	0.0338	0.0396	0.0427	0.0459	0.0594	0.343	0.0775	1784	2237	2894	3403	3846	Santee/ACSS/TW

TABLE 1-37
SHAPED WIRE CONCENTRIC-LAY COMPACT ALUMINUM CONDUCTOR STEEL SUPPORTED (ACSS/TW)
Area Equal to Standard ACSS Sizes

Code Word	Conductor Size kcmil	Type No.	Cross-Sectional Area Sq. in		Layers of Al	Stranding		Diameter		Weight/1000 feet		
			Aluminum	Total		No. of Aluminum Wires	No. & Dia. Individual Steel Wire	Steel Core in	Complete Conductor in	Aluminum lbs	Steel lbs	Total lbs
Linnet/ACSS/TW	336.4	16	0.2641	0.3070	2	18	7 x 0.0884	0.2652	0.659	315.9	145.5	461.4
Oriole/ACSS/TW	336.4	23	0.2642	0.3258	2	16	7 x 0.1059	0.3177	0.693	316.1	208.7	524.8
Flicker/ACSS/TW	477	13	0.3747	0.4233	2	18	7 x 0.0940	0.2820	0.776	447.8	164.5	612.3
Hawk/ACSS/TW	477	16	0.3746	0.4356	2	18	7 x 0.1053	0.3159	0.789	448.1	206.4	654.5
Hen/ACSS/TW	477	23	0.3747	0.4621	2	16	7 x 0.1261	0.3783	0.825	448.3	296.0	744.3
Parakeet/ACSS/TW	556.5	13	0.4371	0.4937	2	18	7 x 0.1015	0.3045	0.835	522.4	191.8	714.2
Dove/ACSS/TW	556.5	16	0.4371	0.5083	2	20	7 x 0.1138	0.3414	0.852	522.9	241.0	763.9
Rook/ACSS/TW	636	13	0.4995	0.5643	2	18	7 x 0.1085	0.3255	0.890	597.0	219.1	816.1
Grosbeak/ACSS/TW	636	16	0.4995	0.5808	2	20	7 x 0.1216	0.3648	0.908	597.6	275.2	872.8
Scoter/ACSS/TW	636	23	0.4995	0.6160	2	18	7 x 0.1456	0.4368	0.953	597.6	394.6	992.2
Tern/ACSS/TW	795	7	0.6244	0.6675	2	17	7 x 0.0886	0.2658	0.960	745.2	146.1	891.3
Puffin/ACSS/TW	795	10	0.6244	0.6919	2	18	7 x 0.1108	0.3324	0.980	745.9	228.5	974.4
Condor/ACSS/TW	795	13	0.6244	0.7053	2	20	7 x 0.1213	0.3639	0.993	746.3	273.9	1020
Drake/ACSS/TW	795	16	0.6244	0.7261	2	20	7 x 0.1360	0.4080	1.010	747.0	344.3	1091
Canary/ACSS/TW	900	13	0.7069	0.7983	2	20	7 x 0.1291	0.3873	1.055	844.9	310.2	1155
Phoenix/ACSS/TW	954	5	0.7493	0.7876	3	30	7 x 0.0837	0.2511	1.044	897.7	130.4	1028
Rail/ACSS/TW	954	7	0.7493	0.8011	3	32	7 x 0.0971	0.2913	1.061	898.6	175.5	1074
Cardinal/ACSS/TW	954	13	0.7493	0.8464	2	20	7 x 0.1329	0.3987	1.084	895.5	328.7	1224
Snowbird/ACSS/TW	1033.5	5	0.8117	0.8534	3	30	7 x 0.0871	0.2613	1.089	972.5	141.2	1114
Ortolan/ACSS/TW	1033.5	7	0.8117	0.8678	3	32	7 x 0.1010	0.3030	1.102	972.5	189.9	1162
Curlew/ACSS/TW	1033.5	13	0.8117	0.9169	2	20	7 x 0.1383	0.4149	1.129	970.1	356.0	1326
Avocet/ACSS/TW	1113	5	0.8742	0.9191	3	30	7 x 0.0904	0.2712	1.129	1047	152.1	1199
Bluejay/ACSS/TW	1113	7	0.8742	0.9347	3	33	7 x 0.1049	0.3147	1.143	1048	204.8	1253
Finch/ACSS/TW	1113	13	0.8743	0.9852	3	38	19 x 0.0862	0.4310	1.185	1051	376.1	1427
Oxbird/ACSS/TW	1192.5	5	0.9366	0.9848	3	30	7 x 0.0936	0.2808	1.167	1122	163.1	1285
Bunting/ACSS/TW	1192.5	7	0.9366	1.0013	3	33	7 x 0.1085	0.3255	1.181	1123	219.1	1342
Grackle/ACSS/TW	1192.5	13	0.9366	1.0554	3	38	19 x 0.0892	0.4460	1.225	1126	402.8	1529
Scissortail/ACSS/TW	1272	5	0.9991	1.0505	3	30	7 x 0.0967	0.2901	1.203	1197	174.0	1371
Bittern/ACSS/TW	1272	7	0.9990	1.0681	3	35	7 x 0.1121	0.3363	1.220	1198	233.9	1432
Pheasant/ACSS/TW	1272	13	0.9990	1.1256	3	39	19 x 0.0921	0.4605	1.264	1201	429.4	1630
Dipper/ACSS/TW	1351.5	7	1.0615	1.1348	3	35	7 x 0.1155	0.3465	1.256	1273	248.3	1521
Martin/ACSS/TW	1351.5	13	1.0615	1.1959	3	39	19 x 0.0949	0.4745	1.300	1276	455.9	1732
Bobolink/ACSS/TW	1431	7	1.1236	1.2017	3	36	7 x 0.1189	0.3567	1.291	1347	263.1	1611
Plover/ACSS/TW	1431	13	1.1239	1.2664	3	39	19 x 0.0977	0.4885	1.337	1351	483.2	1834
Lapwing/ACSS/TW	1590	7	1.2488	1.3351	3	36	7 x 0.1253	0.3759	1.358	1498	292.2	1790
Falcon/ACSS/TW	1590	13	1.2488	1.4071	3	42	19 x 0.1030	0.5150	1.408	1501	537.0	2038
Chukar/ACSS/TW	1780	8	1.3986	1.5126	3	37	19 x 0.0874	0.4370	1.445	1675	386.7	2062
Bluebird/ACSS/TW	2156	8	1.6933	1.8309	4	64	19 x 0.0961	0.4805	1.608	2045	467.5	2512

Notes:
(1) Data based on a nominal cable manufactured in accordance with ASTM B 857.
(2) Resistance and ampacity based on an aluminum conductivity of 63% IACS at 20°C and a steel conductivity of 8% IACS at 20°C.
(3) Ampacity based on reference conductor temperature, 25°C ambient temperature, 2 ft/sec wind, in sun, with an emissivity of 0.5 and a coefficient of solar absorption of 0.5, at sea level.
(4) Rated strengths for standard core based on Class A Galfan coated steel core wire in accordance with ASTM B 802.
(5) Rated Strength for high strength core based on Class A Galfan coated high strength steel core wire in accordance with ASTM B 803.
(6) The final design of a shaped wire compact conductor is contingent upon several factors such as: layer diameter, wire width and wire thickness. The actual configuration of a given size may vary between manufacturers. This may result in a slight variation in the number of wires, number of layers and dimensions of individual wires from that shown in the chart.

TABLE 1-37
SHAPED WIRE CONCENTRIC-LAY COMPACT ALUMINUM CONDUCTORS STEEL SUPPORTED (ACSS/TW)
Area Equal to Standard ACSS Sizes

Rated Breaking Strength			Resistance					Reactance @ 1 ft Spacing 60 Hz		Ampacity					
				ac-60 Hz											
Standard Strength lbs	High Strength lbs	HS285 Strength lbs	dc @ 20°C Ω/mile	@ 25°C Ω/mile	@ 50°C Ω/mile	@ 75°C Ω/mile	GMR ft	Inductive Xa Ω/mile	Capacitive X'a MΩ-mile	@ 75°C amps	@ 100°C amps	@ 150°C amps	@ 200°C amps	@ 250°C amps	Code Word
11,200	12,300	14,400	0.2588	0.2647	0.2914	0.3182	0.0221	0.463	0.1066	523	638	801	921	1021	Linnet/ACSS/TW
14,800	16,300	19,100	0.2565	0.2622	0.2887	0.3151	0.0238	0.454	0.1051	533	650	816	940	1043	Oriole/ACSS/TW
13,000	14,200	16,400	0.1831	0.1877	0.2066	0.2255	0.0257	0.444	0.1017	648	793	998	1151	1279	Flicker/ACSS/TW
15,600	17,100	19,800	0.1825	0.1870	0.2059	0.2247	0.0264	0.441	0.1013	652	799	1005	1159	1289	Hawk/ACSS/TW
21,000	22,700	26,700	0.1809	0.1853	0.2039	0.2225	0.0283	0.432	0.1000	663	813	1024	1181	1315	Hen/ACSS/TW
15,200	16,600	19,100	0.1569	0.1612	0.1774	0.1935	0.0277	0.435	0.0994	713	874	1102	1271	1415	Parakeet/ACSS/TW
18,200	19,900	23,100	0.1564	0.1606	0.1767	0.1928	0.0286	0.431	0.0991	719	881	1111	1282	1427	Dove/ACSS/TW
17,300	19,000	21,900	0.1373	0.1413	0.1554	0.1696	0.0296	0.427	0.0978	775	951	1200	1386	1544	Rook/ACSS/TW
20,700	22,400	26,000	0.1369	0.1407	0.1548	0.1689	0.0305	0.423	0.0971	781	958	1210	1398	1557	Grosbeak/ACSS/TW
27,400	29,700	35,000	0.1357	0.1393	0.1533	0.1672	0.0328	0.415	0.0957	795	976	1234	1427	1591	Scoter/ACSS/TW
14,200	15,200	17,400	0.1105	0.1147	0.1260	0.1373	0.0312	0.421	0.0955	878	1080	1366	1580	1762	Tern/ACSS/TW
18,900	20,600	23,700	0.1101	0.1139	0.1252	0.1365	0.0323	0.417	0.0949	886	1090	1378	1595	1778	Puffin/ACSS/TW
21,700	23,300	26,900	0.1098	0.1136	0.1248	0.1361	0.0331	0.414	0.0945	890	1095	1386	1604	1789	Condor/ACSS/TW
25,900	28,000	32,600	0.1095	0.1130	0.1243	0.1355	0.0339	0.411	0.0940	896	1103	1396	1616	1803	Drake/ACSS/TW
24,600	26,400	30,500	0.0970	0.1007	0.1106	0.1205	0.0359	0.404	0.0927	962	1185	1501	1739	1942	Canary/ACSS/TW
14,200	15,200	17,100	0.0927	0.0970	0.1087	0.1190	0.0343	0.409	0.0928	967	1189	1503	1740	1940	Phoenix/ACSS/TW
16,700	18,000	20,400	0.0926	0.0967	0.1084	0.1187	0.0349	0.407	0.0925	972	1196	1512	1750	1953	Rail/ACSS/TW
26,000	28,000	32,300	0.0915	0.0952	0.1045	0.1138	0.0362	0.403	0.0919	997	1229	1558	1806	2016	Cardinal/ACSS/TW
15,400	16,400	18,500	0.0856	0.0899	0.1007	0.1101	0.0356	0.405	0.0917	1016	1251	1584	1834	2048	Snowbird/ACSS/TW
18,100	19,500	22,000	0.0854	0.0896	0.1004	0.1098	0.0363	0.402	0.0914	1021	1257	1592	1843	2058	Ortolan/ACSS/TW
28,200	30,300	35,000	0.0845	0.0881	0.0967	0.1053	0.0377	0.398	0.0906	1048	1293	1641	1903	2126	Curlew/ACSS/TW
16,300	17,500	19,500	0.0794	0.0838	0.0938	0.1025	0.0369	0.400	0.0906	1063	1310	1661	1925	2150	Avocet/ACSS/TW
19,500	21,000	23,800	0.0793	0.0835	0.0935	0.1022	0.0376	0.398	0.0903	1068	1317	1669	1935	2161	Bluejay/ACSS/TW
30,400	33,200	38,700	0.0789	0.0826	0.0925	0.1012	0.0399	0.391	0.0891	1084	1336	1695	1965	2196	Finch/ACSS/TW
17,500	18,700	20,900	0.0741	0.0786	0.0879	0.0960	0.0382	0.396	0.0896	1108	1367	1735	2013	2249	Oxbird/ACSS/TW
20,900	22,500	25,400	0.0740	0.0783	0.0875	0.0956	0.0390	0.394	0.0893	1114	1374	1744	2023	2261	Bunting/ACSS/TW
32,600	35,500	41,500	0.0737	0.0773	0.0866	0.0947	0.0412	0.387	0.0883	1130	1395	1771	2055	2298	Grackle/ACSS/TW
18,700	20,000	22,300	0.0695	0.0741	0.0827	0.0902	0.0394	0.392	0.0888	1152	1423	1807	2098	2346	Scissortail/ACSS/TW
22,300	24,000	27,100	0.0694	0.0737	0.0824	0.0899	0.0403	0.390	0.0884	1159	1431	1817	2110	2360	Bittern/ACSS/TW
34,100	37,300	43,000	0.0691	0.0728	0.0814	0.0890	0.0426	0.383	0.0874	1176	1452	1846	2143	2398	Pheasant/ACSS/TW
23,700	25,500	28,800	0.0653	0.0697	0.0778	0.0849	0.0415	0.386	0.0874	1202	1485	1888	2194	2455	Dipper/ACSS/TW
36,200	39,600	45,600	0.0650	0.0688	0.0768	0.0839	0.0438	0.377	0.0865	1220	1508	1918	2228	2494	Martin/ACSS/TW
25,100	27,000	30,500	0.0617	0.0662	0.0738	0.0804	0.0427	0.383	0.0867	1243	1538	1958	2276	2549	Bobolink/ACSS/TW
38,400	41,900	48,300	0.0614	0.0652	0.0728	0.0795	0.0451	0.376	0.0860	1263	1562	1989	2313	2590	Plover/ACSS/TW
27,900	29,600	33,500	0.0555	0.0602	0.0670	0.0729	0.0449	0.377	0.0851	1324	1640	2092	2435	2730	Lapwing/ACSS/TW
42,600	46,600	53,700	0.0553	0.0592	0.0660	0.0719	0.0476	0.370	0.0841	1346	1668	2127	2477	2777	Falcon/ACSS/TW
35,300	38,200	43,900	0.0495	0.0542	0.0601	0.0654	0.0482	0.368	0.0832	1421	1764	2255	2630	2952	Chukar/ACSS/TW
42,100	45,500	51,700	0.4117	0.0464	0.0503	0.0543	0.0538	0.355	0.0801	1601	1999	2573	3014	3396	Bluebird/ACSS/TW

TABLE 1-38

ALUMINUM CONDUCTOR ALUMINUM-CLAD STEEL SUPPORTED (ACSS/AW)

Code Word	Conductor Size kcmil	Stranding (Al/AW)	Cross-Sectional Area Sq. in: Aluminum	Cross-Sectional Area Sq. in: Total	Layers of Al	Diameter, Individual Wires: Aluminum in	Diameter, Individual Wires: AW in	Diameter: AW Core in	Diameter: Complete Cable in	Weight/1000 ft: Aluminum lbs	Weight/1000 ft: AW lbs	Weight/1000 ft: Total lbs	Metal Content % Weight: Aluminum %	Metal Content % Weight: AW %
Partridge/ACSS/AW	266.8	26/7	0.2095	0.2436	2	0.1013	0.0788	0.2364	0.642	251.3	97.9	349.2	72.0	28.0
Junco/ACSS/AW	266.8	30/7	0.2095	0.2584	2	0.0943	0.0943	0.2829	0.660	251.3	140.2	391.5	64.2	35.8
Ostrich/ACSS/AW	300	26/7	0.2355	0.2738	2	0.1074	0.0835	0.2505	0.680	282.5	110.0	392.5	72.0	28.0
Woodcock/ACSS/AW	336.4	22/7	0.2644	0.2903	2	0.1237	0.0687	0.2061	0.701	317.1	74.4	391.5	81.0	19.0
Linnet/ACSS/AW	336.4	26/7	0.2640	0.3070	2	0.1137	0.0884	0.2652	0.720	316.6	123.2	439.8	72.0	28.0
Oriole/ACSS/AW	336.4	30/7	0.2642	0.3259	2	0.1059	0.1059	0.3177	0.741	317.7	176.9	494.6	64.2	35.8
Ptarmigan/ACSS/AW	397.5	20/7	0.3123	0.3339	2	0.1410	0.0627	0.1881	0.752	374.5	62.0	436.5	85.8	14.2
Brant/ACSS/AW	397.5	24/7	0.3122	0.3527	2	0.1287	0.0858	0.2574	0.772	374.4	116.1	490.5	76.3	23.7
Ibis/ACSS/AW	397.5	26/7	0.3120	0.3628	2	0.1236	0.0961	0.2883	0.783	374.1	145.6	519.7	72.0	28.0
Lark/ACSS/AW	397.5	30/7	0.3121	0.3849	2	0.1151	0.1151	0.3453	0.806	375.3	208.9	584.2	64.2	35.8
Tailorbird/ACSS/AW	477	20/7	0.3745	0.4004	2	0.1544	0.0686	0.2058	0.823	449.1	74.2	523.3	85.8	14.2
Flicker/ACSS/AW	477	24/7	0.3747	0.4233	2	0.1410	0.0940	0.2820	0.846	449.4	139.4	588.8	76.3	23.7
Hawk/ACSS/AW	477	26/7	0.3744	0.4354	2	0.1354	0.1053	0.3159	0.858	449.0	174.9	623.9	72.0	28.0
Hen/ACSS/AW	477	30/7	0.3747	0.4621	2	0.1261	0.1261	0.3783	0.883	450.4	250.8	701.2	64.2	35.8
Sapsucker/ACSS/AW	556.5	22/7	0.4368	0.4797	2	0.1590	0.0883	0.2649	0.901	523.9	123.0	646.9	81.0	19.0
Parakeet/ACSS/AW	556.5	24/7	0.4372	0.4938	2	0.1523	0.1015	0.3045	0.914	524.3	162.5	686.8	76.3	23.7
Dove/ACSS/AW	556.5	26/7	0.4371	0.5083	2	0.1463	0.1138	0.3414	0.927	524.2	204.2	728.4	72.0	28.0
Eagle/ACSS/AW	556.5	30/7	0.4371	0.5391	2	0.1362	0.1362	0.4086	0.953	525.5	292.6	818.1	64.2	35.8
Peacock/ACSS/AW	605	24/7	0.4753	0.5370	2	0.1588	0.1059	0.3177	0.953	570.0	176.9	746.9	76.3	23.7
Squab/ACSS/AW	605	26/7	0.4749	0.5522	2	0.1525	0.1186	0.3558	0.966	569.5	221.8	791.3	72.0	28.0
Wood Duck/ACSS/AW	605	30/7	0.4751	0.5860	2	0.1420	0.1420	0.4260	0.994	571.2	318.0	889.2	64.2	35.8
Teal/ACSS/AW	605	30/19	0.4751	0.5834	2	0.1420	0.0852	0.4260	0.994	571.2	311.4	882.6	64.7	35.3
Goldfinch/ACSS/AW	636	22/7	0.4994	0.5484	2	0.1700	0.0944	0.2832	0.963	598.9	140.5	739.4	81.0	19.0
Rook/ACSS/AW	636	24/7	0.4996	0.5643	2	0.1628	0.1085	0.3255	0.977	599.1	185.7	784.8	76.3	23.7
Grosbeak/ACSS/AW	636	26/7	0.4995	0.5808	2	0.1564	0.1216	0.3648	0.990	599.0	233.2	832.2	72.0	28.0
Scoter/ACSS/AW	636	30/7	0.4995	0.6160	2	0.1456	0.1456	0.4368	1.019	600.5	334.3	934.8	64.2	35.8
Egret/ACSS/AW	636	30/19	0.4995	0.6135	2	0.1456	0.0874	0.4370	1.019	600.5	327.6	928.1	64.7	35.3
Flamingo/ACSS/AW	666.6	24/7	0.5238	0.5917	2	0.1667	0.1111	0.3333	1.000	628.2	194.7	822.9	76.3	23.7
Gannet/ACSS/AW	666.6	26/7	0.5234	0.6086	2	0.1601	0.1245	0.3735	1.014	627.7	244.5	872.2	72.0	28.0
Stilt/ACSS/AW	715.5	24/7	0.5622	0.6350	2	0.1727	0.1151	0.3453	1.036	674.2	208.9	883.1	76.3	23.7
Starling/ACSS/AW	715.5	26/7	0.5620	0.6535	2	0.1659	0.1290	0.3870	1.051	674.0	262.4	936.4	72.0	28.0
Redwing/ACSS/AW	715.5	30/19	0.5617	0.6897	2	0.1544	0.0926	0.4630	1.081	675.3	367.8	1043.1	64.7	35.3
Cuckoo/ACSS/AW	795	24/7	0.6244	0.7053	2	0.1820	0.1213	0.3639	1.092	748.8	232.1	980.9	76.3	23.7
Drake/ACSS/AW	795	26/7	0.6247	0.7264	2	0.1749	0.1360	0.4080	1.108	749.1	291.7	1040.8	72.0	28.0
Macaw/ACSS/AW	795	42/7	0.6246	0.6567	3	0.1376	0.0764	0.2292	1.055	749.0	92.1	841.1	89.1	10.9
Tern/ACSS/AW	795	45/7	0.6242	0.6674	3	0.1329	0.0886	0.2658	1.063	748.6	123.8	872.4	85.8	14.2
Condor/ACSS/AW	795	54/7	0.6240	0.7049	3	0.1213	0.1213	0.3639	1.092	748.4	232.1	980.5	76.3	23.7
Mallard/ACSS/AW	795	30/19	0.6245	0.7669	2	0.1628	0.0977	0.4885	1.140	750.7	409.4	1160.1	64.7	35.3
Ruddy/ACSS/AW	900	45/7	0.7066	0.7555	3	0.1414	0.0943	0.2829	1.131	847.4	140.2	987.6	85.8	14.2
Canary/ACSS/AW	900	54/7	0.7069	0.7985	3	0.1291	0.1291	0.3873	1.162	847.7	262.9	1110.6	76.3	23.7
Corncrake/ACSS/AW	954	20/7	0.7492	0.8010	2	0.2184	0.0971	0.2913	1.165	898.5	148.7	1047.2	85.8	14.2
Redbird/ACSS/AW	954	24/7	0.7495	0.8466	2	0.1994	0.1329	0.3987	1.196	898.8	278.6	1177.4	76.3	23.7
Rail/ACSS/AW	954	45/7	0.7492	0.8010	3	0.1456	0.0971	0.2913	1.165	898.5	148.7	1047.2	85.8	14.2
Towhee/ACSS/AW	954	48/7	0.7495	0.8157	3	0.1410	0.1097	0.3291	1.175	898.8	189.8	1088.6	82.6	17.4
Cardinal/ACSS/AW	954	54/7	0.7491	0.8462	3	0.1329	0.1329	0.3987	1.196	898.3	278.6	1176.9	76.3	23.7
Canvasback/ACSS/AW	954	30/19	0.7491	0.9199	2	0.1783	0.1070	0.5350	1.248	900.5	491.1	1391.6	64.7	35.3
Snowbird/ACSS/AW	1033.5	42/7	0.8121	0.8539	3	0.1569	0.0872	0.2616	1.203	973.9	119.9	1093.8	89.0	11.0
Ortolan/ACSS/AW	1033.5	45/7	0.8112	0.8673	3	0.1515	0.1010	0.3030	1.212	972.8	160.9	1133.7	85.8	14.2
Curlew/ACSS/AW	1033.5	54/7	0.8112	0.9164	3	0.1383	0.1383	0.4149	1.245	972.8	301.7	1274.5	76.3	23.7
Bluejay/ACSS/AW	1113	45/7	0.8745	0.9350	3	0.1573	0.1049	0.3147	1.259	1048.7	173.5	1222.2	85.8	14.2
Finch/ACSS/AW	1113	54/19	0.8746	0.9855	3	0.1436	0.0862	0.4310	1.293	1053.9	318.7	1372.6	76.8	23.2
Moose/ACSS/AW	1043.3	54/7	0.8194	0.9256	3	0.1390	0.1390	0.4170	1.251	982.7	304.7	1287.4	76.3	23.7
Bunting/ACSS/AW	1192.5	45/7	0.9367	1.0014	3	0.1628	0.1085	0.3255	1.302	1123.4	185.7	1309.1	85.8	14.2
Grackle/ACSS/AW	1192.5	54/19	0.9365	1.0552	3	0.1486	0.0892	0.4460	1.338	1128.6	341.3	1469.9	76.8	23.2
Bittern/ACSS/AW	1272	45/7	0.9987	1.0678	3	0.1681	0.1121	0.3363	1.345	1197.7	198.2	1395.9	85.8	14.2
Diver/ACSS/AW	1272	48/7	0.9992	1.0873	3	0.1628	0.1266	0.3798	1.357	1198.3	252.8	1451.1	82.6	17.4
Pheasant/ACSS/AW	1272	54/19	0.9993	1.1259	3	0.1535	0.0921	0.4605	1.382	1204.3	363.8	1568.1	76.8	23.2
Dipper/ACSS/AW	1351.5	45/7	1.0614	1.1347	3	0.1733	0.1155	0.3465	1.386	1272.9	210.4	1483.3	85.8	14.2
Martin/ACSS/AW	1351.5	54/19	1.0614	1.1958	3	0.1582	0.0949	0.4745	1.424	1279.1	386.3	1665.4	76.8	23.2
Bobolink/ACSS/AW	1431	45/7	1.1236	1.2013	3	0.1783	0.1189	0.3567	1.427	1347.5	223.0	1570.5	85.8	14.2
Plover/ACSS/AW	1431	54/19	1.1241	1.2665	3	0.1628	0.0977	0.4885	1.465	1354.6	409.4	1764.0	76.8	23.2
Nuthatch/ACSS/AW	1510	45/7	1.1862	1.2682	3	0.1832	0.1221	0.3663	1.466	1422.5	235.1	1657.6	85.8	14.2
Parrot/ACSS/AW	1510	54/19	1.1856	1.3357	3	0.1672	0.1003	0.5015	1.505	1428.8	431.5	1860.3	76.8	23.2
Ratite/ACSS/AW	1590	42/7	1.2492	1.3134	3	0.1946	0.1081	0.3243	1.492	1498.1	184.3	1682.4	89.0	11.0
Lapwing/ACSS/AW	1590	45/7	1.2492	1.3355	3	0.1880	0.1253	0.3759	1.504	1498.1	247.6	1745.7	85.8	14.2
Falcon/ACSS/AW	1590	54/19	1.2489	1.4072	3	0.1716	0.1030	0.5150	1.544	1505.0	455.0	1960.0	76.8	23.2
Chukar/ACSS/AW	1780	84/19	1.3986	1.5126	4	0.1456	0.0874	0.4370	1.602	1685.4	327.6	2013.0	83.7	16.3
Mockingbird/ACSS/AW	2034.5	72/7	1.5979	1.6670	4	0.1681	0.1121	0.3363	1.681	1925.7	198.2	2123.9	90.7	9.3
Roadrunner/ACSS/AW	2057	76/19	1.6152	1.7032	4	0.1645	0.0768	0.3840	1.700	1946.5	253.0	2199.5	88.5	11.5
Bluebird/ACSS/AW	2156	84/19	1.6931	1.8309	4	0.1602	0.0961	0.4805	1.762	2040.4	396.1	2436.5	83.7	16.3
Kiwi/ACSS/AW	2167	72/7	1.7022	1.7758	4	0.1735	0.1157	0.3471	1.735	2051.4	211.1	2262.5	90.7	9.3
Thrasher/ACSS/AW	2312	76/19	1.8155	1.9144	4	0.1744	0.0814	0.4070	1.802	2187.9	284.2	2472.1	88.5	11.5
Joree/ACSS/AW	2515	76/19	1.9750	2.0826	4	0.1819	0.0849	0.4245	1.880	2380.1	309.2	2689.3	88.5	11.5

Notes: (1) Data based on a nominal cable manufactured in accordance with ASTM B 856.

(2) Resistance and ampacity based on an aluminum conductivity of 63% IACS at 20°C and an aluminum-clad steel conductivity of 20.3% IACS at 20°C.

(3) Ampacity based on a 75°C conductor temperature, 25°C ambient temperature, 2 ft/sec wind, in sun, with an emissivity of 0.5 and a coefficient of solar absorption of 0.5, at sea level.

(4) Rated strengths based on aluminum-clad steel core wire in accordance with ASTM B 502.

TABLE 1-38
ALUMINUM CONDUCTOR ALUMINUM-CLAD STEEL SUPPORTED (ACSS/AW)

Rated Strength lbs	Resistance dc @ 20°C Ω/mile	Resistance ac-60 Hz @ 25°C Ω/mile	Resistance ac-60 Hz @ 50°C Ω/mile	Resistance ac-60 Hz @ 75°C Ω/mile	GMR ft	Reactance @ 1 ft Spacing 60 Hz Inductive Xa Ω/mile	Reactance @ 1 ft Spacing 60 Hz Capacitive X'a MΩ-mile	Ampacity @ 75°C Amps	Ampacity @ 100°C Amps	Ampacity @ 150°C Amps	Ampacity @ 200°C Amps	Ampacity @ 250°C Amps	Code Word
8,360	0.3169	0.3240	0.3567	0.3894	0.0217	0.4648	0.1074	470	573	719	826	916	Partridge/ACSS/AW
11,200	0.3108	0.3176	0.3496	0.3816	0.0227	0.4592	0.1066	478	584	732	842	934	Junco/ACSS/AW
9,390	0.2819	0.2883	0.3173	0.3464	0.0230	0.4578	0.1057	506	618	775	892	989	Ostrich/ACSS/AW
7,210	0.2566	0.2626	0.2891	0.3156	0.0232	0.4566	0.1048	534	652	819	943	1046	WoodCock/ACSS/AW
10,500	0.2514	0.2572	0.2831	0.3090	0.0243	0.4508	0.1040	544	664	834	961	1067	Linnet/ACSS/AW
14,200	0.2465	0.2521	0.2774	0.3028	0.0255	0.4451	0.1031	553	677	850	980	1089	Oriole/ACSS/AW
6,760	0.2191	0.2247	0.2473	0.2700	0.0246	0.4495	0.1027	588	719	904	1042	1158	Ptarmigan/ACSS/AW
10,400	0.2150	0.2202	0.2424	0.2646	0.0259	0.4436	0.1019	598	732	921	1061	1180	Brant/ACSS/AW
12,400	0.2127	0.2179	0.2398	0.2617	0.0265	0.4407	0.1015	604	739	930	1072	1192	Ibis/ACSS/AW
16,700	0.2086	0.2135	0.2350	0.2564	0.0277	0.4350	0.1007	615	753	948	1093	1216	Lark/ACSS/AW
8,100	0.1826	0.1876	0.2064	0.2253	0.0270	0.4385	0.1000	659	808	1017	1174	1306	Tailorbird/ACSS/AW
12,500	0.1791	0.1838	0.2022	0.2207	0.0283	0.4325	0.0992	671	822	1036	1196	1331	Flicker/ACSS/AW
14,900	0.1773	0.1818	0.2001	0.2183	0.0290	0.4296	0.0988	677	830	1047	1208	1345	Hawk/ACSS/AW
20,100	0.1739	0.1782	0.1960	0.2139	0.0304	0.4239	0.0980	689	846	1067	1232	1372	Hen/ACSS/AW
11,900	0.1551	0.1596	0.1755	0.1915	0.0298	0.4262	0.0974	732	899	1134	1310	1460	Sapsucker/ACSS/AW
14,600	0.1535	0.1579	0.1736	0.1894	0.0306	0.4232	0.0969	739	907	1145	1324	1475	Parakeet/ACSS/AW
17,500	0.1520	0.1561	0.1717	0.1873	0.0313	0.4202	0.0965	746	916	1157	1337	1490	Dove/ACSS/AW
22,900	0.1490	0.1530	0.1682	0.1835	0.0328	0.4146	0.0957	759	933	1179	1363	1520	Eagle/ACSS/AW
15,900	0.1412	0.1454	0.1599	0.1744	0.0319	0.4181	0.0957	779	957	1209	1398	1559	Peacock/ACSS/AW
19,000	0.1398	0.1438	0.1581	0.1725	0.0327	0.4152	0.0953	786	966	1221	1412	1574	Squab/ACSS/AW
24,400	0.1371	0.1408	0.1549	0.1689	0.0342	0.4095	0.0944	800	984	1244	1440	1606	Wood Duck/ACSS/AW
25,000	0.1373	0.1411	0.1551	0.1692	0.0342	0.4095	0.0944	799	983	1243	1439	1605	Teal/ACSS/AW
13,600	0.1357	0.1400	0.1539	0.1678	0.0319	0.4181	0.0954	796	978	1237	1430	1595	Goldfinch/ACSS/AW
16,700	0.1344	0.1384	0.1522	0.1660	0.0327	0.4151	0.0950	803	988	1249	1445	1611	Rook/ACSS/AW
19,900	0.1330	0.1369	0.1505	0.1642	0.0335	0.4122	0.0946	811	997	1261	1459	1627	Grosbeak/ACSS/AW
25,100	0.1304	0.1341	0.1474	0.1608	0.0351	0.4065	0.0937	826	1016	1285	1488	1661	Scoter/ACSS/AW
26,300	0.1306	0.1343	0.1477	0.1610	0.0351	0.4065	0.0937	825	1015	1284	1487	1659	Egret/ACSS/AW
17,500	0.1282	0.1322	0.1453	0.1585	0.0335	0.4122	0.0943	827	1018	1287	1490	1662	Flamingo/ACSS/AW
20,900	0.1269	0.1307	0.1437	0.1567	0.0343	0.4093	0.0939	835	1027	1300	1505	1679	Gannet/ACSS/AW
18,800	0.1194	0.1234	0.1356	0.1478	0.0347	0.4079	0.0932	865	1064	1348	1561	1742	Stilt/ACSS/AW
22,000	0.1182	0.1220	0.1341	0.1462	0.0355	0.4050	0.0928	873	1075	1361	1577	1760	Starling/ACSS/AW
29,500	0.1161	0.1196	0.1315	0.1434	0.0372	0.3994	0.092	888	1094	1386	1606	1794	Redwing/ACSS/AW
20,900	0.1075	0.1113	0.1223	0.1333	0.0365	0.4015	0.0917	923	1138	1443	1672	1868	Cuckoo/ACSS/AW
24,400	0.1064	0.1100	0.1209	0.1318	0.0375	0.3986	0.0912	932	1149	1457	1690	1888	Drake/ACSS/AW
11,400	0.1102	0.1146	0.1265	0.1381	0.0346	0.4080	0.0927	899	1107	1401	1622	1810	Macaw/ACSS/AW
13,500	0.1096	0.1138	0.1257	0.1372	0.0352	0.4062	0.0925	904	1113	1409	1631	1820	Tern/ACSS/AW
20,900	0.1075	0.1113	0.1252	0.1372	0.0368	0.4006	0.0917	911	1121	1418	1641	1831	Condor/ACSS/AW
32,900	0.1045	0.1079	0.1185	0.1292	0.0392	0.3929	0.0904	949	1170	1485	1722	1925	Mallard/ACSS/AW
15,300	0.0968	0.1010	0.1114	0.1216	0.0374	0.3987	0.0906	976	1204	1526	1769	1976	Ruddy/ACSS/AW
23,200	0.0949	0.0987	0.1109	0.1214	0.0392	0.3930	0.0898	984	1213	1537	1781	1989	Canary/ACSS/AW
16,200	0.0913	0.0956	0.1048	0.1141	0.0381	0.3964	0.0897	1015	1253	1592	1848	2067	Corncrake/ACSS/AW
24,600	0.0896	0.0934	0.1025	0.1116	0.0400	0.3905	0.0890	1034	1277	1623	1884	2108	Redbird/ACSS/AW
16,200	0.0913	0.0956	0.1053	0.1149	0.0385	0.3951	0.0897	1012	1249	1585	1838	2055	Rail/ACSS/AW
19,000	0.0908	0.0948	0.1040	0.1140	0.0391	0.3932	0.0895	1018	1256	1595	1850	2068	Towhee/ACSS/AW
24,600	0.0896	0.0934	0.1048	0.1147	0.0404	0.3895	0.0890	1020	1258	1596	1850	2067	Cardinal/ACSS/AW
39,400	0.0871	0.0903	0.0992	0.1080	0.0430	0.3819	0.0877	1062	1313	1670	1941	2172	Canvasback/ACSS/AW
14,800	0.0848	0.0892	0.0978	0.1071	0.0395	0.3921	0.0888	1057	1306	1659	1926	2154	Snowbird/ACSS/AW
17,600	0.0843	0.0886	0.0976	0.1063	0.0401	0.3903	0.0886	1063	1313	1668	1937	2166	Ortolan/ACSS/AW
26,100	0.0827	0.0865	0.0970	0.1061	0.0420	0.3847	0.0878	1072	1324	1681	1951	2181	Curlew/ACSS/AW
18,900	0.0783	0.0826	0.0909	0.0990	0.0416	0.3858	0.0874	1113	1376	1750	2034	2276	Bluejay/ACSS/AW
28,800	0.0772	0.0810	0.0908	0.0993	0.0436	0.3801	0.0867	1120	1384	1759	2043	2286	Finch/ACSS/AW
26,300	0.0819	0.0858	0.0961	0.1051	0.4220	0.3841	0.0876	1079	1332	1691	1963	2195	Moose/ACSS/AW
20,800	0.0731	0.0775	0.0852	0.0927	0.0431	0.3816	0.0864	1160	1436	1829	2127	2382	Bunting/ACSS/AW
30,800	0.0721	0.0760	0.0850	0.0929	0.0451	0.3759	0.0856	1168	1445	1839	2137	2393	Grackle/ACSS/AW
21,600	0.0685	0.0730	0.0802	0.0872	0.0445	0.3777	0.0855	1207	1495	1906	2218	2486	Bittern/ACSS/AW
25,300	0.0681	0.0724	0.0795	0.0865	0.4520	0.3758	0.0852	1214	1505	1918	2233	2502	Diver/ACSS/AW
32,800	0.0676	0.0715	0.0799	0.0873	0.0466	0.3720	0.0847	1215	1505	1917	2230	2498	Pheasant/ACSS/AW
23,000	0.0645	0.0691	0.0758	0.0824	0.0459	0.3740	0.0846	1251	1552	1981	2307	2587	Dipper/ACSS/AW
34,900	0.0636	0.0676	0.0755	0.0824	0.0480	0.3684	0.0838	1261	1563	1993	2319	2600	Martin/ACSS/AW
24,300	0.0609	0.0656	0.0719	0.0781	0.0472	0.3705	0.0837	1295	1608	2054	2394	2687	Bobolink/ACSS/AW
36,900	0.0601	0.0641	0.0716	0.0781	0.0494	0.3649	0.0829	1305	1619	2067	2408	2700	Plover/ACSS/AW
25,700	0.0577	0.0625	0.0685	0.0743	0.0485	0.3673	0.0829	1337	1662	2125	2479	2783	Nuthatch/ACSS/AW
38,900	0.0569	0.0611	0.0680	0.0742	0.0508	0.3616	0.0821	1348	1674	2139	2493	2798	Parrot/ACSS/AW
22,700	0.0551	0.0603	0.0657	0.0711	0.0490	0.3660	0.0824	1370	1704	2182	2547	2861	Ratite/ACSS/AW
27,000	0.0548	0.0597	0.0654	0.0709	0.0497	0.3641	0.0822	1379	1715	2196	2563	2879	Lapwing/ACSS/AW
41,100	0.0540	0.0583	0.0649	0.0707	0.0521	0.3585	0.0814	1391	1728	2210	2578	2895	Falcon/ACSS/AW
33,600	0.0490	0.0540	0.0587	0.0635	0.0534	0.3554	0.0803	1480	1846	2371	2774	3122	Chukar/ACSS/AW
26,500	0.0434	0.0494	0.0535	0.0577	0.0553	0.3513	0.0789	1573	1967	2536	2975	3354	Mockingbird/ACSS/AW
30,300	0.0428	0.0486	0.0527	0.0568	0.0562	0.3494	0.0785	1591	1990	2566	3009	3394	Roadrunner/ACSS/AW
40,700	0.0404	0.0460	0.0498	0.0537	0.0588	0.3439	0.0775	1652	2067	2668	3132	3534	Bluebird/ACSS/AW
28,200	0.0407	0.0470	0.0508	0.0547	0.0570	0.3475	0.0779	1630	2041	2635	3094	3493	Kiwi/ACSS/AW
34,100	0.0380	0.0443	0.0478	0.0514	0.0595	0.3423	0.0768	1698	2129	2753	3237	3657	Thrasher/ACSS/AW
37,100	0.0350	0.0415	0.0447	0.0480	0.0621	0.3372	0.0755	1777	2233	2895	3410	3858	Joree/ACSS/AW

TABLE 1-39

VR™ CONDUCTOR

WIND INDUCED - MOTION RESISTANT CONDUCTORS

Code Word	Closest Standard Conductor Size kcmil or AWG	Conductor Description			Stranding		Cross-Sectional Area Sq. in			Outside Dimensions in	Equivalent Diameter in
		Type	Construction	Total kcmil	Aluminum	Steel	Aluminum	Steel	Total		
Rose / VR	1	AAC	2 x #4 - 7	83.5	7 x 0.0772	- - -	0.0656	- - -	0.0656	0.232 x 0.464	0.380
Swan / VR		ACSR	2 x #4 - 6/1	83.5	6 x 0.0834	1 x 0.0834	0.0656	0.0109	0.0765	0.250 x 0.500	0.409
Lily / VR	1/0	AAC	2 x #3 - 7	105.2	7 x 0.0867	- - -	0.0827	- - -	0.0827	0.260 x 0.520	0.426
Swallow / VR		ACSR	2 x #3 - 6/1	105.2	6 x 0.0937	1 x 0.0937	0.0826	0.0138	0.0964	0.281 x 0.562	0.460
Iris / VR	2/0	AAC	2 x #2 - 7	132.7	7 x 0.0974	- - -	0.1044	- - -	0.1044	0.292 x 0.584	0.478
Sparrow / VR		ACSR	2 x #2 - 6/1	132.7	6 x 0.1052	1 x 0.1052	0.1044	0.0174	0.1218	0.316 x 0.631	0.517
Pansy / VR	3/0	AAC	2 x #1 - 7	167.4	7 x 0.1093	- - -	0.1314	- - -	0.1314	0.328 x 0.656	0.537
Robin / VR		ACSR	2 x #1 - 6/1	167.4	6 x 0.1181	1 x 0.1181	0.1314	0.0220	0.1534	0.355 x 0.710	0.581
Poppy / VR	4/0	AAC	2 x #1/0 - 7	211.2	7 x 0.1228	- - -	0.1658	- - -	0.1658	0.368 x 0.736	0.602
Raven / VR		ACSR	2 x #1/0 - 6/1	211.2	6 x 0.1327	1 x 0.1327	0.1660	0.0276	0.1936	0.398 x 0.796	0.652
Aster / VR	266.8	AAC	2 x #2/0 - 7	266.2	7 x 0.1379	- - -	0.2090	- - -	0.2090	0.414 x 0.828	0.678
Quail / VR		ACSR	2 x #2/0 - 6/1	266.2	6 x 0.1489	1 x 0.1489	0.2090	0.0348	0.2438	0.447 x 0.894	0.732
Phlox / VR	336.4	AAC	2 x #3/0 - 7	335.6	7 x 0.1548	- - -	0.2635	- - -	0.2635	0.464 x 0.928	0.760
Pigeon / VR		ACSR	2 x #3/0 - 6/1	335.6	6 x 0.1672	1 x 0.1672	0.2634	0.0440	0.3074	0.502 x 1.004	0.822
Oxlip / VR	397.5	AAC	2 x #4/0 - 7	423.2	7 x 0.1739	- - -	0.3324	- - -	0.3324	0.552 x 1.104	0.855
Penguin / VR		ACSR	2 x #4/0 - 6/1	423.2	6 x 0.1878	1 x 0.1878	0.3324	0.0554	0.3878	0.563 x 1.126	0.922
Jaeger / VR		ACSR	2 x 228.2 - 18/1	456.4	18 x 0.1126	1 x 0.1126	0.3585	0.0199	0.3784	0.563 x 1.126	0.922
Daisy / VR	477	AAC	2 x 266.8 - 7	533.6	7 x 0.1953	- - -	0.4194	- - -	0.4194	0.586 x 1.172	0.959
Waxwing / VR	AND	ACSR	2 x 266.8 - 18/1	533.6	18 x 0.1217	1 x 0.1217	0.4188	0.0232	0.4420	0.609 x 1.218	0.997
Partridge / VR	556.5	ACSR	2 x 266.8 - 26/7	533.6	26 x 0.1013	7 x 0.0788	0.4191	0.0682	0.4873	0.642 x 1.264	1.051
Tulip / VR	636	AAC	2 x 336.4 - 19	672.8	19 x 0.1331	- - -	0.5284	- - -	0.5284	0.666 x 1.332	1.090
Merlin / VR		ACSR	2 x 336.4 - 18/1	672.8	18 x 0.1367	1 x 0.1367	0.5284	0.0294	0.5578	0.684 x 1.368	1.120
Linnet / VR		ACSR	2 x 336.4 - 26/7	672.8	26 x 0.1137	7 x 0.0884	0.5280	0.0860	0.6140	0.721 x 1.442	1.180
Canna / VR	795	AAC	2 x 397.5 - 19	795	19 x 0.1447	- - -	0.6244	- - -	0.6244	0.724 x 1.448	1.185
Chickadee / VR		ACSR	2 x 397.5 - 18/1	795	18 x 0.1486	1 x 0.1486	0.6244	0.0346	0.6590	0.743 x 1.486	1.216
Ptarmigan / VR		ACSR	2 x 397.5 - 20/7	795	20 x 0.1401	7 x 0.0626	0.6246	0.0432	0.6678	0.752 x 1.504	1.231
Ibis / VR		ACSR	2 x 397.5 - 26/7	795	26 x 0.1236	7 x 0.0961	0.6240	0.1016	0.7256	0.783 x 1.566	1.282
Cosmos / VR	954	AAC	2 x 477.0 - 19	954	19 x 0.1584	- - -	0.7492	- - -	0.7492	0.793 x 1.586	1.298
Pelican / VR		ACSR	2 x 477.0 - 18/1	954	18 x 0.1628	1 x 0.1628	0.7494	0.0416	0.7910	0.814 x 1.628	1.333
Tailorbird / VR		ACSR	2 x 477.0 - 20/7	954	20 x 0.1544	7 x 0.0686	0.7490	0.0518	0.8008	0.823 x 1.647	1.347
Flicker / VR		ACSR	2 x 477.0 - 24/7	954	24 x 0.1410	7 x 0.0940	0.7494	0.0972	0.8466	0.846 x 1.692	1.305
Hawk / VR		ACSR	2 x 477.0 - 26/7	954	26 x 0.1354	7 x 0.1053	0.7488	0.1220	0.8708	0.858 x 1.716	1.405
Hen / VR		ACSR	2 x 477.0 - 30/7	954	30 x 0.1261	7 x 0.1261	0.7494	0.1748	0.9242	0.883 x 1.766	1.445
Cosmos / VR	1033.5	AAC	2 x 477.0 - 19	954	19 x 0.1584	- - -	0.4792	- - -	0.4792	0.793 x 1.586	1.298
Dahlia / VR		AAC	2 x 556.5 - 19	1113	19 x 0.1711	- - -	0.6742	- - -	0.8742	0.856 x 1.712	1.401
Nightingale / VR		ACSR	2 x 517.0 - 18/1	1033.5	18 x 0.1694	1 x 0.1694	0.8114	0.0450	0.8564	0.847 x 1.694	1.387
Creeper / VR		ACSR	2 x 517.0 - 20/7	1033.5	20 x 0.1607	7 x 0.0714	0.8114	0.0460	0.8675	0.857 x 1.714	1.403
Dahlia / VR	1113	AAC	2 x 556.5 - 19	1113	19 x 0.1711	- - -	0.8742	- - -	0.8742	0.856 x 1.712	1.401
Osprey / VR		ACSR	2 x 556.5 - 18/1	1113	18 x 0.1758	1 x 0.1758	0.8742	0.0486	0.9228	0.879 x 1.758	1.439
Tody / VR		ACSR	2 x 556.5 - 20/7	1113	20 x 0.1667	7 x 0.0741	0.8730	0.0604	0.9334	0.889 x 1.778	1.455
Parakeet / VR		ACSR	2 x 556.5 - 24/7	1113	24 x 0.1523	7 x 0.1015	0.8744	0.1132	0.9876	0.914 x 1.828	1.496

Notes: (1) ACSR data based on a nominal cable manufactured in accordance with ASTM Specifications B 911/B 911M.

(2) Resistance and ampacity based on an aluminum conductivity of 61.2% IACS at 20°C and a steel conductivity of 8% IACS at 20°C.

(3) Ampacity based on a 75°C conductor temperature, 25°C ambient temperature, 2 ft/sec wind, in sun, with an emissivity of 0.5 and a coefficient of solar absorption of 0.5, at sea level.

(4) Rated strengths for ACSR based on Class A galvanized steel core wire in accordance with ASTM B 498.

(5) Cable assembly twisted in the left hand lay direction with a 9 ft, +/- 1 ft, length of lay.

TABLE 1-39

VR™ CONDUCTOR

WIND INDUCED - MOTION RESISTANT CONDUCTORS

Weight/1000 feet				Resistance				Reactance @ 1 ft Spacing 60 Hz					
					ac-60 Hz			Inductive		Capacitive			
Aluminum lbs	Steel lbs	Total lbs	Rated Strength lbs	dc @ 20°C Ω/mile	@ 25°C Ω/mile	@ 50°C Ω/mile	@ 75°C Ω/mile	GMR_S ft	Xa Ω/mile	GMR_E ft	X'a MΩ-mile	Ampacity @ 75°C Amps	Code Word
78.3	- - -	78.3	1,762	1.0930	1.1153	1.2257	1.3361	0.0116	0.5403	0.0136	0.1273	222	Rose / VR
78.3	36.8	115.1	3,700	1.0757	1.0976	1.3231	1.3884	0.0087	0.5758	0.0147	0.1251	221	Swan / VR
98.7	- - -	98.7	2,180	0.8675	0.8854	0.9730	1.0606	0.0130	0.5265	0.0153	0.1240	256	Lily / VR
98.4	46.6	145.0	4,600	0.8538	0.8713	1.0080	1.0978	0.0098	0.5616	0.0166	0.1216	256	Swallow / VR
124.5	- - -	124.5	2,700	0.6877	0.7021	0.7715	0.8409	0.0147	0.5124	0.0172	0.1205	296	Iris / VR
124.0	58.7	182.7	5,660	0.6769	0.6909	0.8662	0.9007	0.0110	0.5474	0.0186	0.1182	291	Sparrow / VR
156.9	- - -	156.9	3,280	0.5452	0.5568	0.6118	0.6668	0.0165	0.4983	0.0193	0.1171	343	Pansy / VR
156.3	73.9	230.2	7,120	0.5366	0.5478	0.7036	0.7292	0.0123	0.5332	0.0209	0.1147	333	Robin / VR
198.1	- - -	198.1	3,980	0.4321	0.4416	0.4851	0.5287	0.0185	0.4843	0.0217	0.1136	397	Poppy / VR
197.3	93.4	290.7	8,760	0.4253	0.4344	0.5734	0.5937	0.0138	0.5194	0.0234	0.1113	380	Raven / VR
249.8	- - -	249.8	5,020	0.3428	0.3507	0.3852	0.4198	0.0208	0.4700	0.0244	0.1102	459	Aster / VR
248.4	117.5	365.9	10,600	0.3374	0.3450	0.4704	0.4879	0.0155	0.5053	0.0263	0.1079	432	Quail / VR
314.8	- - -	314.8	6,080	0.2719	0.2786	0.3059	0.3333	0.0233	0.4562	0.0273	0.1068	531	Phlox / VR
313.2	148.2	461.4	13,200	0.2677	0.2740	0.3865	0.4036	0.0175	0.4912	0.0296	0.1044	489	Pigeon / VR
397.2	- - -	397.2	7,660	0.2157	0.2214	0.2431	0.2648	0.0262	0.4419	0.0308	0.1033	614	Oxlip / VR
395.1	187.0	582.1	16,700	0.2122	0.2177	0.3174	0.3349	0.0196	0.4773	0.0332	0.1010	554	Penguin / VR
428.2	67.2	495.4	12,100	0.1992	0.2046	0.2246	0.2446	0.0291	0.4287	0.0332	0.1010	652	Jaeger / VR
501.0	- - -	501.0	9,660	0.1710	0.1763	0.1934	0.2105	0.0300	0.4253	0.0345	0.0998	710	Daisy / VR
500.3	78.5	578.8	13,700	0.1675	0.1725	0.1892	0.2059	0.0317	0.4190	0.0359	0.0987	725	Waxwing / VR
503.1	231.4	734.5	22,600	0.1700	0.1748	0.1918	0.2089	0.0341	0.4099	0.0378	0.0971	730	Partridge / VR
631.6	- - -	631.6	12,300	0.1356	0.1406	0.1541	0.1677	0.0342	0.4098	0.0392	0.0960	823	Tulip / VR
631.2	99.1	730.3	17,400	0.1351	0.1399	0.1534	0.1669	0.0356	0.4049	0.0403	0.0953	831	Merlin / VR
633.8	291.2	925.0	28,200	0.1349	0.1393	0.1527	0.1662	0.0383	0.3959	0.0425	0.0937	844	Linnet / VR
746.5	- - -	746.5	14,200	0.1148	0.1198	0.1311	0.1425	0.0371	0.3996	0.0427	0.0936	912	Canna / VR
745.9	117.1	863.0	19,900	0.1144	0.1191	0.1304	0.1418	0.0386	0.3948	0.0438	0.0928	921	Chickadee / VR
749.0	146.4	895.4	22,200	0.1148	0.1194	0.1308	0.1422	0.0392	0.0930	0.0443	0.0924	923	Ptarmigan / VR
749.0	334.1	1093.1	32,600	0.1141	0.1184	0.1298	0.1411	0.0416	0.3858	0.0461	0.0912	936	Ibis / VR
894.6	- - -	894.6	16,700	0.0957	0.1007	0.1101	0.1196	0.0407	0.3886	0.0467	0.0909	1020	Cosmos / VR
895.2	140.5	1035.7	23,600	0.0953	0.1001	0.1094	0.1189	0.0423	0.3837	0.0480	0.0901	1031	Pelican / VR
899.1	175.3	1074.4	26,200	0.0957	0.1003	0.1098	0.1192	0.0428	0.3824	0.0506	0.0885	1032	Tailorbird / VR
898.8	329.2	1228.0	34,400	0.0953	0.0997	0.1091	0.1185	0.0447	0.3771	0.0498	0.0890	1026	Flicker / VR
898.8	413.1	1311.9	39,000	0.0951	0.0994	0.1088	0.1182	0.0456	0.3747	0.0506	0.0885	1048	Hawk / VR
901.7	592.5	1494.2	47,600	0.0949	0.0990	0.1084	0.1178	0.0473	0.3702	0.0520	0.0877	1058	Hen / VR
894.6	- - -	894.6	16,700	0.0957	0.1007	0.1101	0.1196	0.0406	0.3889	0.0467	0.0909	1020	Cosmos / VR
1043.8	- - -	1043.8	19,500	0.0820	0.0873	0.0953	0.1033	0.0439	0.3793	0.0504	0.0886	1120	Dahlia / VR
969.3	152.1	1121.4	25,400	0.0880	0.0928	0.1014	0.1101	0.0440	0.3789	0.0499	0.0889	1082	Nightingale / VR
973.9	189.9	1163.8	28,400	0.0883	0.0930	0.1017	0.1104	0.0447	0.3772	0.0505	0.0886	1084	Creeper / VR
1043.8	- - -	1043.8	19,500	0.0820	0.0873	0.0953	0.1033	0.0439	0.3793	0.0504	0.0886	1120	Dahlia / VR
1043.9	163.8	1207.7	27,400	0.0817	0.0866	0.0946	0.1026	0.0457	0.3744	0.0518	0.0878	1132	Osprey / VR
1048.0	204.6	1252.6	30,600	0.0820	0.0868	0.0948	0.1029	0.0464	0.3727	0.0524	0.0875	1134	Tody / VR
1048.8	383.6	1432.4	39,600	0.0817	0.0862	0.0942	0.1022	0.0483	0.3678	0.0538	0.0867	1146	Parakeet / VR

TABLE 1-40
COPPERWELD COPPER CONDUCTORS

Size of Conductor				Copper Equiv. kcmil or AWG	Rated Strength lbs	Weight lbs/1000 ft	GMR ft	Resistance					Reactance @ 1 ft Spacing 60 Hz		Ampacity @ 75°C Amps
Size and Type Desig-nation	No. and Diameter of Wires		Outside Diam-eter in					dc			ac-60 Hz		Inductive Xa Ω/mile	Capacitive X'a MΩ-mile	
	Copper-weld	Copper						@ 20°C Ω/mile	@ 25°C Ω/mile	@ 50°C Ω/mile	@ 25°C Ω/mile	@ 50°C Ω/mile			
350 E	7x0.1576	12x0.1576	0.788	350	32,420	1403	0.02200	0.1660	0.1658	0.1812	0.1812	0.204	0.463	0.1014	660
350 EK	4x0.1470	15x0.1470	0.735	350	23,850	1238	0.02450	0.1660	0.1658	0.1812	0.1705	0.188	0.450	0.1034	680
300 E	7x0.1459	12x0.1459	0.729	300	27,770	1203	0.02040	0.1936	0.1934	0.211	0.2090	0.235	0.473	0.1037	600
300 EK	4x0.1361	15x0.1361	0.680	300	20,960	1061	0.02270	0.1936	0.1934	0.211	0.1981	0.219	0.460	0.1057	610
250 E	7x0.1332	12x0.1332	0.666	250	23,920	1002	0.01859	0.2323	0.232	0.254	0.248	0.279	0.484	0.1064	540
250 EK	4x0.1242	15x0.1242	0.621	250	17,840	884.2	0.02070	0.2323	0.232	0.254	0.237	0.261	0.471	0.1084	540
4/0 E	7x0.1225	12x0.1225	0.613	4/0	20,730	848.3	0.01711	0.2745	0.274	0.300	0.290	0.326	0.493	0.1088	480
4/0 G	2x0.1944	5x0.1944	0.583	4/0	15,640	789.4	0.01409	0.2745	0.273	0.299	0.298	0.342	0.517	0.1103	460
4/0 EK	4x0.1143	15x0.1143	0.571	4/0	15,370	748.4	0.01903	0.2745	0.274	0.300	0.279	0.308	0.481	0.1109	490
4/0 F	1x0.1833	6x0.1833	0.550	4/0	12,290	710.2	0.01558	0.2745	0.273	0.299	0.287	0.322	0.505	0.1120	470
3/0 E	7x0.1091	12x0.1091	0.545	3/0	16,800	672.7	0.01521	0.3462	0.346	0.378	0.361	0.407	0.508	0.1123	420
3/0 J	3x0.1851	4x0.1851	0.555	3/0	16,170	706.7	0.01156	0.3462	0.344	0.377	0.372	0.428	0.541	0.1118	410
3/0 G	2x0.1731	5x0.1731	0.519	3/0	12,860	626.0	0.01254	0.3462	0.344	0.377	0.369	0.423	0.531	0.1137	400
3/0 EK	4x0.1018	15x0.1018	0.509	3/0	12,370	593.5	0.01697	0.3462	0.346	0.378	0.351	0.386	0.495	0.1143	420
3/0 F	1x0.1632	6x0.1632	0.490	3/0	9,980	563.2	0.01388	0.3462	0.344	0.377	0.358	0.401	0.519	0.1155	410
2/0 K	4x0.1780	3x0.1780	0.534	2/0	17,600	645.9	0.00912	0.4364	0.434	0.475	0.466	0.535	0.570	0.1129	360
2/0 J	3x0.1648	4x0.1648	0.494	2/0	13,430	560.6	0.01029	0.4364	0.434	0.475	0.462	0.530	0.555	0.1152	350
2/0 G	2x0.1542	5x0.1542	0.463	2/0	10,510	496.6	0.01119	0.4364	0.434	0.475	0.459	0.525	0.545	0.1171	350
2/0 F	1x0.1454	6x0.1454	0.436	2/0	8,094	446.8	0.01235	0.4364	0.434	0.475	0.448	0.501	0.533	0.1189	350
1/0 K	4x0.1585	3x0.1585	0.475	1/0	14,490	512.0	0.00812	0.5507	0.548	0.599	0.579	0.664	0.584	0.1164	310
1/0 J	3x0.1467	4x0.1467	0.440	1/0	10,970	444.3	0.00917	0.5507	0.548	0.599	0.576	0.659	0.569	0.1186	310
1/0 G	2x0.1373	5x0.1373	0.412	1/0	8,563	393.6	0.00996	0.5507	0.548	0.599	0.573	0.654	0.559	0.1206	310
1/0 F	1x0.1294	6x0.1294	0.388	1/0	6,536	354.1	0.01099	0.5507	0.548	0.599	0.562	0.627	0.547	0.1224	310
1 N	5x0.1546	2x0.1546	0.464	1	15,410	481.3	0.00638	0.6943	0.691	0.755	0.726	0.832	0.614	0.1171	280
1 K	4x0.1412	3x0.1412	0.423	1	11,900	406.2	0.00723	0.6943	0.691	0.755	0.722	0.825	0.598	0.1198	270
1 J	3x0.1307	4x0.1307	0.392	1	9,000	352.5	0.00817	0.6943	0.691	0.755	0.719	0.820	0.583	0.1221	270
1 G	2x0.1222	5x0.1222	0.367	1	6,956	312.2	0.00887	0.6943	0.691	0.755	0.716	0.815	0.573	0.1240	260
1 F	1x0.1153	6x0.1153	0.346	1	5,266	280.9	0.00980	0.6943	0.691	0.755	0.705	0.786	0.561	0.1258	270
2 P	6x0.1540	1x0.1540	0.462	2	16,870	471.1	0.00501	0.8754	0.871	0.952	0.909	1.040	0.643	0.1172	250
2 N	5x0.1377	2x0.1377	0.413	2	12,680	381.7	0.00568	0.8754	0.871	0.952	0.906	1.035	0.627	0.1205	240
2 K	4x0.1257	3x0.1257	0.377	2	9,730	322.1	0.00644	0.8754	0.871	0.952	0.902	1.028	0.612	0.1232	240
2 J	3x0.1164	4x0.1164	0.349	2	7,322	279.5	0.00727	0.8754	0.871	0.952	0.899	1.022	0.598	0.1255	230
2 A	1x0.1699	2x0.1699	0.366	2	5,876	256.8	0.00763	0.8754	0.869	0.950	0.882	0.979	0.592	0.1241	240
2 G	2x0.1089	5x0.1089	0.327	2	5,626	247.6	0.00790	0.8754	0.871	0.952	0.896	1.016	0.587	0.1275	230
2 F	1x0.1026	6x0.1026	0.308	2	4,233	222.8	0.00873	0.8754	0.871	0.952	0.885	0.985	0.575	0.1292	230
3 P	6x0.1371	1x0.1371	0.411	3	13,910	373.6	0.00445	1.1035	1.098	1.200	1.136	1.296	0.657	0.1207	220
3 N	5x0.1226	2x0.1226	0.368	3	10,390	302.7	0.00506	1.1035	1.098	1.200	1.133	1.289	0.641	0.1239	210
3 K	4x0.1120	3x0.1120	0.336	3	7,910	255.5	0.00574	1.1035	1.098	1.200	1.129	1.281	0.626	0.1266	210
3 J	3x0.1036	4x0.1036	0.311	3	5,955	221.7	0.00648	1.1035	1.098	1.200	1.126	1.275	0.611	0.1289	200
3 A	1x0.1513	2x0.1513	0.326	3	4,810	203.6	0.00679	1.1035	1.096	1.198	1.109	1.229	0.606	0.1275	210
4 P	6x0.1221	1x0.1221	0.366	4	11,420	296.3	0.00397	1.3918	1.385	1.514	1.423	1.616	0.671	0.1241	190
4 N	5x0.1092	2x0.1092	0.328	4	8,460	240.0	0.00451	1.3918	1.385	1.514	1.420	1.610	0.655	0.1274	180
4 D	2x0.1615	1x0.1615	0.348	4	7,340	225.5	0.00566	1.3918	1.382	1.511	1.399	1.542	0.628	0.1256	190
4 A	1x0.1347	2x0.1347	0.290	4	3,938	161.5	0.00604	1.3918	1.382	1.511	1.395	1.545	0.620	0.1310	180
5 P	6x0.1087	1x0.1087	0.326	5	9,311	234.9	0.00353	1.7376	1.747	1.909	1.785	2.020	0.685	0.1275	160
5 D	2x0.1438	1x0.1438	0.310	5	6,035	178.9	0.00504	1.7376	1.742	1.905	1.759	1.939	0.642	0.1290	160
5 A	1x0.1200	2x0.1200	0.258	5	3,193	128.1	0.00538	1.7376	1.742	1.905	1.755	1.941	0.634	0.1345	160
6 D	2x0.1281	1x0.1281	0.276	6	4,942	141.8	0.00449	2.1912	2.200	2.400	2.22	2.44	0.656	0.1325	140
6 A	1x0.1068	2x0.1068	0.230	6	2,585	101.6	0.00479	2.1912	2.200	2.400	2.21	2.44	0.648	0.1379	140
6 C	1x0.1046	2x0.1046	0.225	6	2,143	97.34	0.00469	2.1912	2.200	2.400	2.21	2.44	0.651	0.1386	130
7 D	2x0.1141	1x0.1141	0.246	7	4,022	112.5	0.00400	2.7625	2.770	3.030	2.79	3.07	0.670	0.1359	120
7 A	1x0.1266	2x0.0895	0.223	7	2,754	93.66	0.00441	2.7625	2.770	3.030	2.78	3.07	0.658	0.1388	120
8 D	2x0.1016	1x0.1016	0.219	8	3,256	89.21	0.00356	3.4837	3.490	3.820	3.51	3.86	0.684	0.1393	110
8 A	1x0.1127	2x0.07969	0.199	8	2,233	74.27	0.00394	3.4837	3.490	3.820	3.51	3.87	0.672	0.1422	100
8 C	1x0.08081	2x0.08336	0.179	8	1,362	60.67	0.00373	3.4837	3.490	3.820	3.51	3.86	0.679	0.1453	100
9 1/2 D	2x0.08081	1x0.08081	0.174	9 1/2	1,743	56.46	0.00283	4.8418	4.910	5.370	4.93	5.42	0.712	0.1462	85

Notes: (1) Data based on ASTM Specifications B229.
(2) Ampacity based on 60 Hz, a conductor temperature of 75°C, ambient temperature of 25°C, with 2 ft/sec wind.
(3) Resistances at 50°C total temperature, based on an ambient of 25°C plus 25°C rise due to the heating effect of current. The approximate magnitude of current necessary to produce the 25°C rise is 75% of the 60 Hz Ampacity at 75°C.

TABLE 1-41

COPPER-CLAD STEEL CONDUCTORS (COPPERWELD®)

Nominal Conductor Size	Number and Wire Size AWG	Conductor Size kcmil	Conductor Area Sq. in	Diameter		Weight lbs/1000 ft	Rated Strength		Resistance			GMR @ 60 Hz ft	Reactance @ 1 ft Spacing 60 Hz		Ampacity @ 125°C Amps
				Individual Wire in	Complete Conductor in		High Strength lbs	Extra High Strength lbs	dc @ 20°C Ω/mile	ac-60 Hz @ 25°C Ω/mile	ac-60 Hz @ 75°C Ω/mile		Inductive Xa Ω/mile	Capacitive X'a MΩ-mile	
30% Conductivity															
7/8"	19 No. 5	628.9	0.4940	0.1819	0.910	1770	55,570	66,910	0.300	0.331	0.499	0.00758	0.592	0.0971	620
13/16"	19 No. 6	498.8	0.3917	0.1620	0.810	1403	45,830	55,530	0.379	0.411	0.605	0.00675	0.606	0.1005	540
23/32"	19 No. 7	395.5	0.3017	0.1443	0.721	1113	37,740	45,850	0.477	0.511	0.737	0.00601	0.621	0.1040	470
21/32"	19 No. 8	313.7	0.2464	0.1285	0.642	883	31,040	37,690	0.602	0.638	0.902	0.00535	0.635	0.1074	410
9/16"	19 No. 9	248.8	0.1954	0.1144	0.572	700	25,500	30,610	0.759	0.798	1.106	0.00477	0.649	0.1109	360
5/8"	7 No.4	292.2	0.2295	0.2043	0.613	819	24,780	29,430	0.644	0.676	0.887	0.00511	0.640	0.1088	410
9/16"	7 No. 5	231.7	0.1820	0.1819	0.546	649	20,470	24,650	0.812	0.847	1.099	0.00455	0.654	0.1122	360
1/2"	7 No. 6	183.8	0.1443	0.1620	0.486	515	16,890	20,460	1.023	1.062	1.364	0.00405	0.668	0.1157	310
7/16"	7 No. 7	145.7	0.1145	0.1443	0.433	408	13,910	16,890	1.290	1.335	1.697	0.00361	0.683	0.1191	270
3/8"	7 No. 8	115.6	0.0908	0.1285	0.385	324	11,440	13,890	1.627	1.678	2.120	0.00321	0.697	0.1126	230
11/32"	7 No. 9	91.65	0.0720	0.1144	0.343	257	9,393	11,280	2.052	2.110	2.640	0.00286	0.711	0.1260	200
5/16"	7 No. 10	72.68	0.0571	0.1019	0.306	204	7,758	9,196	2.587	2.660	3.300	0.00255	0.725	0.1294	170
—	3 No. 5	99.31	0.0780	0.1819	0.392	278	9,262	11,860	1.890	1.938	2.350	0.00457	0.654	0.1221	220
—	3 No. 6	78.75	0.0619	0.1620	0.349	220	7,639	9,754	2.383	2.440	2.950	0.00407	0.668	0.1255	190
—	3 No. 7	62.45	0.0491	0.1443	0.311	175	6,291	7,922	3.005	3.070	3.710	0.00363	0.682	0.1289	160
—	3 No. 8	49.53	0.0389	0.1285	0.277	139	5,174	6,282	3.789	3.870	4.660	0.00323	0.696	0.1324	140
—	3 No. 9	39.28	0.0309	0.1144	0.247	110	4,250	5,129	4.778	4.880	5.860	0.00288	0.710	0.1358	120
—	3 No. 10	31.15	0.0245	0.1019	0.220	87	3,509	4,160	6.026	6.150	7.380	0.00257	0.724	0.1392	110
—	2	66.37	0.05213	0.2576	—	184.2	5,682	6,412	2.805	2.860	3.400	0.00427	0.662	0.1345	160
—	4	41.74	0.03278	0.2043	—	115.8	3,934	4,672	4.460	4.540	5.400	0.00339	0.690	0.1415	120
—	6	26.24	0.02061	0.1620	—	72.9	2,680	3,247	7.091	7.220	8.580	0.00269	0.718	0.1483	90
40% Conductivity															
7/8"	19 No. 5	628.9	0.4940	0.1819	0.910	1770	50,240	—	0.225	0.254	0.391	0.01175	0.539	0.0971	690
13/16"	19 No. 6	498.8	0.3917	0.1620	0.810	1403	41,600	—	0.284	0.314	0.472	0.01046	0.553	0.1005	610
23/32"	19 No. 7	395.5	0.3017	0.1443	0.721	1113	34,390	—	0.358	0.390	0.573	0.00931	0.567	0.1040	530
21/32"	19 No. 8	313.7	0.2464	0.1285	0.642	883	28,380	—	0.451	0.485	0.698	0.00829	0.582	0.1074	470
9/16"	19 No. 9	248.8	0.1954	0.1144	0.572	700	23,390	—	0.569	0.605	0.853	0.00739	0.595	0.1109	410
5/8"	7 No.4	292.2	0.2295	0.2043	0.613	819	22,310	—	0.483	0.512	0.680	0.00792	0.587	0.1088	470
9/16"	7 No. 5	231.7	0.1820	0.1819	0.546	649	18,510	—	0.609	0.640	0.840	0.00705	0.601	0.1122	410
1/2"	7 No. 6	183.8	0.1443	0.1620	0.486	515	15,330	—	0.768	0.802	1.040	0.00628	0.615	0.1157	350
7/16"	7 No. 7	145.7	0.1145	0.1443	0.433	408	12,670	—	0.968	1.006	1.291	0.00559	0.629	0.1191	310
3/8"	7 No. 8	115.6	0.0908	0.1285	0.385	324	10,460	—	1.221	1.264	1.606	0.00497	0.644	0.1126	270
11/32"	7 No. 9	91.65	0.0720	0.1144	0.343	257	8,616	—	1.539	1.588	2.00	0.00443	0.658	0.1260	230
5/16"	7 No. 10	72.68	0.0571	0.1019	0.306	204	7,121	—	1.941	1.998	2.50	0.00395	0.671	0.1294	200
—	3 No. 5	99.31	0.0780	0.1819	0.392	278	8,373	—	1.418	1.457	1.77	0.00621	0.617	0.1221	250
—	3 No. 6	78.75	0.0619	0.1620	0.349	220	6,934	—	1.787	1.833	2.22	0.00553	0.631	0.1255	220
—	3 No. 7	62.45	0.0491	0.1443	0.311	175	5,732	—	2.254	2.31	2.79	0.00492	0.645	0.1289	190
—	3 No. 8	49.53	0.0389	0.1285	0.277	139	4,730	—	2.842	2.91	3.51	0.00439	0.659	0.1324	160
—	3 No. 9	39.28	0.0309	0.1144	0.247	110	3,898	—	3.584	3.66	4.41	0.00391	0.673	0.1358	140
—	3 No. 10	31.15	0.0245	0.1019	0.220	87	3,221	—	4.519	4.62	5.55	0.00348	0.687	0.1392	120
—	3 No. 12	19.59	0.0154	0.0808	0.174	55	2,236	—	7.186	7.34	8.78	0.00276	0.715	0.1462	90
—	2	66.37	0.05213	0.2576	—	184.2	5,056	—	2.104	2.14	2.54	0.00645	0.612	0.1345	190
—	4	41.74	0.03278	0.2043	—	115.8	3,541	—	3.346	3.41	4.05	0.00506	0.640	0.1415	140
—	6	26.24	0.02061	0.1620	—	72.9	2,433	—	5.322	5.42	6.44	0.00403	0.669	0.1483	100

Notes: (1) Data based on a nominal cable manufactured in accordance with ASTM B 227 anc B 228.
(2) Ampacity based on a conductor temperature of 125°C, an ambient of 25°C, a 2ft/sec wind, sun and emissivity of 0.5.
(3) Resistance at 75°C is based on the magnitude of current equal to approximately 75% of the 60 Hz Ampacity at 125°C.
(4) Modulus of Elasticity: Strand, 23 x 10^6 psi
Wire, 24 x 10^6 psi
(5) Coefficient of Linear Expansion: 7.2 x 10^{-6} per degree F
13 x 10^{-6} per degree C
(6) Temperature Coefficient of Resistance: 0.0021 per degree F
0.0038 per degree C

Copperweld® is a registered trademark of Copperweld Corporation

TABLE 1-42
ALUMINUM-CLAD STEEL CONDUCTORS (ALUMOWELD®)

Conductor Size			Diameter				Resistance				Reactance @ 1 ft Spacing 60 Hz		
								ac-60 Hz					
No. and Size of Wires	kcmil	Area Sq. in	Individual Wire in	Complete Conductor in	Weight lbs/1000 ft	Rated Strength lbs	dc @ 20°C Ω/mile	@ 25°C Ω/mile	@ 75°C Ω/mile	GMR ft	Inductive Xa Ω/mile	Capacitive X'a MΩ-mile	Ampacity @ 125°C Amps
37 No. 5	1225	0.96152	0.1819	1.270	2802.0	142,800	0.2242	0.2327	0.4788	0.00688	0.604	0.0872	640
37 No. 6	971.3	0.76264	0.1620	1.130	2222.0	120,200	0.2828	0.2935	0.5546	0.00612	0.618	0.0907	575
37 No. 7	770.3	0.60509	0.1443	1.010	1762.0	100,700	0.3566	0.3702	0.6524	0.00547	0.632	0.0940	520
37 No. 8	610.9	0.47984	0.1285	0.899	1398.0	84,200	0.4496	0.4667	0.7679	0.00487	0.646	0.0974	465
37 No. 9	484.4	0.38032	0.1144	0.801	1108.0	66,770	0.5671	0.5886	0.9144	0.00434	0.660	0.1009	420
37 No. 10	384.2	0.30174	0.1019	0.713	879.0	52,950	0.7149	0.7409	1.0930	0.00386	0.674	0.1043	370
19 No. 5	628.9	0.49438	0.1819	0.910	1430.0	73,350	0.4342	0.4507	0.7585	0.00493	0.645	0.0971	485
19 No. 6	498.8	0.39163	0.1620	0.810	1134.0	61,700	0.5475	0.5683	0.8886	0.00439	0.658	0.1005	425
19 No. 7	395.5	0.31073	0.1443	0.721	899.5	51,730	0.6906	0.7171	1.064	0.00390	0.673	0.1040	380
19 No. 8	313.7	0.24641	0.1285	0.642	713.5	43,240	0.8707	0.9038	1.280	0.00348	0.687	0.1074	335
19 No. 9	248.8	0.19530	0.1144	0.572	565.8	34,290	1.0977	1.140	1.554	0.00310	0.701	0.1109	295
19 No. 10	197.3	0.15495	0.1019	0.509	448.7	27,190	1.3844	1.437	1.896	0.00276	0.715	0.1133	260
7 No. 5	231.7	0.18193	0.1819	0.546	524.9	27,030	1.1954	1.240	1.669	0.00296	0.707	0.1122	280
7 No. 6	183.8	0.14435	0.1620	0.486	416.3	22,730	1.4800	1.536	2.01	0.00263	0.721	0.1157	250
7 No. 7	145.7	0.11448	0.1443	0.433	330.0	19,060	1.8665	1.937	2.47	0.00234	0.735	0.1191	220
7 No. 8	115.6	0.09077	0.1285	0.385	261.8	15,930	2.3538	2.44	3.06	0.00209	0.749	0.1226	190
7 No. 9	91.65	0.07198	0.1144	0.343	207.6	12,630	2.9679	3.08	3.80	0.00186	0.763	0.1260	160
7 No. 10	72.68	0.05708	0.1019	0.306	164.7	10,020	3.7425	3.88	4.73	0.00166	0.777	0.1294	140
3 No. 5	99.31	0.07796	0.1819	0.392	224.5	12,230	2.7335	2.78	3.56	0.00294	0.707	0.1221	170
3 No. 6	78.75	0.06185	0.1620	0.349	178.1	10,280	3.4468	3.51	4.41	0.00262	0.721	0.1255	150
3 No. 7	62.45	0.04905	0.1443	0.311	141.2	8,621	4.3465	4.42	5.47	0.00233	0.735	0.1289	130
3 No. 8	49.53	0.03890	0.1285	0.277	112.0	7,206	5.4806	5.58	6.82	0.00208	0.749	0.1324	110
3 No. 9	39.28	0.03085	0.1144	0.247	88.8	5,715	6.9115	7.04	8.52	0.00185	0.763	0.1358	100
3 No. 10	31.15	0.02446	0.1019	0.220	70.4	4,532	8.7173	8.87	10.67	0.00165	0.777	0.1392	80

Notes:
(1) All data based on a nominal cable manufactured in accordance with ASTM B 416.
(2) The resistance at 75°C is based on an ambient temperature of 25°C plus 50°C rise due to the heating effect of current. The approximate magnitude of current necessary to produce the 50°C rise is 75% of the 60 Hz Ampacity at 125°C .
(3) The Ampacity is based on a strand temperature of 125°C and an ambient of 25°C.
(4) Modulus of Elasticity: Strand, 23 x 10^6 psi
Wire, 23.5 x 10^6 psi
(5) Coefficient of Linear Expansion: 7.2 x 10^{-6} per degree F
13 x 10^{-6} per degree C
(6) Temperature Coefficient of Resistance: 0.0020 per degree F
0.0036 per degree C

® Alumoweld is a trademark of United States Alumoweld Company

TABLE 1-43

HIGH-STRENGTH, UTILITIES GRADE, AND EXTRA-HIGH-STRENGTH STEEL CONDUCTORS

Size	Stranding	Area	Diameter		Weight	Rated	Resistance					Reactance @ 1 ft Spacing, 60 Hz				
							dc	ac-60 Hz				Inductive, Xa				Capacitive
in		Sq. in	Strand in	Complete Cable in	lbs/1000 ft	Strength lbs	20°C Ω/mile	10 Amps Ω/mile	40 Amps Ω/mile	70 Amps Ω/mile	100 Amps Ω/mile	10 Amps Ω/mile	40 Amps Ω/mile	70 Amps Ω/mile	100 Amps Ω/mile	X'a MΩ-mile
High-Strength Steel-Class A Galvanizing (1)																
5/8"	7	0.2356	0.207	0.621	813	29,600	2.15	2.22	2.27	2.44	2.60	0.97	1.12	1.26	1.39	0.1083
1/2"	7	0.1497	0.165	0.495	517	18,800	3.38	3.48	3.57	3.84	4.08	1.01	1.24	1.44	1.63	0.1150
7/16"	7	0.1156	0.145	0.435	399	14,500	4.38	4.51	4.62	4.97	5.28	1.04	1.33	1.58	1.81	0.1188
3/8"	7	0.0792	0.120	0.360	273	10,800	6.40	6.59	6.75	7.26	7.71	1.10	1.50	1.84	2.15	0.1244
Utilities Grade Steel (1) (2)																
7/16"	7	0.1156	0.145	0.435	399	18,000	4.53	4.66	4.74	5.01	5.36	1.00	1.28	1.52	1.74	0.1188
3/8"	7	0.0792	0.120	0.360	273	11,500	6.40	6.59	6.75	7.26	7.71	1.10	1.50	1.84	2.15	0.1244
Extra-High-Strength Steel-Class A Galvanizing (1)																
5/8"	7	0.2356	0.207	0.621	813	42,400	2.22	2.28	2.32	2.46	2.62	0.93	1.08	1.21	1.35	0.1083
1/2"	7	0.1497	0.165	0.495	517	26,900	3.50	3.60	3.66	3.87	4.13	0.98	1.20	1.39	1.57	0.1150
7/16"	7	0.1156	0.145	0.435	399	20,800	4.53	4.66	4.74	5.01	5.36	1.00	1.28	1.52	1.74	0.1188
3/8"	7	0.0792	0.120	0.360	273	15,400	6.63	6.81	6.93	7.33	7.83	1.06	1.44	1.77	2.07	0.1244
Extra-High-Strength Steel-Class C Galvanizing (1)																
7/16"	7	0.1156	0.145	0.435	399	20,800	—	3.68	3.69	3.73	3.77	0.83	0.85	0.89	0.95	0.1188
3/8"	7	0.0792	0.120	0.360	273	15,400	—	5.42	5.43	5.45	5.51	0.82	0.84	0.88	0.93	0.1244
5/16"	7	0.0595	0.104	0.312	205	11,200	—	6.84	6.84	6.87	6.89	0.85	0.87	0.94	1.04	0.1287

Notes: (1) Data based on a nominal cable manufactured in accordance with ASTM A 363 and ASTM A 475.

(2) Data accumulated by EPRI, and published in "The EPRI Transmission Line Reference Book - 345 KV and Above / Second Edition."

(3) — Data not available.

2 SAG AND TENSION OF CONDUCTORS

Overhead lines must be designed and constructed such that the conductors remain at safe distances from buildings, objects, people and vehicles passing beneath the line. Overhead conductors, elongate over time, with temperature and tension, thereby changing their original sag after installation. To ensure safety, the shape of the terrain along the right-of-way, the height and lateral position of the conductor support points, and the position of the conductor between support points under all wind, ice and temperature conditions must be known.

Bare overhead transmission and distribution conductors are typically flexible and uniform in weight along their length. Because of these characteristics, they take the form of a catenary[1][2] between support points. The shape of the catenary changes with conductor temperature, ice and wind loading, and time. To ensure adequate vertical and horizontal clearance under all weather and electrical loadings, and to ensure that the breaking strength of the conductor is not exceeded, the behavior of the conductor catenary under all conditions must be incorporated into the line design. The required prediction of the future behavior of the conductor is determined through calculations commonly referred to as sag-tension calculations.

Sag-tension calculations predict the behavior of conductors based on recommended tension limits under varying load conditions. These tension limits may be specified as percentages of the conductor's rated breaking strength (RBS) that are not to be exceeded upon installation or during the life of the line. These conditions, along with the elastic and permanent elongation properties of the conductor, provide the basis for the determination of the amount of resulting sag during installation and long-term operation of the line.

Accurately determined initial sag limits are essential in the line design process. Final sags and tensions depend on initial installed sags and tensions and on proper handling during installation. The final sag of conductors is used to select support point heights and span lengths so that the minimum clearances will be maintained over the life of the line. If the conductor is damaged or the initial sags are incorrect, the line clearances may be violated or the conductor may break during heavy ice or wind loadings.

FIGURE 2-1
CATENARY VARIATIONS WITH CONDUCTOR TEMPERATURE, ICE AND WIND LOADS, AND TIME AFTER INSTALLATION, WHERE T_{MAX} IS THE MAXIMUM CONDUCTOR TEMPERATURE

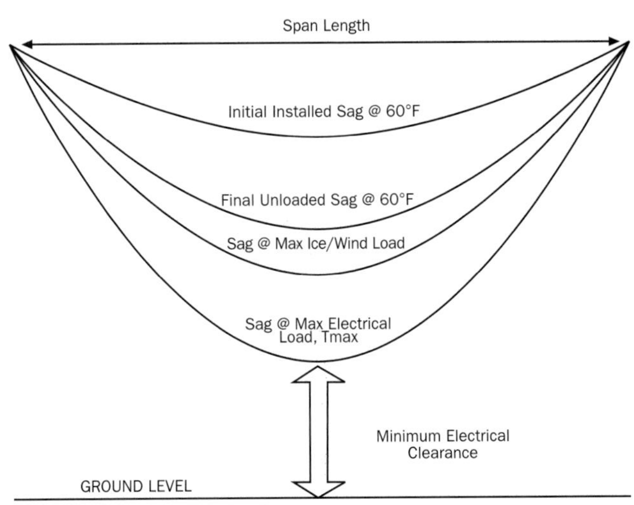

2.1 THE CATENARY OF A SINGLE SPAN

A bare stranded overhead conductor is normally held clear of objects, people and other conductors by periodic attachment to structures. The elevation differences between the supporting structures affect the shape of the conductor catenary. The catenary's shape has a distinct effect on the sag and tension of the conductor, which can be determined using well-defined mathematical equations.

2.1.1 *Level Spans*

The shape of the catenary is a function of the conductor weight per unit length, *w*, the horizontal component of tension, *H*, span length, *S*, and the sag of the conductor, *D*. Sag is measured vertically from a line through the points of conductor support to a line tangent to the conductor. Conductor sag and span length are illustrated in Figure 2-2 for a level span.

FIGURE 2-2
THE CATENARY CURVE FOR LEVEL SPANS

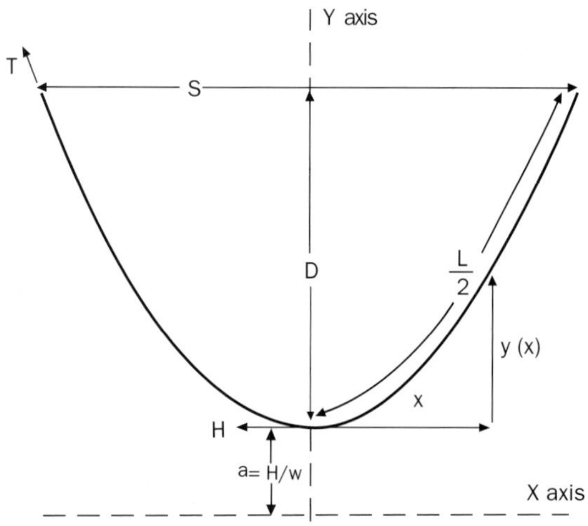

The exact catenary equation uses hyperbolic functions. Relative to the low point of the catenary curve shown in Figure 2-2, the height of the conductor, *y(x)*, is given by the following equation:

$$y(x) = \frac{H}{w}\left[\cosh\left(\frac{wx}{H}\right) - 1\right] \cong \frac{wx^2}{2H} \qquad (2\text{-}1)$$

Note that *x* is positive in either direction from the low point of the catenary. The expression on the right of Equation 2-1 is an approximate parabolic equation based upon a MacLaurin expansion of the hyperbolic cosine. This simpler parabolic equation is valid as long as:

$$\frac{x^2 w^2}{12 H^2} << 1 \qquad (2\text{-}2)$$

For a level span, the low point is in the center, and the sag, *D*, is found by substituting *S/2* for *x* in Equation 2-1. The equations for the catenary and parabolic approximation for sag become the following:

$$D = \frac{H}{w}\left[\cosh\left(\frac{wS}{2H}\right) - 1\right] \cong \frac{wS^2}{8H} \qquad (2\text{-}3)$$

The ratio, *H/w*, which appears in all of the preceding equations, is commonly referred to as the catenary constant. An increase in the catenary constant causes the catenary curve to become more shallow and the sag to decrease. Although it varies with conductor temperature, and ice and wind loading, the catenary constant typically has a value in the range of several thousand feet for most transmission-line catenaries.

The parabolic approximation for sag, *D*, is accurate as long as the sag is less than 5% of the span length.

As an example, consider a 1000 feet span of 795 kcmil - 26/7 ACSR "Drake" conductor (w = 1.094 lbs/ft) installed at a tension of 6300 lbs. The catenary constant equals 5759 feet. The calculated sag is 21.72 feet using the catenary equation and 21.71 feet using the parabolic equation. Both estimates indicate a sag-to-span ratio of 2.17% and the sag difference is only 0.12 inches (0.01 feet).

2.1.2 Conductor Length

Application of calculus to the catenary equation allows the calculation of the conductor length, $L(x)$, measured along the conductor from the low point of the catenary in either direction.

The resulting equation becomes:

$$L(x) = \frac{H}{w} \sinh\left(\frac{wx}{H}\right) \cong x\left(1 + \frac{x^2 w^2}{6H^2}\right) \qquad (2\text{-}4)$$

For a level span, the conductor length corresponding to $x = S/2$ is half of the total conductor length, L:

$$L = \left(\frac{2H}{w}\right) \sinh\left(\frac{Sw}{2H}\right) \cong S\left(1 + \frac{S^2 w^2}{24H^2}\right) \qquad (2\text{-}5)$$

A parabolic equation for conductor length can also be expressed as a function of sag, D, by substitution of the parabolic equation for sag into Equation 2-5:

$$L \cong S + \frac{8D^2}{3S} \qquad (2\text{-}6)$$

For the example in Section 2.1.1, the total conductor length is:

$$L = \left(\frac{(2)(6300)}{1.094}\right) \sinh\left(\frac{(1000)(1.094)}{(2)(6300)}\right) = 1001.257 \text{ ft}$$

In this case the total conductor length is only 1.257 feet (about 0.126%) longer than the horizontal span length of 1000 feet.

The parabolic approximation gives a total conductor length of 1001.256 feet.

2.1.3 Conductor Slack

The difference between the conductor length, L, and the span length, S, is called slack. The equation for slack may be found by equating and rearranging the preceding equations for conductor length, L:

$$L - S = \left(\frac{2H}{w}\right) \sinh\left(\frac{wS}{2H}\right) - S \qquad (2\text{-}7)$$

While slack has units of length, it may also be expressed as a percentage of the span length. For a 1000 feet span of ACSR "Drake" (1.094 lbs/ft) installed at a horizontal tension of 6300 lbs, the exact length of the conductor is 1001.257 feet and the slack is therefore 1.257 feet or 0.126% of the span.

As with previous equations, a parabolic equation in terms of sag, D, can be developed for slack. This offers some additional insight into the interaction of slack, span length and tension that the exact equation does not. For a level span:

$$L - S \cong S^3\left(\frac{w^2}{24H^2}\right) \cong D^2\left(\frac{8}{3S}\right) \qquad (2\text{-}8)$$

Note that slack is related to the cube of the span length for a given H/w ratio and to the square of sag for a given span. For a series of spans having the same H/w ratio, the total slack is largely determined by the longest spans. It is for this reason that the ruling span is nearly equal to the longest span, rather than the average span, in a series of suspension spans.

Equation 2-8 can be solved to obtain other mathematical relationships, which show the dependence of sag, D, and tension, H, upon slack, $(L - S)$:

$$D \cong \sqrt{\frac{3S(L-S)}{8}} \qquad (2\text{-}9)$$

and

$$H \cong \frac{wS}{2}\sqrt{\frac{S}{6(L-S)}} \qquad (2\text{-}10)$$

As can be seen from the preceding equation, small changes in slack typically yield large changes in conductor sag and horizontal tension.

FIGURE 2-3
STRESS AND SAG vs. % SLACK FOR BARE ACSR "DRAKE" IN A 1000 FEET SPAN INSTALLED TO A TENSION OF 20% RBS (6300 LBS) AT 60°F

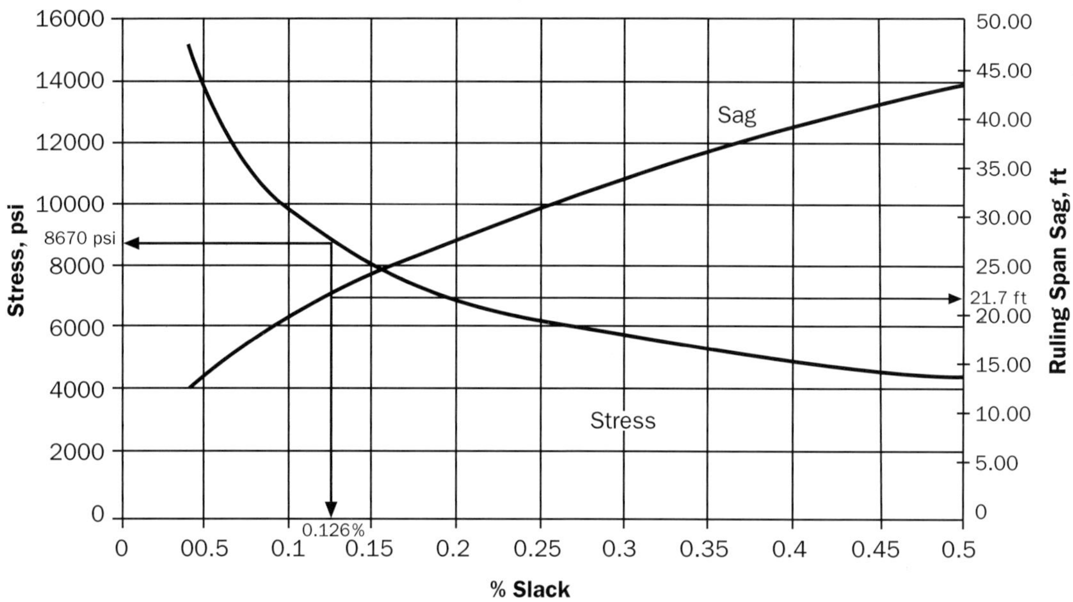

As an example of the impact of small changes in slack on tension and sag, consider the preceeding 1000 feet level span example of 795 kcmil - 26/7 ACSR "Drake" conductor with 1.257 feet of slack. If only 3 inches is added to the total conductor length, the sag and tension are approximately equal to:

$$D = \sqrt{\frac{(3)(1000)(1.257 + 0.25)}{8}} = 23.77 \ ft$$

$$H = \frac{(1.094)(1000)}{2}\sqrt{\frac{1000}{(6)(1.257 + 0.25)}}$$

$$= 5752.5 \ lbs$$

Note that in this example, a relatively small change in the conductor length (e.g. 3 inches, only 0.25% of the conductor length), yields a relatively large sag increase of 2.05 feet and a relatively large reduction in tension of 547.5 lbs, a change of approximately 9.2%. The slack concept is thus very useful in explaining how very small changes in conductor length, such as those caused by thermal, elastic or plastic elongation, have major effects on the sag of long spans.

Given span length, S, and conductor weight per unit length, w, the catenary relationship between slack (or length) and horizontal tension, H, is found with Equation 2-7 and the relationship between sag and tension is found with Equation 2-3. For the sample problem described previously with a span length of 1000 feet, the bare ACSR "Drake" conductor is installed with a horizontal tension of 6300 lbs (6300/0.7264 = 8670 psi). At this tension, the sag of the bare conductor is 21.72 feet, which corresponds to a slack of 0.126%. The variation of conductor sag and conductor stress with % slack (100 × slack/span) is shown in Figure 2-3.

2.1.4 Total and Horizontal Tension

The horizontal component of tension, H, is equal to the total conductor tension, T, at the point in the catenary where the conductor slope is horizontal. At every other location within the span, including the end supports, the total conductor tension is somewhat greater than the horizontal component.

For a level span, at the supports, the vertical component of tension, V, is equal to half the weight of the conductor in the span:

$$V = w\frac{L}{2} = H \sinh\left(\frac{wS}{2H}\right) \qquad (2\text{-}11)$$

At the end supports, the total tension is the vector sum of the horizontal and vertical tension components:

$$T^2 = H^2 + \left(\frac{wL}{2}\right)^2 \qquad (2\text{-}12)$$

Substituting the equation for length, *L*, from Section 2.1.2, and taking the square root of both sides of this equation, we obtain:

$$T = H\sqrt{1 + \sinh^2\left(\frac{wS}{2H}\right)} = H \cosh\left(\frac{wS}{2H}\right) \qquad (2\text{-}13)$$

To relate the total tension to the sag in this level span, the equation can be rearranged as:

$$T = H + H \cosh\left(\frac{wS}{2H}\right) - H \qquad (2\text{-}14)$$

$$T = H + w\left(\frac{H}{w}\cosh\left(\frac{wS}{2H}\right) - \frac{H}{w}\right) \qquad (2\text{-}15)$$

Substituting Equation (2-3) for sag, gives:

$$T \cong H + wD \qquad (2\text{-}16)$$

Given the conditions in the preceding example calculation for a 1000 feet level span of ACSR "Drake," the tension at the attachment points exceeds the horizontal component of tension by 33 lbs, less than 0.4% of the horizontal tension.

2.1.5 Inclined Spans

Inclined spans may be analyzed using essentially the same equations that were used for level spans. The catenary equation for the conductor height above the low point in the span is the same. However, the span is considered to consist of two separate sections, one to the right of the low point and the other to the left as shown in Figure 2-4. The shape of the catenary relative to the low point is unaffected by the difference in suspension point elevation (span inclination).

FIGURE 2-4
INCLINED CATENARY SPAN

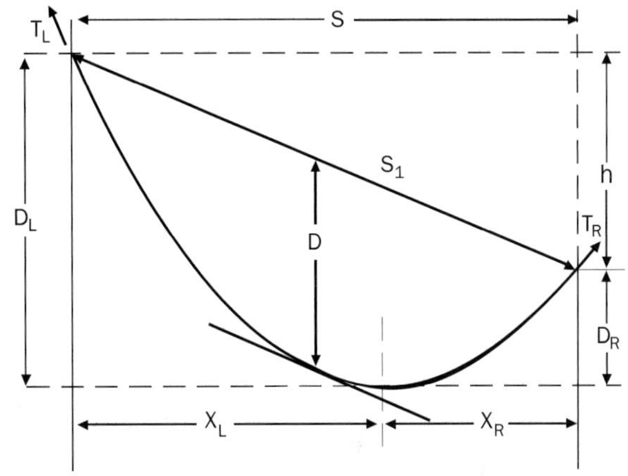

In each direction from the low point, the conductor elevation, *y(x)*, relative to the low point is given by Equation 2-1.

$$y(x) = \frac{H}{w}\left[\cosh\left(\frac{wx}{H}\right) - 1\right] \cong \frac{wx^2}{2H} \qquad (2\text{-}17)$$

and

$$D_L = D_R + h$$

Of course, the low point in an inclined span is not in the center of the horizontal span as it is with a level span. In fact the "low point" may even be beyond the end of the span depending on the difference in support elevations or conductor catenary constant.

The position of the low point in the inclined span may be calculated[3] exactly by Equation 2-18:

$$x_R = \frac{S}{2} - \frac{H}{w} \sinh^{-1}\left[\frac{\frac{h}{2}}{\frac{H}{w} \sinh \frac{S/2}{H/w}}\right] \tag{2-18}$$

Consider the following example, where:

$H = 6300\ lbs$
$h = 30\ ft$
$S = 1000\ ft$
$w = 1.094\ lbs/ft$
$H/w = 6300/1.094 = 5759\ ft$

Then the example calculation becomes:

$$x_R = \frac{1000}{2} - (5759) \sinh^{-1}\left[\frac{\frac{30}{2}}{(5759) \sinh\left(\frac{1000/2}{5759}\right)}\right]$$

$$= 327.48\ ft$$

The elevation of any point in the inclined span can be calculated using the basic catenary equation with being the horizontal distance from the low point, which for the example is 327.48 feet into the span from the right support and 672.52 feet from the left.

In the example span, as the difference in support heights increases, the low point of the catenary moves further toward the low support, eventually going beyond it as shown in Table 2-1.

In field situations, where the sag, D, has been calculated for a level span, the sag-tension data can be applied to an inclined span using the following approximate method.

The horizontal distance, x_L, from the left support point to the low point and the horizontal distance, x_R, from the right support point to the low point of the catenary is:

$$x_R \cong \frac{S}{2}\left(1 - \frac{h}{4D}\right) \qquad x_L \cong \frac{S}{2}\left(1 + \frac{h}{4D}\right) \tag{2-19}$$

where:

S = *horizontal distance between support points*
h = *vertical distance between support points*
D = *sag measured vertically from the midpoint of a line between support points of a conductor to a line tangent to the conductor*

TABLE 2-1
SAG & VERTICAL FORCE IN A 1000 FEET INCLINED SPAN
WITH ACSR "DRAKE" AT H = 6300 LBS

h [ft]	**0.00**	**10.00**	**20.00**	**30.00**	**40.00**	**50.00**	**60.00**	**70.00**	**80.00**	**90.00**
x_R [ft]	500.00	442.49	384.98	327.48	270.00	212.55	155.12	97.73	40.37	–16.94
x_L [ft]	500.00	557.51	615.02	672.52	730.00	787.45	844.88	902.27	959.63	1016.94
D_R [ft]	21.72	17.01	12.87	9.31	6.33	3.92	2.09	0.83	0.14	0.02
D_L [ft]	21.72	27.01	32.87	39.31	46.33	53.92	62.09	70.83	80.14	90.02
V_R [lbs]	548	485	421	358	295	233	170	107	44	–19
V_L [lbs]	548	611	674	737	801	864	928	991	1055	1118
T_R [lbs]	6324	6319	6314	6310	6307	6304	6302	6301	6300	6300
T_L [lbs]	6324	6330	6336	6343	6351	6359	6368	6377	6388	6398

Note:

1. *As the difference in elevation increases from 0 to 90 feet, the vertical force at the left support (V_L) increases from 548 lbs to 1118 lbs and the downward vertical force on the right support (V_R) becomes negative (uplift).*
2. *The total tension at either support (T_R or T_L) never exceeds the horizontal component by more than 98 lbs (1.6%).*
3. *When the difference in elevation equals slightly less than 90 feet, the low point of the catenary occurs at the right support point.*
4. *At greater differences in elevation, the low point position of the catenary occurs beyond the right support point.*

The midpoint sag, D, in an inclined span is approximately equal to the sag in a horizontal span whose length is equal to the inclined span length.

Knowing the horizontal distance from the low point to the support point in each direction, the preceding equations for $y(x)$, L, D and T can be applied to each side of the inclined span. The total conductor length, L, in the inclined span is equal to the sum of the lengths in the and sub-span sections:

$$L \cong S + S^3 \left(\frac{w^2}{6H^2} \right) \qquad (2\text{-}20)$$

where:

$S = x_R + x_L$

For each sub-span, the low point sag relative to the corresponding support point elevation can be calculated using the parabolic approximation in Equation 2-1 and substituting and for x.

$$D_R \cong \frac{w x_R^2}{2H} \qquad D_L \cong \frac{w x_L^2}{2H} \qquad (2\text{-}21)$$

where:

D_R = *sag relative to the right support*

D_L = *sag relative to the left support*

or in terms of sag, D, and the vertical distance between support points, h, D_R and D_L can be found by substituting Equation 2-1 into Equation 2-19:

$$D_R \cong D\left(1 - \frac{h}{4D}\right)^2 \qquad D_L \cong D\left(1 + \frac{h}{4D}\right)^2 \qquad (2\text{-}22)$$

and the maximum tensions at the support points are:

$$T_R \cong H + wD_R \qquad T_L \cong H + wD_L \qquad (2\text{-}23)$$

where:

T_R = *tension in right sub-span section*

T_L = *tension in left sub-span section*

or in terms of upper and lower support points:

$$T_u = T_l + wh \qquad (2\text{-}24)$$

where:

T_u = *tension in conductor at upper support*

T_l = *tension in conductor at lower support*

The horizontal conductor tension is equal at both supports. The vertical component of conductor tension is greater at the upper support, and the resultant tension, T_u, is also greater.

2.2 RULING SPAN CONCEPT

Transmission lines are normally designed in line sections consisting of multiple coupled spans, with each end of the section terminated by a strain structure (one that allows no longitudinal movement of the conductor at the attachment point)[2]. Structures within each line section are typically suspension structures that support the conductor vertically, but allow free movement of the conductor attachment point either longitudinally or transversely.

Sag-tension calculations are normally performed for a single span, called the ruling span, which is unique to each line section. The variation in tension with temperature, ice and wind loadings for the ruling span is assumed to apply to all the suspension spans in the line section. This simplifies clearance calculations for line design (tower spotting) and construction (stringing sags). In mountainous terrain with great variations in span length, the ruling span method may not be sufficiently accurate. In such cases, a detailed force balance assessment can be performed at the support points using finite element methods.

2.2.1 Ruling Span Calculation

Sag-tension can be found for a series of suspension spans in a line section by use of the ruling span concept[1][2]. The ruling span for the line section is defined by the following equation:

$$RS = \sqrt{\frac{S_1^3 + S_2^3 + \cdots + S_n^3}{S_1 + S_2 + \cdots + S_n}} \qquad (2\text{-}25)$$

where:

- RS = *Ruling span for the line section containing n suspension spans*
- S_1 = *Span length of first suspension span*
- S_2 = *Span length of second suspension span*
- S_n = *Span length of nth suspension span*

Alternatively, a generally satisfactory method for estimating the ruling span is to take the sum of the average suspension span length plus two-thirds of the difference between the maximum span and the average span. However, some judgment must be exercised in using this method because a large difference between the average and maximum span may cause a substantial error in the ruling span value.

As discussed, suspension spans are supported by suspension insulators that are free to move in the direction of the conductor axis. This freedom of movement allows the assumption that the tension is each suspension span is the same and equal to that calculated for the ruling span. This assumption is valid for the suspension spans and ruling span under the same conditions of temperature and load, for both initial and final sags. For level spans, sag in each suspension span is given by the parabolic sag equation:

$$D_i \cong \frac{w\,(S_i^2)}{8H_{RS}} \qquad (2\text{-}26)$$

where:

- w = *weight of conductor, lbs/ft*
- D_i = *sag in the ith span, ft*
- S_i = *span length of the ith span, ft*
- H_{RS} = *tension from ruling span sag-tension calculations, lbs*

The sag in level suspension spans may also be calculated using the ratio:

$$D_i = D_{RS}\left(\frac{S_i}{S_{RS}}\right)^2 \qquad (2\text{-}27)$$

Suspension spans vary in length, though typically not over a large range. If the sag in any suspension span exceeds approximately 5% of the span length, a correction factor should be added to the sags obtained from Equation 2-27, or the sag should be calculated using the catenary Equation 2-29. This correction factor may be calculated as follows:

$$Correction = D^2\left(\frac{w}{6H}\right) \qquad (2\text{-}28)$$

where:

- D = *sag obtained from parabolic equation, ft*
- w = *weight of conductor, lbs/ft*
- H = *horizontal tension, lbs*

The catenary equation for calculating the sag in a suspension or stringing span is:

$$D = \frac{H}{w}\left[\cosh\left(\frac{wS}{2H}\right) - 1\right] \qquad (2\text{-}29)$$

where:

- D = *sag in suspension span, ft*
- S = *span length of suspension span, ft*

2.2.2 Stringing Sag Tables

Conductors are typically installed in line section lengths consisting of multiple spans. The conductor is pulled from the conductor reel at a point near one strain structure, progressing through travelers attached to each suspension structure, to a point near the next strain structure. After stringing, the conductor tension is increased until the sag in one or more suspension spans reaches the appropriate stringing sags based on the ruling span for the line section. The calculation of stringing sags is based on Equation 2-29.

TABLE 2-2

STRINGING SAG TABLE FOR 795 KCMIL - 26/7 ACSR "DRAKE" 600 FEET RULING SPAN

600 ft Ruling Span

Controlling Design Conditions
15% RBS at 60°F, No Ice or Wind, Final

NESC Heavy Loading District

Horizontal Tension lbs	6493	6193	5910	5645	5397	5166	4952	4753	4569
Temp,°F	**20**	**30**	**40**	**50**	**60**	**70**	**80**	**90**	**100**
Spans	**Sag, ft-in**	**Sag, ft-in**	**Sag, ft-in**	**Sag, ft-in**	**Sag, ft-in**	**Sag, ft-in**	**Sag, ft-in**	**Sag, ft-in**	**Sag, ft-in**
400	3-4	3-6	3-8	3-11	4-1	4-3	4-5	4-7	4-9
410	3-6	3-9	3-11	4-1	4-3	4-5	4-8	4-10	5-0
420	3-9	3-11	4-1	4-3	4-6	4-8	4-10	5-1	5-3
430	3-11	4-1	4-3	4-6	4-8	4-11	5-1	5-4	5-6
440	4-1	4-3	4-6	4-8	4-11	5-2	5-4	5-7	5-10
450	4-3	4-6	4-8	4-11	5-2	5-4	5-7	5-10	6-1
460	4-5	4-8	4-11	5-2	5-4	5-7	5-10	6-1	6-4
470	4-8	4-11	5-1	5-4	5-7	5-10	6-1	6-4	6-7
480	4-10	5-1	5-4	5-7	5-10	6-1	6-4	6-8	6-11
490	5-1	5-4	5-7	5-10	6-1	6-4	6-8	6-11	7-2
500	5-3	5-6	5-9	6-1	6-4	6-7	6-11	7-2	7-6
510	5-6	5-9	6-0	6-4	6-7	6-11	7-2	7-6	7-9
520	5-8	6-0	6-3	6-7	6-10	7-2	7-6	7-9	8-1
530	5-11	6-2	6-6	6-10	7-1	7-5	7-9	8-1	8-5
540	6-2	6-5	6-9	7-1	7-5	7-9	8-1	8-5	8-9
550	6-4	6-8	7-0	7-4	7-8	8-0	8-4	8-8	9-1
560	6-7	6-11	7-3	7-7	7-11	8-4	8-8	9-0	9-5
570	6-10	7-2	7-6	7-10	8-3	8-7	9-0	9-4	9-9
580	7-1	7-5	7-9	8-2	8-6	8-11	9-4	9-8	10-1
590	7-4	7-8	8-1	8-5	8-10	9-3	9-7	10-0	10-5
600	7-7	7-11	8-4	8-9	9-1	9-6	9-11	10-4	10-9
610	7-10	8-3	8-7	9-0	9-5	9-10	10-3	10-9	11-2
620	8-1	8-6	8-11	9-4	9-9	10-2	10-7	11-1	11-6
630	8-4	8-9	9-2	9 -7	10-1	10-6	11-0	11-5	11-11
640	8-8	9-1	9-6	9-11	10-5	10-10	11-4	11-9	12-3
650	8-11	9-4	9-9	10-3	10-9	11-2	11-8	12-2	12-8
660	9-2	9-7	10-1	10-7	11-1	11-6	12-0	12-6	13-1
670	9-5	9-11	10-5	10-11	11-5	11-11	12-5	12-11	13-5
680	9-9	10-3	10-8	11-2	11-9	12-3	12-9	13-4	13-10
690	10-0	10-6	11-0	11-6	12-1	12-7	13-2	13-8	14-3
700	10-4	10-10	11-4	11-11	12-5	13-0	13-6	14-1	14-8

Table 2-2 shows a typical stringing sag table for a 600 feet ruling span of ACSR "Drake" with suspension spans ranging from 400 to 700 feet and conductor temperatures of 20°F to 100°F. All values in this stringing table are calculated from the ruling span initial tensions shown in Table 2-3 using the parabolic sag equation.

2.2.3 Tension Differences for Adjacent Dead-End Spans

Table 2-4 contains initial and final sag-tension data for 700 feet and 1000 feet dead-end spans of ACSR "Drake" conductor initially installed to a 6300 lbs tension limit at 60°F. Note that the difference between final tensions at 60°F is approximately 260 lbs[4]. Also note that the initial tensions for the spans are equal at 60°F, but differ by almost 900 lbs at -20°F and 600 lbs at 167°F.

2.2.4 Tension Equalization by Suspension Insulators

At a suspension structure, the conductor attachment point on the bottom of the insulator assembly supports the conductor vertically, but allows limited movement in the direction of the line along the conductor axis. Because of the relatively free axial movement of the support points, any difference in conductor tension between suspension spans is typically small and the difference in sags from span to span is proportional to the span lengths squared.

TABLE 2-3

SAG TENSION DATA FOR 795 KCMIL - 26/7 ACSR "DRAKE" 600 FEET RULING SPAN

795 kcmil-26/7 ACSR "Drake"
Area = 0.7264 in^2
Dia. = 1.108 in
Weight = 1.094 lbs/ft
Strength = 31,500 lbs
Creep is ***not*** a factor

Span = 600 ft

NESC Heavy Loading District

Temp. °F	**Ice in**	**Wind lbs/ft^2**	**K lbs/ft**	**Resultant Weight lbs/ft**	**Final Sag ft**	**Final Tension lbs**	**Initial Sag ft**	**Initial Tension lbs**
0	0.50	4.00	0.30	2.509	11.14	10153	11.14	10153
32	0.50	0.00	0.00	2.094	11.54	8185	11.09	8512
–20	0.00	0.00	0.00	1.094	6.68	7372	6.27	7855
0	0.00	0.00	0.00	1.094	7.56	6517	6.89	7147
30	0.00	0.00	0.00	1.094	8.98	5490	7.95	6197
60	0.00	0.00	0.00	1.094	10.44	4725*	9.12	5402
90	0.00	0.00	0.00	1.094	11.87	4157	10.36	4759
120	0.00	0.00	0.00	1.094	13.24	3727	11.61	4248
167	0.00	0.00	0.00	1.094	14.29	3456	13.53	3649
212	0.00	0.00	0.00	1.094	15.24	3241	15.24	3241

* Design Condition

TABLE 2-4
TENSION DIFFERENCES IN ADJACENT STRAIN SPANS

795 kcmil-26/7 ACSR "Drake" **Span = 700 ft**
Area = 0.7264 in^2
Dia. = 1.108 in
Weight = 1.094 lbs/ft
Strength = 31,500 lbs
Creep is ***not*** a factor *NESC Heavy Loading District*

Temp. °F	Ice in	Wind lbs/ft^2	K lbs/ft	Resultant Weight lbs/ft	Final		Initial	
					Sag ft	Tension lbs	Sag ft	Tension lbs
0	0.50	4.00	0.30	2.509	13.61	11318	13.55	11361
32	0.50	0.00	0.00	2.094	13.93	9224	13.33	9643
-20	0.00	0.00	0.00	1.094	8.22	8161	7.60	8824
0	0.00	0.00	0.00	1.094	9.19	7301	8.26	8115
30	0.00	0.00	0.00	1.094	10.75	6242	9.39	7142
60	0.00	0.00	0.00	1.094	12.36	5429	10.65	6300*
90	0.00	0.00	0.00	1.094	13.96	4809	11.99	5596
120	0.00	0.00	0.00	1.094	15.52	4330	13.37	5020
167	0.00	0.00	0.00	1.094	16.97	3960	15.53	4326
212	0.00	0.00	0.00	1.094	18.04	3728	17.52	3837

* Design Condition

795 kcmil-26/7 ACSR "Drake" **Span = 1000 ft**
Area = 0.7264 in^2
Dia. = 1.108 in
Weight = 1.094 lbs/ft
Strength = 31,500 lbs
Creep is ***not*** a factor *NESC Heavy Loading District*

Temp. °F	Ice in	Wind lbs/ft^2	K lbs/ft	Resultant Weight lbs/ft	Final		Initial	
					Sag ft	Tension lbs	Sag ft	Tension lbs
0	0.50	4.00	0.30	2.509	25.98	12116	25.98	12116
32	0.50	0.00	0.00	2.094	26.30	9990	25.53	10290
-20	0.00	0.00	0.00	1.094	18.72	7318	17.25	7940
0	0.00	0.00	0.00	1.094	20.09	6821	18.34	7469
30	0.00	0.00	0.00	1.094	22.13	6197	20.04	6840
60	0.00	0.00	0.00	1.094	24.11	5689	21.76	6300*
90	0.00	0.00	0.00	1.094	26.04	5271	23.49	5839
120	0.00	0.00	0.00	1.094	27.89	4923	25.20	5444
167	0.00	0.00	0.00	1.094	30.14	4559	27.82	4935
212	0.00	0.00	0.00	1.094	31.47	4369	30.24	4544

* Design Condition

2.3 CONDUCTOR TENSION LIMITS AND LOADING

The most widely accepted tension limit recommendations are found in the National Electrical Safety Code (NESC)[5]. These tension limits, specified as a percentage of rated breaking strength, are specified for each loading district as defined in the NESC.

Other tension limit recommendations can be found in the RUS Bulletin 1724E-200[6]. The key difference in RUS (previously REA) and NESC limits is that RUS tension limits are all applied at the temperature of the loading district. This results in the RUS limits being slightly more conservative than the NESC limits.

NESC limits are the industry "standard" and will be used throughout this book. However, it should be kept in mind that the NESC limits are minimum design requirements. Many utilities, while adopting the NESC as their general "standard," have even greater restrictions on tension limits. Utilities may also specify tension limits for extreme ice and wind loading conditions.

Utilities often restrict final unloaded tensions, especially in areas subject to aeolian vibration (AV). It is not uncommon to see final unloaded tension restrictions of 15% to 20% of RBS, along with installation of vibration dampeners, in line sections subject to severe vibration. The appropriate tension limits depend on type of conductor, terrain,

TABLE 2-5
NESC RECOMMENDED MAXIMUM TENSION LIMITS

NESC Tension Limits	Light Loading		Medium Loading		Heavy Loading	
	Temp. °F	Tension %RBS	Temp. °F	Tension %RBS	Temp. °F	Tension %RBS
Initial Loaded	30	60	15	60	0	60
Initial Unloaded	60	35	60	35	60	35
Final Unloaded	60	25	60	25	60	25

TABLE 2-6
RUS RECOMMENDED MAXIMUM TENSION LIMITS

REA/Alcoa Tension Limits	Light Loading		Medium Loading		Heavy Loading	
	Temp. °F	Tension %RBS	Temp. °F	Tension %RBS	Temp. °F	Tension %RBS
Initial Loaded	30	50	15	50	0	50
Initial Unloaded	30	33.3	15	33.3	0	33.3
Final Unloaded	30	25	15	25	0	25

span length and damper effectiveness. Some conductor types, such as ACSS, ACSS/TW, SDC and VR™, have superior self-dampening properties and may be installed to the maximum recommended tension limit, if so desired.

The use of an "everyday" tension limit to control AV to safe levels, expressed as a percentage of the conductor's rated breaking strength, has recently been questioned in a series of CIGRE technical reports [7][8][9]. Using the %RBS criteria as a vibration constraint does not consider the following important fatigue issues:

- The effect of wind speed persistency and direction relative to the line section;
- Terrain covers of trees, shrubs and buildings and the terrain itself, which have a marked effect on the degree of turbulence;
- Occurrence of high tensions at very low-temperature operation during winter months;
- The effect of conductor stranding and the content of steel on mass and strength in ACSR;
- Conductor diameter, which influences the energy induced by the wind and the frequency of vortex shedding.

In light of the above issues, it has been suggested that a tension-over-weight per unit length criterion, H/w, be used instead of %RBS, and that terrain and span length be considered when choosing safe conductor tensions. Recommended maximum values of H/w range from 3000 to 5000 feet. As can be seen in Table 2-7, for ACSR "Drake" in a 1000 feet span, this would correspond to an unloaded initial tension of between 10% and 15% RBS. If dampers are used, then a higher H/w level would be acceptable.

H/w limits for conventional ACSR may not be applicable to special conductors such as ACSS, GTACSR and SDC since these conductors have better self-damping characteristics than standard ACSR. Also, these limits may not apply to VR™ conductors that have non-circular cross-sections.

Given the complexity of the calculation process for estimating aeolian vibration levels, and given the availability of well documented calculation methods for the prediction of aeolian vibration levels under known terrain, span lengths, conductor type, tension, etc., it is not reasonable to use rules of thumb for setting tension limits when designing new lines in areas where there is no experience regarding aeolian vibration damage. The process of selecting suitable tension limits in conjunction with damper suppliers is the most sensible approach.

2.3.1 *Ice and Wind Loading*

In addition to tensions, the NESC also provides guidelines for ice and wind loading. When a conductor is covered with ice and/or is exposed to wind, the effective conductor weight per unit length increases. During occasions of heavy ice and/or wind load, the conductor tension increases dramatically along with the loads on angle and dead-end structures. Both the conductor and its supports can fail unless these high tension conditions are considered in the line design.

The NESC suggests certain combinations of ice and wind corresponding to heavy, medium and light loading regions of the United States. Figure 2-5 is a map of the United States indicating those areas. The combinations of ice and wind corresponding to each loading district are listed in Table 2-8.

The values shown in Table 2-8 are recommended minimum values. Certain utilities in very heavy ice areas use glaze ice thicknesses of 2 inches or more to calculate iced conductor weight. Similarly, utilities in regions where hurricane winds occur may use wind loads as high as 42 lbs/ft^2. These loads must be accounted for in the line design process to prevent a detrimental effect on the line. Some of the effects of both the individual and combined components of ice and wind loads are discussed in the following sub-sections.

TABLE 2-8
DEFINITIONS OF ICE AND WIND LOAD FOR NESC LOADING AREAS

	Loading Districts			
	Heavy	**Medium**	**Light**	**Extreme wind loading**
Radial thickness of ice (in) (mm)	0.50 12.5	0.25 6.5	0 0	0 0
Horizontal wind pressure (lbs/ft^2) (Pa)	4 190	4 190	9 430	See Fig 2-6
Temperature (°F) (°C)	0 –20	+15 –10	+30 –1	+60 +15
NESC safety factors (K_{NESC}) to be added to the resultant (lbs/ft) (N/m)	0.30 4.40	0.20 2.90	0.05 0.73	0.0 0.0

TABLE 2-7
H/W VALUES FOR ACSR "DRAKE" IN A 1000 FEET SPAN

Initial unloaded conductor tension @ 60°F %RBS	Initial unloaded H/w @ 0°F (average temperature for coldest month) ft	Tension @ 0°F with 0.5" ice and 4 lbs/ft^2 wind load %RBS	Tension @ max ice and wind load lbs	Final sag @ 212°F ft
10	3040	22.6	7,115	49.1
15	4860	31.7	9,980	37.0
20	6825	38.4	12,090	31.5
25	8730	43.5	13,710	28.3

Ice Loading

The formation of ice on overhead conductors is usually in the form of glaze ice or rime ice. The impact of lower density rime ice formation is usually considered in the design of line sections at high altitudes. The formation of ice on overhead conductors can influence:

- The maximum vertical conductor loads that structures and foundations must withstand
- The maximum transverse loads on structures since the ice build-up increases the conductor profile to the wind
- The maximum sags and the permanent increase in sag

Ice loads for use in designing lines are normally derived on the basis of past experience, code requirements, state regulations and analysis of historical weather data. Mean recurrence intervals for heavy ice loadings are a function of local conditions along various routings. The impact of varying assumptions concerning ice loading can be investigated with line design software.

The calculation of ice loads on conductors is normally done with an assumed glaze ice density of 57 lbs/ft^3. If rime ice is assumed, the density is 37 lbs/ft^3. Using a density of 57 lbs/ft^3, the weight of ice per unit length is calculated with the following equation:

$$w_i = 1.244t\ (D_c + t) \tag{2-30}$$

where:

t = *thickness of ice, in*

D_c = *conductor outside diameter, in*

w_i = *resultant weight of ice, lbs/ft*

The ratio of iced weight to bare weight depends strongly upon conductor diameter. As shown in Table 2-9, for three different conductors covered with 0.5 inches of radial glaze ice, this ratio ranges from 4.8 for 1/0 AWG to 1.6 for 1590 kcmil conductors. As a result, in climates prone to high ice loads, small diameter conductors may need to have higher elastic modulus and higher tensile strength than large conductors for the same span lengths.

FIGURE 2-5
ICE AND WIND LOAD AREAS OF THE UNITED STATES

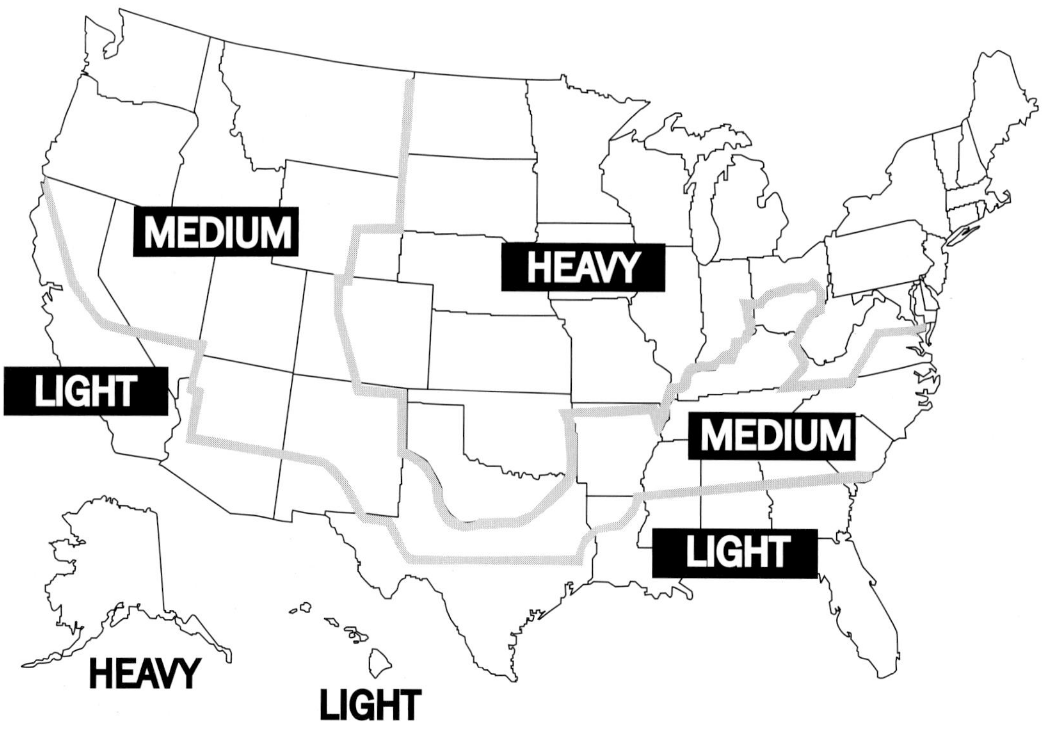

TABLE 2-9
RATIO OF ICED TO BARE CONDUCTOR WEIGHT

ACSR Conductor	D_c in	w_{bare} lbs/ft	w_{ice} lbs/ft	$\frac{w_{bare} + w_{ice}}{w_{bare}}$
#1/0 AWG-6/1 "Raven"	0.398	0.1451	0.559	4.8
477 kcmil-26/7 "Hawk"	0.858	0.6553	0.845	2.3
1590 kcmil-54/19 "Falcon"	1.545	2.042	1.272	1.6

Wind Loading

Wind loadings on overhead conductors influence line design in a number of ways:

- The maximum span between structures may be limited by the need for horizontal clearance to edge of right-of-way during moderate winds.
- The maximum transverse loads for tangent and small angle suspension structures may be determined by infrequent high wind-speed loadings.
- Permanent increases in conductor sag may occur due to wind loading in areas of light ice load.

A component of wind loading per unit length of conductor is wind pressure load. Wind pressure load is calculated:

$$P_W = 0.00256(V_W)^2 \tag{2-31}$$

where:

P_w = *wind pressure load on the conductor, lbs/ft²*

0.00256 = *ambient air density value, the mass density of air for the standard atmosphere, i.e., temperature of 15°C (59°F) and sea level pressure of 760 mm (29.92 in) of mercury*

V_w = *wind speed per NESC, mph*

Wind loading per unit length of conductor is calculated one of two ways based on the height of the structure. The determining factor is whether any portion of a structure, or its supported facilities, exceeds 60 feet above ground or water level.

For those structures that are below 60 feet:

$$w_W = P_W \left[\frac{D_c + 2t}{12} \right] \tag{2-32}$$

Where a structure, or its supported facilities, exceeds 60 feet above ground or water level, additional extreme wind load conditions must be taken into account in accordance per NESC requirements:

$$w_W = P_W\, k_z\, G_{RF}\, I\, C_d \left[\frac{D_c + 2t}{12} \right] \tag{2-33}$$

where:

w_W = *wind load per unit length of conductor, lbs/ft*

k_z = *velocity pressure exposure coefficient, as defined in Rule 250C1, Table 250-2*

G_{RF} = *gust response factor, as defined in Rule 250C2*

I = *importance factor, 1.0 for utility structures and their supported facilities*

C_d = *shape factor, as defined in Rule 252B*

D_c = *conductor diameter, in*

t = *ice thickness, if ice exists, in*

Values for k_z, V, G_{RF}, I and C_d are specified in the NESC, Section 25, Rule 250C and 252B2. The wind pressure parameters (k_z, V and G_{RF}) are based on open terrain with scattered obstructions (Exposure Category C as defined in ASCE 7-05[10]. Topographical features such as ridges, hills and escarpments may increase the wind loads. A Topographical Factor, K_{zt}, from ASCE 7-05 may be used in these special cases.

FIGURE 2-6
WIND PRESSURE DESIGN VALUES IN THE UNITED STATES

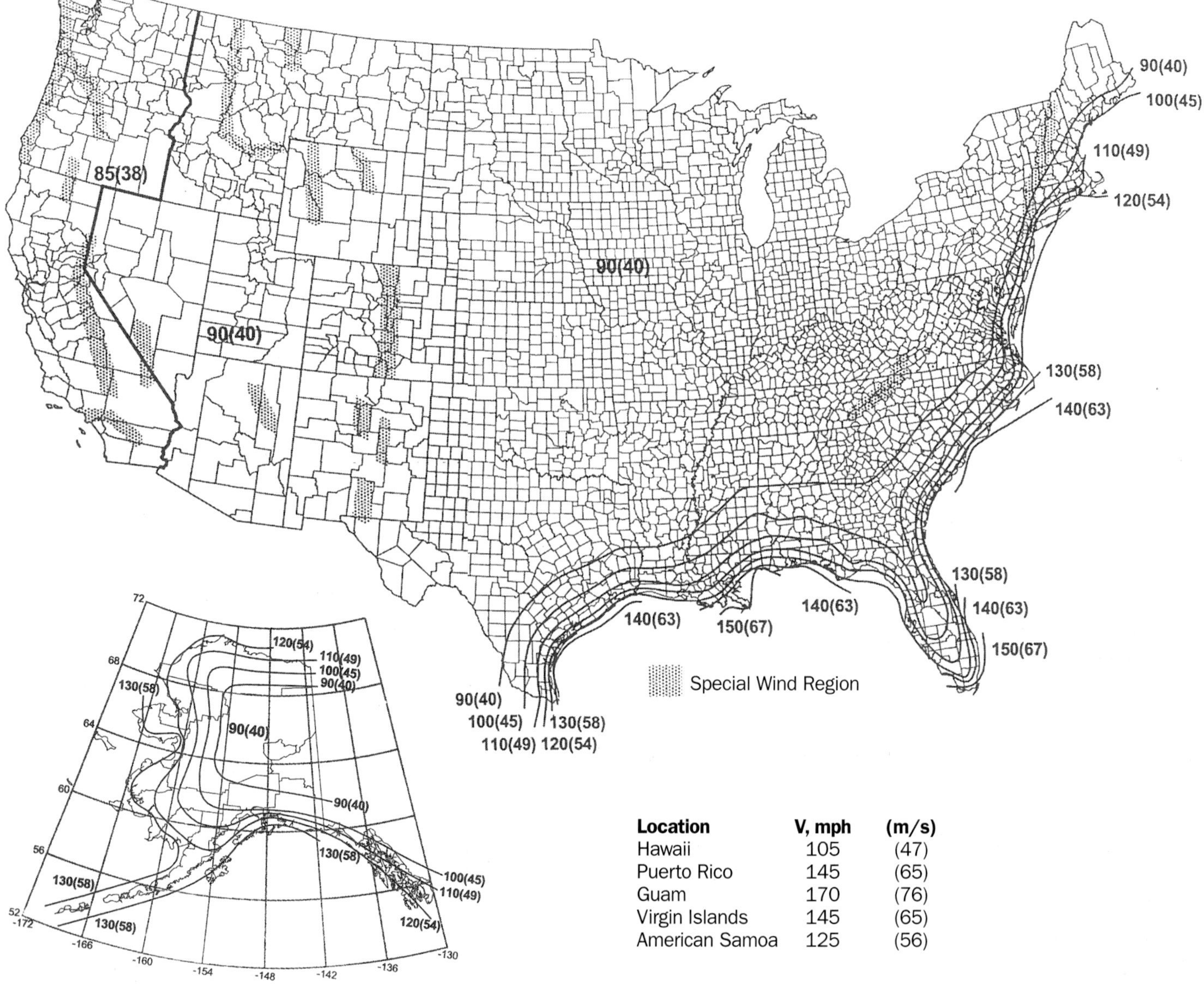

Location	V, mph	(m/s)
Hawaii	105	(47)
Puerto Rico	145	(65)
Guam	170	(76)
Virgin Islands	145	(65)
American Samoa	125	(56)

Notes:

1. *Values are nominal design 3-second gust wind speeds in miles per hour (m/s) at 33 ft (10 m) above ground for Exposure C category.*
2. *Linear interpolation between wind contours is permitted.*
3. *Islands and coastal areas outside the last contour shall use the last wind speed contours of the coastal area.*
4. *Mountainous terrain, gorges, ocean promontories, and special wind regions shall be examined for unusual wind conditions.*

Resultant Ice and Wind Loading

If the conductor weight is to include both ice and wind loading, the resultant magnitude of the loads must be determined. The weight of a conductor under both ice and wind loading is given by the following equation:

$$w_{w+i} = \sqrt{(w + w_i)^2 + (w_w)^2} + K_{NESC} \qquad (2\text{-}34)$$

where:

w_{w+i} = *total load per unit length of conductor, including ice and wind loads, lbs/ft*

w = *bare conductor weight per unit length, lbs/ft*

w_i = *weight of ice per unit length, lbs/ft*

w_w = *wind load per unit length, lbs/ft*

K_{NESC} = *NESC loading constant from Table 2-8, lbs/ft*

The factor K_{NESC} is a safety factor prescribed by the NESC. It is dependent upon loading district and is added to the resultant ice and wind loadings when performing sag and tension calculations.

2.4 SIMPLIFIED SAG-TENSION CALCULATIONS

Sag-tension calculations, for a range of conductor temperatures and loading conditions, are usually performed with the aid of a computer. These calculations involve the simultaneous effects from temperature, elastic and plastic elongation, which requires an iterative process for their solution. However, if the simultaneous effects of these parameters are ignored, certain basic calculations can be made with a handheld calculator or spreadsheet. Such simplified calculations provide insight into the calculation of sags and tensions that is not possible with the use of computer programs.

For example, calculate the sag and slack for a 600 feet level span of 795 kcmil-26/7 ACSR "Drake" conductor. The bare conductor weight per unit length, w, is 1.094 lbs/ft. The conductor is installed with a horizontal tension, H, of 6300 lbs, which is equal to 20% of its rated breaking strength of 31,500 lbs.

By use of the parabolic Equation 2-3, the sag for this level span is:

$$D = \frac{1.094\,(600)^2}{(8)\,6300} = 7.81 \text{ ft}$$

The length of the conductor between the support points is determined using Equation 2-6:

$$L = 600 \; \frac{8\,(7.81)^2}{3\,(600)} = 600.27 \text{ ft}$$

The conductor slack is the conductor length (600.27 feet) minus the span length (600 feet), 0.27 feet . It is useful to describe the slack as a "strain" equal to slack divided by span. For a slack of 0.27 feet and a span of 600 feet, the slack strain is 0.000453. Alternatively, the strain can be expressed as 0.0453% or 453 "microstrain" (microstrain = strain × 10^{-6}).

2.4.1 Sag-Tension Change with Conductor Temperature

Most conventional conductors elongate with increasing conductor temperature. The rate of linear thermal expansion for a composite ACSR conductor is less than that of an AAC conductor because the steel strands in the ACSR elongate at approximately half the rate of aluminum. The composite coefficient of linear thermal expansion of a non-homogenous conductor, such as ACSR "Drake," may be calculated by[11]:

$$\alpha_{AS} = \alpha_A \left(\frac{E_A}{E_{AS}}\right)\left(\frac{A_A}{A_{AS}}\right) + \alpha_S \left(\frac{E_S}{E_{AS}}\right)\left(\frac{A_S}{A_{AS}}\right) \qquad (2\text{-}35)$$

The modulus of elasticity for an aluminum-steel composite can be calculated by:

$$E_{AS} = E_A \left(\frac{A_A}{A_{AS}} \right) + E_S \left(\frac{A_S}{A_{AS}} \right) \quad (2\text{-}36)$$

where:

E_A = *modulus of elasticity of aluminum, psi*

E_S = *modulus of elasticity of steel, psi*

E_{AS} = *composite modulus of elasticity of aluminum-steel, psi*

A_A = *area of aluminum strands, square units*

A_S = *area of steel strands, square units*

A_{AS} = *total cross-sectional area, square units*

α_A = *aluminum coefficient of thermal elongation, per °F*

α_S = *steel coefficient of thermal elongation, per °F*

α_{AS} = *composite aluminum-steel coefficient of thermal elongation, per °F*

The modulus of elasticity for solid aluminum wire is 10 million psi, and 30 million psi for solid steel wire. For stranded wire the modulus of elasticity is reduced. The modulus for stranded aluminum is assumed to be 8.6×10^6 psi for all strandings, however the modulus for the steel core wire varies with stranding:

- 27.5×10^6 for single strand core
- 27.0×10^6 for 7 strand core
- 26.5×10^6 for 19 strand core

The coefficient of thermal elongation of aluminum is 0.0000128 per °F and 0.0000064 per °F for steel. Therefore aluminum expands at twice the rate of steel. In sag-tension calculations for non-homogeneous conductors, such as ACSR, the expansion of each component must be considered as shown in Equation 2-35. The composite coefficient of thermal elongation can be calculated as follows:

For example, for a 795 kcmil - 26/7 ACSR "Drake," the composite modulus of elasticity may be calculated using Equation 2-36, assuming elastic moduli of 8.6×10^6 and 27.0×10^6 psi for aluminum and steel respectively:

$$E_{AS} = (8.6 \times 10^6) \left(\frac{0.6247}{0.7264} \right) + (27.0 \times 10^6) \left(\frac{0.1017}{0.7264} \right) = 11.2 \times 10^6 \text{ psi}$$

and the composite coefficient of linear thermal expansion is calculated using Equation 2-35:

$$\alpha_{AS} = (12.8 \times 10^{-6}) \left(\frac{8.6 \times 10^6}{11.2 \times 10^6} \right) \left(\frac{0.6247}{0.7264} \right) + 6.4 \times 10^{-6} \left(\frac{27.0 \times 10^6}{11.2 \times 10^6} \right) \left(\frac{0.1017}{0.7264} \right) = 10.6 \times 10^{-6} \text{ /°F}$$

If the conductor temperature changes from a reference temperature, T_{REF}, to another temperature, T, the approximate change in conductor length, L, can be calculated by the product of the conductor's composite coefficient of linear thermal elongation, α_{AS}, and the change in temperature, $T - T_{REF}$, as shown as:

$$L_T = L_{T_{REF}} [1 + \alpha_{AS} (T - T_{REF})] \quad (2\text{-}37)$$

For example, consider the ACSR "Drake" in the example of Section 2.4 with temperature increasing from 60°F (15°C) to 212°F (100°C). The length increases by 0.97 feet from 600.27 feet to 601.24 feet:

$$L_{(212°F)} = 600.27(1 + (10.6 \times 10^{-6})(212 - 60)) = 601.24 \text{ ft}$$

Ignoring for the moment any change in length due to change in tension, the sag at 212°F (100°C) may be calculated for the conductor length of 601.24 feet using Equation 2-9:

$$D = \sqrt{\frac{3(600)(1.24)}{8}} = 16.70 \ ft$$

Using a rearrangement of Equation 2-3, this increased sag is found to correspond to a decreased tension of:

$$H = \frac{wS^2}{8D} = \frac{1.094(600^2)}{8(16.70)} = 2948 \ lbs$$

However, this is a simplified calculation, which ignores the effect of tension on the conductor length. These values of sag and tension for a conductor temperature of 212°F (100°C) are not correct for a real conductor, which would experience additional change in length due to the decreased tension. In the previous example, the ACSR "Drake" conductor has an area of 0.7264 in^2, and a composite modulus of elasticity, E_{AS}, of 11.2 × 10^6 psi as calculated in the previous example. Every 100 lbs change in tension equates to a tension change, in psi, of:

$$\frac{100}{0.7264} = 137.67 \ psi$$

Using the composite modulus for the ACSR "Drake," the change in elongation per 100 lbs of change in tension can be calculated:

$$\frac{137.67}{11.2 \times 10^6} = 0.0000123$$

Thus the ACSR "Drake" conductor will contract (shorten) elastically approximately 12 microstrain for every 100 lbs of tension relieved, or approximately 396 microstrain total in changing from a tension of 6300 lbs to 2948 lbs. In the preceding example only the length change (increase) due to thermal elongation was used to calculate the sag. This would result in an overestimation of the sag because the elastic change in length (decrease) would offset the thermal elongation. Therefore, ignoring the elastic effects results in an overestimation of both the increase in sag and the decrease in tension that occur with an increase in conductor temperature.

2.4.2 Sag-Tension Change with Ice or Wind Loading

Consider the calculation of sag and tension for the 600 feet ACSR "Drake" span with the addition of 0.5 inches of radial ice. Using Equation 2-30, the weight of the conductor increases by:

$$w_i = 1.244(0.5)(1.108 + 0.5) = 1.000 \ lbs/ft$$

As in Section 2.1.3, conductor tension is a function of slack and elastic elongation. If elongation of the conductor due to increased tension is ignored, then the sag of the iced conductor remains the same and the tension increases in direct proportion to the iced weight. If elongation due to tension is not ignored, then, as in the case for thermal elongation, the tension change resulting from the increased weight per unit length of conductor is reduced by the conductor modulus.

The effect due to wind loading is identical to ice loading. Assuming a wind load of 22 lbs/ft^2 and no ice, the increased conductor weight of the ACSR "Drake," can be calculated using Equation 2-32, assuming no portion of the structures is above 60 feet:

$$w_w = 22\left(\frac{1.108}{12}\right) = 2.031 \ lbs/ft$$

Small distribution conductors, such as the 1/0 AWG ACSR in Table 2-9, experience a larger iced to bare conductor weight ratio (4.8), so the conductor sag under maximum wind and ice load may exceed the sag at moderately higher temperatures.

2.4.3 Sag-Tension Change of a Simple Elastic Conductor

Bare stranded overhead conductors are typically made up of aluminum strands with or without a steel reinforcing core. Steel is essentially "elastic" (its length increases linearly with tension and is reversible) over the normal stress ranges seen in an overhead line. While the aluminum strands are not elastic, assuming them to be can provide useful insight into the behavior of the conductor. Assuming purely elastic moduli for the aluminum and steel, the conductor will return to its un-stressed length if an applied tension is brought to zero, no matter how high the tension or how long it has been applied. Essentially, the bare stranded conductor is modeled as a simple linear spring. However, ignoring the plastic elongation of aluminum stranded overhead conductors does not yield sag-tension estimates that are sufficiently accurate for line design.

The complex stress-strain behavior of composite conductors usually requires a sophisticated, computer-aided approach. For illustration purposes, however, the effect of permanent elongation of the conductor on sag and tension calculations will be ignored and a simplified elastic conductor assumed. This idealized conductor is assumed to elongate linearly with load and to undergo no permanent increase in length regardless of loading or temperature. For such a conductor, the relationship between tension and length is:

$$L_H = L_{H_{REF}}\left(1 + \frac{H - H_{REF}}{E_C A}\right) \tag{2-38}$$

where:

L_H = *length of conductor under horizontal tension H*

$L_{H_{REF}}$ = *length of conductor under horizontal reference tension* H_{REF}

E_C = *modulus of elasticity of the conductor, psi*

A = *cross-sectional area,* in^2

With a modulus of elasticity of approximately 8.6×10^6 psi for all-aluminum stranded conductor and 11×10^6 psi for moderately strong ACSR, all-aluminum and ACSR conductors elongate approximately 0.012% and 0.009%, respectively, for every 1000 psi change in tension. In the example presented in Section 2.4.1, the increase in temperature caused an increase in length and sag and a decrease in tension, but the effect of tension change on length was ignored.

Based on actual test data for ACSR "Drake," a linear elastic modulus of 11.2×10^6 psi and a thermal elongation coefficient of 10.6×10^{-6} per °F will be used. This idealized conductor is assumed to elongate linearly with load and to undergo no permanent increase in length regardless of loading or temperature.

Consider the example presented in Section 2.4 for ACSR "Drake" in a 600 feet level span. Assuming an initial conductor temperature 60°F, the conductor length under tension is 600.27 feet, and the composite linear modulus, E_{AS}, *is* 11.2×10^6 *psi. Using Equation 2-38, the reduction of the initial tension from 6300 lbs to zero yields a zero tension length, ZTL, at 60°F of:*

$$ZTL_{(60°F)} = 600.27\left(1 + \frac{0 - 6300}{(11.2 \times 10^6)\,0.7264}\right)$$

$$= 599.81\ ft$$

Keeping the tension zero and increasing the conductor temperature to 212°F yields a purely thermal elongation. The zero tension length at 212°F can be calculated using Equation 2-37:

$$ZTL_{(60°F)} = 599.81\left(1 + (10.6 \times 10^{-6})(212 - 60)\right)$$

$$= 600.78\ ft$$

According to Equations 2-9 and 2-3, this length corresponds to a sag of 13.2 feet and a horizontal tension of 3736 lbs. However, this length is for zero tension. The conductor will also elongate elastically under tension, thus increasing the length and the sag. The actual conductor sag-tension is determined by a process of iteration, which is best done through the use of a computer program.

2.5 PLASTIC CONDUCTOR ELONGATION

Relatively small changes in the length of overhead conductors can cause rather large changes in sag. In the preceding calculation, a thermal elongation of 0.16% (11.5 inches) in a 600 feet span produced an increase in sag from 7.8 to 13.2 feet and the inclusion of a further elastic elongation of 0.05% (3.5 inches) increased the sag to 15.4 feet. In this particular case, the conductor sag changes about a foot for every 2 inch increase in conductor length in a 600 feet span. In a similar fashion, small amounts of plastic elongation can have a large effect on the accuracy of sag-tension calculations.

This section concerns the proper modeling of plastic elongation for sag-tension calculations. Most of the plastic elongation in bare overhead stranded conductors occurs in the aluminum strands. Steel elongates linearly under tension load and exhibits very little plastic elongation. For non-homogeneous conductors, the calculation of plastic elongation is complex. Plastic elongation in the aluminum layers increases the fraction of the total conductor tension carried by the steel core.

There are three primary sources of plastic elongation in the stranded aluminum strand layers of bare overhead conductor. They are strand settlement, deformation and creep elongation. Plastic elongation due to strand settlement and deformation occurs when a newly manufactured conductor is first tensioned. Creep elongation occurs throughout the life of an overhead line so long as the aluminum strand layers are under tension. However, the rate of creep decreases exponentially with time.

2.5.1 Conductor Elongation Models

At the time of installation, the initial conductor sag and tension are set. Sag and tension typically change as a result of the following factors:

- Tension change due to ice and wind loading (Section 2.4.1)
- Length change with conductor temperature driven by changes in weather and electrical current (Section 2.4.3)
- Elastic length change associated with conductor tension changes
- Permanent increases in aluminum strand length over time depending on operating temperature and tension history

In order to calculate sag and tension the change in conductor length due to each of these factors must be modeled.

In this section, three conductor elongation models will be considered. These models differ primarily in how plastic conductor elongation is included in sag-tension calculations. These elongation models are the Linear Elongation Model, the Simplified Plastic Elongation Model and the Experimental Plastic Elongation Model. Note that the three elongation models differ in how plastic elongation is calculated, but that all three use the same linear stress-strain model for elastic strain calculation.

Linear Elongation Model (LEM)

Also know as the Linear Elastic Model, this model does not include any consideration for plastic elongation of the conductor. The conductor is modeled as being purely elastic. As conductor tension increases, the strain increases linearly. As tension decreases, the same stress strain relationship applies. There is no allowance for the plastic elongation that may have occurred.

This model is included here as it is the basis for the simplified plastic elongation model, which follows. This method should never be used for actual sag-tension calculations, but is useful in discussing the concepts of elastic and plastic elongation.

FIGURE 2-7
ILLUSTRATION OF THE LINEAR ELONGATION MODEL

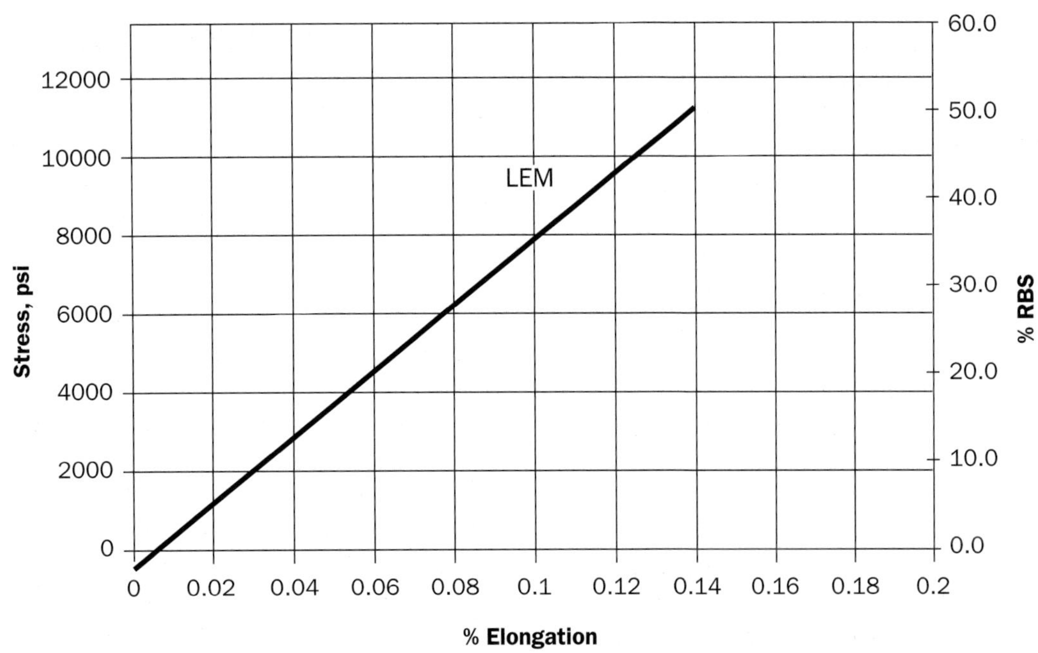

Simplified Plastic Elongation Model (SPE)
This model is basically the LEM with an allowance for plastic elongation of the conductor. However, the amount of plastic elongation is not calculated as a function of tension or temperature over time, but rather is a value based on engineering experience.

In Figure 2-8, the two straight lines, representing the initial and final stress-strain curves, are separated by a constant amount of plastic strain (elongation). The permanent plastic elongation due to creep and strand settlement is represented by the horizontal shift of the linear stress-strain curves. Short-term plastic elongation of the conductor due to strand settlement and deformation is ignored. Long-term creep elongation is represented by a permanent elongation value (e.g. 450 microstrain) based on engineering experience. This creep elongation value is commonly considered to be equivalent to the amount of elongation due to an increase in conductor temperature (e.g. 35°F).

The attractiveness of this method is its simplicity and intuitiveness, however they are several drawbacks:

- It over estimates structure tensions under high ice and wind loads.
- It does not represent the high temperature thermal elongation of ACSR properly.
- It does not allow the estimation of plastic elongation for conductor and line designs for which engineering experience is limited.

This method has mainly been used overseas. Its use is currently declining in favor of the more analytical Experimental Plastic Elongation Model. In a document developed by the CIGRE WG 11[9], a thorough description of this method, with typical conductor data and the basic formulas, is presented.

Experimental Plastic Elongation Model (EPE)
The Experimental Plastic Elongation Model is based upon laboratory conductor test data. The stress-strain and creep data from these tests are used to calculate plastic conductor elongation as a function of tension and temperature over time. With non-homogeneous conductors, such as ACSR, the tension, elastic and plastic elongation can be calculated for each component (i.e. aluminum and steel). As a result, the Experimental Plastic Elongation Model overcomes most of the limitations noted with the Simplified Plastic Elongation Model. However, the EPE is a more complex, numerical method typically requiring a computerized solution.

Sag-tension methods that incorporate the EPE are the Varney Graphical Method[12] and the Strain Summation Method developed at Ontario Hydro[13]. Both methods use

FIGURE 2-8
ILLUSTRATION OF THE SIMPLIFIED PLASTIC ELONGATION MODEL

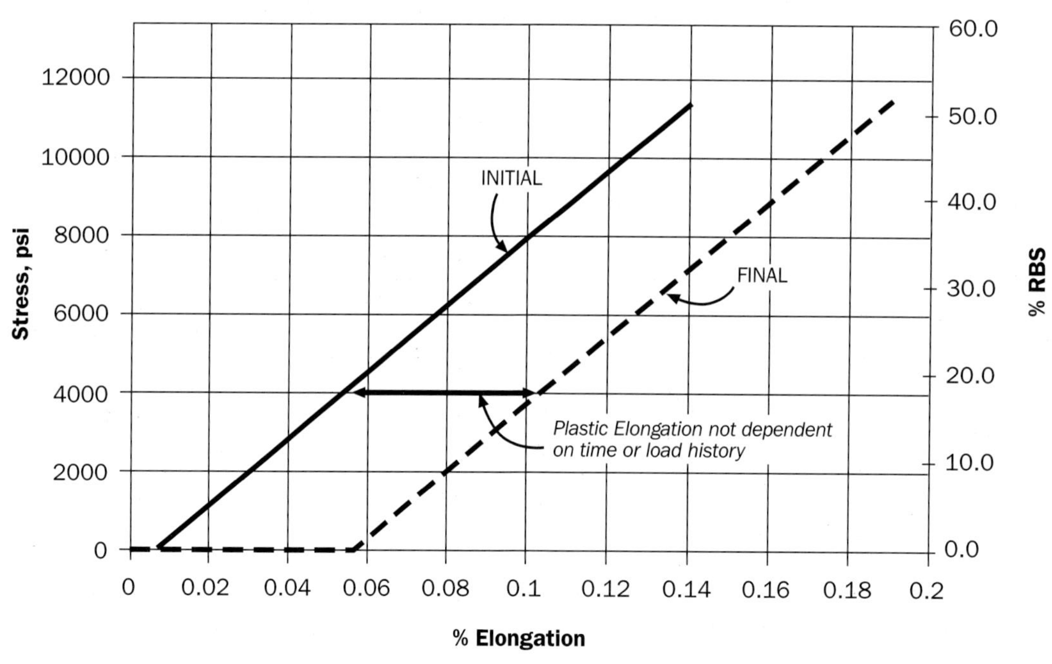

experimental conductor data and both are complex enough to require implementation by numerical methods.

In both the SPE and the EPE the elastic elongation of the conductor is modeled using an elastic modulus whose response to changes in tension does not change over the life of the conductor. However, in the EPE, plastic elongation is modeled as consisting of two components which are not additive:

- Strand settlement and deformation that occurs when tension is first applied (1-hour) to the conductor.

- "Metallurgical" creep elongation that occurs over the life of the line, the amount of which depends on the tension history of the line and the conductor design.

Strand settlement and deformation are modeled by non-linear "initial" stress-strain curves (usually expressed as 4th order polynomials) for both the aluminum layers and steel core, when present[14]. "Metallurgical" creep elongation is also modeled by polynomials, but the curves used to derive them cannot be generated from a single test.

Figure 2-9 shows an example of the EPE Model. It should be noted that:

- The EPE INITIAL curve is derived from a standardized stress-strain test procedure[15]. The difference between the LEM INITIAL curve and the EPE INITIAL curve is plastic elongation due to strand settlement and deformation that occurs during the initial application of tension. As shown, if the tension reaches 50% RBS and persists for an hour, then the plastic elongation is approximately 0.055% (i.e. 550 microstrain).

- The CREEP curve summarizes the plastic strain that occurs as a result of long-term creep at a series of different everyday tensions. For example, if the conductor is held at a tension of 4000 psi for 10 years, then the plastic elongation due to metallurgical creep at this moderate tension is approximately 0.07% (i.e. 700 microstrain).

- Physically, the plastic elongation due to high loads and the plastic elongation due to metallurgical creep over long time periods are not additive. In the Varney graphical method, the total plastic elongation of a conductor exposed to both load events (i.e. 10 years at 4000 psi and an hour at 11,300 psi) is not the sum of the two, but is approximated as the larger of the two.

2.5.2 Experimental Plastic Elongation Model Data

With any of the models of conductor elongation, the initial length of the conductor is assumed to be that produced by the manufacturer (see "Manufactured length" in Figure 2-10), as stranded and wound on the reel, at very low tension.

FIGURE 2-9
ILLUSTRATION OF THE EXPERIMENTAL PLASTIC ELONGATION MODEL

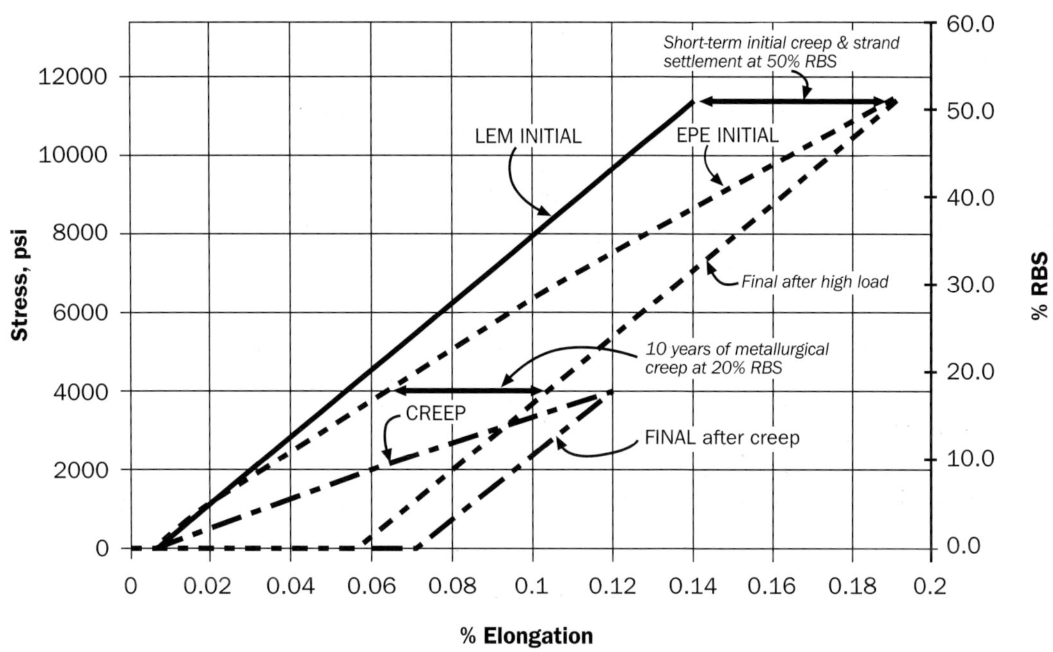

If the conductor has a steel core, then the "Manufactured length" of the steel core and the surrounding aluminum layers are assumed equal (Rawlins [16] has suggested that there may be certain residual stresses in the aluminum layers as a result of the stranding process, but these residual stresses are only important in calculating the conductor elongation at high temperatures).

Stranded aluminum conductors undergo plastic elongation as a result of unreeling and applying tension during stringing and sagging ("Settlement & strand deformation," Figure 2-10). Subsequent to clipping the conductor into its support hardware, conductors experience further gradual plastic elongation due to aluminum creep strain (also referred to as metallurgical creep).

Over the life of the line, the conductor will undergo creep elongation at "everyday" tension levels. At certain times, the conductor tension will increase due to ice and wind loading events. The combined effect of creep elongation at continuous everyday tension and occasional accelerated creep during loading events is shown as "Long-term aluminum creep strain" in Figure 2-10.

Normal changes in temperature and tension cause reversible changes in length as represented by the strain components "Thermal strain" and "Elastic strain" in Figure 2-10.

Stress-Strain Tests

To predict the elongation of conductors under tension, two types of test are performed. The first is a stress-strain test to measure the short time behavior of a conductor under varying tensile loads, and the second is a series of creep tests to measure the long-term non-elastic behavior of a conductor under tension. Details of these test methods are provided by The Aluminum Association[15].

To perform a stress-strain test, a sample of new conductor, with an active gauge length of approximately 500 inches, is placed in a long bed tensile test machine with the conductor ends gripped in such a manner as to assure positive holding of the conductor strands during the test. An initial bias tension load of 1000 lbs or 8% of the conductor rated breaking strength (RBS), whichever is smaller, is applied to the conductor to remove any looseness in the conductor. Leveling jack supports are placed under the cable to keep the conductor straight.

Tension is then applied three consecutive times, at increasing tension levels of 30%, 50% and 70% of RBS, measuring the elongation of the conductor during each loading series. The tension is held for 30 minutes at the 30% tension level, and one hour at both the 50% and 70% tension levels. At the end of each hold period, the tension is reduced to the starting bias tension. The conductor is loaded a fourth, and final time, to 75% of its calculated

FIGURE 2-10
ELONGATION DIAGRAM OF STRANDED CONDUCTOR IN A CATENARY

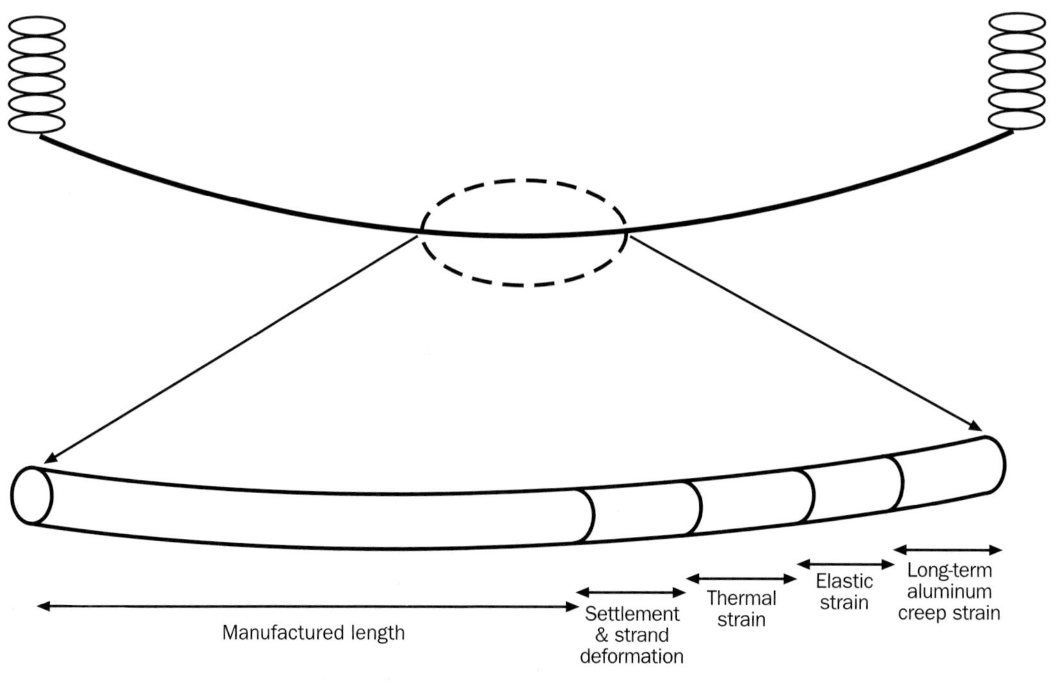

RBS. The extensometers, elongation measuring devices, are removed from the sample and the conductor is pulled to tensile failure.

A fitted curve is drawn through the endpoints of each hold period. This curve is typically referred to as the one hour modulus of the composite conductor. A straight line is drawn tangent to the one hour modulus curve between a stress of approximately 8% and 35% of the conductor RBS. The slope of this line is referred to as the initial modulus. The slopes of the stress-strain curves during unloading, at each tension level, are essentially the same. This slope is called the final modulus of the composite conductor. Typical results of such a test are shown in Figure 2-11.

If the conductor being tested is all-aluminum, only one stress-strain test is required. If the conductor is a non-homogenous conductor, such as an ACSR, having a core with different tensile and elongation characteristics, a second stress-strain test is required to predict the stress-strain behavior of the core. The stress-strain characteristic of the core is then subtracted from the composite stress-strain curve to determine the initial and final stress-strain characteristics of the aluminum.

Creep Tests

Next, a series of creep tests are performed to determine the long-term non-elastic elongation of a conductor under constant tension. These tests consist of loading a sample of new conductor into a long-bed tensile test machine and gripping the ends to prevent slippage of the conductor strands relative to each other. A pre-determined tension, typically between 10% and 50% of the conductor RBS, is applied to the conductor and held constant for the duration of the test cycle. The elongation of the conductor is monitored over time. A creep test cycle lasts for 1000 hours, approximately 42 days. The room temperature must be held constant during the test to prevent thermally induced elongation.

Upon completion of the test, the percent strain (percent elongation of original sample length) between one hour and 1000 hours is recorded on a log-log plot, and the curve is extrapolated to 100,000 hours. Multiple creep tests are typically performed at increasing tension levels to predict the creep characteristic of a conductor. A representative creep plot is shown in Figure 2-12.

The creep strain is added to the stress-strain curve to indicate the total stress-strain-creep characteristic of a conductor.

FIGURE 2-11
TYPICAL STRESS-STRAIN TEST PLOT
THE ONE HOUR MODULUS CURVE, THE INITIAL AND FINAL MODULUS CURVES

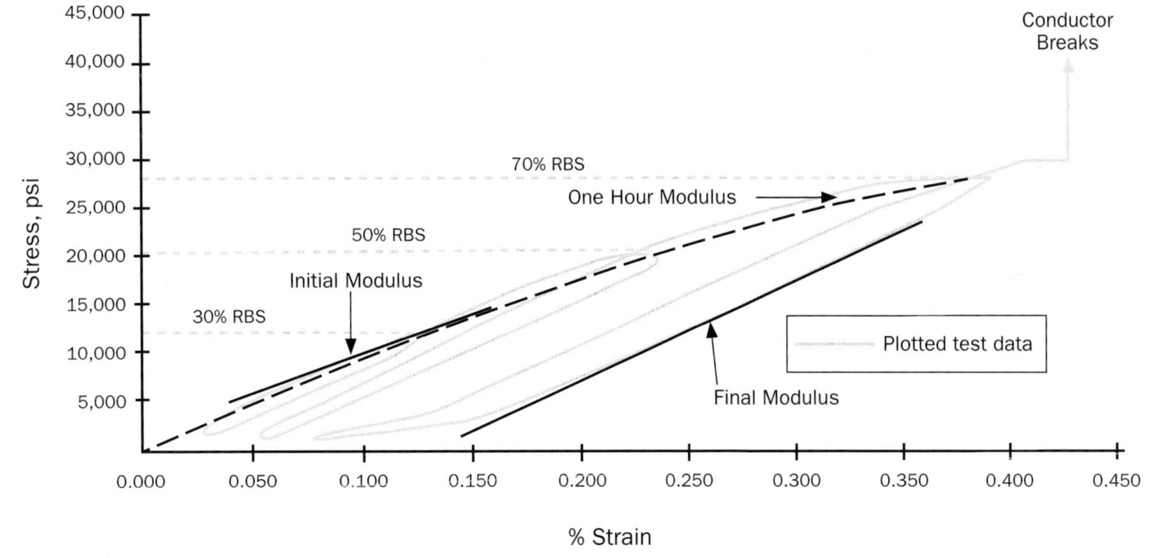

Note:

1. The conductor length increases due to non-elastic elongation during each hold period and the final modulus (slope) after each hold period is approximately the same.
2. The curve drawn through the end of the hold periods is called the ONE HOUR MODULUS curve of the initial composite stress-strain curve.
3. The slope of the curves during each unloading, decreasing tension level, for each hold period is essentially parallel.
4. Since the INITIAL composite curve is non-linear, it is normally expressed as a polynomial equation. The FINAL curve is linear and is normally expressed as a simple linear modulus.

Stress, the conductor tension divided by the total area of the conductor strands, expressed in psi, is plotted on the vertical axis. Conductor strain, expressed as percent of conductor length, is plotted on the horizontal axis.

Stress-strain characteristics are typically grouped by stranding (i.e., 7, 19, 37, 61, etc) for all-aluminum conductors, and by steel to aluminum ratio for ACSR and ACSS strandings, such as 26/7, 54/9, etc. Stress-strain curves are available for all stranded conductors.

2.5.3 Formulas for Predicting Creep Elongation

Creep test data is difficult and expensive to obtain. Reference [17] presents formulas for estimation of creep elongation, based upon extensive testing by Alcoa. A review of these formulas allows for some understanding of creep elongation. The rate of creep is a function of the type of material, temperature, and tensile stress. For a given material, the creep rate increases with temperature and tensile stress. Under similar conditions, the creep rates of aluminum and aluminum alloys are significantly greater than steel. It is evident, therefore, that the creep rates of AAC, AAAC and ACAR conductors will be considerably greater than that of ACSR.

Any increase in the elongation of an overhead conductor, regardless of whether it is temporary or permanent, increases line sag and decreases line clearance. Therefore, to maintain minimum clearances, the line designer must take into account the creep characteristics of the conductor, and the effects of conductor temperature and mechanical stress on these characteristics.

AAC, AAAC and ACAR Conductors

All-aluminum conductors are composed of strands of 1350-H19, all-aluminum alloy conductors (AAAC) of 6201-T81, and aluminum conductor, aluminum-alloy reinforced (ACAR) aluminum conductors of a combination of the two materials. Although the creep rates of individual strands of these materials can be determined, no acceptable correlation between the creep rates of individual strands and multi-strand cables has been established. It has been necessary, therefore, to test various constructions of these conductors to empirically derive equations

FIGURE 2-12
CREEP PLOT

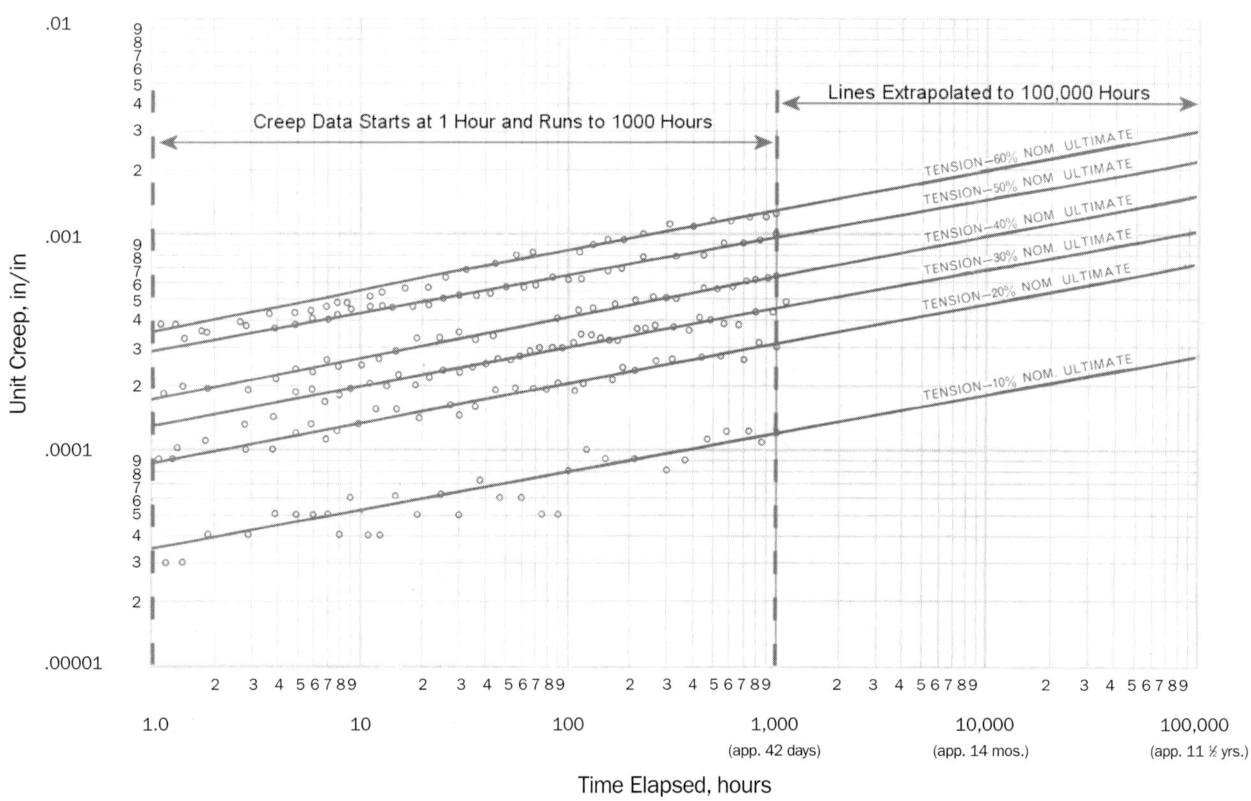

Courtesy of Alcoa Conductor Products Company.

to predict cable elongation due to creep at room temperature. These equations, developed by Harvey and Larson[18], are presented in Table 2-10, along with the coefficient values in Table 2-11. Although different coefficient values for *K* are shown, based on the method of manufacture, it is noted that the great majority of redraw rod from which 1350 aluminum wire is produced today is manufactured by a continuous-cast process.

TABLE 2-10
CREEP ELONGATION EQUATIONS FOR ALL ALUMINUM CONDUCTORS

Conductor Types	Room Temperature
AAC	$\epsilon_c = K\,\sigma^{1.3}\,t^{0.16}$
AAAC (6201)	$\epsilon_c = G\,\sigma^{1.3}\,t^{0.16}$
ACAR	$\epsilon_c = (0.0003 + 0.0021\,\frac{A_{1350}}{\Sigma A})(\sigma^{1.3}\,t^{0.16})$

where:

ϵ_c = *creep strain, micro-units/unit*

t = *elapsed time, hours*

σ = *tensile stress (total tensile load divided by the conductor cross-sectional area), psi*

A_{1350} = *area of 1350-H19 aluminum strands*

ΣA = *total cross-sectional area of conductor*

K_1, K_2 and G are constants listed in Table 2-11

Creep of ACSR Conductors

ACSR conductors are composed of 1350-H19 aluminum strands spiraled around a core comprised of steel strand(s). The creep rate at room temperature of these conductors is typically less than that of AAC, AAAC and ACAR conductors. The lower creep rate of ACSR results from the tensile load carried by the aluminum strands being transferred to the steel core after the aluminum strands have undergone a relatively small increase in elongation. For ACSR with steel cross-sectional areas of 7.5%, or more, elevated temperatures of the magnitude experienced in transmission and distribution lines cause no significant increase in creep rate over that normally experienced at room temperature. For constructions with less steel, the increased creep rate at elevated temperatures should be considered.

Creep at Room Temperature

Creep is a function of conductor stress, temperature, the process used to manufacture the aluminum rod, the elapsed time and, for ACSR and ACAR, the area ratio between the component materials. The following equations for predicting creep of ACSR conductors were derived by Harvey and Larson.

For an ACSR conductor with 1350-H19 strands drawn from hot-rolled rod:

$$\epsilon_c = 2.4\,(\%RBS)^{1.3}\,t^{0.16} \quad (2\text{-}39)$$

where:

$\%RBS$ = *the ratio of tension to rated breaking strength expressed as a percentage*

TABLE 2-11
CONDUCTOR CREEP EQUATION COEFFICIENTS

	STRANDING			
# Strands	**7**	**19**	**37**	**61**
K_1	0.0021	0.0020	0.0019	0.0018
K_2	0.0013	0.0012	0.0012	0.0011
G	0.0011	0.0010	0.0012	0.00094

Subscript "1" denotes 1350-H19 wire drawn from hot-rolled rod.
Subscript "2" denotes 1350-H19 wire drawn from continuous-cast rod.

For an ACSR conductor with 1350-H19 strands drawn from continuous-cast rod:

$$\epsilon_c = 1.4\,(\%RBS)^{1.3}\; t^{0.16} \qquad (2\text{-}40)$$

Careful examination of these equations leads to a number of useful observations:

- For any period of time, when conductor tension and temperature are held constant, 70% of the creep occurs in the first 10% of the period.
- Conductors made with hard drawn aluminum strands drawn from continuous-cast rod creep only about 60% as much as conductors manufactured with aluminum strands drawn from hot-rolled rod.
- Increasing the conductor temperature by 50% increases creep by 76% for AAC, AAAC and ACAR, by 50% for low strength strandings of ACSR (18/1, 45/7, etc.), but has no effect on the creep of higher strength strandings of ACSR (54/7, 26/7, etc.).
- At room temperature, increasing the conductor tension by 50% increases creep by 70% for AAC, ACAR, AAAC and ACSR.

2.5.4 Knee-Point

Due to the differing temperature coefficients of expansion, at some increasing temperature, the tension in the aluminum is reduced to zero and the total tension load is shifted to the steel core. This temperature is commonly referred to as the "knee-point" temperature and it corresponds to the point at which the rate of increase of sag with temperature changes. This transition appears as a "knee" on a plot of sag versus temperature. Beyond this temperature, the ACSR conductor elongates with temperature at the rate of the steel core. When the temperature of the conductor decreases, the aluminum strands once again come under tension and the conductor behaves according to the composite modulus and thermal elongation equations previously presented.

As shown in Figure 2-13, creep elongation of the aluminum strands causes the "knee-point" to occur at a lower temperature.

The "knee-point temperature" of the example sag-tension calculation for ACSR "Drake" in a 1000 feet span occurs at approximately 135°C when the conductor is first installed, and decreases to about 60°C after aluminum creep has occurred for 10 years.

However, tests of multi-layer ACSR indicate that the "full hard" 1350-H19 aluminum layers support a modest level of compression beyond the "knee-point" temperature. Several theories have been proposed to model the

FIGURE 2-13
"KNEE-POINT TEMPERATURE" AS AFFECTED BY CREEP ELONGATION OF ALUMINUM STRANDS

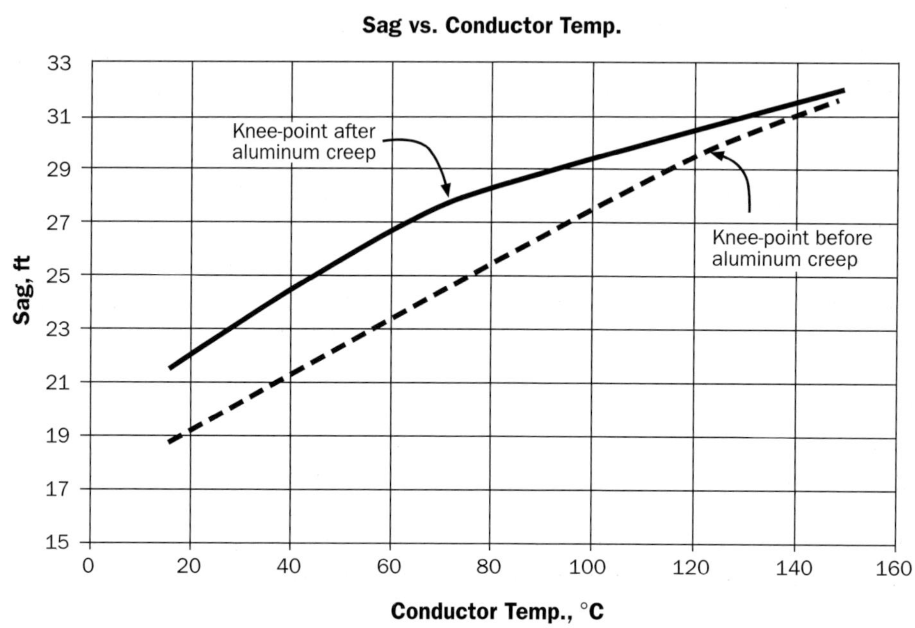

composite conductor thermal elongation in this transition region beyond the knee-point temperatures, two of which are Rawlins and Barrett.

Rawlins[16] found that the increased sag at temperatures above the knee-point is due to a combination of residual stress in the aluminum layers (a result of "multi-pass" manufacturing stranding process) and interference between the helical layers trying to expand radially (lending the aluminum component layers a small compressive modulus). Barrett[19] suggested that the aluminum layers can support up to 1500 psi of compression which can be increased by the relatively high radial temperature gradients that can occur at high current densities. Barrett notes a high level of random variation in the observed levels of thermal elongation beyond the knee-point. Rawlins' study found the compressive aluminum stresses beyond the knee-point temperature to be insignificant. In Barrett's studies of compression levels in ACSR, he found that the aluminum layers supported between 1000 and 3000 psi.

Given the variability of sag calculation at high temperatures, a conservative approach may be to account for aluminum compression. Additional sag due to aluminum compression for several conductors is summarized in Table 2-12.

TABLE 2-12
MAXIMUM EXCESS SAG (FEET) DUE TO THE COMPRESSIVE ALUMINUM STRESS AS PROPOSED BY RAWLINS, AT TEMPERATURES ABOVE THE KNEE-POINT, ESTIMATED WITH THE ALCOA SAG10 COMPUTER PROGRAM

ACSR Conductor	Span Length, ft		
	650	1300	1950
"Tern" – 45/7	1.0	1.0	-
"Drake" – 26/7	1.5	2.3	3.3
"Mallard" – 30/19	1.3	1.6	2.2

2.5.5 Examples of Typical Stress-Strain Conductor Curves

Sag-tension calculations are normally done numerically using a computer program allowing the user to enter many different loading and conductor temperature conditions. Both initial and final conditions are calculated and multiple tension constraints can be specified. The complex stress-strain behavior of ACSR-type conductors can be modeled numerically, including both temperature and elastic and plastic effects.

Stress-strain curves for bare overhead conductors include a minimum of an initial curve and a final curve over a range of elongations from 0 to a minimum of 0.40%. For conductors consisting of two materials, an initial and final curve for each is included. Creep curves for various lengths of time are typically included as well[14].

Overhead conductors are not purely elastic. They stretch with tension, but when the tension is reduced to zero, they do not return to their initial length. The change in conductor length cannot be expressed with a simple linear equation. The permanent length increase that occurs in overhead conductors yields the difference in initial and final sag-tension data calculated by most computer programs.

Figure 2-14 is a stress-strain curve for all-aluminum 37-strand conductors ranging in size from 250 kcmil to 1033.5 kcmil. Because the conductor is made entirely of aluminum, there is only one initial and final curve.

Figure 2-15 shows a typical stress-strain curve[14] for a 26/7 ACSR conductors ranging in conductor sizes from 266.8 to 795 kcmil. A 795 kcmil-26/7 ACSR "Drake" conductor has a rated breaking strength of 31,500 lbs and an area of 0.7264 in^2 thus it fails at an average stress of 43,000 psi. The stress-strain curve illustrates that a stress equal to 50% of the conductor's breaking strength (21,500 psi), the elongation is less than 0.3%. This translates to an elongation of 1.8 feet in a 600 feet span.

From prior examples, with the ACSR "Drake" at a tension of 6300 lbs, the length of the conductor in the 600 feet span was found to be 0.27 feet longer than the span. This tension corresponds to a stress of 8600 psi. From the stress-strain curve in Figure 2-15, this corresponds to an initial elongation of 0.105% (0.63 feet). As in the calculation in Section 2.4.3, if the conductor is reduced to zero tension, its unstressed length would be less than the span length.

ACSS is very similar to ACSR in construction, however the use of fully annealed 1350-O aluminum results in very different stress-strain relationships. Figure 2-16 is a stress strain curve for a 26/7 ACSS. When compared to that of an equivalent ACSR, Figure 2-15, several observations can be made:

- The initial (elastic) modulus for the ACSS is approximately the same as that of the ACSR. However, since this modulus is only valid before the aluminum goes into yield, the initial modulus for the ACSS is valid over a much lower tension range than that for the ACSR.

- The initial aluminum curve for the ACSS goes into yield much quicker, resulting in a relatively flat upper initial aluminum curve. Thus the aluminum strength contribution to the initial composite curve is limited, resulting in more conductor elongation at a given stress compared to an equivalent ACSR.

- The relatively low yield, and resulting flat initial aluminum curve, for ACSS also translates to practically no creep. At relatively low tension in the aluminum, less than 5000 psi in this graph, the aluminum goes into yield, thus the aluminum rapidly elongates as much as it ever will, as soon as a new tension level is applied. This can be verified by observing that the slope of the initial composite curve closely matches that of the initial steel curve after the aluminum goes into yield.
- After experiencing heavy loading, the ACSS conductor will respond to tension changes per the final steel curve – it is said to "ride" the steel curve – as long as the tension does not exceed that of the heavy loading event. This is a benefit for the ACSS in that this results in the conductor thermally elongating at the rate of steel. Thus, the ACSS will have a lower rate of sag change in response to temperature increases compared to an ACSR.

FIGURE 2-14
STRESS-STRAIN CURVES FOR 37-STRAND AAC

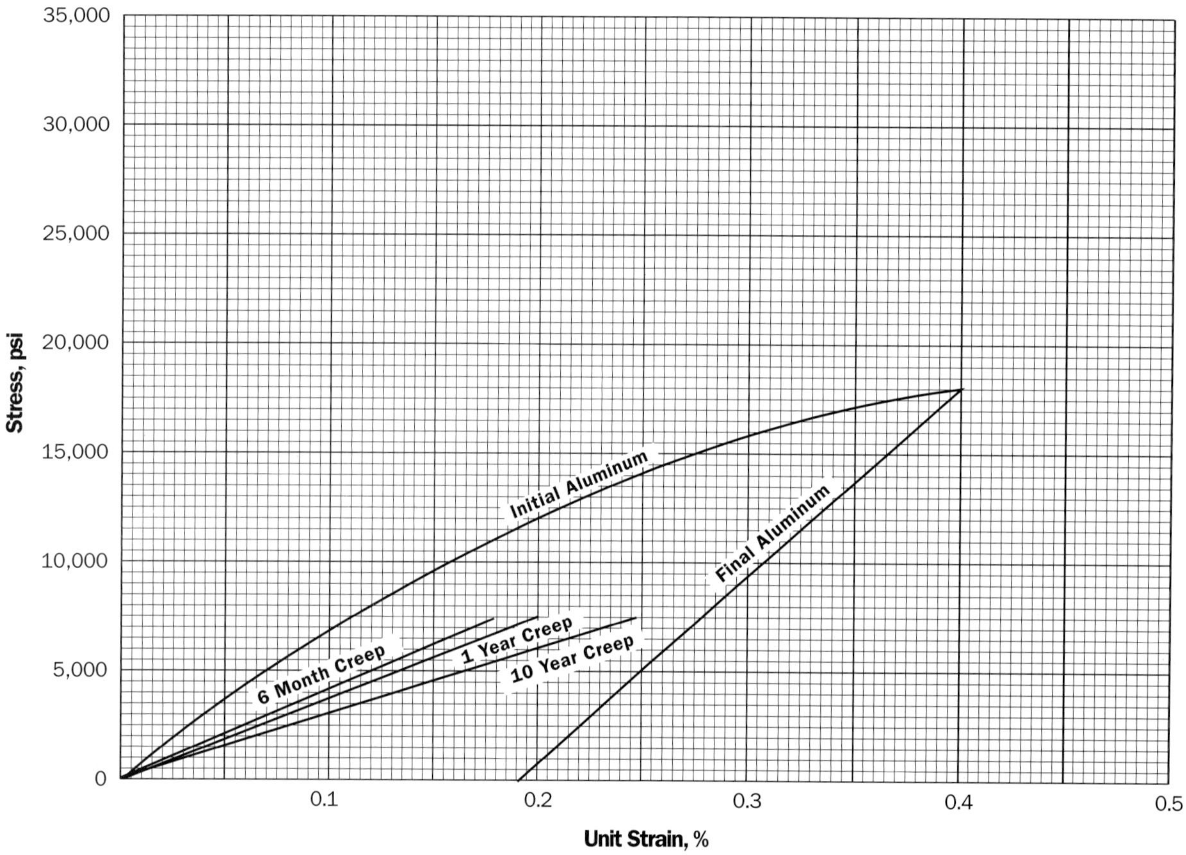

Equations for Curves (X = unit strain in %: Y = stress in psi):

Initial Aluminum: $X = -5.31 \times 10^{-3} + (1.74 \times 10^{-5})Y - (6.17 \times 10^{-10})Y^2 + (5.05 \times 10^{-14})Y^3$

$Y = 136 + (7.46 \times 10^4)X - (8.51 \times 10^4)X^2 + (2.33 \times 10^4)X^3$

Final Aluminum: $Y = (85.20X - 16.14) \times 10^3$

6 Month Creep: $Y = (42.30 \times 10^3)X$

1 Year Creep: $Y = (38.20 \times 10^3)X$

10 Year Creep: $Y = (30.60 \times 10^3)X$

Test Temperature 70°F to 75°F

Once a conductor has been installed to an initial tension, it can elongate further. Such elongation results from two phenomena: permanent elongation due to high tension levels resulting from ice and wind loads, and creep elongation.

Both Figures 2-14 and 2-15 indicate that when the conductor is initially installed, it elongates non-linearly. If the conductor tension increases to a relatively high level under ice and wind loading the conductor will elongate. When the wind and ice loads abate, the conductor elongation will reduce along a curve parallel to the final curve, but will never return to its original length.

Referring to Figure 2-14 and assuming that a newly strung 795 kcmil 37-strand AAC "Arbutus" conductor has an everyday tension of 2780 lbs, a conductor area of 0.6245 in^2, then the everyday stress is 4450 psi and the elongation is 0.062%. Following an extremely heavy ice and wind load event, assume that the conductor stress reaches 18,000 psi. When the conductor tension decreases back to everyday levels, the conductor elongation will be

FIGURE 2-15
STRESS-STRAIN CURVES FOR 26/7 ACSR

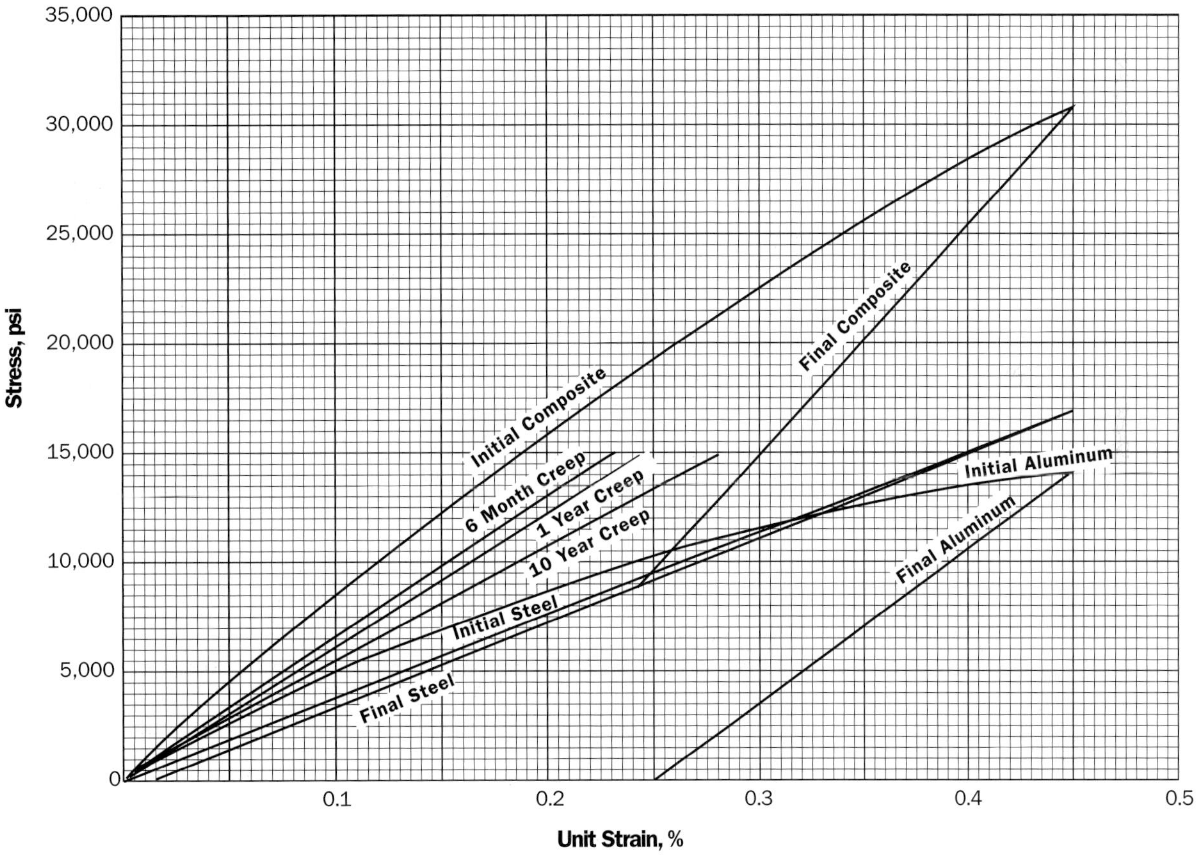

Equations for Curves (X = unit strain in %: Y = stress in psi):

Initial Composite:	X	$= 4.07 \times 10^{-3} + (1.28 \times 10^{-5})\,Y - (1.18 \times 10^{-10})\,Y^2 + (5.64 \times 10^{-15})\,Y^3$
	Y	$= -512 + (8.617 \times 10^{4})\,X - (1.18 \times 10^{4})\,X^2 - (5.76 \times 10^{-4})\,X^3$
Initial Steel:	Y	$= (37.15 \times 10^{3})\,X$
Initial Aluminum:	Y	$= -512 + (4.902 \times 10^{4})\,X - (1.18 \times 10^{4})\,X^2 - (5.76 \times 10^{4})\,X^3$
Final Composite:	Y	$= (107.55X - 17.65) \times 10^{3}$
Final Steel:	Y	$= (38.60X - 0.65) \times 10^{3}$
Final Aluminum:	Y	$= (68.95X - 17.00) \times 10^{3}$
6 Month Creep:	Y	$= (64.75 \times 10^{3})\,X$
1 Year Creep:	Y	$= (60.60 \times 10^{3})\,X$
10 Year Creep:	Y	$= (53.45 \times 10^{3})\,X$

Test Temperature 70°F to 75°F

permanently increased by approximately 0.2%. Also the sag under everyday conditions will be correspondingly greater, and the tension will be less. In most numerical sag-tension methods, final sags and tensions are calculated including the permanent elongation due to heavy loading conditions.

With ACSR, creep elongation in the aluminum strands reduces the fraction of the total conductor tension carried in the aluminum strands and increases the fraction carried in the steel core strands. If the creep strain is sufficiently large, the tension in the aluminum strands can reach zero.

2.6 SAG-TENSION CALCULATIONS WITH EXPERIMENTAL ELONGATION

Several commercial sag-tension computer programs are available in the United States and Canada. Each of these programs is based on the experimental elongation model. The sag-tension output varies by vendor, but all programs provide the basic initial and final sag-tension output data. Typically sag-tension calculations are based on initial and final loading conditions. Initial sags and tensions are simply the sags and tensions at the time the line is built before any mechanical (ice and/or wind) loading occurs. Final

FIGURE 2-16
STRESS-STRAIN CURVE FOR 26/7 ACSS

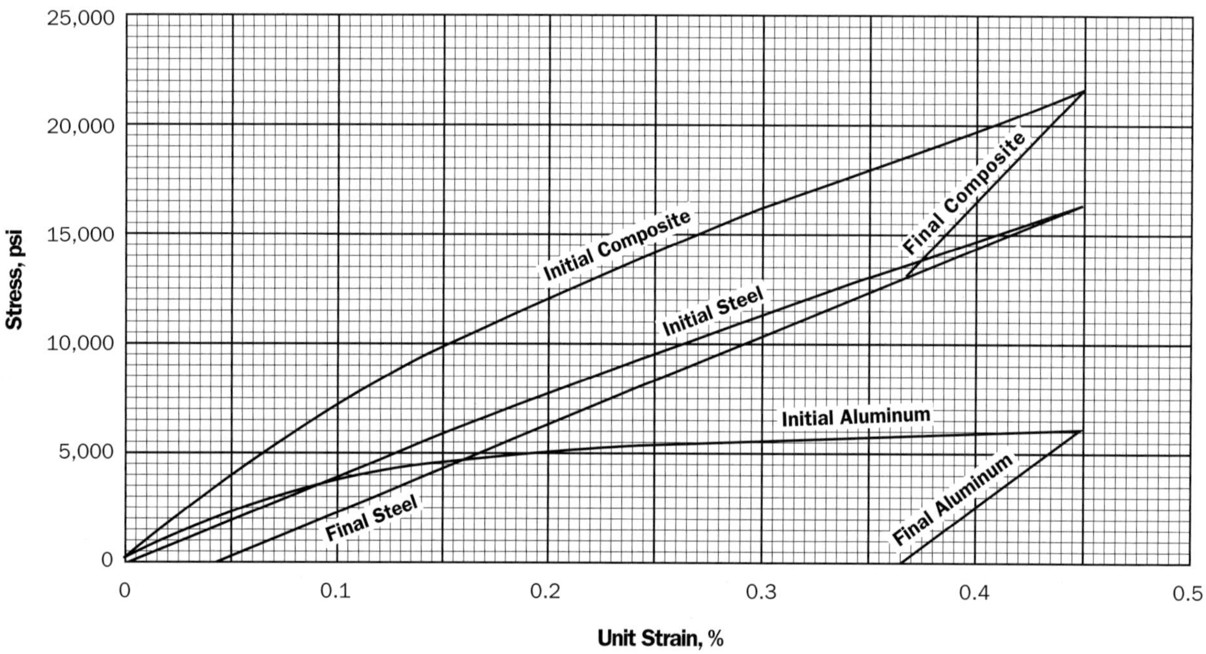

Equations for Curves (X = unit strain in %: Y = stress in kpsi):

Initial Composite: $Y = 133.67 + (8.73 \times 10^4)X - (1.97 \times 10^5)X^2 + (3.46 \times 10^5)X^3 - (2.37 \times 10^5)X^4$

Initial Steel: $Y = -21.97 + (3.90 \times 10^4)X - (1.53 \times 10^4)X^2 + (2.50 \times 10^4)X^3 - (2.77 \times 10^4)X^4$

Initial Aluminum: $Y = 155.64 + (4.83 \times 10^4)X - (1.82 \times 10^5)X^2 + (3.21 \times 10^5)X^3 - (2.10 \times 10^5)X^4$

Final Composite: $Y = (107.55X - 27.07) \times 10^3$

Final Steel: $Y = (38.225X - 1.6116) \times 10^3$

Final Aluminum: $Y = (69.325X - 25.46) \times 10^3$

sags and tensions are usually calculated assuming both the specified ice and/or wind loading has occurred, and the conductor has experienced 10 years of predicted creep elongation at a conductor temperature of 60°F at the user-specified initial tension. Unusual or extreme ice and/or wind loading may be included in a design consideration. Determination of whether the unusual or extreme ice and/or wind load will be considered a random event or should be included in the loading of the design parameters must be considered when preparing the calculations.

Sag-tension calculations for ACSR conductors are more complex than such calculations with AAC, AAAC or ACAR conductors. The complexity results from the different behavior of steel and aluminum strands in response to tension and temperature. Steel wires do not exhibit creep elongation, and plastic elongation is negligible in the range of conductor operating tensions. Aluminum wires do creep and respond plastically to high stress levels. Also, they elongate at approximately twice the rate of steel wires in response to changes in temperature.

Table 2-13 presents various initial and final sag-tension values for a 600 feet span of an ACSR "Drake" conductor under heavy loading conditions. Note that the tensions in the aluminum and steel components are given separately. In particular some useful observations are:

(1) At 60°F, without ice or wind, the tension level in the aluminum strands decreases with time as the strands permanently elongate due to creep.

(2) Both the initial and final tension in the aluminum strands decrease with increasing conductor temperature, reaching zero tension at 212°F and 167°F for initial and final conditions, respectively.

(3) At the highest temperature (212°F), where all the tension is in the steel core, the initial and final sag-tensions are nearly the same, illustrating the fact that the steel core is practically elastic.

2.7 LINE DESIGN SAG-TENSION PARAMETERS

In laying out a transmission line, the first step is to survey the route and draw up a plan-profile of the selected right-of-way. The plan-profile drawings serve an important function in linking together the various stages involved in the design and construction of the line. These drawings, based on the route survey, show the location and elevation of all natural and man-made obstacles to be traversed by, or adjacent to, the proposed line. These plan-profiles are drawn to scale and provide the basis for tower spotting and line design work.

Once the plan-profile is completed, one or more estimated ruling spans for the line may be selected. Based on these estimated ruling spans and the maximum design tensions, sag-tension data may be calculated providing initial and final sag values. From this data, sag templates may be constructed to the same scale as the plan-profile for each ruling span, and used to graphically spot structures. Other parameters, such as wind span, weight span and uplift, are used in the design of structures.

2.7.1 Catenary Constants

Using the Parabolic Equation 2-3, the sag in a ruling span is approximately equal to the weight per unit length, w, times the spans length, S, squared, divided by 8 times the horizontal component of the conductor tension, H. The ratio of conductor horizontal tension, H, to weight per unit length, w, is the catenary constant, H/w. For a ruling span sag-tension calculation using eight loading conditions, a total of 16 catenary constant values could be defined, one for initial and final tension under each loading condition.

Catenary constants can be defined for each loading condition of interest and are used in locating structures. Some typical uses of catenary constants for locating structures are to avoid overloading, assure ground clearance is sufficient at all points along the right-of-way and minimize blowout or uplift under cold weather conditions. To do this, catenary constants are typically found for; (1) the maximum line temperature; (2) heavy ice and wind loading; (3) wind blowout; and (4) minimum conductor temperature. Under any of these loading conditions, the catenary constant allows sag calculation at any point within the span.

2.7.2 Wind Span

The wind span of any structure is equal to the sum of the horizontal distances measured from a structure to the center of each adjacent span. Traditional line design practice defines the wind span for a structure as the average of the two adjacent spans. The wind span is used to determine the maximum horizontal, or transverse, force a structure must withstand under high wind conditions. The transverse wind load on a structure is the product of the horizontal wind load per unit length of cable times the wind span. The wind span is not dependent on conductor sag or tension, only on horizontal span length.

TABLE 2-13
SAG-TENSION OF AN ACSR "DRAKE" SHOWING COMPONENT TENSIONS

Span = 600 ft

Creep is ***not*** a factor — *NESC Heavy Loading District*

Temp. °F	Ice in	Wind lbs/ft²	K lbs/ft	Resultant Weight lbs/ft	Final		Initial	
					Sag ft	Tension lbs	Sag ft	Tension lbs
0	0.50	4.00	0.30	2.509	11.14	10153	11.14	10153
						5414 Al		5414 Al
						4738 St		4738 St
32	0.50	0.00	0.00	2.094	44.54	8185	11.09	8512
						3819 Al		4343 Al
						4366 St		4169 St
−20	0.00	0.00	0.00	1.094	6.68	7372	6.27	7855
						3871 Al		4465 Al
						3501 St		3390 St
0	0.00	0.00	0.00	1.094	7.56	6517	6.89	7147
						3111 Al		3942 Al
						3406 St		3205 St
30	0.00	0.00	0.00	1.094	8.98	5490	7.95	6197
						2133 Al		3201 Al
						3357 St		2996 St
60	0.00	0.00	0.00	1.094	10.44	4725 *	9.12	5402
						1321 Al		2526 Al
						3404 St		2875 St
90	0.00	0.00	0.00	1.094	11.87	4157	10.36	4759
						634 Al		1922 Al
						3522 St		2837 St
120	0.00	0.00	0.00	1.094	13.24	3727	11.61	4248
						35 Al		1379 Al
						3692 St		2869 St
167	0.00	0.00	0.00	1.094	14.29	3456	13.53	3649
						0 Al		626 Al
						3456 St		3022 St
212	0.00	0.00	0.00	1.094	15.24	3241	15.24	3241
						0 Al		0 Al
						3241 St		3239 St

* Design Condition

2.7.3 Weight Spans

The weight span of a structure is a measure of the maximum vertical force a structure must be designed to withstand. The weight span, of a structure, is equal to the sum of the horizontal distances between the low points of sag of the two adjacent spans. The maximum weight span for a structure is dependent upon the design ice and wind loading condition. The maximum vertical load transferred by a conductor to a structure is the product of the vertical load per unit length of cable times the weight span of the structure. When the elevations of adjacent structures are the same, the wind and weight spans are equal. If the weight span is found to be negative, the structure will experience uplift.

2.7.4 Uplift at Suspension Structures

Conductor uplift, shown in Figure 2-17, occurs when the weight span of a structure is negative. On steeply inclined spans, the low point of sag may fall beyond the lower support. This indicates that the conductor in the uphill span is exerting a negative, or upward, force on the lower tower. The amount of this upward force is equal to the weight of the conductor from the lower tower to the low point in the sag. If the upward pull of the uphill span is greater than the downward load of the next adjacent span, actual uplift will occur and the conductor will swing free of the tower. This usually occurs under minimum temperature conditions and must be dealt with by adding weights to the insulator suspension string or using a strain structure.

FIGURE 2-17
CONDUCTOR UPLIFT

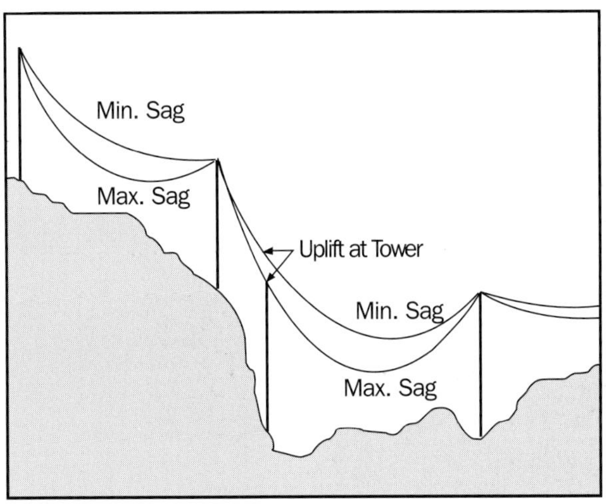

2.7.5 Tower Spotting

Given sufficiently detailed plan-profile drawings, structure heights, wind/weight spans, catenary constants, and minimum ground clearances, structure locations can be chosen such that ground clearance is maintained and structure loads are acceptable. Tower spotting can be performed using a sag template, plan-profile drawing and structure heights, or by using one of several commercial computer programs. Tower spotting and optimization of line design parameters are discussed in Chapter 5.

2.8 CONDUCTOR INSTALLATION

Installation of a bare overhead conductor can present complex problems. Careful planning and a thorough understanding of stringing procedures are needed to prevent damage to the conductor during the stringing operations. The selection of stringing sheaves, tensioning method and measurement techniques are critical factors in obtaining the desired sagging results. Conductor stringing and sagging equipment and techniques are discussed in detail in the 524™ *IEEE Guide to the Installation of Overhead Transmission Line Conductors*, IEEE Std 524-2003[20]. Some basic factors concerning installation are covered in this section. Because the terminology used for equipment and installation procedures for overhead conductor varies throughout the utility industry, a limited glossary of terms and equipment definitions excerpted from IEEE[20] is provided in the Chapter 2 Appendix.

2.8.1 Conductor Stringing Methods

There are two basic methods of stringing conductors, categorized as either slack or tension stringing. There are as many variations of these methods as there are organizations installing conductors. The selected method, however, depends primarily on the terrain and the sensitivity to conductor surface damage.

Slack or Layout Stringing Method

Slack stringing of conductor is normally limited to lower voltage lines and smaller conductors where some surface damage is tolerable. The conductor reel(s) are placed on reel stands or "jack stands" at the beginning of the stringing location. The conductor is unreeled from the shipping reel and dragged along the ground by means of a vehicle or pulling device. When the conductor is dragged past a supporting structure, pulling is stopped and the conductor placed in stringing sheaves attached to the structure. The conductor is then reattached to the pulling equipment and the pull is continued to the next structure.

FIGURE 2-18
TENSION STRINGING EQUIPMENT SETUP

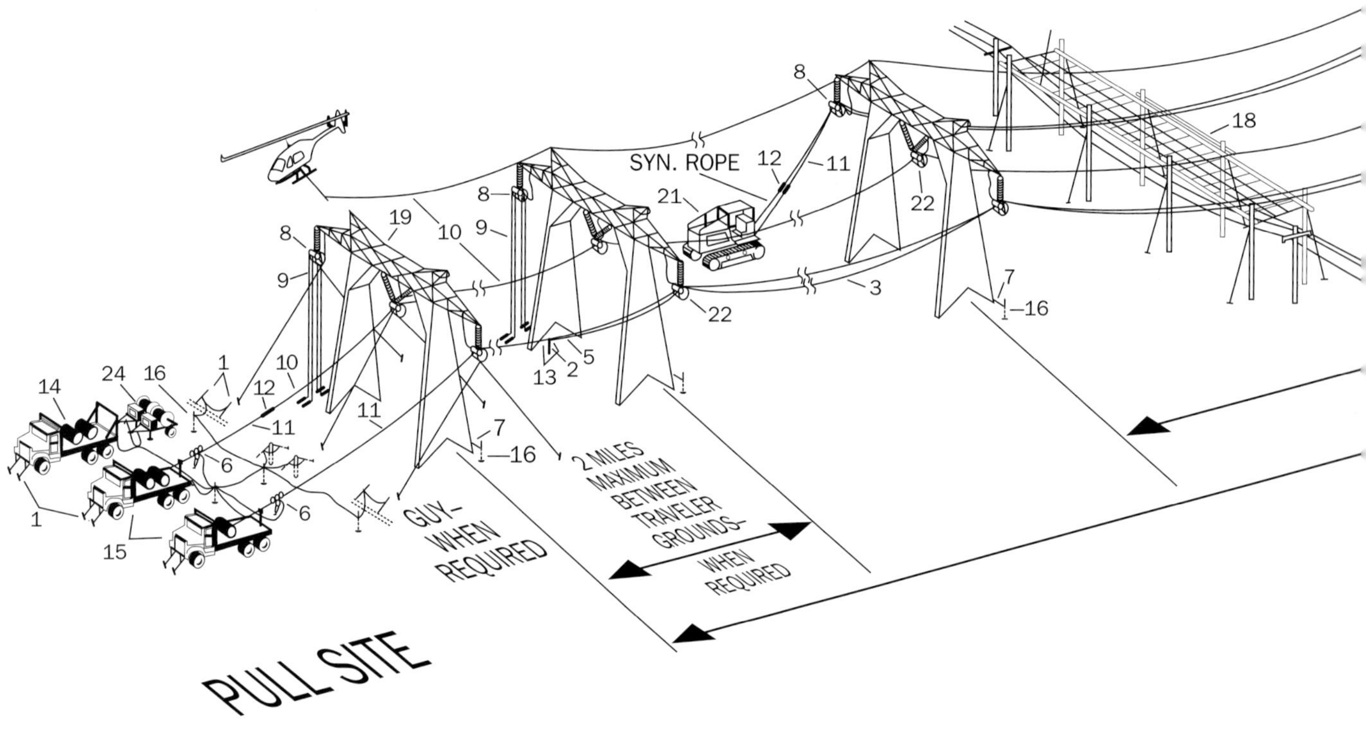

1. Anchor
2. Running board
3. Bundled conductor
4. Ground grid
5. Woven wire grip
6. Running ground
7. Structure base ground
8. Traveler ground
9. Finger line
10. Pilot line
11. Pulling line
12. Connector link
13. Swivel link
14. Bullwheel puller
15. Drum puller
16. Ground rod
17. Reel stand
18. Crossing structure
19. Snub structure
20. Bullwheel tensioner
21. Crawler tractor
22. Traveler
23. Pilot line winder
24. Reel winder

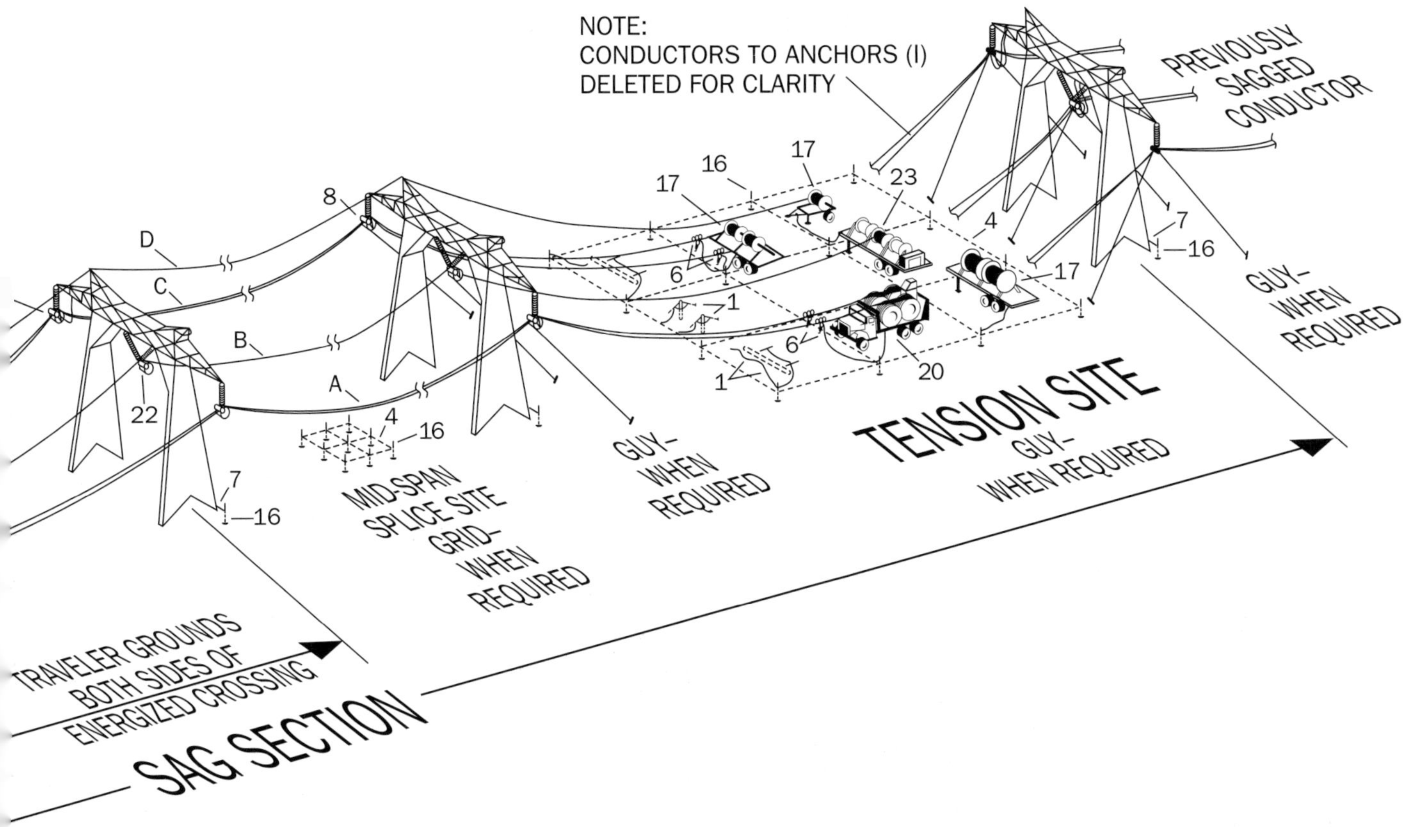

A. Typical stringing arrangement
B. Typical pulling line installation with pilot line winder
C. Typical installation of pulling lines with tractor and finger lines
D. Typical installation of pilot lines with helicopter

This stringing method is typically used during construction of new lines in areas where the right-of-way is readily accessible to vehicles used to pull the conductor. However, slack stringing may be used for repair or maintenance of transmission lines where rugged terrain limits use of pulling and tensioning equipment. The slack stringing method is seldom used in urban areas or where there is any danger of contact with high voltage conductors.

Tension Stringing

A tension stringing method is normally employed when installing transmission conductors. Using this method, the conductor is unreeled under tension and is not allowed to contact the ground. In a typical tension stringing operation, travelers are attached to each structure. A pilot line is pulled through the travelers and is used, in turn, to pull in heavier pulling line. This pulling line is then used to pull the conductor from the reels and through the travelers. The tension on the conductor is controlled by the tension puller at the pulling end, and the bullwheel tension retarder at the conductor payout end of the installation. Tension stringing is preferred for all transmission installations. This installation method keeps the conductor off the ground, minimizing the possibility of surface damage and limiting problems at roadway crossings. It also limits damage to the right-of-way by minimizing heavy vehicular traffic.

2.8.2 Tension Stringing Equipment and Setup

Stringing equipment typically includes bullwheel or drum pullers for back-tensioning the conductor during stringing and sagging; travelers (stringing blocks) attached to every phase conductor and shield wire attachment point on every structure; a bullwheel or crawler tractor for pulling the conductor through travelers; and various other special items of equipment. Figure 2-18 illustrates a typical stringing and sagging setup for a stringing section and the range of stringing equipment required. Provision for conductor splicing prior to sagging must be made at the tension site or midspan sites to avoid pulling splices through the travelers.

During the stringing operation, it is necessary to use proper tools to grip the strands of the conductor evenly to avoid damaging the outer layer of wires. Two basic types or categories of grips are normally used in transmission construction. The first is a type of grip referred to as a pocketbook, suitcase, bolted, etc., that hinges to completely surround the conductor and incorporates a bail for attaching to the pulling line. The second type grip shown in Figure 2-19, is similar to a Chinese finger grip and often referred to as a basket or Kellems® grip. It is often used because its flexibility and small size make it easily pulled through sheaves during the stringing operation. Whatever type of gripping device is used, a swivel should be installed between the pulling grip and pulling line, or running board, to allow free rotation of both the conductor and the pulling line.

A traveler consists of a sheave or pulley wheel enclosed in a frame to allow it to be suspended from structures or insulator strings. The frame must have some type of latching mechanism to allow insertion and removal of the conductor during the stringing operation. Travelers are designed for a maximum safe working load. This safe working load must not be exceeded during the stringing operation. Sheaves are often lined with neoprene or urethane materials to prevent scratching of conductor in high voltage applications, however, unlined sheaves are also available for special applications.

Travelers used in tension stringing must be free-rolling and capable of withstand high running or static loads without damage. Proper maintenance is essential. Very high longitudinal tension loads can develop on transmission structures if a traveler should "freeze" during tension stringing, possibly causing conductor and/or structure damage. Significant levels of rotation resistance will also yield tension differences between spans, resulting in incorrect sag.

Proper selection of travelers is important to assure that travelers operate correctly during tension stringing and sagging. The sheave diameter and the groove radius must be matched to the conductor.

Figure 2-20 illustrates the minimum sheave diameter for typical stringing and sagging operations. Larger diameter sheaves may be required if particularly severe installation conditions exist.

FIGURE 2-19
BASKET GRIP PULLING DEVICE

FIGURE 2-20
RECOMMENDED SHEAVE DIMENSIONS

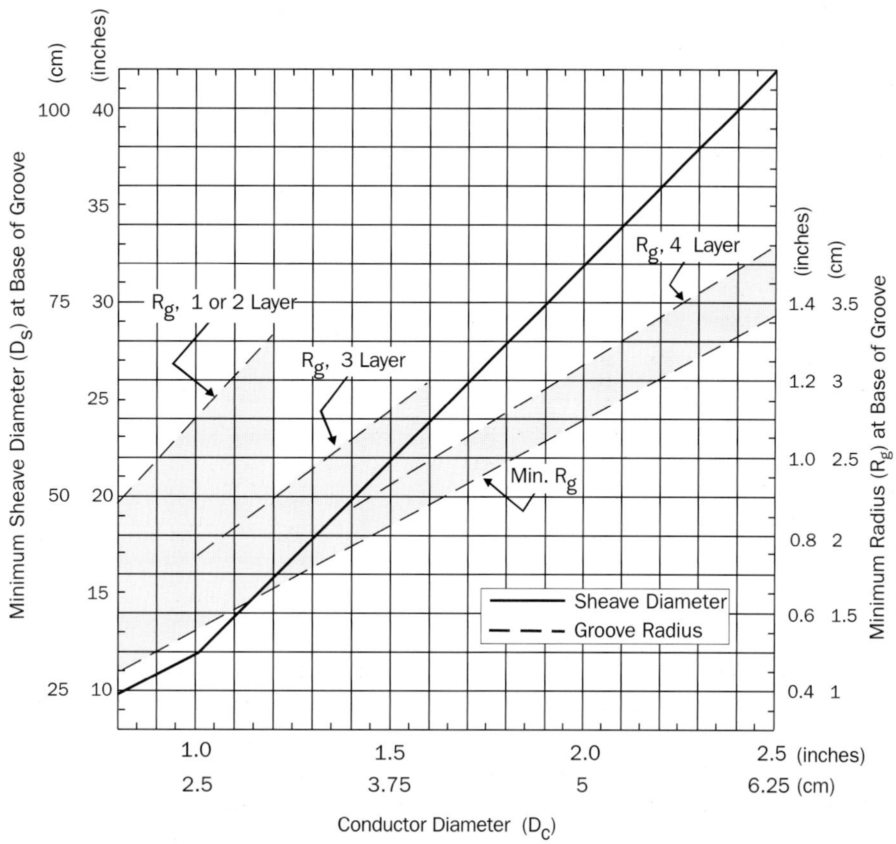

2.8.3 Sagging Procedure

It is important that the conductors be properly sagged, at the correct stringing tension and reference temperature for the design ruling span. A series of several spans, called a line section, is usually sagged in one operation. To obtain the correct sags, and to insure the suspension insulators hang vertically, the horizontal tension in all spans must be equal. Figures 2-22 through 2-27 presented in the Chapter 2 Appendix depict typical parabolic methods and computations required for sagging conductors. Factors that must be considered for sagging conductors are discussed in the following sub-sections.

Creep Elongation During Stringing

Upon completion of conductor stringing, a time of up to several days may elapse before the conductor is tensioned to design sag. Since the conductor tension during the stringing process is normally well below the initial sagging tension, and because the conductor remains in the stringing sheaves for only a few days or less, any elongation due to creep is normally neglected. The conductor should be sagged to the initial stringing sags listed in the sag tables. However, if the conductor tension is excessively high during stringing, or the conductor is allowed to remain in the blocks for an extended period of time, an abnormal amount of creep will occur. If this occurs, the stringing sag tables should be corrected to compensate for the additional creep elongation. Using the creep estimation formulas given in Chapter 3, the creep strain can be estimated and adjustments made accordingly.

Pre-stressing Conductor

Pre-stressing is sometimes used to stabilize the elongation of a conductor for some defined period of time. The pre-stressing tension is normally much higher than the unloaded design tension for a conductor. The degree of stabilization is dependent upon the time maintained at the pre-stressing tension. After pre-stressing, the tension on the conductor is reduced to stringing or design tension limits. At this reduced tension, the creep or plastic elongation of the conductor has been temporarily halted, reducing the permanent elongation due to strain and creep for a defined period of time. By tensioning an ACSR a conductor to levels approaching 50% of its rated breaking

TABLE 2-14

TIME-SAG TABLE FOR STOPWATCH METHOD

Return of Wave											
Sag in	**3rd Time sec**	**5th Time sec**	**Sag in**	**3rd Time sec**	**5th Time sec**	**Sag in**	**3rd Time sec**	**5th Time sec**	**Sag in**	**3rd Time sec**	**5th Time sec**
5	1.9	3.2	55	6.4	10.7	105	8.8	14.7	155	10.7	17.9
6	2.1	3.5	56	6.5	10.8	106	8.9	14.8	156	10.8	18.0
7	2.3	3.8	57	6.5	10.9	107	8.9	14.9	157	10.8	18.0
8	2.4	4.1	58	6.6	11.0	108	9.0	15.0	158	10.9	18.1
9	2.6	4.3	59	6.6	11.1	109	9.0	15.0	159	10.9	18.1
10	2.7	4.6	60	6.7	11.1	110	9.1	15.1	160	10.9	18.2
11	2.9	4.8	61	6.7	11.2	111	9.1	15.2	161	11.0	18.2
12	3.0	5.0	62	6.8	11.3	112	9.1	15.2	162	11.0	18.2
13	3.1	5.2	63	6.9	11.4	113	9.2	15.3	163	11.0	18.4
14	3.2	5.4	64	6.9	11.5	114	9.2	15.4	164	11.1	18.4
15	3.3	5.6	65	7.0	11.6	115	9.3	15.4	165	11.1	18.5
16	3.5	5.8	66	7.0	11.7	116	9.3	15.5	166	11.1	18.5
17	3.6	5.9	67	7.1	11.8	117	9.3	15.6	167	11.2	18.6
18	3.7	6.1	68	7.1	11.9	118	9.4	15.6	168	11.2	18.7
19	3.8	6.3	69	7.2	12.0	119	9.4	15.7	169	11.2	18.7
20	3.9	6.4	70	7.2	12.0	120	9.5	15.8	170	11.3	18.8
21	4.0	6.6	71	7.3	12.1	121	9.5	15.8	171	11.3	18.8
22	4.0	6.7	72	7.3	12.2	122	9.5	15.9	172	11.3	18.9
23	4.1	6.9	73	7.4	12.3	123	9.6	16.0	173	11.4	18.9
24	4.2	7.0	74	7.4	12.4	124	9.6	16.0	174	11.4	19.0
25	4.3	7.2	75	7.5	12.5	125	9.7	16.1	175	11.4	19.0
26	4.4	7.3	76	7.5	12.5	126	9.7	16.2	176	11.4	19.1
27	4.5	7.5	77	7.6	12.6	127	9.7	16.2	177	11.5	19.1
28	4.6	7.6	78	7.6	12.7	128	9.8	16.3	178	11.5	19.2
29	4.6	7.7	79	7.7	12.8	129	9.8	16.3	179	11.5	19.3
30	4.7	7.9	80	7.7	12.9	130	9.8	16.4	180	11.6	19.3
31	4.8	8.0	81	7.8	13.0	131	9.9	16.5	181	11.6	19.4
32	4.9	8.1	82	7.8	13.0	132	9.9	16.5	182	11.6	19.4
33	5.0	8.3	83	7.9	13.1	133	10.0	16.6	183	11.7	19.5
34	5.0	8.4	84	7.9	13.2	134	10.0	16.7	184	11.7	19.5
35	5.1	8.5	85	8.0	13.3	135	10.0	19.7	185	11.7	19.6
36	5.2	8.6	86	8.0	13.3	136	10.1	16.8	186	11.8	19.6
37	5.3	8.8	87	8.1	13.4	137	10.1	16.8	187	11.8	19.7
38	5.3	8.9	88	8.1	13.5	138	10.1	16.8	188	11.8	19.7
39	5.4	9.0	89	8.1	13.6	139	10.2	17.0	189	11.9	19.8
40	5.5	9.1	90	8.2	13.7	140	10.2	17.0	190	11.9	19.8
41	5.5	9.2	91	8.2	13.7	141	10.3	17.1	191	11.9	19.9
42	5.6	9.3	92	9.3	13.8	142	10.3	17.1	192	12.0	19.9
43	5.7	9.4	93	8.3	13.9	143	10.3	17.2	193	12.0	20.0
44	5.7	9.5	94	8.4	14.0	144	10.4	17.3	194	12.0	20.0
45	5.8	9.7	95	8.4	14.0	145	10.4	17.3	195	12.1	20.1
46	5.9	9.8	96	8.5	14.1	146	10.4	17.4	196	12.1	20.1
47	5.9	9.9	97	8.5	14.2	147	10.5	17.4	197	12.1	20.2
48	6.0	10.0	98	8.5	14.2	148	10.5	17.5	198	12.1	20.2
49	6.0	10.1	99	8.6	14.3	149	10.5	17.6	199	12.2	20.3
50	6.1	10.2	100	8.6	14.4	150	10.6	17.6	200	12.2	20.3
51	6.2	10.3	101	8.7	14.5	151	10.6	17.7	201	12.2	20.4
52	6.2	10.4	102	8.7	14.5	152	10.6	17.7	202	12.3	20.5
53	6.3	10.5	103	8.8	14.6	153	10.7	17.8	203	12.3	20.5
54	6.3	10.6	104	8.8	14.7	154	10.7	17.9	204	12.3	20.6

Note: To calculate the time of return of other waves, multiply the time in seconds for one wave return by the number of wave returns or, more simply, select the combinations of values from the table that represents the number of wave returns desired. For example, the time of return of the 8th wave is the sum of the 3rd and 5th, while for the 10th wave it is twice the time of the 5th.

The approximate formula giving the relationship between sag and time is given as:

$$D = 12.075 \left(\frac{T}{N}\right)^2 \text{ inches}$$

$$D = 1.00325 \left(\frac{T}{N}\right)^2 \text{ feet}$$

where:

D = sag, in

T = time, sec

N = number of return waves counted

strength for 20-30 hours, creep elongation will be temporarily halted[21]. For an ACSS conductor, the held time is substantially less. This simplifies concerns about creep during subsequent installation but presents both equipment, and safety problems.

Sagging by Stopwatch Method

A mechanical pulse imparted to a tensioned conductor moves at a speed proportional to the square root of tension divided by weight per unit length. By initiating a pulse on a tensioned conductor and measuring the time required for the pulse to move to the nearest termination and back, the tension and sag of the conductor can be determined. This stopwatch method has come into wide use even for long spans and large conductors[22].

This process is started by giving the conductor a sharp blow near one support and simultaneously starting a stopwatch. A mechanical wave moves from the point where the conductor was struck to the next support point, where it will be partially reflected. If the initiating blow is sharp, the wave will travel up and down the span many times before dying out. Time-sag tables, such as the one shown in Table 2-14, are available from many sources. Specially designed sagging stopwatches are also available.

The reflected wave can be detected by lightly touching the conductor, but the procedure is more likely to be accurate if the wave is both initiated and detected with a light rope over the conductor. Normally, the time for the return of the 3rd or 5th wave is monitored.

Traditionally, the transit sagging method has been considered to be more accurate for sagging than the stopwatch method. However, many transmission-line constructors use the stopwatch method exclusively, even with large conductors.

Sagging by Transit Methods

IEEE Standard 524[20] lists three methods of sagging conductor with a transit: "Calculated Angle of Sight," "Calculated Target Method" and "Horizontal Line of Sight." The method best suited to a particular line sagging situation may vary with terrain and line design.

Sagging Accuracy

Sagging a conductor during construction of a new line, or in the reconductoring of an old line, involves many variables that can lead to a small degree of error. IEEE Standard 524[20] suggests that all sags be within 6 inches of the stringing sag values. Additionally, possible errors in terrain measurement, variations in conductor properties, loading conditions and hardware installation have led some utilities to allow up to a 3 feet safety buffer in addition to the required minimum ground clearance.

Clipping Offsets

If conductor is to be sagged in a series of suspension spans, with span lengths approximately equal and the terrain reasonably level, the conductor is sagged using conventional stringing sag tables and the conductor is clipped into suspension clamps that replace the travelers. If conductor is to be sagged in a series of suspension spans where span lengths vary widely, or more commonly, where the terrain is steep, then clipping offset calculations may need to be employed in order to yield vertical suspension strings after installation.

Clipping offsets are illustrated in Figure 2-21[20] showing a series of steeply inclined spans terminated in a "snub" structure at the bottom and a "deadend" structure at the top. The vector diagram illustrates a balance of total conductor tension in the travelers, but an imbalance in the horizontal component of tension.

FIGURE 2-21
CLIPPING OFFSET ILLUSTRATION

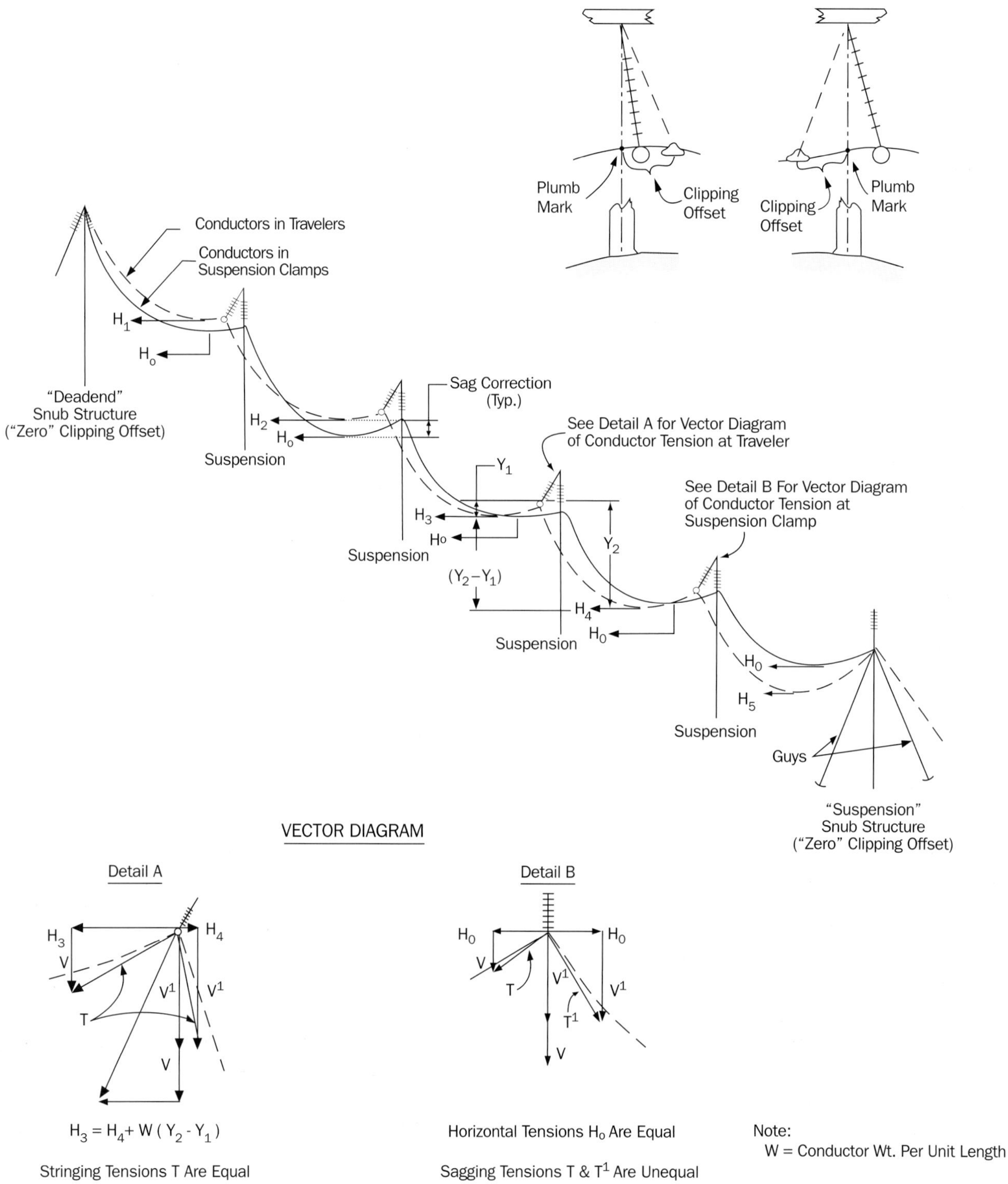

2.9 REFERENCES

[1] Ehrenburg, D.O., "Transmission Line Catenary Calculations," AIEE Paper, Committee on Power Transmission & Distribution, July 1935.

[2] Winkelman, P.F., "Sag-Tension Computations and Field Measurements of Bonneville Power Administration," AIEE Paper 59-900, June 1959.

[3] CIGRE SCB2/WG12 "Sag-Tension Calculation Methods for Overhead Lines," Technical Brochure, 2007.

[4] Aluminum Company of America, *Graphic method for the Sag Tension Calculation for ACSR and Other Conductors*, 1961.

[5] *National Electrical Safety Code®*, C2-2002, 2002 Edition, The Institute of Electrical and Electronic Engineers, Inc.

[6] RUS Bulletin 1724E-200, "Design Manual for High Voltage Transmission Lines," United States Department of Agriculture, Rural Utilities Service, September 2004.

[7] CIGRE SC22/WG11.04 "Safe Design Tension with Respect to Aeolian Vibrations. - Part I: Single unprotected conductors," ELECTRA, No. 186, pp. 53-87, October 1999.

[8] CIGRE SC22/WG11.04 "Safe Design Tensions with Respect to Aeolian Vibrations. – Part II – Damped Single Conductors with Dampers" ELECTRA, No. 198, October 2001.

[9] CIGRE SCB2/WG 11.04 "Overhead Conductor Safe Design Tension with Respect to Aeolian Vibrations" with Part III "Bundled Conductor Lines," ELECTRA, No. 220, pp. 48-59, June 2005. Technical Brochure No. 273, 2005.

[10] American Society of Civil Engineers, *Minimun Design Loads for Buildings and Other Structures*, ASCE Standard SEI/ASCE 7-05.

[11] Fink, D.G., and Beaty, H.W., *Standard Handbook for Electrical Engineers*, 13th Edition, McGraw Hill.

[12] Varney T., Aluminium Company of America, *Graphic Method for Sag Tension Calculations for ACSR and Other Conductors*, Pittsburg, 1927.

[13] Barrett, J.S., Dutta S., and Nigol, O., "A New Computer Model of A1/S1A (ACSR) Conductors," IEEE Trans., Vol. PAS-102, No. 3, pp. 614-621, March 1983.

[14] The Aluminum Association, "Stress-Strain-Creep Curves for Aluminum Electrical Conductors," A Technical Report of the Aluminum Association's Technical Committee on Electrical Conductors, Updated May 1997.

[15] The Aluminum Assocoation, "A Method of Stress-Strain Testing of Aluminium Conductor and ACSR" and "A Test Method for Determining the Long Time Tensile Creep of Aluminium Conductors in Overhead Lines," January, 1999.

[16] Rawlins, C.B., "Some Effects of Mill Practice on the Stress-Strain Behaviour of ACSR," IEEE WPM 1998, Tampa, FL, February 1998.

[17] Harvey, J.R., "Creep of Transmission Line Conductors," IEEE Trans., Vol. PAS-88, No. 4, pp. 281-285, April 1969.

[18] Harvey, J.R., Larson, R.E. "Creep Equations of Conductors for Sag-Tension Calculations," 1972 IEEE/PES Winter Meeting Conference Paper No. C72 190-2.

[19] Barrett, J.S., and Nigol, O., "Characteristics of A1/S1A (ACSR) Conductors as High Temperatures and Stresses," IEEE Trans., Vol. PAS-100, No. 2, pp. 485-493, February 1981.

[20] *524™ IEEE Guide to the Installation of Overhead Transmission Line Conductors*, IEEE Std 524™-2003, Published by The Institute of Electrical and Electronics Engineers, Inc., New York, NY.

[21] Cahill, T., "Development of Low-Creep ACSR Conductor," *Wire Journal*, July 1973.

[22] Overend, P.R., and Smith, S., "Impulse Time Method of Sag Measurement."

2.10 ADDITIONAL PUBLICATIONS

[1] Dwight, H.B., "Sag Calculations for Transmission Lines," AIEE Transactions, pp. 796-805, May 1926.

[2] Lummis, J., Ficher, H.D., "Practical Application of Sag and Tension Calculations to Transmission Line Design," AIEE Trans. Paper 54-501.

[3] Martin, J.S., *Sag Calculations by the use of Martin's Table,* Copperweld Steel Co., 1961.

[4] Healy, E.S., Wright, A.J., "Unbalanced Conductor Tensions," AIEE Transactions, pp. 1064-1070, September 1926.

[5] IEEE Working Group, "Limitations of the Ruling Span Method for Overhead Line Conductors at High Operating Temperatures". Report of IEEE WG on Thermal Aspects of Conductors, of the Towers, Poles and Conductors Subcommittee, IEEE Transactions on Power Delivery, Vol. 14, No. 2, April 1999.

[6] CIGRE WG 22.05, "PA Practical Method of Conductor Creep Determination" Electra, No. 24, pp. 105-137, October 1972.

[7] CIGRE WG 22.05 (12), "Permanent Elongation of Conductors. Predictor Equations and Evaluation Methods," Electra, No. 75, pp. 63-98, March 1981.

[8] IEC 61395, "Overhead Electrical Conductors – Creep Test Procedures for Stranded Conductors."

[9] CIGRE Technical Brochure 244, "Conductors for the Uprating of Overhead Lines," 2004.

[10] Seppa, Tapani O., "Factors Influencing the Accuracy of High Temperature Sag Calculations," IEEE/PES Summer Power Meeting Paper No. 93 SM 440-8 PWRD, 1993.

[11] Harvey, John R., "Use of Elevated-Temperature Creep Data in Sag-Tension Calculations," IEEE Trans. Power Apparatus and Systems, Vol. PAS-89, No. 3, March 1970.

[12] Chen, S.L., Black, W.Z., and Fancher, M.L., "High-Temperature Sag Model for Overhead Conductors," IEEE Trans. Power Delivery, Vol. 18, No. 1, January 2003.

[13] Black, W.Z., Collins, S.S., and Hall, J.F., "Theoretical Model for Temperature Gradients within Bare Overhead Conductors," IEEE Trans. Power Delivery, Vol. 3, pp. 707-715, April 1988.

CHAPTER 2
APPENDIX

FIGURE 2-22

NOMOGRAPH FOR DETERMINING LEVEL SPAN EQUIVALENTS ON NON-LEVEL SPANS

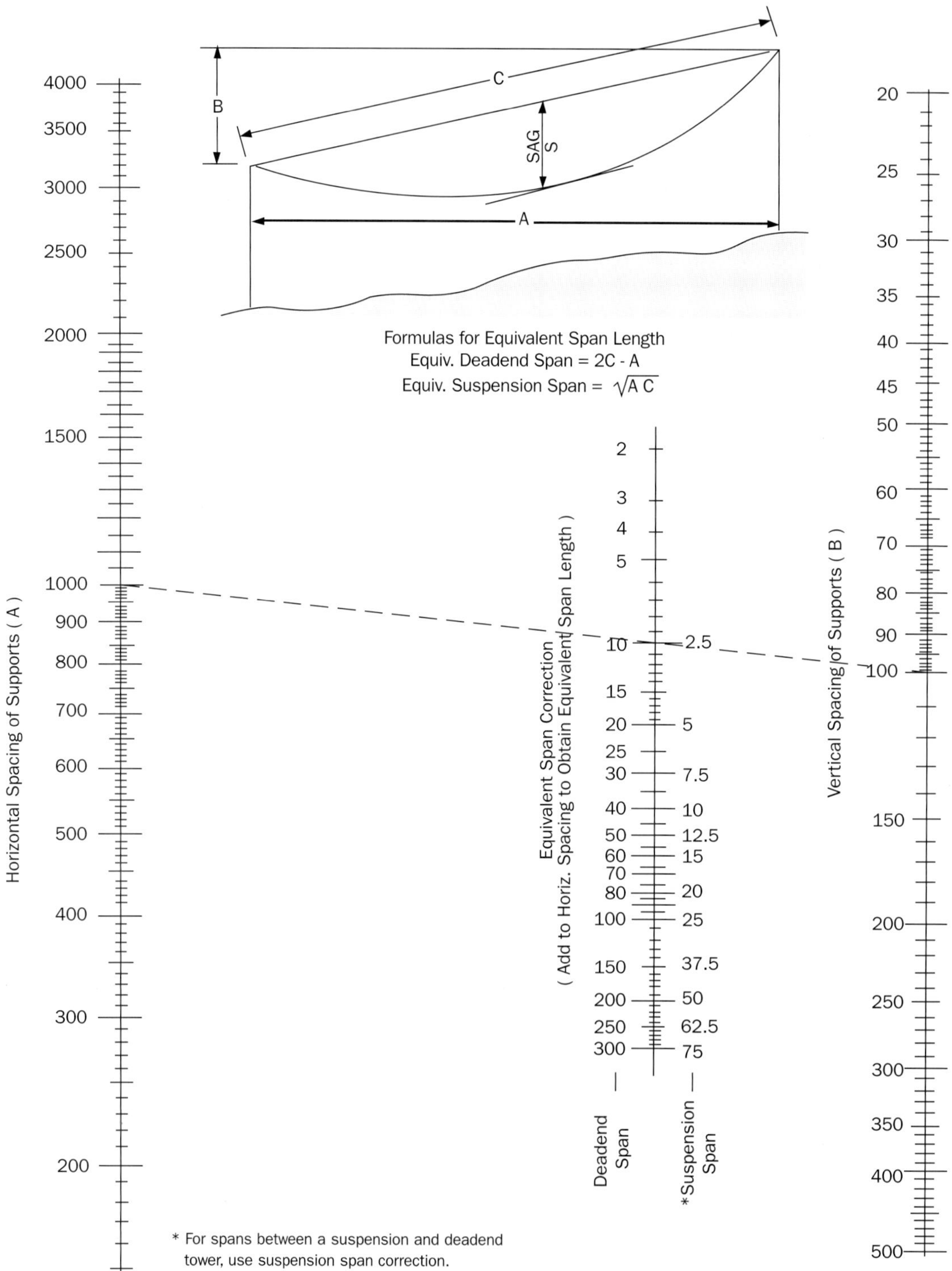

* For spans between a suspension and deadend tower, use suspension span correction.

Example: Assume span with A = 1000 ft, B = 100 ft, if deadend span, correction = 10 ft (see above). If suspension span, correction = 2.5 ft (see above). Equivalent span = 1000 ft + correction. Read chart sag for equivalent span length.

Sag is based on parabolic functions.
If sag exceeds 5% of span, do not use this chart.

FIGURE 2-23

NOMOGRAPH FOR DETERMINING CONTROL FACTOR FOR CONDUCTOR SAGGING

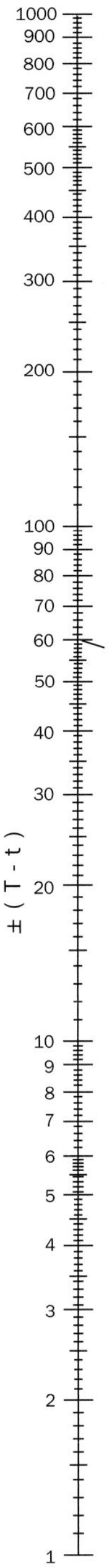

Procedure

Determine from nomograph the control factor of transit "setup" used in sagging the conductor (see examples on the right).

For most accurate result in sagging the conductor this value of control factor should not be below the curve shown below.

In all cases a control factor of 1.00 is ideal (For T = t).

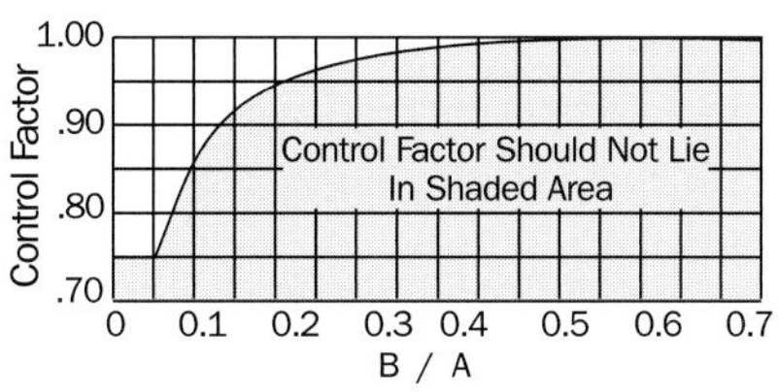

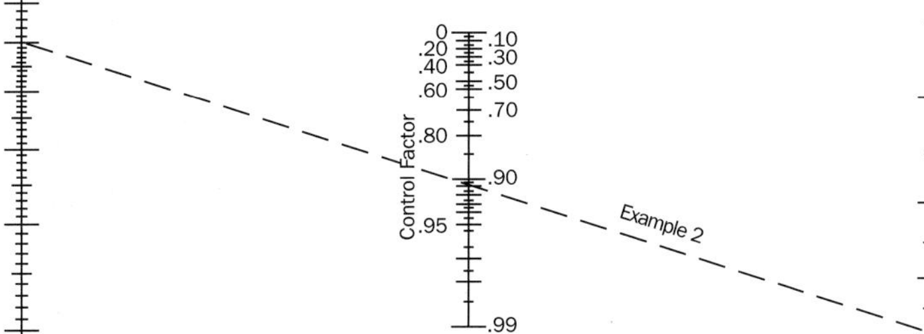

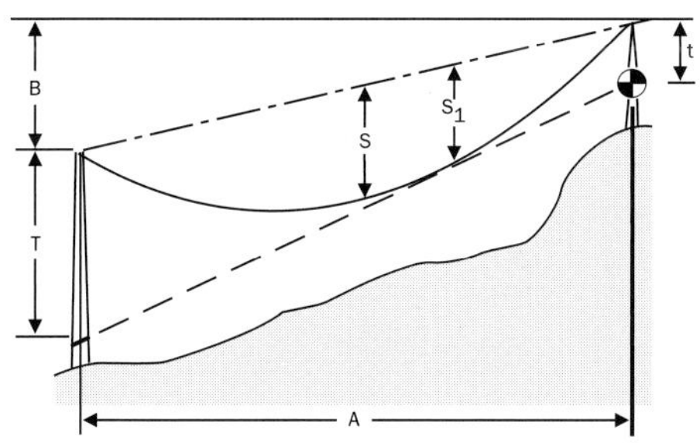

$$\text{Control Factor} = \frac{S_1}{S} = \frac{\Delta S_1}{\Delta S} = 1 - \frac{(T-t)^2}{(4\,S)^2}$$

T = Distance transit is set below conductor support.

t = Corresponding distance target is set below opposite support.

S = Conductor sag determined from stringing charts.

S_1 = Corresponding sag at point of tangency of conductor and line of sight.

ΔS = Change of sag "S".

ΔS_1 = Change of sag "S_1".

Sag is based on parabolic functions.

If sag exceeds 5% of span, do not use this chart.

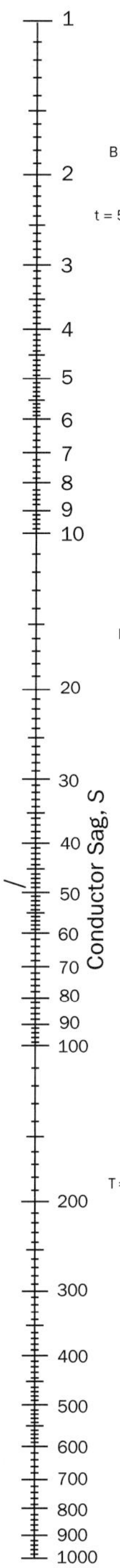

Example 1

When sagging by calculated target setting. (See Figure 2-25)

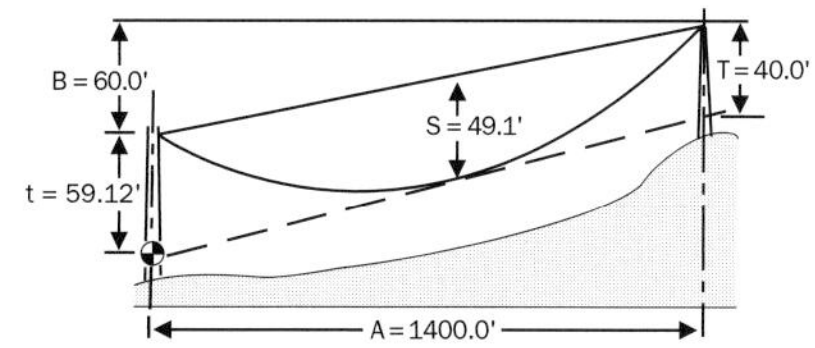

$(T-t) = -19.12'$

$S = 49.1'$

Control factor = 0.99 (From nomograph)

Example 2

When sagging by horizontal line of sight. (See Figure 2-26)

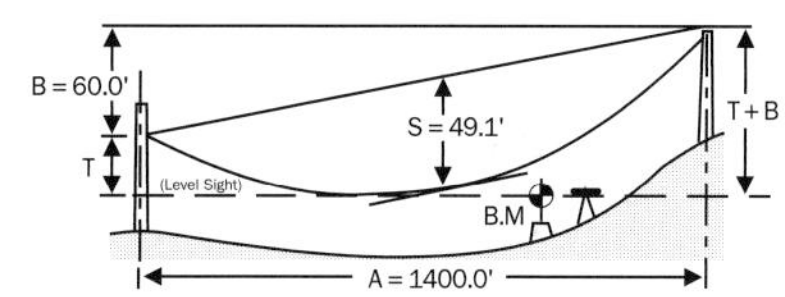

$(T-t)$ = "B" for horizontal line of sight.

= 60.0'

$S = 49.1'$

Control factor = 0.91 (From nomograph)

Example 3

When sagging by calculated angle of sight. (See Figure 2-24)

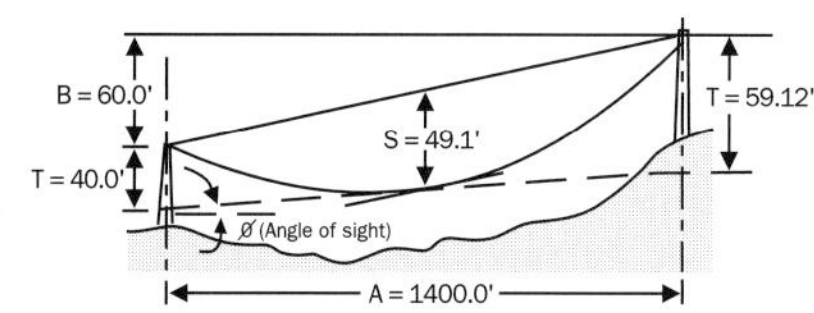

$(T-t)$ = A tan Ø − (±B)

Ø = Angle of sight.

+Ø = When angle is above horizontal.

−Ø = When angle is below horizontal.

B = Vertical distance between points of support.

+B = When support ahead is higher.

−B = When support ahead is lower.

In example, Ø = +1°40'21" or tan Ø = +0.02920

A = 1400.0'

B = +60.0'

S = 49.1'

Then (T − t) = 1400.0 (+0.02920) − (+60.0) = −19.12

Control factor = 0.99 (From nomograph)

FIGURE 2-24
CONDUCTOR SAGGING BY CALCULATED ANGLE OF SIGHT

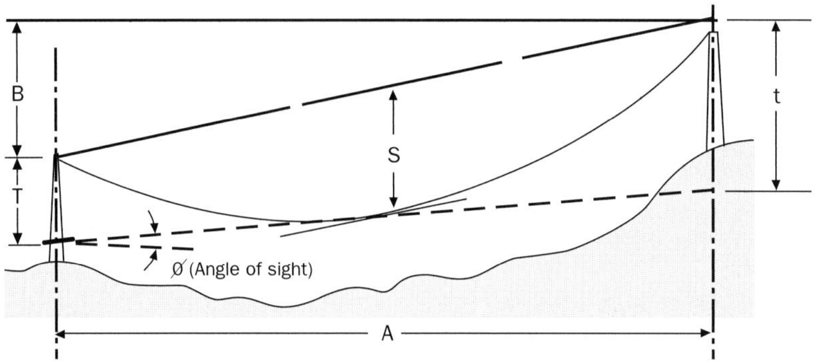

Method 1: $\text{Tan } Ø = \frac{T \pm B - t}{A}$

Method 2: $\text{Tan } Ø = \frac{B + 2T - S(2 + M)}{A}$

Ø = Angle of sight.
  + Ø when angle is above horizontal.
  – Ø when angle is below horizontal.
t = Vertical distance below support to line of sight. (See Figure 2-25)
T = Vertical distance below support for transit.
S = Sag.
A = Horizontal distance between points of support–obtained from structure list or plan and profile.
B = Vertical distance between points of support–obtained from plan and profile, tower site data sheets or field measurement.
  + B when support ahead is higher.
  – B when support ahead is lower.
M = Determined from curve on Figure 2-25.

Examples

Given:

A = 1400.0'	S = 49.1' @ 60°F
B = 60.0'	S = 51.2' @ 90°F
T = 40.0'	t = 59.12' @ 60°F
	t = 63.76' @ 90°F

Method 1

$\text{Tan } Ø = \frac{T \pm B - t}{A}$

$\text{Tan } Ø_{60°F} = \frac{40.0 + 60.0 - 59.12}{1400.0} = 0.02920$

$Ø_{60°F} = +\ 1°\ 40'\ 21"$

$\text{Tan } Ø_{90°F} = \frac{40.0 + 60.0 - 63.76}{1400.0} = 0.02589$

$Ø_{90°F} = +\ 1°\ 28'\ 59"$

Change in angle Ø for 5°F = $(1°\ 40'\ 21" - 1°\ 28'\ 59")\left(\frac{5}{30}\right) = 0°\ 1'\ 54"$

Method 2

$\text{Tan } Ø = \frac{B + 2T - S(2 + M)}{A}$

$\text{Tan } Ø_{60°F} = \frac{60.0 + (40.0)(2) - (49.1)(2 + 0.019)}{1400.0} = 0.02919$

$Ø_{60°F} = +\ 1°\ 40'\ 19"$

$\text{Tan } Ø_{90°F} = \frac{60.0 + (40.0)(2) - (51.2)(2 + 0.027)}{1400.0} = 0.02587$

$Ø_{90°F} = +\ 1°\ 28'\ 55"$

Change in angle Ø for 5°F = $(1°\ 40'\ 19" - 1°\ 28'\ 55")\left(\frac{5}{30}\right) = 0°\ 1'\ 54"$

Sag is based on parabolic functions. If sag exceeds 5% of span, do not use this chart.

FIGURE 2-25
CONDUCTOR SAGGING BY CALCULATED TARGET METHOD

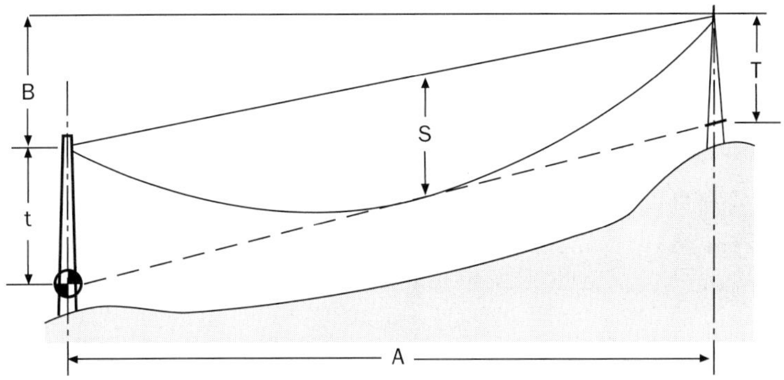

Method 1: $t = \left(2\sqrt{S} - \sqrt{T}\right)^2$

Method 2: $t = 2S - T + SM$

t = Vertical distance below support for target.
T = Vertical distance below support for transit.
S = Sag.
A = Horizontal distance between structures - obtained from structure list or plan & profile.
B = Vertical distance between points of support - obtained from plan & profile, tower site data sheets or field measurement.
M = Determined from curve below.

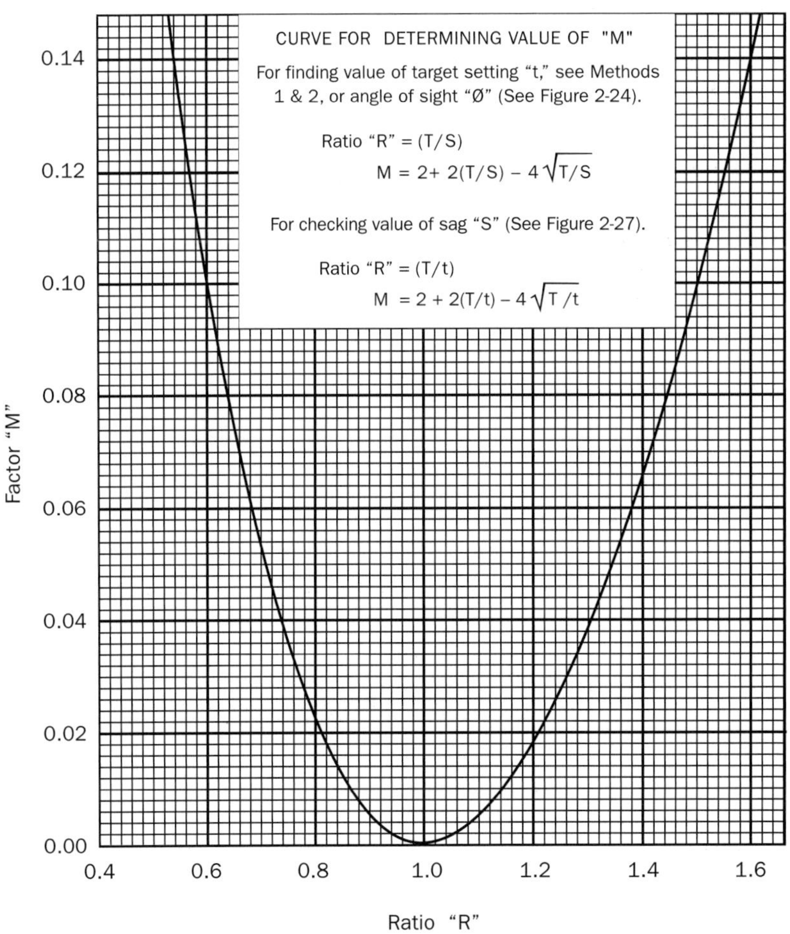

Sag is based on parabolic functions. If sag exceeds 5% of span, do not use this chart.

Examples

Given:

A = 1400.0'
B = 60.0'
T = 40.0'
S = 49.1' @ 60°F
S = 51.2' @ 90°F

Method 1

$t = \left(2\sqrt{S} - \sqrt{T}\right)^2$

$\sqrt{T} = 6.325$

$\sqrt{S_{60°F}} = 7.007$

$2\sqrt{S_{60°F}} = 14.014$

$t_{60°F} = 59.12'$

$\sqrt{S_{90°F}} = 7.155$

$2\sqrt{S_{90°F}} = 14.310$

$t_{90°F} = 63.76'$

Change in "t" for 5°F $= (63.76 - 59.12)\left(\frac{5}{30}\right) = 0.77'$

Method 2

$t = 2S - T + SM$

$T/S_{60°F} = 0.815$

$M_{60°F} = 0.019$

$2S_{60°F} = 98.2'$

$t_{60°F} = 59.13'$

$T/S_{90°F} = 0.781$

$M_{90°F} = 0.027$

$2S_{90°F} = 102.4'$

$t_{90°F} = 63.78'$

Change in "t" for 5°F $= (63.78 - 59.13)\left(\frac{5}{30}\right) = 0.78'$

FIGURE 2-26
CONDUCTOR SAGGING BY HORIZONTAL LINE OF SIGHT

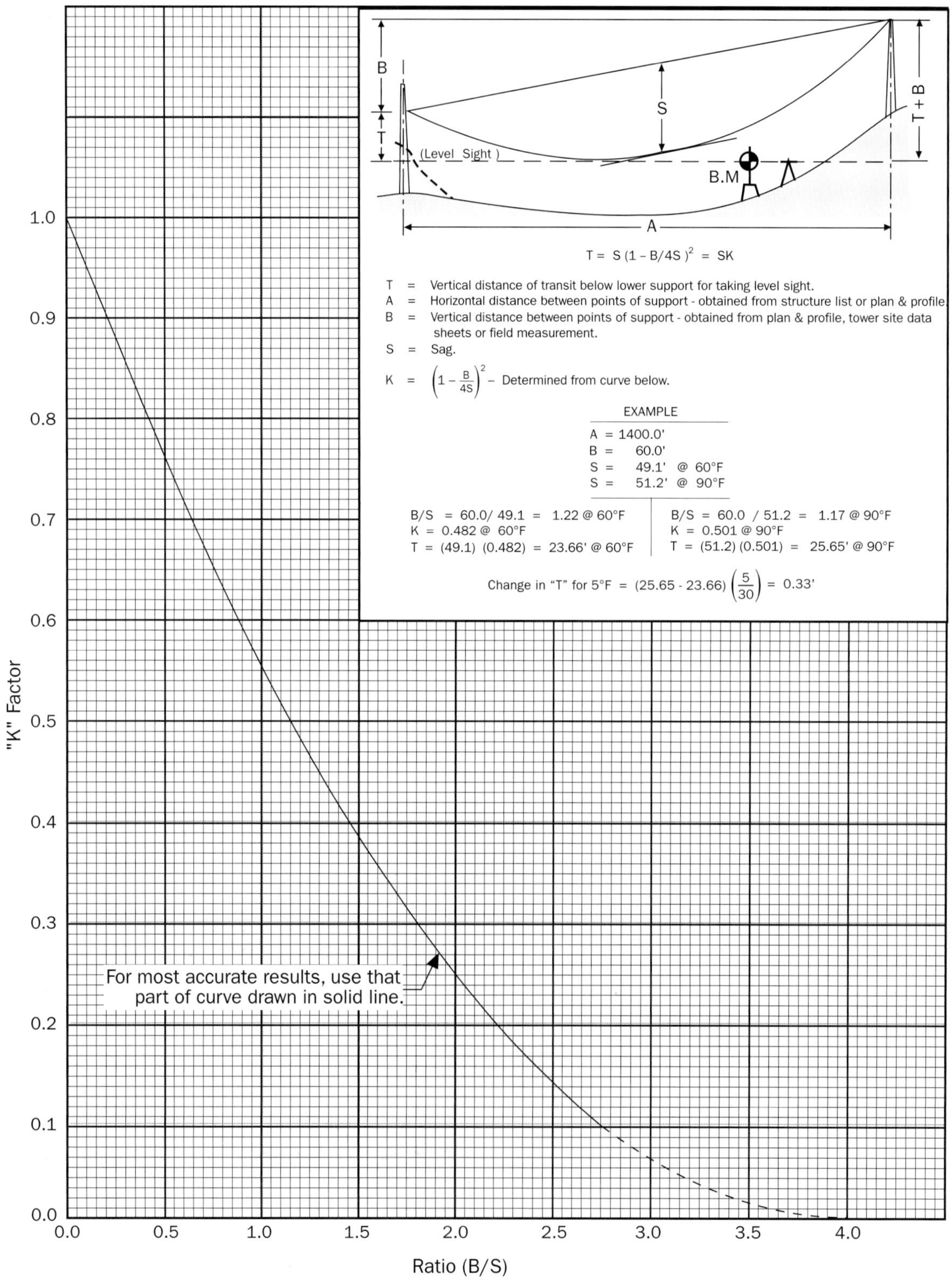

Sag is based on parabolic functions. If sag exceeds 5% of span, do not use this chart.

FIGURE 2-27
CONDUCTOR SAGGING FOR CHECKING SAG, *S*

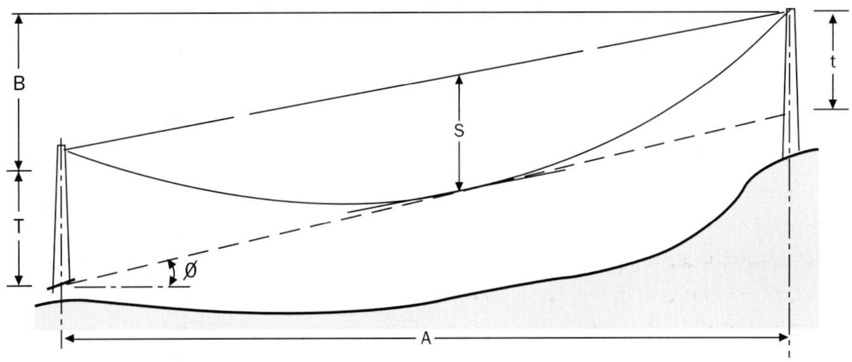

Method 1: $S = \left(\frac{\sqrt{T} + \sqrt{t}}{2}\right)^2$

Method 2: $S = \frac{T}{2} + \frac{t}{2} - \frac{tM}{8}$

S = Sag
t = Vertical distance below support for line of sight
  = T ± B – A tan Ø when angle Ø is above horizontal
  = T ± B + A tan Ø when angle Ø is below horizontal

T = Vertical distance below support for transit
B = Vertical distance between points of support–obtained from plan and profile, tower site data sheets or field measurement.
  + B when support ahead is higher.
  – B when support ahead is lower.
A = Horizontal distance between points of support–obtained from structure list or plan and profile
Ø = Angle of sight
M = Determined from curve on Figure 2-25

Examples

Given:

A = 1400.0'	T = 40.0'
B = 60.0'	Ø = + 1° 40' 21" @ 60°F (Field Measured)

Method 1

$S = \left(\frac{\sqrt{T} + \sqrt{t}}{2}\right)^2$

t = 40.0 + 60.0 – 1400.0 tan 1° 40' 21"
  = 59.12'

$\sqrt{t}$ = 7.689

$\sqrt{T}$ = 6.325

$S_{60°F}$ = 49.1'

Method 2

NOTE: *When using Method 2, value of "T" should lie between 3/4 "S" & 4/3 "S"*

$S = \frac{T}{2} + \frac{t}{2} - \frac{tM}{8}$

t = 59.12'

t/2 = 29.56'

T/2 = 20.0'

M = 0.061

$S_{60°F} = 20.0 + 29.56 - \frac{(59.12)(0.061)}{8}$

$S_{60°F}$ = 49.1'

Sag is based on parabolic functions. If sag exceeds 5% of span, do not use this chart.

APPENDIX 2.2
GLOSSARY OF EQUIPMENT TERMS

block: A device designed with one or more single sheaves, a wood or metal shell, and an attachment hook or shackle. When rope is reeved through two of these devices, the assembly is commonly referred to as a *block and tackle.* A *set of 4s* refers to a block and tackle arrangement utilizing two 4-inch double-sheave blocks to obtain four load bearing lines. Similarly, a *set of 5s* or a *set of 6s* refers to the same number of load bearing lines obtained using two 5-inch or two 6-inch double-sheave blocks, respectively.
Synonyms: set of 4s, set of 5s, set of 6s.

bullwheel: A wheel incorporated as an integral part of a bullwheel puller or tensioner to generate pulling or braking tension on conductors or pulling lines, or both, through friction. A puller or tensioner normally has one or more pairs arranged in tandem incorporated in its design. The physical size of the wheels will vary for different designs, but 17 in (43 cm) face widths and diameters of 5 ft (150 cm) are common. The wheels are power driven or retarded and lined with single or multiple groove neoprene or urethane linings. Friction is accomplished by reeving the pulling line or conductor around the groove of each pair.

clipping-in: The transferring of sagged conductors from the traveler to their permanent suspension positions and the installing of the permanent suspension clamps.
Synonyms: clamping, clipping.

clipping offset: A calculated distance, measured along the conductor from the plum mark to a point on the conductor at which the center of the suspension clamp is to be placed. When stringing in rough terrain, clipping offset may be required to balance the horizontal forces on each suspension structure.

grip, conductor: A device designed to permit the pulling of conductor without splicing on fittings, eyes, etc. It permits the pulling of a *continuous* conductor where threading is not possible. The designs of these grips vary considerably. Grips such as the Klein (Chicago) and Crescent utilize an open sided rigid body with opposing jaws and swing latch. In addition to pulling conductors, this type is commonly used to tension guys and, in some cases, pull wire rope. The design of the come-along (pocketbook, suitcase, four bolt, etc.) incorporates a bail attached to the body of a clamp which folds to completely surround and envelope the conductor. Bolts are then used to close the clamp and obtain a grip.
Synonyms: buffalo, Chicago grip, come-along, Crescent, four bolt, grip, Klein, pocketbook, seven bolt, six bolt, slip-grip, suitcase.

line, pilot: A lightweight line, normally synthetic fiber rope, used to pull heavier pulling lines which in turn are used to pull the conductor. Pilot lines may be installed with the aid of finger lines or by helicopter when the insulators and travelers are hung.
Synonyms: lead line, leader, P-line, straw line.

line, pulling: A high strength line, normally synthetic fiber rope or wire rope, used to pull the conductor. However, on reconstruction jobs where a conductor is being replaced, the old conductor often serves as the pulling line for the new conductor. In such cases, the old conductor must be closely examined for any damage prior to the pulling operations.
Synonyms: bull line, hard line, light line, sock line.

puller, bullwheel: A device designed to pull pulling lines and conductors during stringing operations. It normally incorporates one or more pairs of urethane- or neoprene-lined, power driven, single or multiple groove bullwheels where each pair is arranged in tandem. Pulling is accomplished by friction generated against the pulling line which is reeved around the grooves of a pair of the bullwheels. The puller is usually equipped with its own engine which drives the bullwheels mechanically, hydraulically or through a combination of both. Some of these devices function as either a puller or tensioner.
Synonym: puller.

puller, drum: A device designed to pull a conductor during stringing operations. It is normally equipped with its own engine which drives the drum mechanically, hydraulically or through a combination of both. It may be equipped with synthetic fiber rope or wire rope to be used as the pulling line. The pulling line is payed out from the unit, pulled through the travelers in the sag section and attached to the conductor. The conductor is then pulled in by winding the pulling line back onto the drum. This unit is sometimes used with synthetic fiber rope acting as a pilot line to pull heavier pulling lines across canyons, rivers, etc.
Synonyms: hoist, single drum hoist, single drum winch, tugger.

puller, reel: A device designed to pull a conductor during stringing operations. It is normally equipped with its own engine which drives the supporting shaft for the reel mechanically, hydraulically or through a combination of both. The shaft, in turn, drives the reel. The application of this unit is essentially the same as that for the drum puller previously described. Some of these devices function as either a puller or tensioner.

reel stand: A device designed to support one or more reels and having the possibility of being skid, trailer or truck mounted. These devices may accommodate rope or conductor reels of varying sizes and are usually equipped with reel brakes to prevent the reels from turning when pulling is stopped. They are used for either slack or tension stringing. The designation of reel trailer or reel truck implies that the trailer or truck has been equipped with a reel stand (jacks) and may serve as a reel transport or *payout* unit, or both, for stringing operations. Depending upon the sizes of the reels to be carried, the transporting vehicles may range from single axle trailers to semi-trucks with trailers having multiple axles.
Synonyms: reel trailer, reel transporter, reel truck.

running board: A pulling device designed to permit stringing more than one conductor simultaneously with a single pulling line. For distribution stringing, it is usually made of lightweight tubing with the forward end curved gently upward to provide smooth transition over pole crossarm rollers. For transmission stringing, the device is either made of sections hinged transversely to the direction of pull or of a hard nose rigid design, both having a flexible pendulum tail suspended from the rear. This configuration stops the conductors from twisting together and permits smooth transition over the sheaves of bundle travelers.
Synonyms: alligator, bird, birdie, monkey tail, sled.

sag section: The section of line between snub structures. More than one sag section may be required in order to sag properly the actual length of conductor which has been strung.
Synonyms: pull, setting, stringing section.

site, pull: The location on the line where the puller, reel winder and anchors (snubs) are located. This site may also serve as the pull or tension site for the next sag section.
Synonyms: reel setup, tugger setup.

site, tension: The location on the line where the tensioner, reel stands and anchors (snubs) are located. This site may also serve as the pull or tension site for the next sag section.
Synonyms: conductor payout station, payout site, reel setup.

snub structure: A structure located at one end of a sag section and considered as a *zero* point for sagging and clipping offset calculations. The section of line between two such structures is the sag section, but more than one sag section may be required in order to sag properly the actual length of conductor which has been strung.
Synonyms: *0* structure, zero structure.

tensioner, bullwheel: A device designed to hold tension against a pulling line or conductor during the stringing phase. Normally, it consists of one or more pairs of urethane- or neoprene-lined, power braked, single or multiple groove bullwheels where each pair is arranged in tandem. Tension is accomplished by friction generated against the conductor which is reeved around the grooves of a pair of the bullwheels. Some tensioners are equipped with their own engines which retard the bullwheels mechanically, hydraulically or through a combination of both. Some of these devices function as either a puller or tensioner. Other tensioners are only equipped with friction type retardation.
Synonyms: retarder, tensioner.

tensioner, reel: A device designed to generate tension against a pulling line or conductor during the stringing phase. Some are equipped with their own engines which retard the supporting shaft for the reel mechanically, hydraulically or through a combination of both. The shaft, in turn, retards the reel. Some of these devices function as either a puller or tensioner. Other tensioners are only equipped with friction type retardation.
Synonyms: retarder, tensioner.

traveler: A sheave complete with suspension arm or frame used separately or in groups and suspended from structures to permit the stringing of conductors. These devices are sometimes bundled with a center drum, or sheave and another traveler, and used to string more than one conductor simultaneously. For protection of conductors that should not be nicked or scratched, the sheaves are often lined with nonconductive or semiconductive neoprene or with nonconductive urethane.

Any one of these materials acts as a padding or cushion for the conductor as it passes over the sheave. Traveler grounds must be used with lined travelers in order to establish an electrical ground. *Synonyms*: block, dolly, sheave, stringing block, stringing sheave, stringing traveler.

winder reel: A device designed to serve as a recovery unit for a pulling line. It is normally equipped with its own engine which drives a supporting shaft for a reel mechanically, hydraulically or through a combination of both. The shaft, in turn, drives the reel. It is normally used to rewind a pulling line as it leaves the bullwheel puller during stringing operations. This unit is not intended to serve as a puller, but sometimes serves this function where only low tensions are involved.

3 THERMAL RATINGS OF BARE OVERHEAD CONDUCTORS

In determining steady-state ratings, a continuous current is calculated that would yield the maximum allowable conductor temperature for certain weather assumptions and conductor parameters. Transient thermal ratings, however, apply to the final current value that yields the maximum allowable conductor temperature in a specified time after an increase in current from some initial value. Traditional conservative weather assumptions used in steady-state ratings allow for considerable increases in current under emergency conditions.

Because conductor temperature indirectly determines allowable conductor current, the factors affecting the actual conductor temperature must be considered. Conductor temperature is impacted by the heat produced from the current flowing through the conductor, the thermal properties of the conductor, and the surrounding ambient conditions. The conductor temperature is raised primarily through heat input from ohmic losses (I^2R), and secondarily from solar radiation. Ohmic losses are a function of conductor resistance and current; solar radiation input is dependent on the conductor's absorptivity.

Convection, the major source of heat loss, is a function of air temperature, wind speed and direction. Conductor radiation, having a lesser effect on conductor temperature, is affected by air temperature and the conductor's emissivity. Solar heating of the conductor varies with altitude, time-of-day, season and latitude. Since overhead lines exist in many different types of climate and terrain, the weather conditions which affect line rating can differ drastically. Some locations may have a winter seasonal average temperature of 25°C, while others may average only 5°C. In locations that are heavily sheltered by forests and terrain, periods of calm may be common whereas in other more exposed regions, the average wind speed may be greater than 10 ft/sec. Selection of weather conditions that are "conservative" is essential to thermal rating accuracy.

Choosing a maximum allowable conductor temperature should include considerations of adequate electrical clearance at increased sag, excessive loss of strength and any permanent increase in sag due to accelerated creep elongation under extended high temperature events. During short-circuit conditions, however, it is permissible to exceed the maximum allowable temperature since short-circuit conditions occur infrequently and last for a very short duration of time.

In the following sections of Chapter 3, three primary aspects of the thermal rating calculation process are considered – rating calculation method; selection of weather conditions; and high temperature operating limits. In Sections 3.1 to 3.3, the IEEE thermal rating calculation method is presented for steady state ratings, transient ratings and fault current ratings. The Southwire computer program for determining these ratings is described in Section 3.4. In Section 3.5, the process of selecting "conservative" weather conditions for line rating calculations is discussed. Finally, in Sections 3.6 to 3.9, the impact of high operating temperatures on conductor and connectors is described.

3.1 STEADY-STATE THERMAL RATINGS

Power system operators limit the electrical current loading of transmission circuits to keep the conductor temperature within acceptable limits. The maximum allowable continuous current that corresponds to the maximum allowable conductor temperature is called the steady-state thermal rating of the line. The method used to calculate this rating assumes the conductor is in thermal equilibrium at the maximum allowable conductor temperature. If values of ambient and conductor variables are measured, calculated, or estimated, the ampacity can be calculated using the steady-state heat balance equation.

3.1.1 Units and Symbols

Units and symbols follow IEEE Standard 738[1] whenever possible and are defined following each equation in this chapter.

3.1.2 Steady-State Calculation Methods

There are a number of methods available to calculate the electrical current corresponding to the maximum allowable conductor temperature. Most of these methods are similar and yield comparable results. One of the earliest models for bare overhead conductors is that of Schurig and Frick, published in 1929[2], and still used by some utility engineers. Today, however, the most widely accepted model is the IEEE method (IEEE Standard 738)[1], based largely on the work of House and Tuttle[3] that was subsequently modified by a group sponsored by the East Central Area Reliability Coordination Agreement (ECAR)[4].

3.1.3 Steady-State Heat Balance

In the steady-state, heat input to the bare overhead conductor from solar radiation and ohmic (I^2R) losses balances the heat lost by convection and radiated cooling. For a unit

FIGURE 3-1
CONVECTION HEAT LOSS vs. PERPENDICULAR WIND SPEED

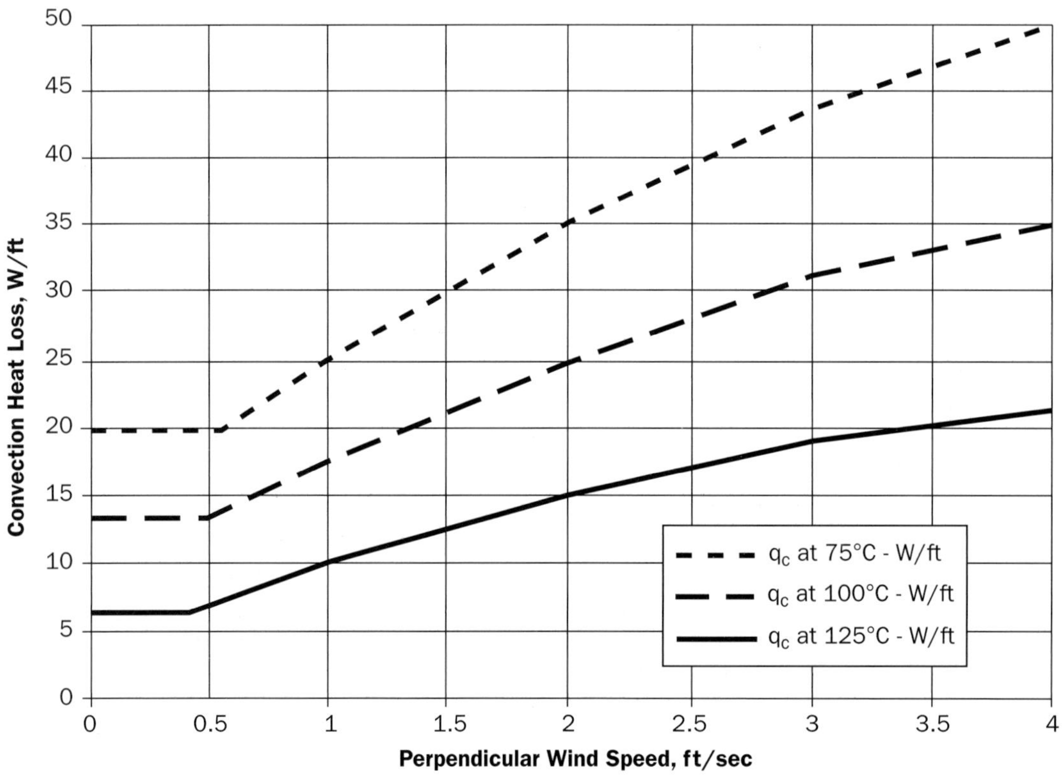

length of bare overhead conductor, the steady-state heat balance equation is:

$$q_c + q_r = q_s + I^2 R_{T_c} \quad (3\text{-}1)$$

Solving for I:

$$I = \sqrt{\frac{q_c + q_r - q_s}{R_{T_c}}} \quad (3\text{-}2)$$

where:

q_c = *convective heat loss rate, W/ft*
q_r = *radiated heat loss rate, W/ft*
q_s = *solar heat gain rate, W/ft*
I = *conductor current, amperes at 60 Hz*
R_{T_c} = *60 Hz ac resistance of conductor at operating temperature T_c, Ω/ft*

A method for the calculation for each of these terms is presented in the following sections.

3.1.4 Convected Heat Loss Rate (q_c)

The major heat loss from an overhead conductor occurs through convection and is typically three or four times as large as radiated heat loss. There are two types of convection applicable to the thermal rating of an overhead conductor - forced and natural. Forced convection heat loss occurs in non-still air and is proportional to the conductor temperature rise above air temperature, the conductor diameter and wind speed. Natural convection occurs in "still air" and is approximately the same as the convection heat loss predicted by the equation for forced convection at wind speeds of 0.5 ft/sec or less. This can be seen in Figure 3-1.

FIGURE 3-2
CONVECTION HEAT LOSS RATE vs. WIND ANGLE RELATIVE TO LINE

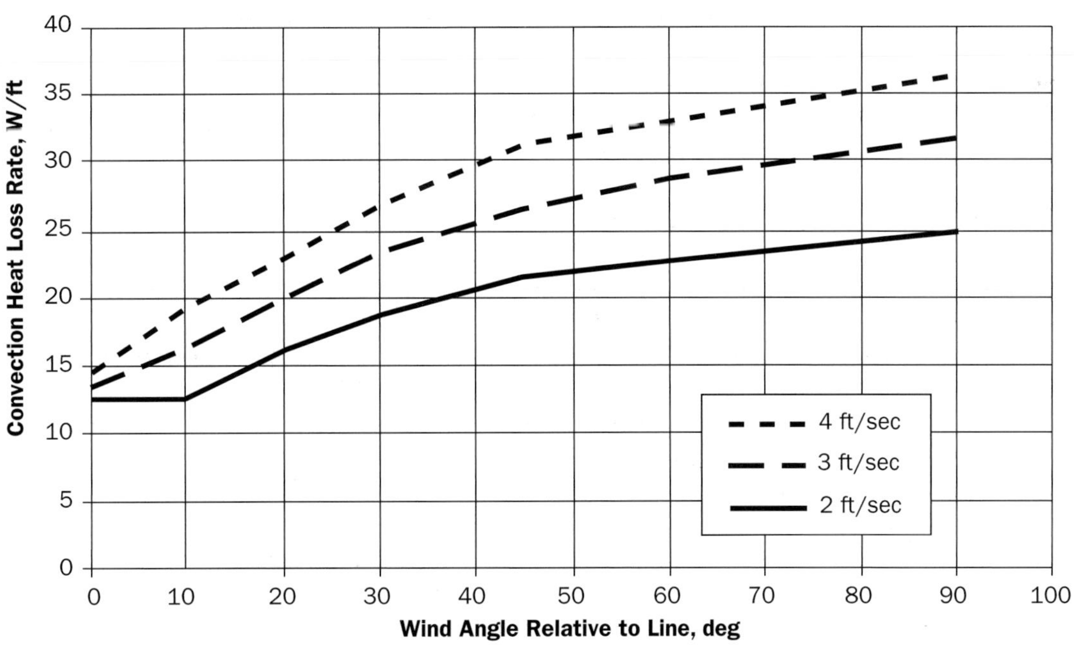

Natural Convection Heat Loss Rate

With zero wind speed, the forced convection goes to zero, but natural convection due to rising hot air still occurs. The natural convection heat loss can be calculated by:

$$q_c = 0.283\ \rho_f^{0.5}\ D^{0.75}\ (T_c - T_a)^{1.25} \qquad (3\text{-}3)$$

where:

- q_c = convection heat loss rate, W/ft
- ρ_f = density of air at temperature T_{film}, lbs/ft^3
- D = conductor diameter, in
- T_c = conductor temperature, °C
- T_a = ambient temperature, °C

$$T_{film} = \frac{T_c + T_a}{2} \qquad (3\text{-}4)$$

In calculating line ratings for low wind speeds, the IEEE 738 Standard suggests taking the convection cooling rate as equal to the larger of the natural or forced convection cooling rates.

Forced Convection Heat Loss Rate

In an attempt to find an accurate model of convection heat loss over the whole range of possible wind speeds, forced convection heat loss is represented by two separate calculations:

$$q_{c1} = \left[1.01 + 0.371\left(\frac{D\rho_f V_w}{\mu_f}\right)^{0.52}\right] k_f (T_c - T_a) \qquad (3\text{-}5)$$

and

$$q_{c2} = 0.1695\left(\frac{D\rho_f V_w}{\mu_f}\right)^{0.6} k_f\ (T_c - T_a) \qquad (3\text{-}6)$$

where:

- q_{c1} = convection heat loss rate at low wind speeds, W/ft
- q_{c2} = convection heat loss rate at high wind speeds, W/ft
- μ_f = absolute viscosity of air at temperature T_{film}, lbs/ft-hr
- k_f = thermal conductivity of air at temperature T_{film}, W/ft-°C
- D = conductor diameter, in
- V_w = wind speed, ft/hr

The first equation, q_{c1}, applies at low winds, but gives convective heat losses that are too low at high speeds. The second equation, q_{c2}, applies at high wind speeds, but predicts heat losses that are too low at low wind speeds. At any wind speed, the convective heat loss is calculated using both equations. The larger of the two values is conservatively chosen for thermal rating calculations.

If the wind is not perpendicular to the conductor, then the convective cooling term is multiplied by the wind direction factor, K_{angle}, as defined by Davis[5] below:

$$K_{angle} = 1.194 - \cos(\phi) + 0.194\cos(2\phi) + 0.368\sin(2\phi) \qquad (3\text{-}7)$$

where:

- ϕ = angle between wind and axis of the conductor, degrees

Equation (3-7) defines the effect of wind direction relative to the conductor orientation. Use of this equation suggests wind blowing parallel to a line can result in approximately 60% lower convective heat loss than a wind of the same speed blowing perpendicular to the line, as shown in Figure 3-2. There can be situations where forced convection from near parallel wind can be less than natural convection.

The values of air density, ρ_f, air viscosity, μ_f, and the coefficient of thermal conductivity of air, k_f, are given by the following equations and are also listed, for convenience, in Table 3-1.

$$\mu_f = \frac{0.00353\ (T_{film} + 273)^{1.5}}{T_{film} + 383.4} \qquad (3\text{-}8)$$

$$\rho_f = \frac{0.080695 - 2.901 \times 10^{-6}\ H_e + 3.7 \times 10^{-11} H_e^2}{1 + 0.00367\ T_{film}} \qquad (3\text{-}9)$$

$$k_f = 7.388 \times 10^{-3} + 2.279 \times 10^{-5}\, T_{film} - 1.343 \times 10^{-9}\, T_{film}^2 \qquad (3\text{-}10)$$

where:

H_e = *elevation of conductor above sea level, ft*

3.1.5 Radiated Heat Loss Rate (q_r)

Radiated heat loss per unit length of conductor is approximately equal to convective loss under still air conditions and is negligible at wind speeds above 10 mph. Over the normal range of conductor temperatures and air temperatures used in rating calculations, the radiated heat loss is dependent upon the conductor diameter, emissivity and conductor temperature rise above ambient and can be determined by using the following equation:

$$q_r = 0.138\, D\, \epsilon \left[\left(\frac{T_c + 273}{100} \right)^4 - \left(\frac{T_a + 273}{100} \right)^4 \right] \qquad (3\text{-}11)$$

where:

q_r = *radiated heat loss rate, W/ft*
ϵ = *emissivity*
D = *diameter of conductor, in*
T_c = *temperature of conductor, °C*
T_a = *ambient temperature, °C*

Emissivity is defined as the ratio of heat radiated to the surrounding air by the conductor to the heat radiated to the surrounding air by a perfect "black body" radiator of the same dimensions. Although solar heat absorption occurs at a different frequency range than heat radiation, the absorptivity and emissivity of an overhead conductor both increase over the life of the conductor as a result of age and atmospheric pollution. Figure 3-3 shows the change of emissivity with age[6].

Emissivity values of approximately 0.2 for new conductors and 0.9 for weathered bare overhead conductors have been reported by House and Tuttle[3]. A value of 0.5, or a more conservative 0.7, has been widely used when the actual conductor surface condition is unknown.

FIGURE 3-3
CONDUCTOR EMISSIVITY AS A FUNCTION OF AGE

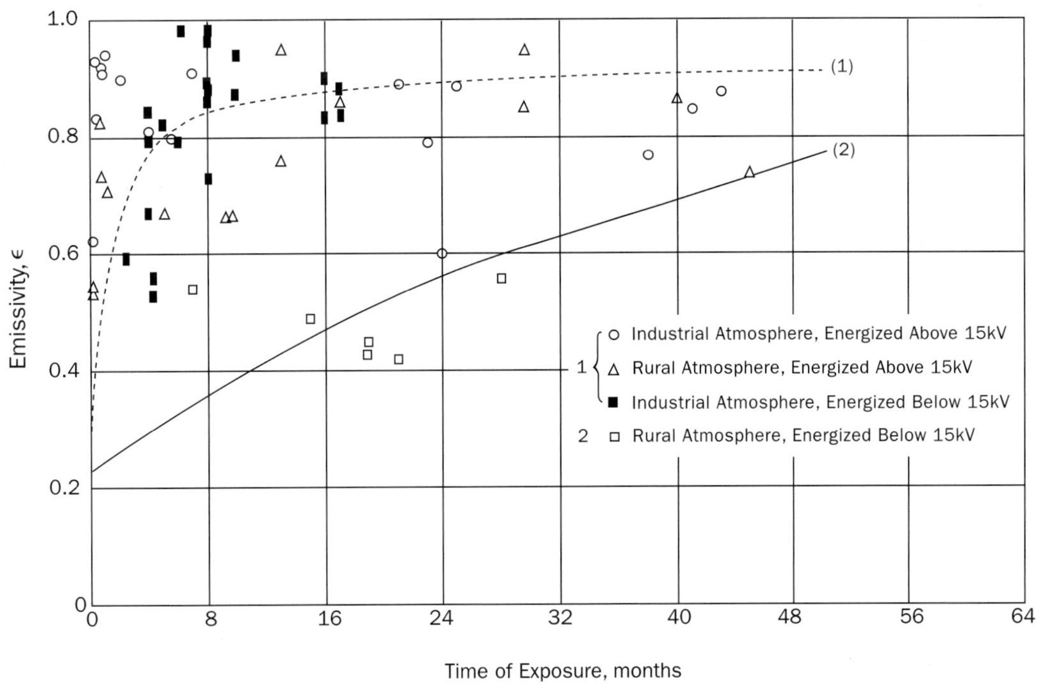

TABLE 3-1
VISCOSITY, DENSITY AND THERMAL CONDUCTIVITY OF AIR

Temperature			$\left(\frac{°K}{100}\right)^4$	Absolute Viscosity, μ_f lbs/ft-hr	Density, ρ_f lbs/ft³				Thermal Conductivity of Air, k_f W/ft-°C
°F	°C	°K			Sea Level	5,000 ft	10,000 ft	15,000 ft	
32	0	273	55.55	0.0415	0.0807	0.0671	0.0554	0.0455	0.00739
41	5	278	59.73	0.0421	0.0793	0.0660	0.0545	0.0447	0.00750
50	10	283	64.14	0.0427	0.0779	0.0648	0.0535	0.0439	0.00762
59	15	288	68.80	0.0433	0.0765	0.0636	0.0526	0.0431	0.00773
68	20	293	73.70	0.0439	0.0752	0.0626	0.0517	0.0424	0.00784
77	25	298	78.86	0.0444	0.0740	0.0616	0.0508	0.0417	0.00795
86	30	303	84.29	0.0450	0.0728	0.0606	0.0500	0.0411	0.00807
95	35	308	89.99	0.0456	0.0716	0.0596	0.0492	0.0404	0.00818
104	40	313	95.98	0.0461	0.0704	0.0586	0.0484	0.0397	0.00830
113	45	318	102.26	0.0467	0.0693	0.0577	0.0476	0.0391	0.00841
122	50	323	108.85	0.0473	0.0683	0.0568	0.0469	0.0385	0.00852
131	55	328	115.74	0.0478	0.0672	0.0559	0.0462	0.0379	0.00864
140	60	333	122.96	0.0484	0.0661	0.0550	0.0454	0.0373	0.00875
149	65	338	130.52	0.0489	0.0652	0.0542	0.0448	0.0367	0.00886
158	70	343	138.41	0.0494	0.0643	0.0535	0.0442	0.0363	0.00898
167	75	348	146.66	0.0500	0.0634	0.0527	0.0436	0.0358	0.00909
176	80	353	155.27	0.0505	0.0627	0.0522	0.0431	0.0354	0.00921
185	85	358	164.26	0.0510	0.0616	0.0513	0.0423	0.0347	0.00932
194	90	363	173.63	0.0515	0.0608	0.0506	0.0418	0.0343	0.00943
203	95	368	183.40	0.0521	0.0599	0.0498	0.0412	0.0338	0.00952
212	100	373	193.57	0.0526	0.0591	0.0492	0.0406	0.0333	0.00966

TABLE 3-2
HEAT-TRANSMISSION FACTOR FOR ALTITUDES ABOVE SEA LEVEL

Elevation Above Sea Level ft	Multiplier for Values in Table 3-4
0	1.00
5,000	1.15
10,000	1.25
15,000	1.30

TABLE 3-3
ALTITUDE, H_C, AND AZIMUTH, Z_C, IN DEGREES OF SUN AT VARIOUS LATITUDES AT A DECLINATION OF 23.0 DEGREES
(NORTHERN HEMISPHERE, JUNE 10 AND JULY 3)

Degrees North Latitude	Local Sun Time					
	10:00 a.m.		12:00 noon		2:00 p.m.	
	H_c	Z_c	H_c	Z_c	H_c	Z_c
20	62	78	87	0	62	282
25	62	88	88	180	62	272
30	62	98	83	180	62	262
35	61	107	78	180	61	253
40	60	115	73	180	60	245
45	57	122	68	180	57	238
50	54	128	63	180	54	232
60	47	137	53	180	47	223
70	40	143	43	180	40	217

TABLE 3-4
TOTAL HEAT RECEIVED BY SURFACE AT SEA LEVEL NORMAL TO SUN'S RAYS

Solar Altitude, H_c Degrees	Q_s W/ft²	
	Clear Atmosphere	Industrial Atmosphere
5	21.7	12.6
10	40.2	22.3
15	54.2	30.5
20	64.4	39.2
25	71.5	46.6
30	77.0	53.0
35	81.5	57.5
40	84.8	61.5
45	87.4	64.5
50	90.0	67.5
60	92.9	71.6
70	95.0	75.2
80	95.8	77.4
90	96.4	78.9

3.1.6 Solar Heat Gain Rate (q_s)

Solar heating normally raises the conductor temperature by 5°C to 10°C above the air temperature, and is usually included in thermal rating calculations. The solar heat gain rate equation is:

$$q_s = \alpha\, Q_s \sin(\theta)\, A' \qquad (3\text{-}12)$$

where:

q_s = solar heat gain rate, W/ft
α = solar absorptivity
Q_s = total solar radiated heat flux, W/ft^2
θ = effective angle of incidence of the sun's rays, degrees
A' = projected area of conductor, per unit length, ft^2/ft

Absorptivity is the ability of a conductor to absorb heat from its surroundings and is dependent upon conductor surface conditions, increasing from about 0.2 to about 0.9 with conductor age[6][7]. The exact rate of increase depends on the level of atmospheric pollution and the line's operating voltage. The absorptivity, α, of the conductor as reported by House and Tuttle[3] is between 0.2 for a new conductor and 0.9 for a weathered conductor. An average value of 0.5, or a more conservative 0.9, is often used when actual conductor surface condition is unknown.

$$\theta = \arccos[\cos(H_c)\cos(Z_c - Z_l)] \qquad (3\text{-}13)$$

The effective angle of incidence, θ, of the sun's rays is calculated as follows:

where:

H_c = altitude of sun, degrees
Z_c = azimuth of sun, degrees
Z_l = azimuth of line, degrees

Values for the solar altitude, H_c, and azimuth of the sun, Z_c, are taken from Table 3-3. The azimuth of the power line, Z_l, is assumed to be 0° (or 180°) for North-South lines and 90° (or 270°) for East-West lines. For conditions not shown in the tables, H_c and Z_c can be calculated using the following equations:

$$H_c = \arcsin[\cos(Lat)\cos(\delta)\cos(\omega) + \sin(Lat)\sin(\delta)] \qquad (3\text{-}14)$$

where:

δ = solar declination, degrees
ω = hour angle, degrees
Lat = north latitude, degrees

The hour angle, ω, is the number of hours from noon time 15 degrees (11 a.m. is –15°, 2 p.m. is +30°).

The solar declination, δ, can be calculated as follows:

$$\delta = 23.4583 \sin\left[360\left(\frac{284 + N}{365}\right)\right] \qquad (3\text{-}15)$$

where:

N = the day of the year (Jan. 21st = 21, Feb. 12th = 43)

Z_c, the solar azimuth, can be calculated as follows:

$$Z_c = C + \arctan(\chi) \qquad (3\text{-}16)$$

where:

C = solar azimuth constant, degrees
χ = solar azimuth variable

The solar azimuth constant, C, is a function of the hour angle, ω, and the solar azimuth variable, χ, as shown in Table 3-5. χ is calculated as follows:

$$\chi = \frac{\sin(\omega)}{\sin(Lat)\cos(\omega) - \cos(Lat)\tan(\delta)} \qquad (3\text{-}17)$$

TABLE 3-5
SOLAR AZIMUTH CONSTANT, C

"Hour Angle," ω, degrees	C if $X \geq 0$, degrees	C if $X < 0$, degrees
$-180 \leq \omega < 0$	0	180
$0 \leq \omega < 180$	180	360

The total solar heat flux, Q_S, at sea level varies with solar altitude. It can reach a maximum of nearly 100 W/ft² in the summer in the U.S. Values for Q_S can be found in Table 3-4 for a wide range of solar altitudes at sea level. For solar altitudes not shown, or elevations other than sea level, Q_S, can be calculated using the following equations.

For clear atmospheres:

$$Q_s = -3.9241 + 5.9276H_c - 1.7856 \times 10^{-1} H_c^2 + 3.223 \times 10^{-3} H_c^3 - 3.3549 \times 10^{-5} H_c^4 + 1.8053 \times 10^{-7} H_c^5 - 3.7868 \times 10^{-10} H_c^6 \quad (3\text{-}18)$$

For industrial atmospheres:

$$Q_s = 4.9408 + 1.3202H_c + 6.1444 \times 10^{-2}H_c^2 - 2.9411 \times 10^{-3}H_c^3 + 5.07752 \times 10^{-5} H_c^4 - 4.03627 \times 10^{-7} H_c^5 + 1.22967 \times 10^{-9} H_c^6 \quad (3\text{-}19)$$

where:

H_c = *altitude of the sun, degrees*

For elevations other than sea level, a correction factor, K_{solar}, can be calculated and applied to Q_s:

$$K_{solar} = 1 + 3.5 \times 10^{-5} H_e - 1.0 \times 10^{-9} H_e^2 \quad (3\text{-}20)$$

where:

H_e = *elevation of conductor above sea level, ft*

Note that the solar heating calculation method described in the foregoing works for any latitude including negative (southern hemisphere) latitudes.

3.1.7 Conductor AC Resistance

The ohmic, or I²R, loss per unit length of conductor is equal to the conductor resistance in ohms per unit length times the square of the rms electrical current in amperes. The I²R loss accounts for a heat gain in a conductor many times larger than the solar input when performing thermal rating calculations.

The ac resistance of a bare, stranded conductor varies with conductor stranding, metal conductivity, frequency, average current density and temperature. Values of ac resistance at temperatures of 25°C, 50°C and 75°C, calculated considering skin effect at 60 Hz, are listed in the appendix of Chapter 1 for most standard bare overhead conductors. To determine thermal ratings at temperatures for which no resistance values are listed, the resistance at the desired temperature may be calculated with the following linear equation:

$$R_{T_c} = R_{T_L} + \left[\frac{R_{T_H} - R_{T_L}}{T_H - T_L} \right] (T_c - T_L) \quad (3\text{-}21)$$

where:

R_{T_c} = *resistance calculated at temperature,* T_c
R_{T_L} = *60 Hz ac resistance at temperature,* T_L
R_{T_H} = *60 Hz ac resistance at temperature,* T_H
T_L = *low conductor temperature at which the resistance,* R_{T_L} *is specified*
T_H = *high conductor temperature at which the resistance,* R_{T_H} *is specified*
T_c = *conductor temperature at which the new 60 Hz resistance is desired*

The resistance of most common metals used in stranded conductors does not increase linearly with temperature. Resistance normally increases at a slightly faster rate than the temperature. Since the above equation assumes a linear relationship between resistance and temperature, resistance values calculated at temperatures between the high and low resistance values specified, will be slightly high. Resistance values calculated at conductor temperatures greater than T_H will be slightly low. Determining the conductor resistance at 175°C using the resistance values for 25°C and 75°C will yield a value that is approximately 1% too low.

As an example calculation, consider the following:

The thermal rating of 795 kcmil - 26/7 ACSR "Drake" is to be calculated at 100°C. Only values of resistance at 25°C, 50°C and 75°C are tabulated in the Chapter 1 Appendix. The resistance at 100°C can be calculated using Equation 3-21, and the tabulated resistances at 25°C and 75°C from the tables.

From the tables:

$$R_{T_L} = R_{25} = 0.1166 \quad \Omega/mile \; (T_L = 25°C)$$

$$R_{T_H} = R_{75} = 0.1390 \quad \Omega/mile \; (T_H = 75°C)$$

Substituting in Equation 3-21 yields:

$$R_{T_C} = R_{100} = 0.1166 + \left(\frac{0.1390 - 0.1166}{75 - 25}\right)(100 - 25)$$

$$= 0.1502 \; \Omega/mile = 2.845 \times 10^{-5} \; \Omega/ft$$

When calculating resistance at very high conductor temperatures, the resistance value R_{T_H} should be selected at a temperature near the desired new resistance value as possible.

As discussed in Chapter 1, the resistance of single and three layer ACSR conductors is dependent on current density due to magnetic effects that occur due to the odd number of layers and the presence of the steel core. The resistance values listed in the conductor tables of Chapter 1 include this effect.

3.1.8 Example of a Steady-State Thermal Rating Calculation

The calculation of steady state thermal rating, given the maximum allowable conductor temperature, weather conditions and conductor characteristics, can be determined using the computer program included with this book. The following example is presented to give some insight into the calculation method used in this program.

Find the steady state thermal rating (ampacity) for a 795 kcmil - 26/7 ACSR "Drake" conductor under the following conditions:

Conductor resistance and diameter are taken from the conductor tables in appendix of Chapter 1.

(1) *Conductor diameter, D, is 1.108 in*

(2) *Conductor ac resistance, R_{Tc}, is:*
$R_{25°C} = 0.1166 \; \Omega/mile$
$R_{75°C} = 0.1390 \; \Omega/mile$

(3) *Maximum allowable conductor temperature, T_c, is 75°C*

(4) *Ambient air temperature, T_a, is 25°C*

(5) *Wind velocity, V_w, is 2 ft/sec at sea level, perpendicular to the conductor*

(6) *Emissivity, ϵ, is 0.5*

(7) *Solar absorptivity, α, is 0.5*

(8) *The line runs in a East-West direction*

(9) *Latitude is 30° N*

(10) *Atmosphere is clear*

(11) *Sun altitude, H_c, 11:00 a.m.*

Convection Heat Loss Rate (q_c)

The natural convection heat loss is calculated by means of Equation 3-3:

$$q_c = 0.283 \, \rho_f^{0.5} \, D^{0.75} \, (T_c - T_a)^{1.25}$$

where:

$D = 1.108$ in

$T_c = 75°C$

$T_a = 25°C$

$T_{film} = 50°C$ (From Equation 3-4)

$\rho_f = 0.0683 \; lbs/ft^3$ (From Table 3-1)

$$q_c = 0.283\,(0.0683)^{0.5}\,(1.108)^{0.75}\,(75\text{-}25)^{1.25}$$
$$= 0.283\,(0.2613)\,(1.0799)\,(132.957)$$
$$= 10.62 \; W/ft$$

Since the wind velocity is greater than 0 ft/sec, the forced convection heat loss for perpendicular wind is calculated according to Equations 3-5 and 3-6 and compared to the natural convection heat loss. The maximum of the heat losses due to both natural and forced convection is then used in calculating the thermal rating:

$$q_{c1} = \left[1.01 + 0.371 \left(\frac{D \rho_f V_w}{\mu_f} \right)^{0.52} \right] k_f \, (T_c - T_a)$$

and

$$q_{c2} = 0.1695 \left(\frac{D \rho_f V_w}{\mu_f} \right)^{0.6} k_f \, (T_c - T_a)$$

where:

D = 1.108 in

V_w = (2 ft/sec) · 3600 (sec/hr) = 7200 ft/hr

T_c = 75°C

T_a = 25°C

μ_f = 0.0473 lbs/ft-hr (From Table 3-1 at 50°C)

ρ_f = 0.0683 lbs/ft^3 (From Table 3-1 at 50°C)

k_f = 0.00852 W/ft-°C (From Table 3-1 at 50°C)

$$q_{c1} = \left[1.01 + 0.0371 \left(\frac{(1.108)(0.0683)(7200)}{0.0473} \right)^{0.52} \right] \times (0.00852)(75 - 25)$$

$$= 20.88 \text{ W/ft}$$

$$q_{c2} = 0.1695 \left[\frac{(1.108)(0.0683)(7200)}{0.0473} \right]^{0.6} (0.00852)(75 - 25)$$

$$= 19.74 \text{ W/ft}$$

Because the wind is perpendicular to the axis of the conductor, the wind direction multiplier, K_{angle}, given in Equation 3-7, is 1.0. Since the maximum of the calculated forced convection heat losses (20.88 W/ft) is greater than the natural convection heat loss, (10.62 W/ft), the larger convection heat loss value of 20.88 W/ft will be used in the calculation of thermal rating.

Radiated Heat Loss Rate (q_r)

Using Equation 3-11, the radiated heat loss rate per unit length of "Drake" at 75°C can be calculated:

$$q_r = 0.138 \, D \epsilon \left[\left(\frac{T_c + 273}{100} \right)^4 - \left(\frac{T_a + 273}{100} \right)^4 \right]$$

where:

D = 1.108 in

ϵ = 0.5

T_c = 75°C

T_a = 25°C

$$q_r = 0.138 \, (1.108) \, 0.5 \left[\left(\frac{75 + 273}{100} \right)^4 - \left(\frac{25 + 273}{100} \right)^4 \right]$$

$$= 5.18 \text{ W/ft}$$

Solar Heat Gain Rate (q_s)

Using Equation 3-12 and 3-13, the solar heat gain per unit length can be calculated:

$$q_s = \alpha \, Q_s \sin(\theta) \, A'$$

and

$$q_s = \arccos\left[\cos(H_c) \cos(Z_c - Z_1)\right]$$

where:

α = 0.5

$$A' = \frac{D}{12} = \frac{1.108}{12} = 0.092 \text{ ft}^2/\text{ft}$$

The sun altitude and azimuth at 11:00 a.m., can be interpolated from values of H_c and Z_c in Table 3-3 at 10:00 a.m. and noon. The average values for 11:00 a.m. are:

H_c @ 10 a.m. = 62°

H_c @ Noon = 83°

$$H_c \text{ @ 11 a.m.} = \frac{83° + 62°}{2} = 72.5°$$

Z_c @ 10 a.m. = 98°

Z_c @ Noon = 180°

$$Z_c @ 11\ a.m. = \frac{98° + 180°}{2} = 139°$$

From Table 3-4 for H_c = 72.5° with a clear atmosphere, Q_s, by interpolation, equals 95.2 W/ft^2.

$$Z_1 = 90° \text{ or } 270°$$

$$\theta = \arccos\left[\cos(72.5°)\cos(139° - 90°)\right] = 78.62°$$

$$q_s = (0.5)(95.2)\sin(78.62°)(0.092)$$

$$= 4.29\ W/ft$$

The values used in the determination of the solar azimuth and the altitude above were interpolated from Tables 3-1 through Table 3-4. The interpolated values provide perfectly adequate estimations for the values used. However, if more accurate values are desired, or if data not listed in the tables is required, Equations 3-8 through 3-10 can be used to determine these values.

In order to calculate the solar altitude, the day of the year, N, must be determined. For this example, June 10 will be used and:

$$N = 31 + 28 + 31 + 30 + 31 + 10 = 161$$

The solar declination, δ, for June 10 is calculated:

$$\delta = 23.4583 \sin\left[360\left(\frac{284 + 161}{365}\right)\right] = 23.02°$$

Using a latitude of 30°, the solar altitude, H_c, is calculated using Equation 3-14:

$$H_c = \arcsin\left[\cos(30°)\cos(23.02°)\cos(-15°) + \sin(30°)\sin(23.02°)\right]$$

$$= 74.8°$$

To calculate the solar azimuth, Z_c, the solar azimuth variable, X, must first be calculated using Equation 3-17:

$$X = \frac{\sin(-15°)}{\sin(30°)\cos(-15°) - \cos(30°)\tan(23.02°)} = -2.24°$$

Referring to Table 3-5, for an hour angle, ω, of –15° and a solar azimuth variable, X, less than zero, the solar azimuth constant, C, is 180 degrees.

The solar azimuth can now be calculated using Equation 3-16:

$$Z_c = 180° + \arctan(-2.24°) = 114°$$

Using these calculated values for H_c and Z_c, the solar heat gain can be calculated with greater precision. First, the effective angle of incidence of the sun's rays, θ, and the total solar heat flux, Q_s, must be calculated. θ is calculated using Equation 3-13:

$$\theta = \arccos\left[\cos(74.8°)\cos(114° - 90°)\right] = 76.1°$$

Q_s is then calculated using Equation 3-18:

$$Q_s = -3.9241 + 5.9276(74.8) - 1.7856 \times 10^{-1}(74.8)^2 + 3.223 \times 10^{-3}(74.8)^3 - 3.3549 \times 10^{-5}(74.8)^4 + 1.8053 \times 10^{-7}(74.8)^5 - 3.7868 \times 10^{-10}(74.8)^6$$

$$= 95.4\ W/ft^2$$

And finally, the solar heat gain rate is calculated using Equation 3-12:

$$q_s = (0.5)(95.4)\sin(76.1°)(0.092) = 4.26\ W/ft$$

Resistance (R_{T_C})

Resistance at 75°C = 0.1390 Ω/mile, (from conductor tables, Chapter 1)

Convert resistance given to Ω/ft:

$$R_{75°C} = (0.1390\ \Omega/mile) \,/\, (5280\ ft/mile)$$
$$= 2.633 \times 10^{-5}\ \Omega/ft$$

If an ampacity is desired at a conductor temperature for which no resistance is given in the tables in Chapter 1, the value of resistance may be calculated using Equation 3-21.

Steady State Ampacity

Using Equation 3-2,

$$I = \sqrt{\frac{q_c + q_r - q_s}{R_{75°C}}}$$

where:

q_c = 20.882 W/ft
q_r = 5.183 W/ft
q_s = 4.293 W/ft
$R_{75°C}$ = 2.633×10^{-5} Ω/ft

$$I = \sqrt{\frac{20.882 + 5.183 - 4.293}{2.633 \times 10^{-5}}} = 909\ amperes$$

It should be noted that given the approximate nature of conductor thermal parameters and minor differences in calculation methods, the thermal rating may vary approximately 1% among different calculation methods.

3.2 TRANSIENT THERMAL RATINGS

Power utilities are required to make every effort to maintain continuity of service to their customers under all operating conditions. When emergency or fault conditions occur, the loading of overhead lines is often allowed to exceed the normal maximum continuous rating for limited periods of time. The maximum permissible conductor temperature and its duration under these emergency operation must be determined in advance to limit the loss of strength and prevent the violation of minimum electrical clearances.

If the time duration allowed is 30 minutes or less, the heat storage capacity of the bare overhead conductor should be considered in the rating calculation as described in the next section.

3.2.1 *Step Increase in Load Current*

Short time emergency loadings are usually calculated for a step increase in load current as shown in Figure 3-4. The current prior to the step increase is typically assumed to be at or near the line's normal thermal rating.

FIGURE 3-4
CONDUCTOR TEMPERATURE RESPONSE TO A STEP CHANGE IN CURRENT @ 40°C AMBIENT TEMPERATURE

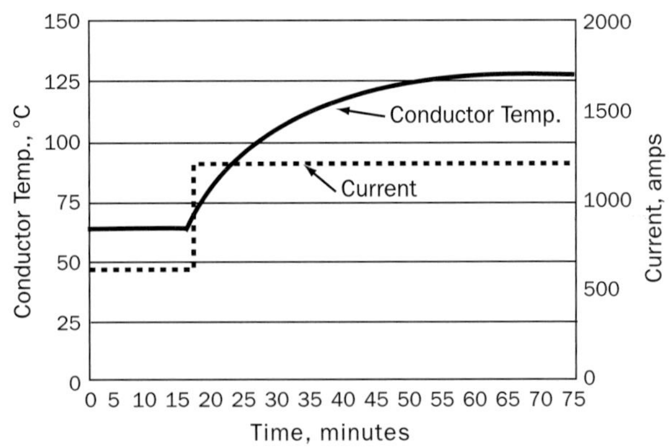

Immediately after a step increase in current, the convection and radiation heat losses remain unchanged and all of the increased heat generated is stored by the conductor which increases its temperature. As the conductor temperature increases, the convection and radiation heat losses also increase, and the rate at which heat is stored by the conductor begins to decrease. Eventually, the conductor temperature approaches a new steady-state temperature and the heat storage rate drops to zero.

The transient heat balance equation is similar to the steady state heat balance Equation 3-1 with an additional term for heat storage:

$$q_c + q_r + mC_p\left(\frac{dT_c}{dt}\right) = q_s + I^2R_{T_c} \qquad (3\text{-}22)$$

and rearranging,

$$\frac{dT_c}{dt} = \frac{1}{mC_p}\,[I^2R_{T_c} + q_s - q_c - q_r] \qquad (3\text{-}23)$$

where:

q_c = *convection heat loss rate, W/ft*

q_r = *radiated heat loss rate, W/ft*

q_s = *solar heat gain rate, W/ft*

m = *mass per unit length, lbs/ft*

C_p = *specific heat of conductor, W-sec/lbs-°C*

I = *conductor current, amperes at 60 Hz*

R_{T_c} = *ac resistance of conductor at temperature T_c, Ω/ft*

T_c = *conductor temperature, °C*

3.2.2 Conductor Heat Capacity

Conductor heat capacity, mC_p, is defined as the product of specific heat and mass per unit length. If the conductor consists of more than one material, such as ACSR and ACAR, the heat capacity of each material is calculated separately and added to obtain the total heat capacity of the conductor.

$$mC_p = \Sigma m_i C_{pi} \qquad (3\text{-}24)$$

where:

m_i = *component weight per unit length of conductor, lbs/ft*

C_{pi} = *specific heat of conductor material, W-sec/lbs-°C*

Although varying slightly, the heat capacities of aluminum and 6201 aluminum alloy are assumed to be the same. Values for the specific heat at 25°C of common conductor materials are listed in Table 3-6.

TABLE 3-6
SPECIFIC HEATS FOR
COMMON CONDUCTOR MATERIALS @ 25°C

MATERIAL	C_p W-sec/kg-°C	C_p W-sec/lbs-°C
Aluminum	955	433
Copper	423	192
Steel	476	216
Alumoweld*	534	242

* *The specific heat of Aluminum Clad Steel depends on the aluminum-to-steel ratio. The values listed are typical for Aluminum Clad Steel wire with a conductivity of 20.3% IACS.*

As an example, the total conductor heat capacity for a 795 kcmil - 26/7 ACSR "Drake" conductor can be calculated as follows:

The weights of the steel core and the outer layers of aluminum, determined from the ACSR table in the Chapter 1 Appendix, are 0.3443 lbs/ft and 0.7491 lbs/ft, respectively. The specific heats for both steel and aluminum are taken from Table 3-6. The component heat capacities for the aluminum and steel are:

$$mC_{pa}\,(alum.\ @\ 25°C) = (0.7491)(433) = 324.4\ \ W\text{-}sec/ft\text{-}°C$$

$$mC_{ps}\,(steel\ @\ 25°C) = (0.3443)(216) = 74.4\ \ W\text{-}sec/ft\text{-}°C$$

The total heat capacity for 795 kcmil - 26/7 ACSR "Drake" conductor is:

$$mC_p = 324.4 + 74.4 = 399\ W\text{-}sec/ft\text{-}°C$$

3.2.3 Example of Transient Thermal Rating Determination

Transient thermal ratings are calculated assuming a step increase in current from a steady-state current loading. The current in the conductor prior to the step increase, the duration of the step increase (typically 5 to 30 minutes), the maximum allowable conductor temperature at the end of the step increase, and all conductor and weather parameters must be specified. The transient rating is defined as the post-step current that causes the conductor temperature to reach the maximum allowable conductor temperature in the specified time duration.

As an example, consider a 795 kcmil - 26/7 ACSR "Drake" conductor, initially carrying a steady-state "pre-step" current of 400 amperes that causes a conductor temperature of 55°C. During both the pre-step and post-step periods, the rating conditions are the same - a 2 ft/sec wind and 40°C ambient temperature with sun. The post-step current that causes the conductor temperature to reach, but not exceed 100°C, based on a 15-minute duration, is the transient rating. The post-step current required to reach the 100°C limit is normally determined by repeating the temperature versus time calculations (after the step) until the post-step current that causes the conductor to reach the maximum defined temperature in the allotted time is found.

FIGURE 3-5
TRANSIENT RESPONSE TO STEP INCREASE CURRENT AT 25°C AMBIENT TEMPERATURE

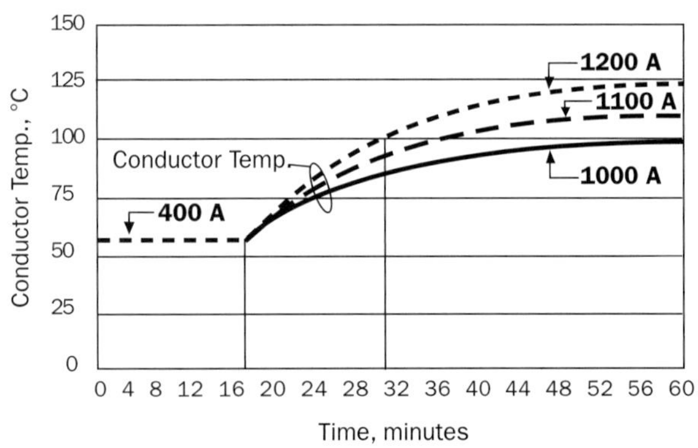

Figure 3-5 shows the increase in conductor temperature versus time corresponding to step increases in currents of 1000, 1100 and 1200 amperes. Prior to the step increase, a steady-state conductor current of 400 amperes with a conductor temperature of 55°C is assumed. With step increases in current of 1000 and 1100 amperes the conductor temperature does not reach 100°C during the 15-minute duration. With a step increase to 1200 amperes, the conductor temperature actually exceeds the maximum allowable temperature of 100°C by a few degrees. Through repeated calculations, a step increase to 1160 amperes is determined to yield a conductor temperature increase to 100°C in exactly 15 minutes. Therefore, the 15-minute transient thermal rating of the conductor is 1160 amperes.

3.2.4 Thermal Time Constant

When the electrical current flowing in a conductor undergoes a sudden "step" increase, the conductor temperature goes through a more gradual change from an initial to a new, higher final temperature. The thermal time constant is the time required for the conductor temperature to attain 63.2% of this ultimate change in temperature. The thermal time constant of an overhead conductor is typically about 5 to 20 minutes. In a period of 3 thermal-time-constants, the conductor temperature will reach 95% of its new steady-state temperature. Figure 3-6[8] shows the influence of conductor size on the time constant for a step current overload of 50% above the 75°C steady-state ampacity rating. As can be seen, the time constant increases with conductor size and decreases with wind speed.

FIGURE 3-6
THERMAL TIME CONSTANT AS A FUNCTION OF WIND VELOCITY FOR TYPICAL ACSR CONDUCTORS

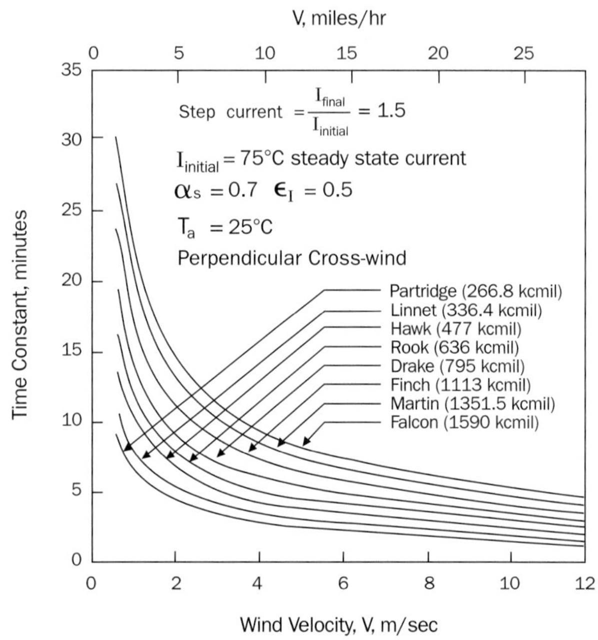

© 1985 IEEE.

3.3 FAULT CURRENT RATINGS

Fault-current ratings are used to determine the adequacy of a conductor for short-circuit currents with a duration measured in cycles (60 Hz), not minutes. For fault-current ratings with all-aluminum conductors, a maximum temperature of 340°C is commonly used. Exposure of aluminum to this temperature for a few seconds has little or no effect on tensile strength. The melting temperature of aluminum, 645°C, is used in steel core conductor fault current calculations because the steel core wires are unaffected by short exposures to such temperatures.

With fault-current thermal ratings of less than 1-minute duration, the heat lost through convection and radiation is normally neglected. Therefore, such ratings are not a function of the weather conditions, and all heat is considered lost to heat storage in the conductor. In an ACSR conductor, 98% of the current flows through the aluminum strands. Therefore, when calculating fault-current ratings with durations of less than 1 minute, the heat storage in the ACSR steel core is neglected, and only the specific heat of aluminum is used to calculate heat capacity. The maximum allowable short-circuit current is typically calculated using the specific heat at 20°C given in Table 3-7. However, an average specific heat for the temperature range of the short-circuit will give more accurate results. The maximum allowable short-circuit (fault) current versus time may be calculated using the following equation:

$$\left(\frac{I_{SC}}{A}\right)^2 t = 247.0 \times 10^{-6} (\rho_m)(C_p) \left(\frac{T_0+\lambda}{\rho_0}\right) \log_{10} \left(\frac{T_2+\lambda}{T_1+\lambda}\right) \quad (3\text{-}25)$$

where:

- I_{SC} = *short-circuit current, amperes*
- A = *effective cross-sectional area of conductor, cmil*
- t = *time of short-circuit, sec*
- ρ_m = *density, g/cm³*
- C_p = *specific heat, cal/g-°C*
- T_0 = *arbitrary temperature of reference (usually 20°C)*
- ρ_0 = *specific resistivity at temperature T_0, μΩ-cm*
- λ = *inferred temperature of zero resistance, °C below zero*
- T_2 = *short-circuit temperature, °C*
- T_1 = *conductor temperature before short-circuit, °C*

If we let:

$$K = 247.0 \times 10^{-6} (\rho_m)(C_p) \left(\frac{T_0 + \lambda}{\rho_0}\right) \quad (3\text{-}26)$$

Then:

$$\left(\frac{I_{SC}}{A}\right)^2 t = K \log_{10} \left(\frac{T_2 + \lambda}{T_1 + \lambda}\right) \quad (3\text{-}27)$$

and K may be read from Table 3-7.

TABLE 3-7
CONDUCTOR MATERIAL PROPERTIES

Material & Conductivity (%IACS)	ρ_m Density @ 20°C g/cm³	C_p Specific Heat @ 20°C cal/g-°C	ρ_0 Resistivity @ 20°C μΩ-cm	λ °C	K
Copper (97%)	8.890	0.092	1.7745	234.0	0.0289
1350-H19 (61.2%)	2.705	0.214	2.8171	228.1	0.0126
6201 (52.5%)	2.690	0.214	3.2841	228.1	0.0107
Steel (8%)	7.800	0.110	13.0	180.0	0.00327
1350-O (63%)	2.705	0.214	2.7367	228.1	0.0130

Figures 3-13 through 3-20, located in the Chapter 3 Appendix, show the conductor fault current limits for time durations commonly encountered in modern systems. The figures were calculated using a 40°C ambient temperature and the specific heat, C_p, taken as the average value between the 40°C ambient and the maximum short circuit temperature (0.2351 cal/g-°C for stranded aluminum and 0.2469 cal/g-°C for ACSR and ACSS). The maximum short circuit temperature is assumed to be 645°C for ACSR and ACSS conductors and 340°C for all-aluminum conductors. The maximum allowable temperature for all-aluminum conductors is set to prevent dangerous annealing of the aluminum. Where rapid multiple reclosures can occur, fault duration time should be taken as the sum of the individual reclosures.

The following simplified equations are the bases of the graphs presented in Figures 3-13 through 3-20 in the Chapter 3 Appendix and may be used as an alternative to the charts.

ACSR:

$$I = 0.0862 \frac{A}{\sqrt{t}} \tag{3-28}$$

AAC:

$$I = 0.0671 \frac{A}{\sqrt{t}} \tag{3-29}$$

AAAC/6201:

$$I = 0.0638 \frac{A}{\sqrt{t}} \tag{3-30}$$

where:

I = *current, amperes*

A = *cross-sectional area, cmil*

t = *time, seconds*

3.4 COMPUTER PROGRAM FOR CALCULATING THERMAL RATINGS

The computer program, SWRate, included with this book to calculate the thermal rating (ampacity) of conductors is based on IEEE Standard 738[1] as presented earlier in this chapter. Careful consideration should be given to all parameters before beginning rating calculations.

The computer program can be used to calculate:

(1) **Steady-State Thermal Rating** – can be calculated, given a maximum allowable conductor temperature, weather conditions, and conductor characteristics.

(2) **Steady-State Conductor Temperature** – can be determined, given the electrical current, the weather conditions, and conductor characteristics.

(3) **Transient Thermal Ratings** – may be determined by calculating conductor temperature versus time curves such as those shown in Figure 3-5. The program calculates a transient thermal rating by performing a series of calculations, adjusting the assumed final current until the conductor temperature just reaches the set temperature requirement in the defined time duration.

(4) **Transient Conductor Temperature** – is the change in conductor temperature over time in response to a step change in electrical current. It can be calculated when either the initial current or the initial conductor temperature is known. Transient conductor temperature calculations for relatively high fault currents for short times can also be performed. Accuracy in the iterative transient calculation requires that the time step chosen be sufficiently small with respect to the thermal time constant. IEEE suggests a step equal to 1% of the thermal time constant is suitable.

3.5 SELECTION OF WEATHER CONDITIONS FOR LINE RATING CALCULATIONS

For a given conductor, the thermal line rating is dependent on the maximum allowable temperature and weather conditions selected for the calculation[9]. The selection of maximum conductor temperature is based on either electrical clearance or annealing concerns. Weather condition selection is based simply upon engineering judgment.

Weather conditions have a major effect on the thermal rating of an overhead line. Wind speed is the most widely varying parameter and the most important determinant of ratings. Careful selection of weather parameters for rating thermal calculations is essential for both public safety and long-term mechanical reliability.

One common problem in choosing weather conditions involves the quality of weather data. For example, weather data can be obtained from local weather bureaus, usually located at airports. Such data, computed hourly, may be available for the past 40 years or more. Unfortunately, wind speeds of less than 3.5 mph (5 ft/sec) from weather bureau records are often inaccurate[10]. This inaccuracy is due to measurement errors with standard cup-type anemometers at low wind speeds (see Figure 3-7) and to the time intervals of measurement of one to three hours.

FIGURE 3-7
NOAA WEATHER DATA – NUMBER OF HOURLY VALUES

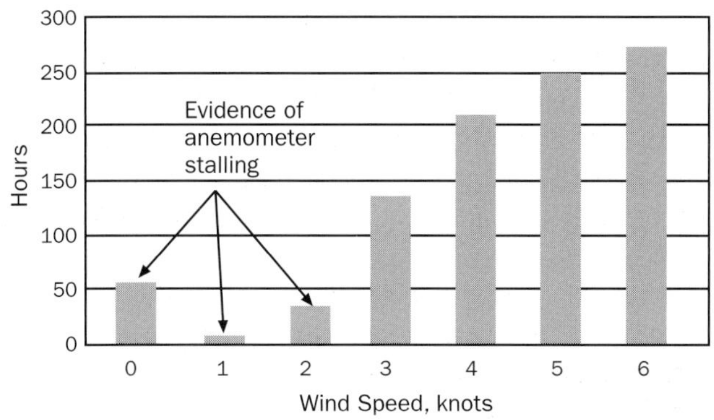

Wind speeds along transmission line right-of-ways may vary substantially from the wind speed measured at airports. The variations in wind speeds typically result from shielding (terrain, trees, structures, etc.), or are due to the inconsistencies in wind speeds over the geographical area.

As an example of the possible consequences of choosing wind speed improperly for line rating calculations, consider the thermal risk for a line with a 795 kcmil - 26/7 ACSR "Drake" conductor with a maximum allowable temperature of 100°C. Traditionally, overhead conductors have been rated using conservative weather conditions with wind speeds of 0 to 2 ft/sec and air temperatures of 25°C to 40°C[11].

The use of less conservative wind speeds can yield increased risk of conductor "over-temperature" that exceeds the line design conductor temperature as shown in Table 3-8.

By assuming a conservative wind speed of 2 ft/sec, the line rating is 990 amps, and the thermal risk (last column) during occasional still air conditions is 30°C. Assuming a less conservative wind speed of 4 ft/sec yields a much larger line rating of 1160 amperes, but the thermal risk during still air periods increases to 60°C. Choosing appropriate weather conditions for line rating calculations is discussed in more detail in Chapter 7.

3.6 CONDUCTOR LOSS OF STRENGTH

Temperature of transmission conductors must be limited in order to avoid reducing the conductor strength. Since some degradation of tensile strength due to conductor heating is inevitable, a maximum 10% loss of strength over the life of the line is allowed for design purposes.

The most often used method of calculating the loss in conductor strength due to annealing of the aluminum strands is to recalculate the total breaking strength using the annealed aluminum tensile strength. For example, consider that over its 30-year life, a "Drake" conductor is exposed to an equivalent time-temperature effect of 500 hours of operation at 150°C. Figure 3-8 indicates that the tensile strength of the aluminum strands would be reduced to 77% of its initial value. This decrease corresponds to a decrease of about 23% in the tensile strength of the aluminum strands and about 11% in the strength of the composite conductor.

3.6.1 Annealing of Individual Strands

Aluminum strands of any temper other than dead soft (O-temper) experience a loss of tensile strength when exposed to elevated temperatures for extended periods of time, as shown in Figures 3-8 and 3-9[12]. This annealing of aluminum occurs for all the common alloys used in overhead conductors and becomes more severe as the temperature and/or time increase. Even special high temperature alloys of aluminum will be annealed if the temperature is sufficiently elevated or the time at temperature adequately extended.

Figure 3-8 shows "percent remaining of initial strength" as a function of the duration of the high temperature exposure for conductor temperatures of 100°C, 125°C and 150°C for individual strands of 1350-H19 aluminum wire. If fully annealed (O-temper) by exposure to high temperatures over a sufficient period of time, it ruptures at only 10,000 psi after an elongation of more than 30%.

Figure 3-9 indicates the change in tensile strength for 1350-H19 and 6201-T81 aluminum alloy during a 30-minute exposure to various temperatures. The rate of decline in tensile strength for 6201-T81 aluminum alloy is much greater than that of 1350-H19 aluminum.

This annealing data for individual strands can be used to

TABLE 3-8
THERMAL RATING RISK

Wind Speed Used in Rating Calculation ft/sec	Rated Ampacity with Wind Amperes	Conductor Temperature at Rated Ampacity with Wind °C	Conductor Temperature at Rated Ampacity in Still Air °C	Conductor Temperature Rise above Rated Temperature in Still Air °C
0	750	100	100	0
2	990	100	130	30
3	1080	100	145	45
4	1160	100	160	60

calculate loss of strength with either homogeneous or non-homogeneous bare, stranded conductor. For non-homogeneous cables, it is the sum of the remaining tensile strengths of its component materials taken at the rupture point of the least ductile material.

3.6.2 Stranded Conductor Loss of Strength

Extensive laboratory research has been done to study the relationship of composite conductor loss of strength as a function of temperature and time[13]. This work has resulted in the derivation of formulas, reproduced herein, to calculate remaining tensile strength after high-temperature excursions. These formulas work well for extended exposure to high temperatures, but there exist some discontinuities at relatively short exposure times for ACSR. If the exposure time is short enough, it is possible to calculate a conductor strength that exceeds the rated strength of the conductor. If the remaining strength calculated is greater than 100% of the rated breaking strength (RBS) of the conductor, the rated strength should be used.

Another issue with ACSR is that measurements of strength indicate that the predicted decrease in breaking strength of ACSR "Drake" does not occur. Studies indicate that exposure of ACSR "Drake" to 150°C for 500 hours yields a loss of strength of only 2%. This difference occurs for two reasons. First the steel core wire does not loose strength at the high temperature. Secondly, the resulting increased ductility of the aluminum reduces, or eliminates, the need to restrict the steel strength to its strength at 1% elongation. This restriction on steel is used in ACSR to prevent failure of the 1350-H19, full-hard, aluminum strands, which have a maximum elongation of approximately 1%. Thus the additional strength made available from the steel offsets the loss of strength in the aluminum strands due to annealing from exposure to the high temperatures.

The measurements of all-aluminum conductors, however, indicate reasonable agreement with calculated loss-of-strength. For example, 795 kcmil AAC, after 430 hours at 125°C, yields a measured loss-of-strength of 11%. Based on the single strand data of Figure 3-8, use of the same temperature-time exposure yields a loss-of-strength equal to 12%.

3.6.3 Calculation of Remaining Strength

Mathematical models for the remaining rated strength of AAC, AAAC, ACAR and ACSR conductors are shown below[13][14]. Comparisons of rated strengths calculated by these models to laboratory measurements of rated strength are shown in Tables 3-9, 3-10, 3-11 and 3-12 for AAC, AAAC, ACAR and ACSR conductors.

1. AAC[1]:

$$RS = (-0.24T + 135)t^{-(0.001T - 0.095)\left(\frac{0.1}{d}\right)} \tag{3-31}$$

2. AAAC[2]:

$$RS = (-0.52T + 176)t^{-(0.0012T - 0.118)\left(\frac{0.1}{d}\right)} \tag{3-32}$$

3. ACAR:

$$RS = RS_{1350}\frac{A_{1350}}{\Sigma A} + RS_{6201}\frac{A_{6201}}{\Sigma A} \tag{3-33}$$

FIGURE 3-8
ANNEALING OF INDIVIDUAL ALUMINUM STRANDS

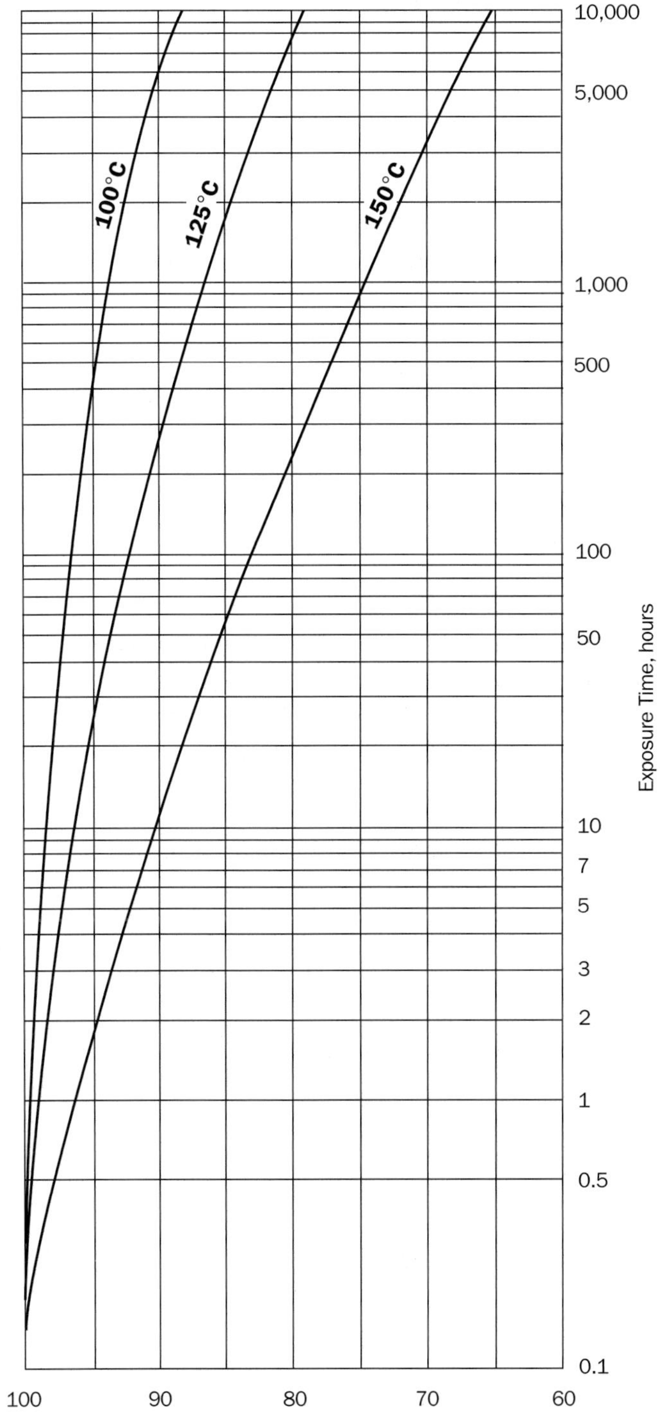

Time-temperature relationship for loss-of-strength for 1350-H19 aluminum strands. Tensile tests made at room temperature after exposure to indicated temperature.

FIGURE 3-9
ANNEALING OF ALUMINUM AND ALUMINUM ALLOY STRANDED CONDUCTORS

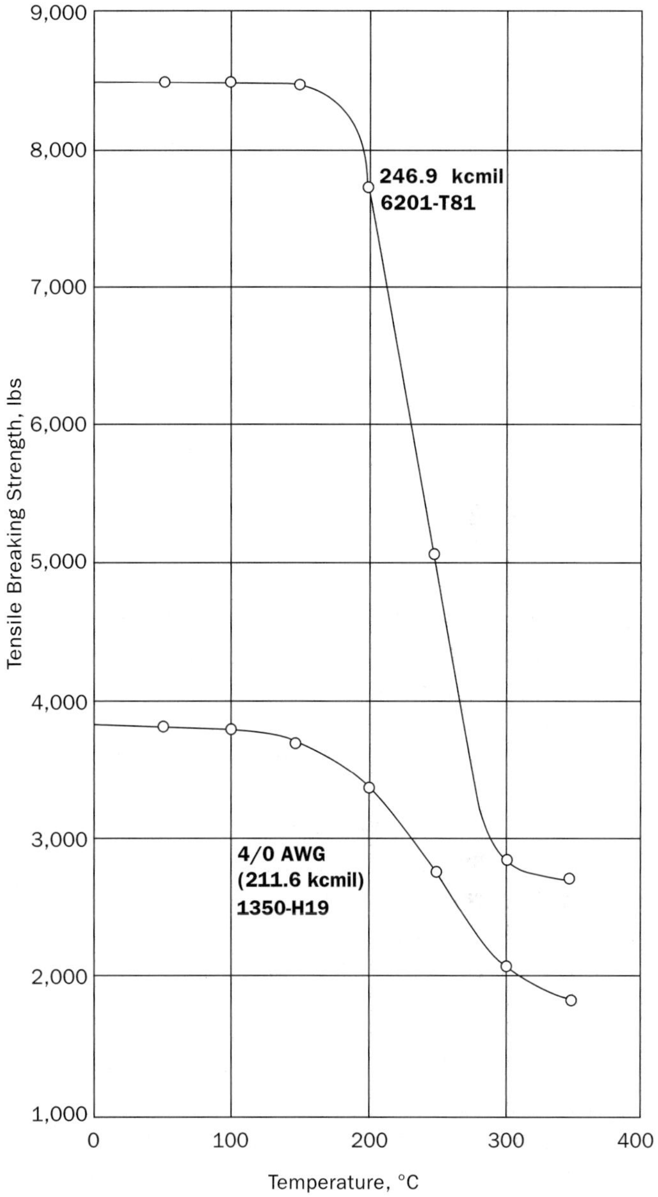

Remaining tensile strength for aluminum and aluminum alloy strand conductors of approximately equivalent conductivity after 1/2 hour exposure to elevated temperature.

4. ACSR:

$$RS = RS_{1350}\left(\frac{STR_{1350}}{STR_T}\right) + 109\left(\frac{STR_{ST}}{STR_T}\right) \quad (3\text{-}34)$$

where:

RS	= *remaining strength as a percentage of initial strength*
RS_{1350}	= *remaining strength of 1350-H19 aluminum strands*
RS_{6201}	= *remaining strength of 6201-T81 aluminum alloy strands*
T	= *exposure temperature, °C*
t	= *elapsed time, hours*
d	= *strand diameter, in*
A_{1350}	= *area of 1350-H19 aluminum strands*
A_{6201}	= *area of 6201-T81 strands*
ΣA	= *total area* = $A_{1350} + A_{6201}$
STR_{1350}	= *calculated initial strength of the 1350-H19 aluminum strands*
STR_{ST}	= *calculated initial strength of the steel core*
STR_T	= *calculated initial strength of the conductor*

[1] *If (−0.24T + 135) >100, use (100)*

[2] *If (−0.52T + 176) >100, use (100)*

These equations may be employed where several episodes of high-temperature operation have occurred. The effects are cumulative for each episode or occurrence. This is illustrated by the following example where an all-aluminum (AAC) conductor, during one year of operation, is exposed to 8000 hours at 75°C, 600 hours at 100°C, 150 hours at 125°C and 10 hours at 150°C. The remaining strength calculated using conversions to 150°C–equivalents is shown below:

- Start with the lowest temperature exposure - 8000 hours at 75°C, the remaining strength, *RS*, found from Equation 3-31 is 100%. This is the equivalent of 0 hours at 150°C.

- 600 hours at 100°C yields a remaining strength of 98%. This is the equivalent of less than 1 hour at 150°C.

- 150 hours at 125°C yields a remaining strength of 90%. This is the equivalent of 10 hours at 150°C.

- Actual time the conductor operated at 150°C is 10 hours.

- The summation of the actual operating time at 150°C plus the equivalent operating times at 150°C is approximately 20 hours. The remaining calculated strength is 88% of the original calculated strength based on 20 hours of operation at 150°C during the year.

3.7 SAG AT HIGH TEMPERATURE

Transmission lines are designed to maintain minimum clearance to ground and to other conductors under all foreseeable operating conditions, including operation at high conductor temperatures that may be generated by emergency loadings. The maximum allowable conductor temperature is specified such that clearances are maintained for any conductor temperature up to this maximum value.

In determining the sag in any given span at high temperature, one must consider momentary reversible thermal elongation, mechanical influences of adjacent spans, and the long-term permanent elongation (creep) that sustained elevated temperatures may cause. For steel core aluminum conductors, the reversible thermal elongation may vary with temperature and radial differences in temperature may affect high temperature sag. These factors are discussed in the following sections.

3.7.1 Thermal Coefficient of Expansion

The thermal coefficients of expansion of common conductor materials differ. In the case of ACSR, aluminum has a coefficient of thermal expansion that is approximately twice that of steel. As a consequence, the faster expansion of the aluminum causes the tensile load of the line to shift to the steel core. The composite cable will then have elongation less than one-half that of an all-aluminum conductor.

Figure 3-10 shows the increase in sag with temperature for both a 795 kcmil - 26/7 ACSR "Drake" and a 795 kcmil 37 - strand AAC "Arbutus" conductor in a 1000 feet span. Although the sags of the two different conductors are the same at 15°C, the sag of the all-aluminum conductor increases more rapidly with temperature. Also note that the thermal elongation rate of ACSR decreases at a temperature between 60°C and 80°C (the "knee-point" temperature) where the aluminum tension reaches zero and the steel becomes the sole tension member of the conductor. Note the relatively large difference – about 4 feet – in sag at 150°C in the AAC and ACSR conductors and that two different ACSR expansion rates, defining the knee point, are shown.

Tests of ACSS and ACSS/TW indicate that beyond the knee-point temperature the conductor elongates thermally at the rate of the steel core alone. Tests of multi-layer

TABLE 3-9
LABORATORY TESTS OF REMAINING STRENGTH
FOR AAC CONDUCTOR

AAC			
Time, Hours	Conductor	Remaining Strength, %	
		Measured	Predicted
100°C			
1	# 2 AWG	100	100
3	# 2 AWG	99	99
10	# 2 AWG	96	99
30	# 2 AWG	99	98
100	# 2 AWG	96	98
125°C			
3	# 2 AWG	94	97
6	# 2 AWG	98	95
10	# 2 AWG	97	93
20	# 2 AWG	91	91
30	# 2 AWG	91	90
71	795 kcmil	92	92
100	# 2 AWG	89	87
125	795 kcmil	92	91
143	# 2 AWG	86	86
164	# 2 AWG	82	85
192	# 2 AWG	84	85
237	795 kcmil	91	89
370	# 2 AWG	84	83
430	795 kcmil	89	88
150°C			
3	# 2 AWG	95	92
3	795 kcmil	93	94
10	# 2 AWG	86	86
10	795 kcmil	92	90
19	336.4 kcmil	81	87
20	# 2 AWG	82	83
30	# 2 AWG	79	81
30	795 kcmil	97	86
74	336.4 kcmil	82	82
100	795 kcmil	83	82
168	336.4 kcmil	79	79
175°C			
3	# 2 AWG	85	84
10	# 2 AWG	78	76

TABLE 3-10
LABORATORY TESTS OF REMAINING STRENGTH
FOR ACAR CONDUCTOR

ACAR			
Time, Hours	Conductor	Remaining Strength, %	
		Measured	Predicted
150°C			
3	649.5 kcmil-18/19	93	93
3	852.6 kcmil-30/7	99	94
24	852.6 kcmil-30/7	90	87
50	649.5 kcmil-18/19	80	82
97	649.5 kcmil-18/19	73	80
120	852.6 kcmil-30/7	84	82
220	852.6 kcmil-30/7	81	81

TABLE 3-11
LABORATORY TESTS OF REMAINING STRENGTH
FOR AAAC CONDUCTOR

AAAC			
Time, Hours	Conductor	Remaining Strength, %	
		Measured	Predicted
100°C			
1	# 2 AWG	100	100
10	# 2 AWG	99	100
30	# 2 AWG	99	100
100	# 2 AWG	100	100
500	# 2 AWG	95	99
125°C			
3	# 2 AWG	99	97
10	# 2 AWG	97	93
24	# 2 AWG	94	91
30	# 2 AWG	90	90
100	# 2 AWG	88	87
260	# 2 AWG	84	84
150°C			
3	# 2 AWG	95	92
3	652.4 kcmil	97	94
10	# 2 AWG	84	86
13	652.4 kcmil	93	90
25	# 2 AWG	79	81
33	652.4 kcmil	87	87
100	652.4 kcmil	80	84
175°C			
3	# 2 AWG	79	77

TABLE 3-12
LABORATORY TESTS OF REMAINING STRENGTH
FOR ACSR CONDUCTOR

ACSR			
Time, Hours	Conductor	Remaining Strength, %	
		Measured	Predicted
125°C			
24	336.4 kcmil-18/1	99	98
50	336.4 kcmil-18/1	98	97
120	336.4 kcmil-18/1	97	96
150°C			
2	336.4 kcmil-18/1	100	99
3	795 kcmil-26/7	100	100
10	795 kcmil-26/7	100	100
30	795 kcmil-26/7	100	100
50	795 kcmil-26/7	101	100
52	336.4 kcmil-18/1	89	91
100	795 kcmil-26/7	98	100
100	795 kcmil-26/7	100	100
120	336.4 kcmil-18/1	87	89
170	336.4 kcmil-18/1	89	88
200	795 kcmil-26/7	98	100
500	795 kcmil-26/7	98	99

ACSR, however, indicate that the "full hard" H19 aluminum layers support a modest level of compression beyond the kneepoint. Several theories have been proposed to model the composite conductor thermal elongation in this transition region beyond the knee-point temperature.

Rawlins[15] proposed that the increased thermal elongation is due to a combination of residual stress in the aluminum layers (a result of manufacturing stranding processes) and to aluminum compression (interference between the helical layers leading to the ability to support up to 2500 psi of compression). Barrett[16] suggested that the aluminum layers can support up to 1500 psi of compression which can be increased by the relatively high radial temperature gradients that can occur at high current densities. In either case, the authors note a high level of random variation in the observed levels of thermal elongation beyond the kneepoint. In Rawlin's study of residual stress in ACSR conductors, residual stress levels ranging from 0 to 4000 psi were found. In Barrett's studies of compression levels in ACSR, it was found that the aluminum layers supported between 1000 and 3000 psi. The effect is small enough that it is nearly impossible to determine which theory is correct.

Given the variability of ACSR sag calculations at high temperatures, and the uncertainty of the details, it seems prudent to take a conservative approach to accounting for aluminum compression. Recommendations for additional sag due to aluminum compression are summarized in Table 3-13.

3.7.2 Suspension Span Interaction at Elevated Temperature

Transmission lines are normally constructed in line sections that terminate at dead-end structures. In between are suspension structures where tension equalization occurs. Sag in any suspension span is a function of average conductor temperature within the line section rather than the conductor temperature at any point in the span.

At normal electrical loads, the conductor temperature is typically 5°C to 20°C above the air temperature, and conductor temperature differences between spans are likely to be no more than a few degrees centigrade. With many suspension spans between dead-end structures, one may simply assume the tension in adjacent suspension spans is equalized by suspension insulator swing. The sags in these spans therefore, may be calculated by normal ruling span methods.

Under heavy electrical loading, the conductor temperature differences between suspension spans may be tens

FIGURE 3-10
SAG vs. TEMPERATURE FOR AAC AND ACSR CONDUCTORS

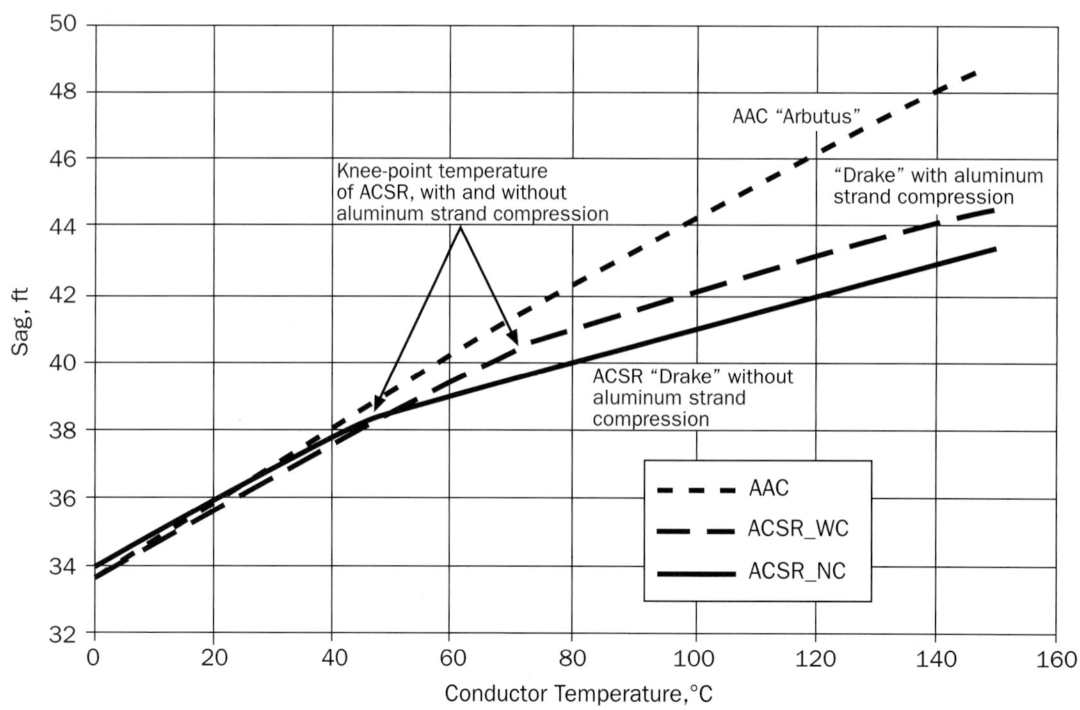

of degrees centigrade. Under such conditions, greater suspension insulator movement may result, but it appears that nearly full tension equalization still occurs between suspension spans in flat or rolling terrain.

Tension equalization between suspension spans relieves the concern regarding excessive sag in a sheltered span where conductor temperatures may greatly exceed those of other less sheltered spans. A simple calculation gives some confidence that the tension equalization/ruling span process limits the sag in this hottest span.

Figure 3-11 shows the sags in a series of four suspension spans in a single dead-end section. In one case, the temperature in all the spans is 120°F. In the next case, the temperature of all the spans is 212°F. Finally, the temperature of the long span is 212°F while the other three spans are at 120°F. Note that, when the long span is at 212°F and other spans are at 120°F, the sag in the long span is less than when all the spans are at the high temperature.

FIGURE 3-11
SAG IN HOTTEST SPAN INFLUENCED BY ADJACENT COOLER SPANS

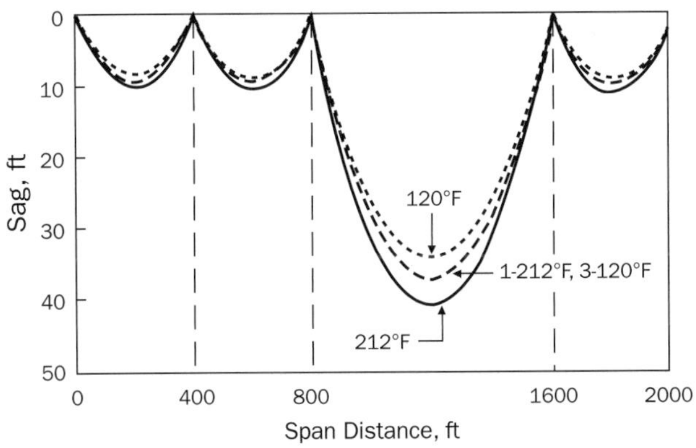

3.7.3 Radial Temperature Difference

Heat generated in the inner strands of a multilayer stranded conductor must reach the conductor surface in order to be convected and radiated away. Thus the core of the conductor is normally somewhat hotter than the surface. At conventional current densities of less than 1 ampere per kcmil, the radial temperature difference is only a few degrees C. As current densities reach levels on the order of 2 or more amperes per kcmil (e.g. >1600 amperes for ACSR "Drake"), the radial temperature difference between core and surface may be 30°C or more as shown in Figure 3-12.

Operating conductors at surface temperatures in the range of 200°C will almost certainly produce significant radial temperature differences which need to be considered in sag calculation procedures. Since ACSR and ACSS conductors are essentially supported by the steel core at high temperature, the sag of these conductors should be calculated on the basis of the core temperature, not the surface temperature of the conductor.

Laboratory testing of "Suwannee/ACSS/TW" conductor sponsored by CenterPoint Energy[17] yielded average radial gradients of between 30°C and 40°C for surface temperatures in the range of 150°C to 240°C. Unlike the smooth increase in radial temperature difference shown in Figure 3-12, there was a great deal of variation in radial temperature difference as a function of the testing variables. The result of three separate tests each of low (L) and high (H) surface temperatures using "Suwannee/ACSS/TW" in a laboratory span are shown in Table 3-14.

Although there is considerable variation in the measured radial temperature differences, the results appear to be consistent with other reported laboratory and field tests differences of 20°C to 50°C.

The various radial temperature difference calculation methods derived from laboratory testing of conductors at

TABLE 3-13
EXCESS SAG DUE TO ALUMINUM COMPRESSION FOR SEVERAL STRANDINGS OF 795 KCMIL ACSR*

ACSR Conductor	Type Designation	Span Length, ft		
		600	1200	1800
795 kcmil – 45/7 "Tern"	7	1.0	1.0	–
795 kcmil – 26/7 "Drake"	16	1.5	2.0	3.0
795 kcmil – 30/19 "Mallard"	23	1.0	1.5	2.0

**Estimated with the Alcoa SAG10 computer program based on the compression model developed by Rawlins.*

lower current levels (surface temperatures below 100°C) appear to underestimate the temperature difference at very high current loading. Apparently, birdcaging of the aluminum strands at these high temperatures reduces metallic contact between the layers and increases the effective radial thermal conductivity.

As a practical matter, depending upon design conditions, a radial temperature difference may exist whenever nonhomogeneous stranded conductors are operated beyond the knee-point temperature and the line current exceeds 1.5 amps/kcmil.

FIGURE 3-12
CALCULATED RADIAL TEMPERATURE DIFFERENCE
FOR $k_r = 1\ W/°C\text{-}m$**

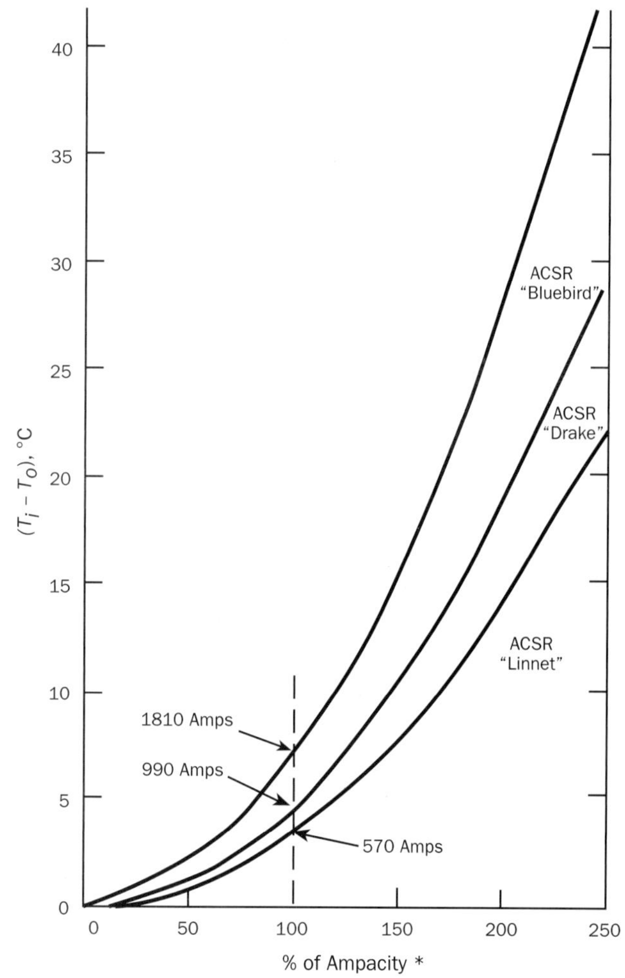

* Current at which $T_O = 75°C$, $T_{amb} = 25°C$, $V_{wind} = 2$ ft/sec perpendicular to conductor

** k_r is the effective radial thermal conductivity of the stranded conductor.

TABLE 3-14
RADIAL TEMPERATURE DIFFERENCE
959.6 KCMIL TYPE-16 ACSS/TW CONDUCTOR

Test No.	Surface Temperature °C	Core Temperature °C	Radial Temp. Difference °C
1L	172	240	68
2L	144	188	44
3L	184	221	37
Average Temps.	167	216	50
1H	217	303	86
2H	236	260	24
3H	220	259	39
Average Temps.	224	274	50

3.8 HIGH-TEMPERATURE CREEP ELONGATION

Permanent elongation of metals normally occurs only after being subjected to a stress that exceeds their yield strengths. If, however, these metals are subjected to stresses less than the yield strength for an extended period of time, a metallurgical phenomenon called creep will also cause an increase in permanent elongation.

The rate of creep is a function of the type of material, temperature, and tensile stress. For a given material, the creep rate increases with temperature and tensile stress. Under similar conditions, the creep rates of aluminum and aluminum alloys are significantly greater than steel. It is evident, therefore, that the creep rates of AAC, AAAC and ACAR conductors will be considerably greater than that of ACSR.

Any increase in the elongation of an overhead transmission or distribution line, regardless of whether it is temporary or permanent, increases line sag and decreases line clearance. To maintain minimum clearances, therefore, the line designer must take into account the creep characteristics of the conductor.

3.8.1 *Creep of AAC, AAAC and ACAR Conductors at High Temperature*

All-aluminum conductors are composed of strands of 1350-H19, all-aluminum-alloy conductors of 6201-T81, and aluminum conductors alloy-reinforced of a combination of the two materials. Although the creep rates of individual strands of these materials can be determined, no acceptable correlation between the creep rates of individual strands and multi-strand cables has been established. It has been necessary, therefore, to test various constructions of these conductors to empirically derive equations to predict cable elongation due to creep at room temperature, as well as at elevated temperatures. Stress-strain curves, such as those provided by The Aluminum Association, normally provide curves to estimate creep elongation at various stress levels for 6 month, 1 year and 10 years, with the conductor assumed to be at room temperature (e.g. 60°F) throughout these periods.

Since creep elongation is a function of temperature, as well as stress and time, the impact of extended periods of high temperature operation on creep must be studied using equations such as those developed by Harvey and Larson[18] as presented in Table 3-15, along with coefficient values found in Table 3-16. Although different coefficient values of AAC are shown as a function of method of manufacture, it is noted that the great majority of redraw rod from which 1350 aluminum wire is produced today is manufactured by a continuous-cast process.

3.8.2 Creep of ACSR Conductors at High Temperature

ACSR conductors are composed of 1350-H19 aluminum strands spiraled around a core comprised of steel strand(s). The creep rate of these conductors at room temperature is typically less than that of AAC, AAAC and ACAR conductors. The lower creep rate results from the tensile load carried by the aluminum strands being transferred to the steel core after the conductor has undergone a relatively small increase in elongation due to thermal elongation and creep. For ACSR with steel cross-sectional areas of 7.5%, or more, elevated temperatures of the magnitude experienced in transmission and distribution lines cause no significant increase in creep rate over that normally experienced at room temperature. For constructions with less steel, the increased creep rate at elevated temperatures must be considered.

The following equations for predicting creep of ACSR conductors were also derived by Harvey and Larson.

Room Temperature

For an ACSR conductor with 1350-H19 strands drawn from hot-rolled rod:

$$\epsilon_c = 2.4\,(\%RBS)^{1.3}\, t^{0.16} \qquad (3\text{-}35)$$

where:

$\%RBS$ = *the ratio of tension to rated breaking strength expressed as a percentage*

ϵ_c = *creep strain, micro-units/unit*

t = *elapsed time, hours*

TABLE 3-15

PREDICTOR EQUATIONS FOR ALL-ALUMINUM CONDUCTOR

Conductor Types	Room Temperature	Elevated Temperature
AAC	$\epsilon_c = K\sigma^{1.3} t^{0.16}$	$\epsilon_c = MT^{1.3}\sigma^{1.3} t^{0.16}$
6201	$\epsilon_c = G\sigma^{1.3} t^{0.16}$	$\epsilon_c = 0.000012T^{1.4}\sigma^{1.3} t^{0.16}$
ACAR	$\epsilon_c = \left(0.0003 + 0.0021\,\frac{A_{1350}}{\Sigma A}\right)(\sigma^{1.3} t^{0.16})$	$\epsilon_c = \left(0.000003 + 0.0021\,\frac{A_{1350}}{\Sigma A}\right)(T^{1.4}\sigma^{1.3} t^{0.16})$

where:

ϵ_c = *creep strain, micro-units/unit*

t = *elapsed time, hours*

σ = *tensile stress (total tensile load divided by the conductor cross-sectional area), psi*

A_{1350} = *area of 1350-H19 aluminum strands*

ΣA = *total cross-sectional area of conductor*

T = *conductor temperature, °C*

K, M and G are constants listed in Table 3-16

TABLE 3-16
CONDUCTOR CREEP EQUATION COEFFICIENTS

	STRANDING			
# Strands	7	19	37	61
K_1	0.0021	0.0020	0.0019	0.0018
K_2	0.0013	0.0012	0.0012	0.0011
M_1	0.000023	0.000022	0.000021	0.000020
M_2	0.000014	0.000014	0.000013	0.000012
G	0.0011	0.0010	0.0012	0.00094

Subscript "1" denotes 1350-H19 wire drawn from hot-rolled rod.
Subscript "2" denotes 1350-H19 wire drawn from continuous-cast rod.

For an ACSR conductor with 1350-H19 strands drawn from continuous-cast rod:

$$\epsilon_c = 1.4\,(\%RBS)^{1.3}\,t^{0.16} \qquad (3\text{-}36)$$

Elevated Temperature
For ACSR with less than 7.5% steel by area, the governing equation is:

$$\epsilon_c = 0.24\,(\%RBS)\,T\,t^{0.16} \qquad (3\text{-}37)$$

As discussed in Chapter 2, the creep elongation calculated with these equations includes initial settling strain and creep elongation to 1 hour. That is, the creep elongation obtained from these equations includes the initial plastic elongation which occurs during stringing, sagging and clipping. To obtain the creep elongation that occurs after installation, the initial plastic elongation must be subtracted from the creep strain obtained above.

This can be done graphically as described in the Alcoa graphical method, or algebraically using the equations for initial plastic elongation provided in the Harvey and Larson paper. High temperature creep is of concern only if it increases the line's final sags and leads to a clearance problem over time. This can only happen if the total creep elongation resulting from a series of high temperature loadings exceeds the room temperature creep elongation for 10 years. A method of high temperature creep calculation is suggested in the following section.

3.8.3 Example of Creep Calculation at High Temperature

The total creep elongation that a conductor experiences during a series of high temperature events cannot be determined by simply adding the creep elongation for each of the high temperature periods. Harvey and Larson[18] suggested the following process:

- Perform a normal sag-tension calculation to find the tension of the conductor as a function of conductor temperature.
- Calculate the creep elongation for 10 years at 60°F.
- Calculate the creep elongation for the high temperature event using the appropriate tension.
- Compare the creep elongation to the normal creep elongation to see if creep elongation at high temperature is significant.

For example, consider an 800 feet span of 795 kcmil - 37 strand AAC "Arbutus" drawn from rolled rod, initially tensioned at 25% of its breaking strength at 60°F. The variation of sag and tension with conductor temperature is shown in Table 3-17.

TABLE 3-17
VARIATION OF SAG AND TENSION WITH
CONDUCTOR TEMPERATURE
800 FT SPAN OF 795 KCMIL - 37 STRAND, AAC "ARBUTUS"

Temperature		Sag	Tension
°F	°C	ft	lbs
0	-17.8	17.73	3372
30	-1.1	19.67	3042
60	15.6	21.53	2780
90	32.2	23.31	2568
120	48.9	25.02	2395
167	75	27.54	2177
212	100	29.80	2014
257	125	31.93	1881
302	150	33.95	1770

As described in Chapter 2, using Tables 3-15 and 3-16, the creep elongation over 10 years at 60°F (15.6°C) for a tension of 2780 lbs is given by the following equation:

$$\epsilon_c = 0.0019 \left(\frac{2780}{0.6244}\right)^{1.3} [(10)(365)(24)]^{0.16}$$

$$= 650 \text{ microstain}$$

Given a single extended emergency electrical loading of the line during this 10 year period where the conductor reaches 167°F (75°C) for 24 hours, the high temperature creep elongation is calculated to be:

$$\epsilon_c = 0.000021\,(75)^{1.4} \left(\frac{2177}{0.6244}\right)^{1.3} (24)^{0.16}$$

$$= 593 \text{ microstrain}$$

Because the creep strain for the high temperature period is less than the normal 10 year creep strain at 60°F (15.6°C) it has no effect on the final sags of the line.

On the other hand for an emergency loading which caused the conductor to reach 212°F (100°C) for 24 hours, the high temperature creep strain is:

$$\epsilon_c = 0.000021\,(100)^{1.4} \left(\frac{2177}{0.6244}\right)^{1.3} (24)^{0.16}$$

$$= 802 \text{ microstrain}$$

And the final sags will be increased due to the additional creep elongation of 152 microstrain. Given the AAC conductor's thermal elongation rate of 0.0000128 per °F, this increased creep elongation is equivalent to an additional conductor temperature of approximately 12°F (6.7°C) or an increase in final sag of about 0.6 feet.

3.9 EFFECT OF ELEVATED TEMPERATURES ON CONNECTORS AND HARDWARE

Temperatures attained by properly installed conductor connectors and line hardware are less than the conductor temperature. The recent IEEE Guide P1283, "Determining the Effects of High-Temperature Operation on Conductors, Connectors, and Accessories"[14] provides a detailed discussion of the conclusions noted in this section.

As compared to the conductor, the lower temperature of properly installed connectors occurs because (1) the mass per unit length of these components is greater than the conductor, allowing them to act as heat sinks, and (2) they have greater surface area per unit length, increasing their cooling rates. The lower temperature of conductor accessories (with the exception of older ferrous type accessories) results from their not carrying current and dissipating heat quickly with their greater surface area.

The major concern in operating hardware at high temperatures is whether it was installed properly. There have been cases where the higher electrical loads resulting from thermal uprating methods have caused splice failures, not because a properly made splice would have failed, but because an improperly installed splice was exposed to higher average operating temperatures. It is crucial that all existing hardware be carefully inspected prior to the thermal uprating of any line.

Clarke[19] presented rather extensive measurements of temperatures for splices, dead-end terminals, ferrous and aluminum clamps.

From this data, Clark concluded the following:

- "Fully cleaned, properly installed dead-end and compression joint fittings will run substantially cooler than the conductor..."

- "...Within the limiting temperature of the steel core wires" of ACSR conductor (220°C according to Clarke), the accessories "should not show any detrimental effects."
- "Ferrous hardware components will run hotter than the conductor..."
- "...Non-ferrous conductor hardware components run much cooler than the conductor and should not be a concern. However, the service life... of damper rubber washer components may be reduced."

In a paper regarding the high temperature performance of accessories, Howitt[20] concluded that "Properly designed fittings are not the limiting factor for the permissible operating temperature of transmission or distribution lines under emergency conditions. The maximum permissible operating temperature is governed by the acceptable loss of strength of the conductor or by the...ground clearance." It should be noted, however, that Howitt's study only considered new hardware and that the issue of proper installation was stated repeatedly.

It is apparent that a careful review of hardware type, condition and installation method must precede the decision to thermally uprate an existing line. All hardware should be evaluated for suitability of operation at the maximum design temperature of the line.

3.10 REFERENCES

[1] *IEEE Standard for Calculating the Current-Temperature Relationship of Bare Overhead Conductors*, IEEE Standard 738- 2005.

[2] Schurig, O. R. and Frick, C. U., "Heating and Current Carrying Capacity of Bare Conductor for Outdoor Service," *General Electric Review*, Vol. 33, No. 3, pp. 141-157, March 1930.

[3] House, H.E. and Tuttle, P.D., "Current Carrying Capacity of ACSR," *IEEE Transactions Power Apparatus and Systems*, pp. 1169-1178, February 1958.

[4] "Transmission Conductors Thermal Ratings," 68-TAP-28. Report by Transmission Advisory Panel, East Central Area Reliability Coordination Agreement.

[5] Davis, M. W., A New Thermal Rating Approach: The real time thermal rating system for strategic overhead conductor transmission lines, Part II", *IEEE Transactions on Power Apparatus and Systems*, Vol. PAS-97, pp. 810-825, March/April 1978.

[6] Taylor, C.S., House, H.E., "Emissivity and its Effects on the Current Carrying Capacity of Stranded Aluminium Conductors," *AIEE Transactions*, Vol. 75, Pt. III, pp. 970-976, October, 1956.

[7] House, H.E., Rigdon, W.S., Grosh, R.J., Cottingham, W.B., "Emissivity of Weathered Conductors after Service in Rural and Industrial Environments," *AIEE Transactions*, pp. 891-896, February, 1963.

[8] Black, W.Z., Rehberg, R.L., "Simplified Model for Steady-State and Real-Time Ampacity of Overhead Conductors," *IEEE Transactions on Power Apparatus and Systems*, Vol. 104, pp. 29-42, October, 1985.

[9] "Guide for Selection of Weather Parameters for Bare Overhead Conductor Ratings," CIGRE Task Force B2.12.6, November, 2005.

[10] Beers, G.M., Gilligan, S.R., Lis, H.W., Schamberger, J.M., *Transmission Conductor Ratings*, IEEE, 63-86, October, 1963.

[11] Seppa, T.O., "A Practical Approach for Increasing the Thermal Capabilities of Transmission Lines," *IEEE Transactions on Power Delivery*, Vol. 8, No. 3, pp. 1536-1550, July, 1993.

[12] *Aluminum Electrical Conductor Handbook*, Second Edition, The Aluminum Association, Washington, DC., 1982.

[13] Harvey, J.R., "Effect of Elevated Temperature Operation on the Strength of Aluminum Conductors," IEEE/PES Winter Meeting Paper T 72-189-4, 1971.

[14] "IEEE Guide for Determining the Effects of High Temperature Operation on Conductors, Connectors, and Accessories," IEEE Standard 1238-2005.

[15] Rawlins, C.B., "Some Effects of Mill Practice on the Stress-Strain Behaviour of ACSR," IEEE WPM 1998, Tampa, FL, February 1998.

[16] Barrett, JS, and Nigol, O., Characteristics of A1/S1A (ACSR) Conductors as High Temperatures and Stresses, IEEE Trans., Vol. PAS-100, No. 2, pp. 485-493, February 1981.

[17] "Corona Performance and Qualification Testing on Suwannee 959.6 kcmil ACSS/TW Conductor at 345 kV", EPRI Report WO-045724, January 2001.

[18] Harvey, J.R., Larson, R.E. "Creep Equations of Conductors for Sag-Tension Calculations," IEEE/PES Winter Meeting Conference Paper No. C72 190-2, 1972.

[19] Clarke, G.J., "Summary Report on the Effects of High Operating Temperatures on Conductors and Hardware Behavior," Ontario Hydro Research Division, Report No. 77-177-H, April 25, 1977.

[20] Howitt, W.B., "Elevated Temperature Performance of Conductor Accessories," Conference Paper presented at Seminar on Effects of Elevated Temperature Operation on Overhead Conductors and Accessories, May 20, 1986.

CHAPTER 3
APPENDIX

FIGURE 3-13
BARE STRANDED ACSR CONDUCTOR,
MAXIMUM FAULT-CURRENT OPERATING LIMIT CURVES

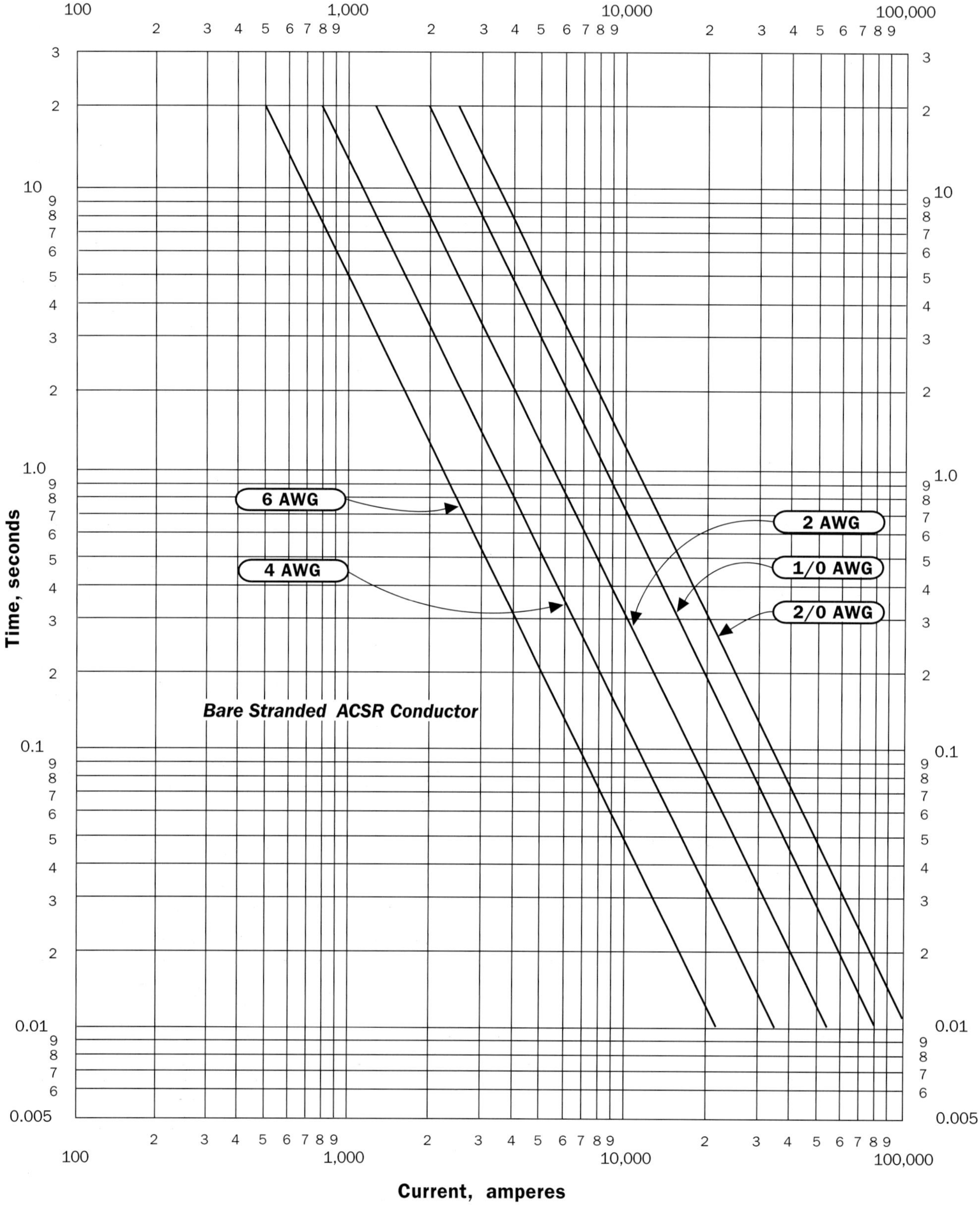

Upper temperature limit - 645°C (aluminum melting point)
Initial temperature - 40°C
Curves assume no conductor heat loss

FIGURE 3-14
BARE STRANDED ACSR CONDUCTOR,
MAXIMUM FAULT-CURRENT OPERATING LIMIT CURVES

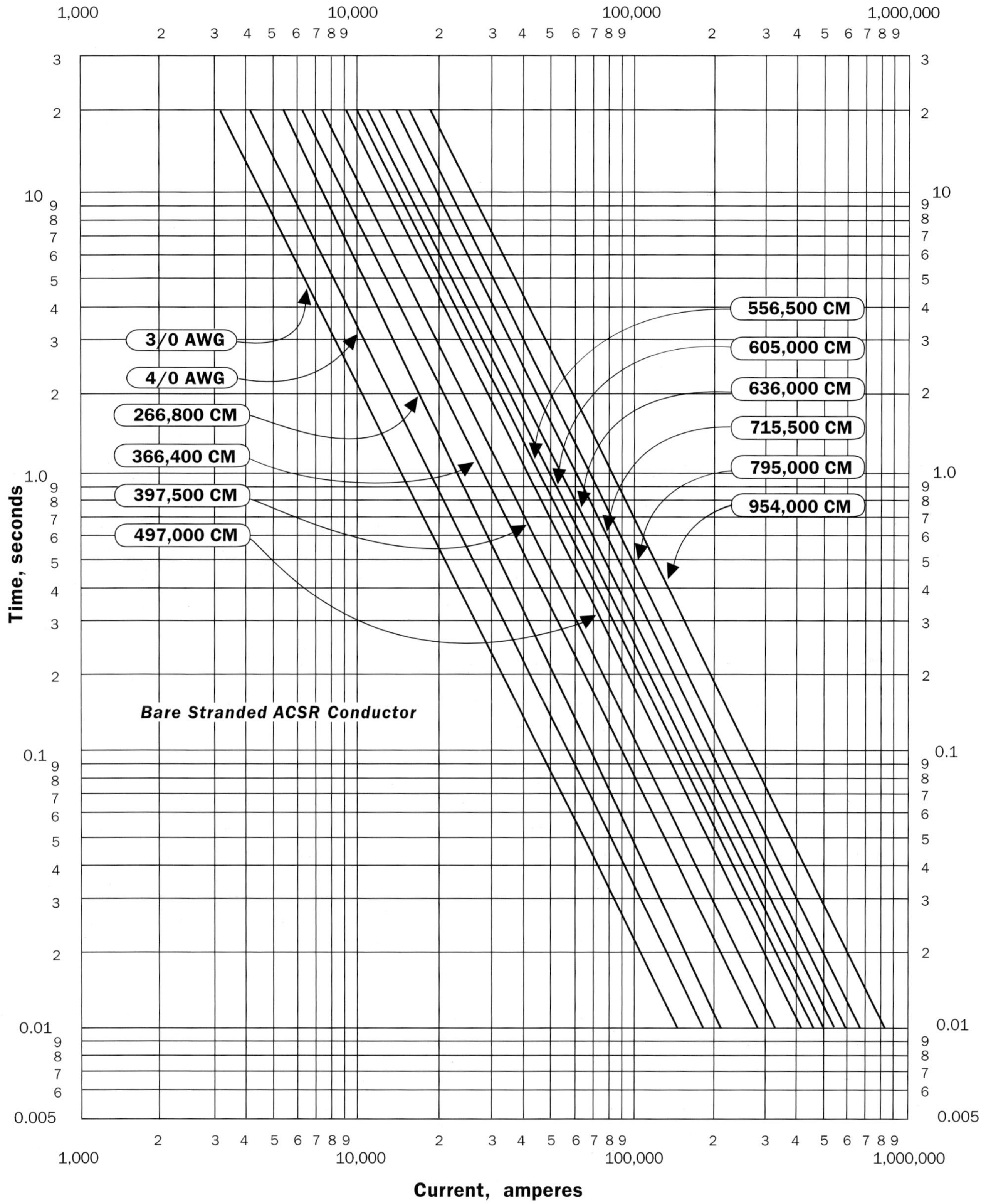

Upper temperature limit - 645°C (aluminum melting point)
Initial temperature - 40°C
Curves assume no conductor heat loss

FIGURE 3-15
BARE STRANDED ACSR CONDUCTOR,
MAXIMUM FAULT-CURRENT OPERATING LIMIT CURVES

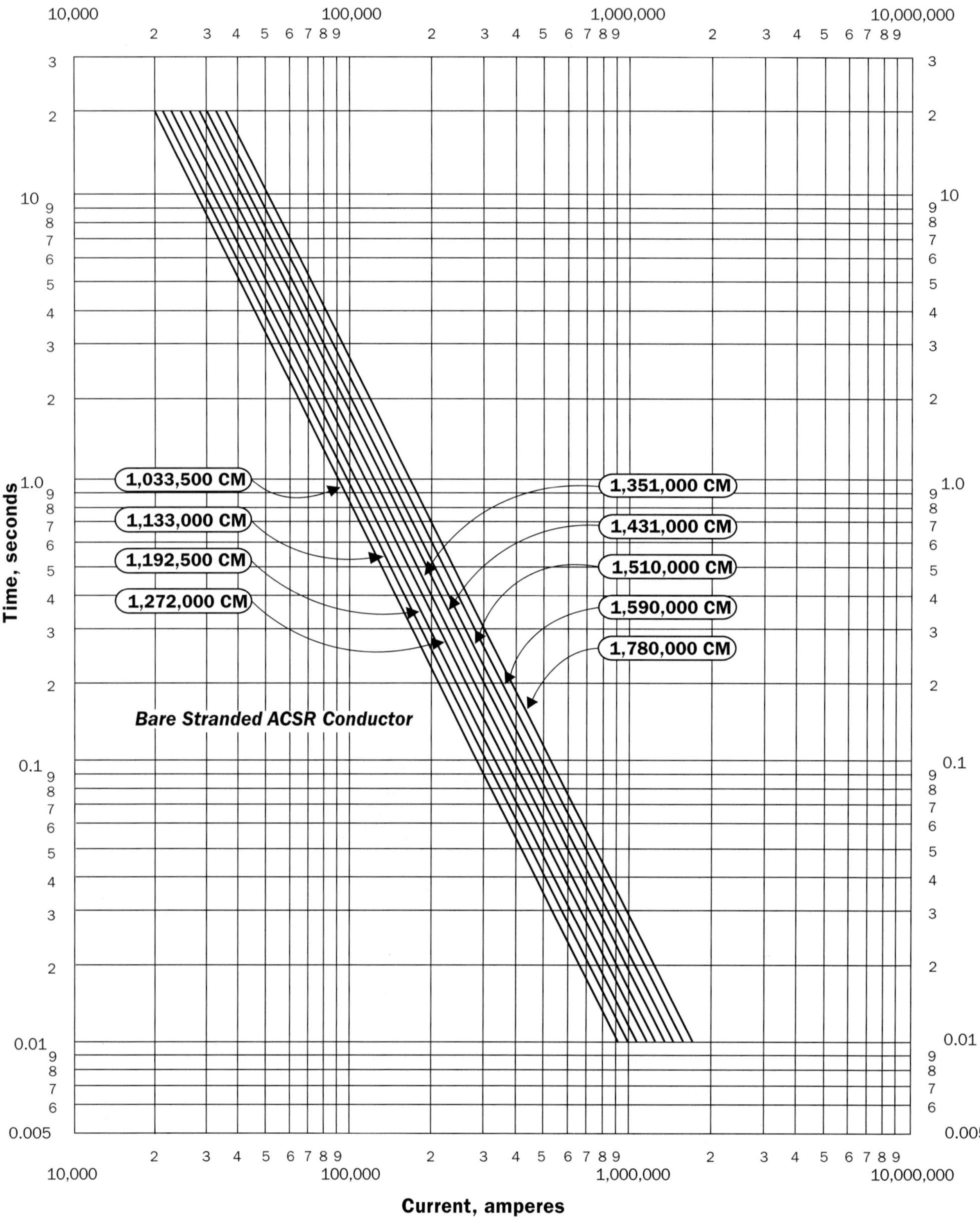

Upper temperature limit - 645°C (aluminum melting point)
Initial temperature - 40°C
Curves assume no conductor heat loss

FIGURE 3-16
BARE STRANDED ALUMINUM CONDUCTOR,
MAXIMUM FAULT-CURRENT OPERATING LIMIT CURVES

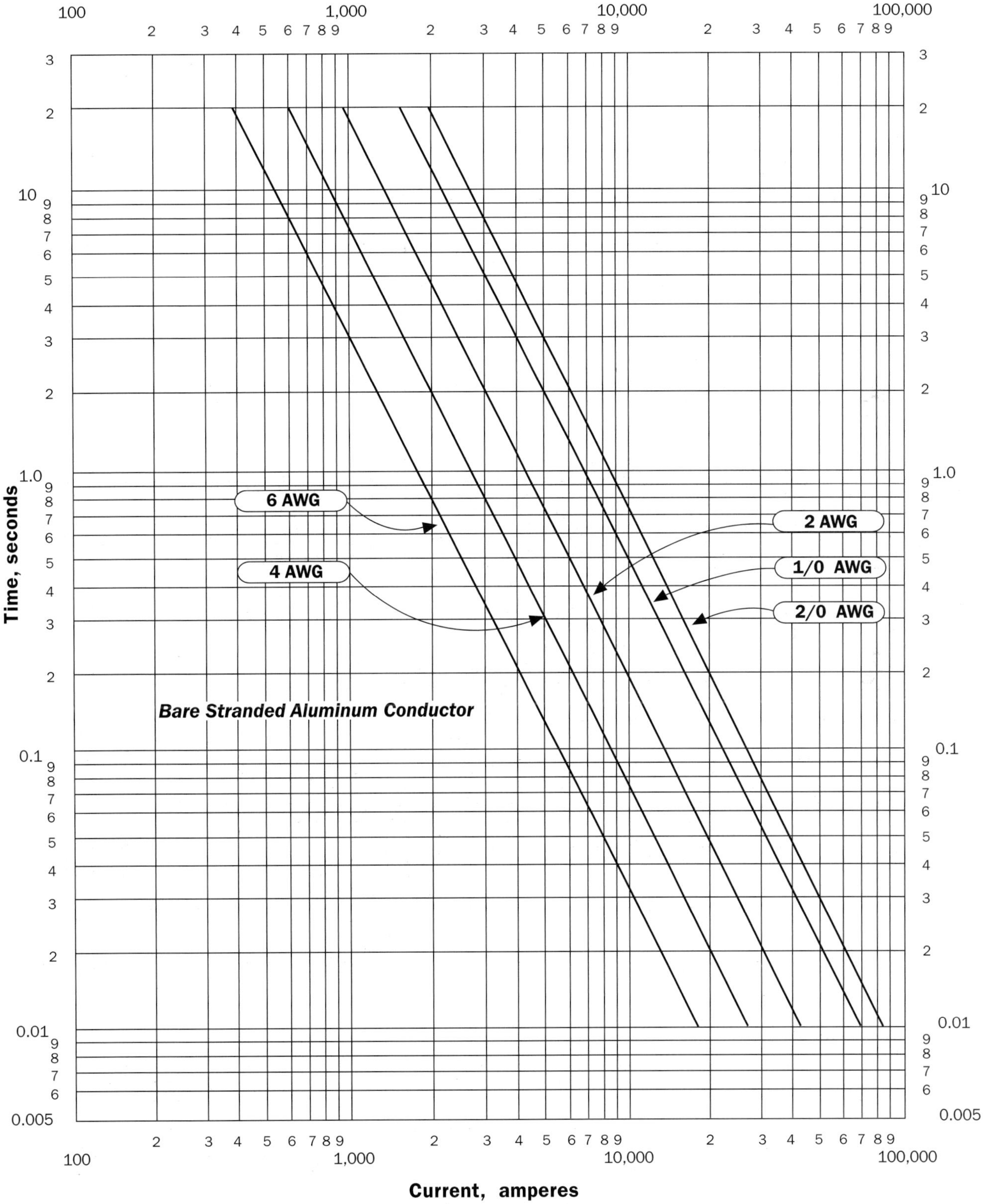

Upper termperature limit - 340°C (point of damaging annealing)
Initial temperature - 40°C
Curves assume no conductor heat loss

FIGURE 3-17
BARE STRANDED ALUMINUM CONDUCTOR,
MAXIMUM FAULT-CURRENT OPERATING LIMIT CURVES

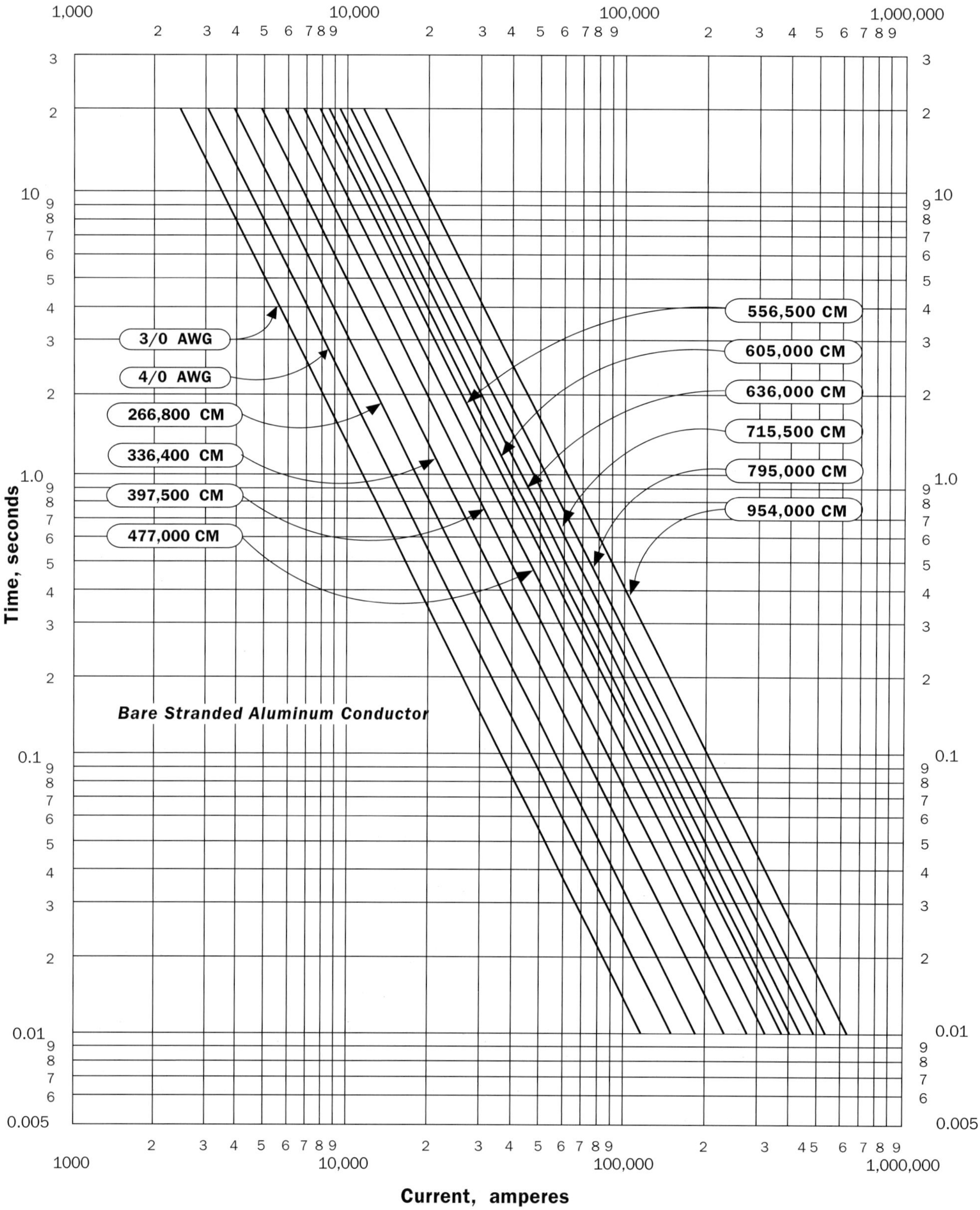

Upper temperature limit - 340°C (point of damaging annealing)
Initial temperature - 40°C
Curves assume no conductor heat loss

FIGURE 3-18
BARE STRANDED ALUMINUM CONDUCTOR, MAXIMUM FAULT-CURRENT OPERATING LIMIT CURVES

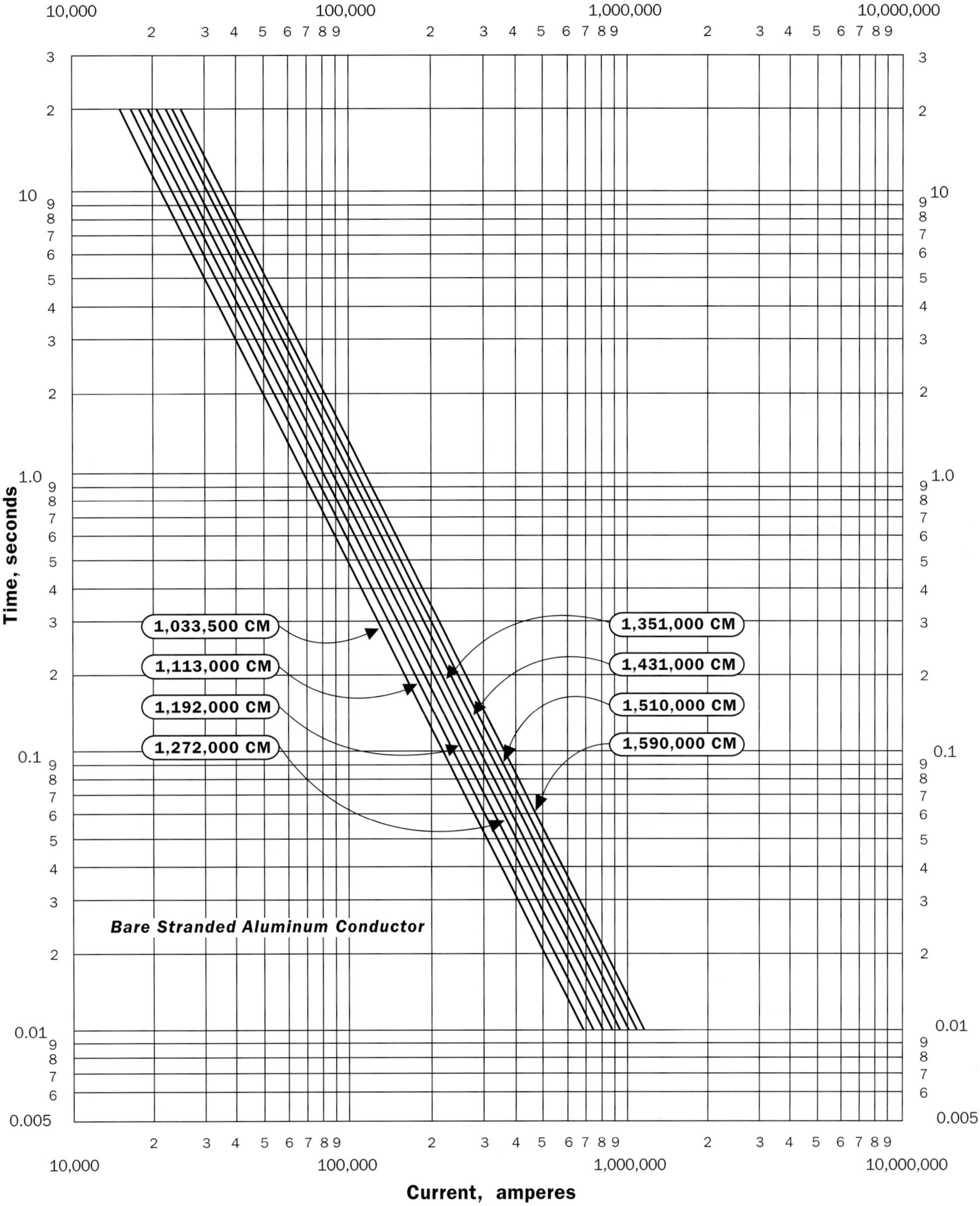

Upper temperature limit - 340°C (point of damaging annealing)
Initial temperature - 40°C
Curves assume no conductor heat loss

FIGURE 3-19
BARE STRANDED 6201 ALUMINUM-ALLOY CONDUCTOR,
MAXIMUM FAULT-CURRENT OPERATING LIMIT CURVES

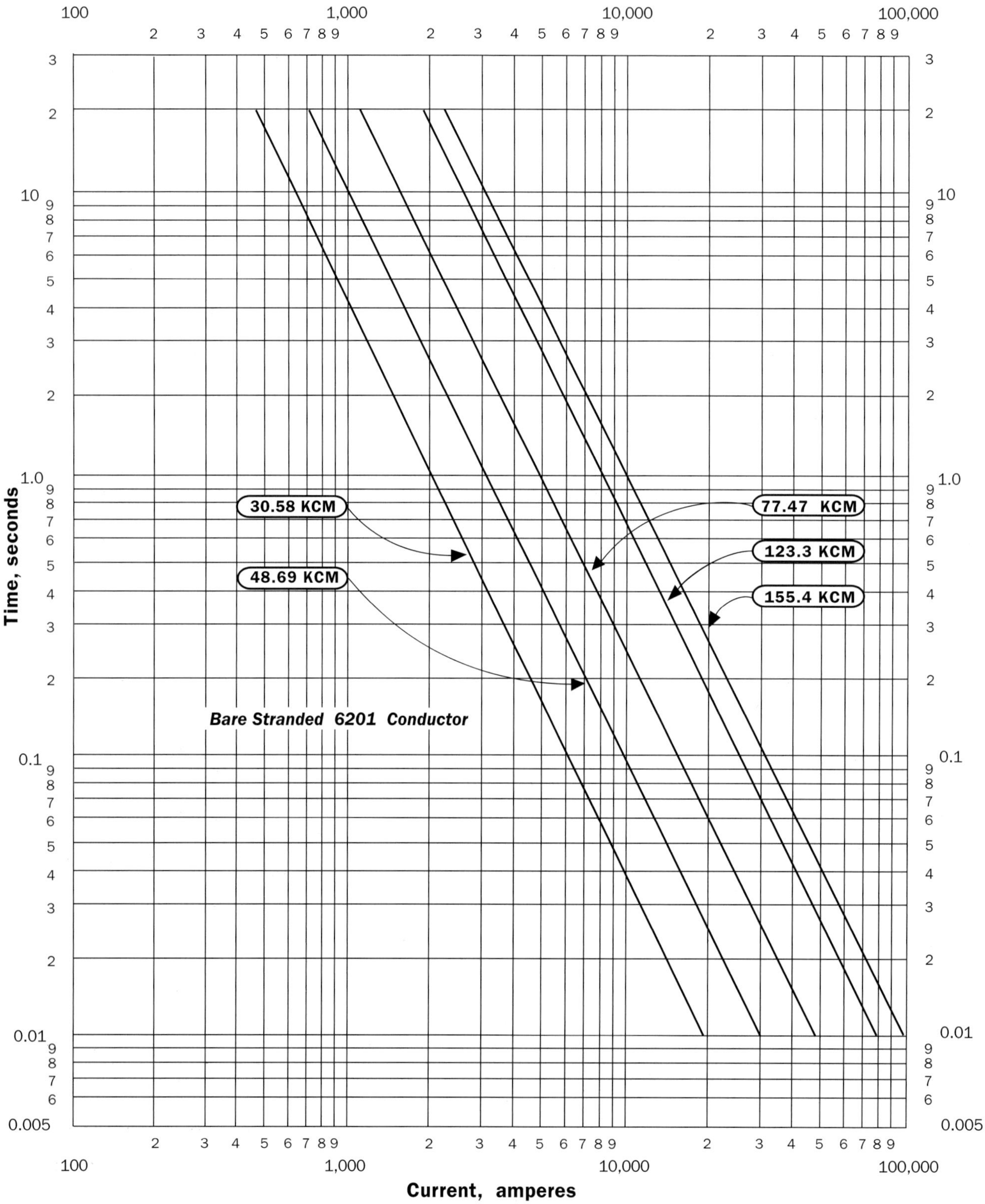

Upper temperature limit - 340°C (point of damaging annealing)
Initial temperature - 40°C
Curves assume no conductor heat loss

FIGURE 3-20
BARE STRANDED 6201 ALUMINUM-ALLOY CONDUCTOR,
MAXIMUM FAULT-CURRENT OPERATING LIMIT CURVES

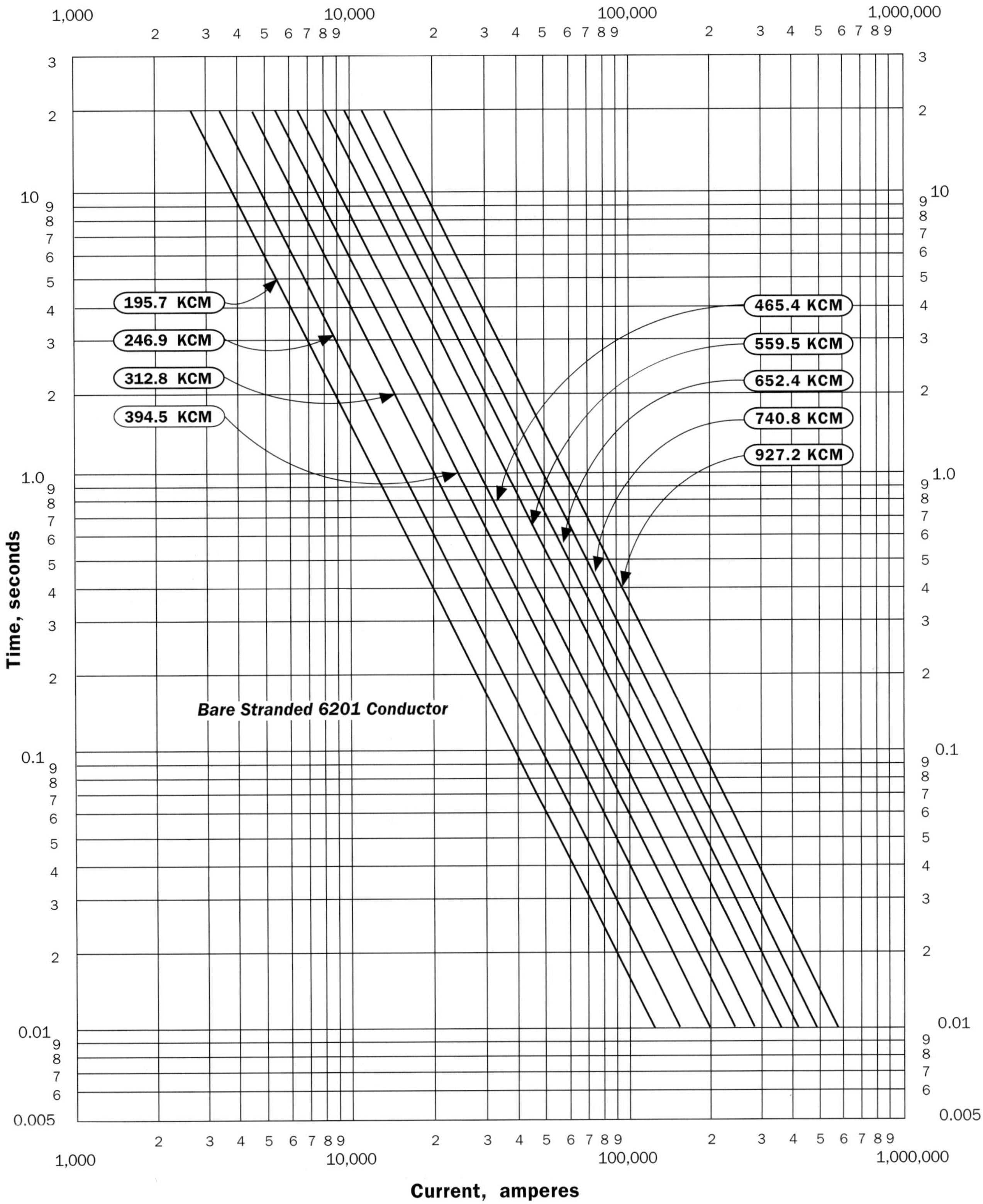

Upper temperature limit - 340°C (point of damaging annealing)
Initial temperature - 40°C
Curves assume no conductor heat loss

4 ELECTRICAL ENVIRONMENTAL EFFECTS

Overhead transmission lines have an impact on their surrounding environment, including people. This impact is caused by two electrical effects: corona-induced noise and coupling due to ground-level electric and magnetic fields[1]. Corona is caused by ionization of the air at the surface of line conductors, insulators and hardware, resulting in radio/TV interference and audible noise. Although difficult to eliminate entirely, corona can be minimized by using an optimum conductor size and properly shaped hardware. Even under the best of circumstances, occasional low levels of radio/TV interference and audible noise are unavoidable. Such noises may be acceptable, however, if radio and television signal strengths and ambient sound levels are sufficiently high.

Electric and magnetic fields are a natural and unavoidable result of transmitting power through overhead transmission lines. These fields can induce voltages and currents into electrically conducting objects near the line. Electric and magnetic fields both may be reduced by raising conductor height, proper phase spacing and geometry, and appropriate right-of-way widths.

4.1 RADIO NOISE

Power transmission and distribution lines may interfere with nearby radio and television reception. The electrical noise may be apparent at different frequencies, and may vary with weather conditions. A 345 kV transmission line, for example, may create large amounts of noise in the AM broadcast band during rain, while a 13.8 kV distribution line may produce "sparkles" in a television picture in dry weather.

Some aspects of radio noise are addressed in the design phase of a line, others in the specification of hardware and fittings, and still others in construction and maintenance considerations. Due to extensive studies and analyzes in the past, recent line designs have been relatively free of radio noise problems. This success, however, does not preclude the need for the line designer to include radio noise considerations in all aspects of design and construction.

4.1.1 Radio Noise Sources

Radio noise from overhead ac power lines is produced by two distinct phenomena, corona and sparking. Corona occurs when the electric field at a particular point reaches a sufficiently high value to cause ionization of the surrounding air. Conductors, as well as insulators and hardware, can serve as sites for corona. The threshold value of the electric field at which corona occurs on conductors depends on conductor diameter, conductor surface condition and atmospheric conditions. Conductor corona noise drops off rapidly with increasing radio frequency, and is primarily of concern in the AM broadcast band of 0.535 to 1.605 MHz.

Corona can also occur on insulators and hardware, particularly if the line is subject to heavy contamination by pollen, dust, etc. The use of glass insulators instead of porcelain or polymer insulators also increases the tendency for corona to occur. In most instances, however, conductor corona noise dominates over noise from insulators and hardware.

Sparking, also called gap discharge, occurs at poorly conducting electrical connections. For example, this can occur at the connections between individual units in a porcelain suspension insulator string supporting a jumper connection. When there is only a small mechanical load on the insulator string, there is a possibility that enough friction will not be generated during joint movement to prevent the buildup of a film of corrosion on the insulator pin. The capacitive voltage distribution across the insulators can then cause sparking across this insulating film.

Other locations where sparking occurs are between tie wires and insulators on distribution lines, and on wood poles where staples make poor contact with ground down leads. Because of the small size of the sparks, they are called microsparks, and the resulting noise is called microsparking noise. Spark noise can extend into the UHF range (above 300 MHz), and is the primary cause of television interference.

Sparking on both transmission and distribution lines can be a serious source of radio and television interference. The minimization of spark discharge noise is more of a maintenance matter than a design consideration. Experience suggests that for a line properly designed with respect to conductor corona, 90-95% of all noise complaints are sparking-related, and can be located and eliminated by normal procedures[2].

4.1.2 Prediction Programs

Conductor corona-induced radio noise is a complex process, involving both noise generation and propagation. Individual corona spots inject noise currents into the power line conductor. These noise currents propagate along the conductor according to the principles of electromagnetic theory and produce noise fields in the surrounding space. The noise heard on a radio is the sum of all noise sources on the line and the propagation of that noise to the radio location.

Two approaches, called the semi-analytical and comparative methods, are used to develop computer programs for predicting conductor corona-related radio noise. The semi-analytical method[3] combines corona noise generation functions developed in laboratory cage tests with an analytical propagation model to predict noise profiles. The comparative method[4] is based on long-term measurements on many test and operating lines. It gives an empirical expression that is based on the measured data.

The semi-analytical method is complex, but has the advantage of using theoretical fundamentals as much as possible. It can be used with confidence for new innovative line designs since the analytical part of the method incorporates the conductor configuration. The comparative method is simple and reliable for line designs that do not depart greatly from existing configurations. In practice, it gives surprisingly good results for new designs as well. Both types of prediction programs are based on transmission lines at sea level. For higher elevations, 1 dB per 300 meters of increased elevation is added to the calculated noise to correct for altitude.

Radio noise prediction programs may be limited to a single frequency, usually 0.5 or 1.0 MHz, or may cover several frequencies throughout the HF, VHF and UHF bands. If a program for a single frequency is available, noise at other frequencies can be determined from published curves of corona noise versus frequency[5].

Experience has shown that transmission lines designed to adequately limit conductor corona radio noise give satisfactory general radio noise performance. Of the complaints that are received, over 90% result from sparking that can be located and eliminated by normal maintenance operations. Radio noise computer prediction programs, therefore, usually predict conductor corona noise only. In some countries, however, additional corrections are made for radio noise from insulators.

4.1.3 Weather Dependence

Radio noise sources are affected differently by changing weather conditions. When the weather is dry, radio noise is found to vary with season and wind velocity. Seasonal variations are due to different amounts of pollen and other organic matter in the air that deposit on the conductors. More of this material passes the conductors when it is windy; so, the corona radio noise increases accordingly. In foul weather, water droplets on the conductor form discontinuities that act as corona-generating noise sources.

The same moisture that increases corona noise can reduce spark noise if it provides a conducting bridge across the sparking surfaces. For example, sparking to a down lead on a distribution pole may only occur when the wood is dry. On the other hand, contamination on insulator surfaces can cause increased noise when the insulator is wet. Observation of how radio noise changes with weather provides a clue to its cause and cure.

Radio noise levels change with time[6]. These changes can be characterized by a complete statistical distribution of these levels obtained by long-term measurements on a single line. For many purposes, however, it is adequate to describe radio noise in terms of exceedence levels. Exceedence levels are stated in terms of L_N, where $_N$ is the percentage of time the noise exceeds the given value. For example, if radio noise is given as 45 dB L_{50} fair weather, it means that the noise is 45 dB above 1 µV/meter 50% of the time during fair weather. L_{50} and L_5 foul weather noise levels are the ones most frequently evaluated. The all-weather statistical distribution is presented in three general portions corresponding to fair weather, foul weather and a transition region. Normally L_{50} and L_5 foul weather statistics include both the foul weather and transition portions of the overall distribution. An alternate description of foul weather radio noise is "average stable foul weather," that is the average of only the purely foul weather portion of the distribution.

Radio noise programs predict some measure of fair weather, and one or more measures of foul weather noise. Some programs try to predict separately the different noise levels for the specific line design; others assume a fixed dB difference between the fair and foul weather levels. The person making the analysis must select the weather condition that is appropriate for the case at hand.

4.1.4 Radio Noise Profiles

Radio noise levels are customarily presented in terms of lateral profiles. These profiles consist of plots of noise level against distance from the center of the line. Figure 4-1 shows calculated lateral profiles of 1 MHz radio noise for a 765 kV transmission line with 4-conductor bundles and horizontal flat conductor configuration. Radio noise levels are compared for three weather conditions: L_{50} fair weather, L_{50} foul weather and L_5 foul weather. Profiles of this type are used to estimate the radio noise performance of proposed transmission line designs.

FIGURE 4-1
TRANSMISSION LINE RADIO NOISE LATERAL PROFILES

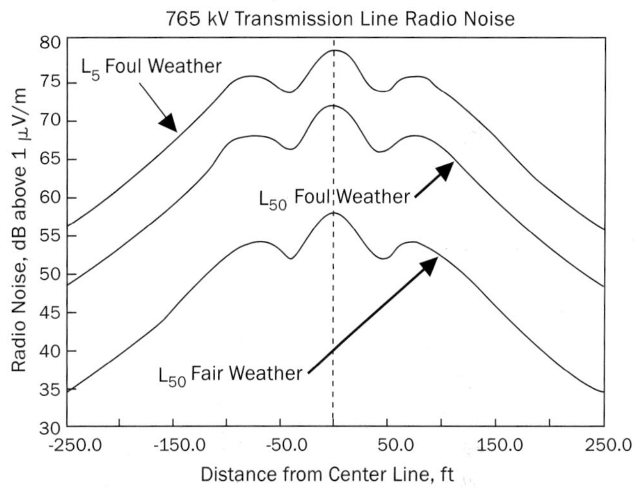

Radio noise from a transmission line is a function of several design variables. Changes in line design that increase the electric field at the conductor surface increase the level of corona activity and the amount of radio noise. The effects of the more common changes are:

- Increased voltage for the same conductor size and phase geometry leads to higher noise.
- Reduced phase spacing, called line compaction, at the same voltage increases the conductor surface electric field which increases radio noise.
- Increasing conductor diameter reduces the conductor surface electric field and reduces radio noise.
- The division of the phase conductor into subconductors, with conductor bundles of 2, 3 or 4 or more subconductors, reduces radio noise. Typically, 2 conductor bundles are used at 345 kV, 3 conductor bundles are used for 500 kV, and 4 conductor bundles for 765 kV.

- Different conductor configurations, such as horizontal, vertical, or delta, result in unique noise profiles.

4.1.5 Evaluation of Radio Noise

Evaluation of radio noise performance of a transmission line requires the consideration of three areas:

- Criteria for acceptable noise level
- Prediction of noise profile
- Evaluation of effect on surrounding environment

It is insufficient merely to calculate a radio noise profile for a proposed line design using a computer program. Some criteria to determine noise levels at which annoyance occurs also must be established.

Some countries, such as Canada[7], have national standards for radio noise from overhead power lines. Others, such as the United States, have no general standards, but rely on understanding of the situation at specific line locations. A good argument can be made for this approach since local land use and weather affect the background noise level in a given location. The following steps are usually taken in the development of radio noise criteria for a specific transmission line[8]:

(1) Ascertain the radio station signal strength that exists at the edge of the right-of-way under different weather conditions. This may be done in several ways: by measurement of signal strengths at the line location; by estimation from radio station coverage maps that are usually given on their advertising rate schedule; or by application of Federal Communication Commission (FCC) rules for signal strength to cover a specific type of terrain[9]. FCC rules specify the following minimum signal strengths in mV/m for different coverage areas.

Area	*Field Strength Groundwave (mV/m)*
City business or factory areas	10 to 50
City residential areas	2 to 10
Rural -all areas during winter or northern areas during summer	0.1 to 0.5
Rural-southern areas during summer	0.25 to 1.0

(2) Determine the signal-to-noise ratio necessary for edge of right-of-way reception under the specific weather conditions. Listener tests of AM radio yield the following[10]:

Signal-to-Noise Ratio in dB	*Reception Quality*
>32	Entirely satisfactory
27-32	Very good, background unobtrusive
22-27	Fairly satisfactory, background plainly evident
16-22	Background very evident, speech easily understood
6-16	Speech understandable with severe concentration
<7	Speech unintelligible

(3) Subtract the appropriate signal-to-noise ratio in dB, listed in the above table, from the radio station signal strength in dB above 1 μV/m to determine the allowable edge of right-of-way radio noise under the specified weather conditions. A commonly used value is 24 dB. Signal and noise strengths are commonly expressed in terms of decibels (dB) above to 1 μV/m.

(4) Compare the developed criteria with the predicted edge of right-of-way radio noise from the line. If the noise is above the criteria, repeat the entire process, including evaluation of the line design and the process for selection of the criteria.

Television noise can be evaluated in the same manner. Care must be taken in evaluating television noise to ensure that proper bandwidth corrections are made to reflect use of the particular measuring receiver, computer program, and signal-to-noise ratio versus viewer reactions. These are often given based on different bandwidths for television, and seemingly disparate numbers can often be traced to different bandwidths used for evaluation.

When this method is properly applied, experience suggests that conductor corona will not be the cause of listener complaints. More than 90% of complaints will be due to sparking sources that can be located and corrected by normal techniques.

4.1.6 Hardware and Insulator Radio Noise

If conductor corona has been properly handled in the design process, hardware and insulator noise have only a secondary effect and are usually ignored. However, when designing an EHV line, a high voltage compact line, or a line different from that previously constructed, it is important to give consideration to hardware corona. For example, a 138 kV compact line may operate at 138 kV, but have an electric field at the conductor surface typical of a 345 kV line. If catalog 138 kV hardware were applied to such a line, the radio noise of the line might be excessive due to corona on the hardware.

Because the electrical stress is at EHV levels on this 138 kV compact line, it may be necessary to specify EHV class hardware to secure desired corona performance. Similar considerations apply to any situation where electric field stress on energized parts is increased as a result of changes in line design. For example, a fiber optic communications cable wrap on a phase wire may be applied with success on a traditionally designed 115 kV transmission line, but could result in radio noise when applied on a 115 kV compact line.

Laboratory test procedures must consider the electric field stress on the hardware and not simply the voltage at which the hardware is to be used. Traditional test procedures involve installation of the device in a testing laboratory to measure radio noise as a function of voltage. If the noise threshold is found to be 20% or more above the intended voltage rating, the hardware is said to perform adequately. This test gives satisfactory results for normally spaced lines, but does not detect noise caused by the increased conductor surface electric field conditions on compact or non-conventional designs. In such cases, it is necessary to specify a test using a conductor surface electric field similar to that anticipated on the transmission line.

4.1.7 Measurement Techniques

Measurement techniques for radio noise are specified in IEEE Standard 430[11]. Noise is generally measured at 0.5 or 1 MHz with a quasi-peak detector having a time constant and bandwidth as specified in the standard. The quasi-peak detector uses a time constant to represent the noise in a way that correlates with annoyance. Also used are field intensity detectors to measure the carrier strength of radio stations and peak detectors to measure television noise levels.

The radio noise receiver may have either a loop or rod antenna. Particular care must be employed, however, when using a rod antenna since (1) the rod antenna is sensitive to the electric field component of the noise electromagnetic field, and (2) the calibration of a radio noise receiver with a rod antenna depends on the capacitance between the receiver case and ground. Thus, there may be some variation in measured noise decibel level, depending upon the height of the receiver above ground and the distance to nearby interfering objects.

Another source of error with rod antennas under EHV lines is the possibility of corona from the tip of the rod causing erroneously high readings. Because of the possible errors associated with the use of rod antennas, many investigators prefer, and strongly recommend, use of the loop antenna. The loop antenna responds to the magnetic field component of the noise electromagnetic field. As a result, the loop antenna is less disturbed than the rod antenna by the presence of nearby objects.

IEEE Standard 430[11] presents procedures and precautions to ensure the measured noise profiles represent the transmission line performance. Any single-time spot measurements are limited in validity to the weather conditions that exist at the time of measurement. For a full characterization of the line, long-term measurements are necessary. Whether spot or long-term measurements are taken, it is necessary to ensure that the noise being measured is actually coming from the transmission line, and not from some other nearby source. A common source of extraneous noise is the signal from a radio station. For example, "splatter" from a strong station at 980 kHz may obscure the noise in a 1 MHz measurement. If so, it may be necessary to change the receiver frequency slightly to avoid the splatter. Whenever taking noise measurements, listening to the noise on the radio is an added means of identification of the source. This is especially important for night and long-term measurements when sky wave radio signals may overpower the noise.

4.1.8 Noise Location Technique

Radio noise in the AM broadcast band propagates great distances from the source and can couple to circuits other than the circuit that contains the noise source. Hence, it is difficult to locate the exact source of the noise with an AM receiver. Fortunately, there are several noise locating techniques[2].

An effective tool used to isolate radio noise sources is a tunable VHF receiver coupled with a yagi antenna. The procedure is to adjust the receiver to the highest frequency where the noise can be heard, then move the yagi antenna around to find the direction of the source. As the receiver is brought closer to the noise source, the volume of the noise increases, allowing identification of the specific source. Adjusting the receiver to higher frequencies as the noise source is approached gives greater location accuracy.

When near a noise source, an ultrasonic detector can be

used to zero in on the offending component. Such detectors, sometimes together with a small UHF radio receiver, can be mounted on the end of a hot stick. The hot stick allows placement of the detector close to a distribution line to identify the specific noise source.

In the past, noise source locations on distribution lines have been narrowed down to a specific pole by hitting the pole with a sledgehammer while listening to the noise on a radio. When the pole with the source is hit, the vibration sometimes changes the noise and alerts the investigator to the presence of the problem. Safety concerns have precluded use of this method in recent years.

4.2 AUDIBLE NOISE

Extensive research has been conducted to determine the mechanisms that produce audible noise and how it is perceived by people. Audible noise first appeared as a problem when 765 kV transmission lines were introduced. At that time, it appeared audible noise was related only to the greater electrical stress produced by the higher transmission voltages. Subsequent studies have shown, however, that the electrical stress on conductors and hardware in compact transmission lines can be equivalent to the stress produced in earlier EHV lines. Audible noise also must be considered, therefore, in the design and construction of these lower voltage lines.

4.2.1 Audible Noise Sources

Audible noise, like radio noise, is produced by corona on ac transmission line conductors. It takes on two forms: a sizzling or crackling sound called random noise and a single pitch tone called hum[3]. Random noise and hum come from two distinct physical mechanisms that may not appear simultaneously. Transmission line random audible noise is rich in high frequency components, giving it its distinctive sound. A typical frequency spectrum of audible noise presented in Figure 4-2 illustrates the dominance of high frequencies.

Random noise results from a multitude of small snapping sounds at corona points on the conductor as the voltage passes through its crest. Sound propagates through air at approximately 1100 ft/sec, or 18 feet during one cycle of a 60 Hz wave. Because the phase conductor is more than 18 feet from the listener's ear, the path length and the phase shift are different from each corona point to the ear. Each sound arrives with a different phase delay and results in the distinctive random noise rich in high frequency components.

Hum results from corona-generated space charge that is trapped by the electric field around the conductor. This space charge is pumped by the alternating electric field that acts as a loudspeaker to produce the hum. In contrast to random noise, which does not correlate with corona loss on the line, hum involves significant power loss and correlates closely with measured corona loss.

FIGURE 4-2
TRANSMISSION LINE AUDIBLE NOISE SPECTRUM

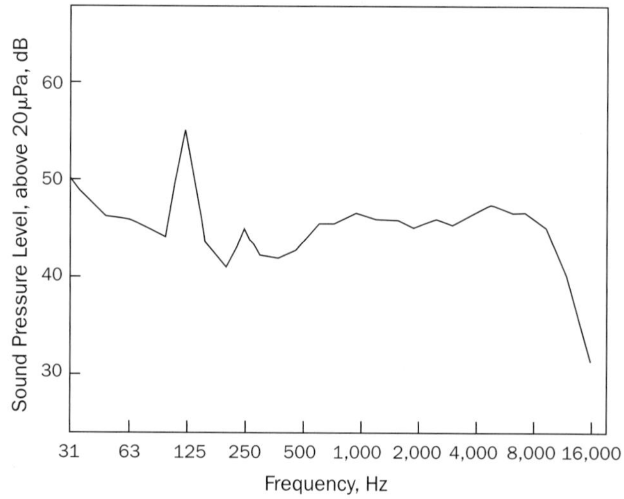

Of the two noise components, random noise is the component predicted by computer programs used for analysis of possible annoyance. Random noise correlates well with human response, while hum is highly variable and is very difficult to predict.

4.2.2 Prediction Programs

Audible noise propagation is less complicated than radio noise because it does not propagate over long distances. It travels directly from the source to the listener's ear. Therefore, audible noise prediction does not require an elaborate propagation model such as that used for radio noise. Some prediction models are based on empirical data obtained from many long-term measurements on experimental and operating transmission lines. Other algorithms are based on cage tests of conductors and noise propagation models in a manner similar to that used in radio noise evaluations[12].

The human ear does not have a linear frequency response. As a result, it is necessary to adjust measured noise levels to obtain correlation with human ear sensitivity. The correlation is provided by frequency response "weighting" curves as shown in Figure 4-3. The "A" weighting curve is used for most community noise and transmission line audible noise studies. However, some psycho-acoustic studies suggest the "B" curve more closely correlates to transmission line annoyance.

Audible noise meters are designed to allow the selection of different weighting curves. Noise calculated or measured with a particular weighting curve is identified with the letter of the curve in parentheses. For example, a noise of 50 dB(A) is based on the "A" weighting curve.

FIGURE 4-3
WEIGHTING CURVES

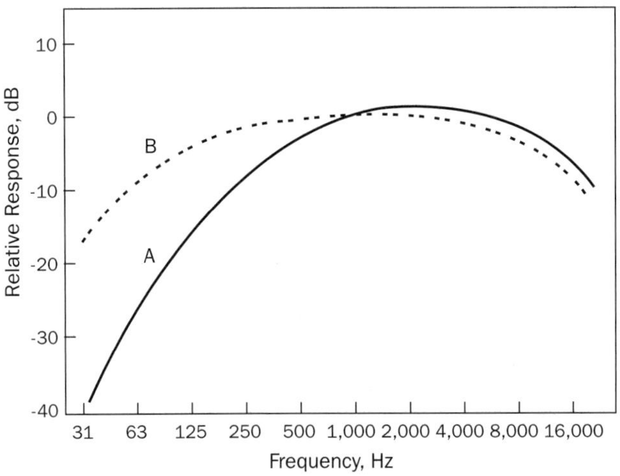

4.2.3 Weather Dependence

In contrast to radio noise, which can be significant in any weather condition, audible noise from ac transmission lines is generally of concern only in wet conditions. Although fair weather audible noise can sometimes be heard, its measurement is difficult, and is not as noticeable, because of the presence of background noise. Even during rain when the highest noise levels occur, the rain itself tends to mask the noise. The critical period is the quiet time immediately following the rain.

Like radio noise, audible noise can be characterized by exceedence levels, typically L_5 and L_{50} foul weather. Some references call these "heavy rain" or "maximum" and "wet conductor" or "average" foul weather conditions, respectively. The L_{50} value is the level most commonly used for audible noise evaluation.

4.2.4 Audible Noise Profiles

Design factors specified for radio noise can be applied to audible noise also. Historically, radio noise was of more concern than audible noise at voltages through 345 kV, whereas audible noise was a greater problem at 765 kV and above. Both radio and audible noise were of concern at intermediate voltages. With compact and uprated transmission lines used today, however, it is possible that either noise may be a significant problem at voltages lower than those previously considered.

The audible noise profiles predicted by computer programs assume no obstructions between the transmission line and the surrounding site. In practice, however, more and more sound-absorbing trees and vegetation are introduced as the distance from the line increases. Consequently, measured sound profiles tend to decrease with distance faster than predicted profiles.

Figure 4-4 shows calculated lateral profiles of audible noise for the same 765 kV transmission line used for the radio noise profile of Figure 4-1. Audible noise levels are compared for two weather conditions: L_{50} foul weather and L_5 foul weather. Profiles of this type are used to estimate the audible noise performance of proposed transmission line designs.

FIGURE 4-4
TRANSMISSION LINE AUDIBLE NOISE LATERAL PROFILES

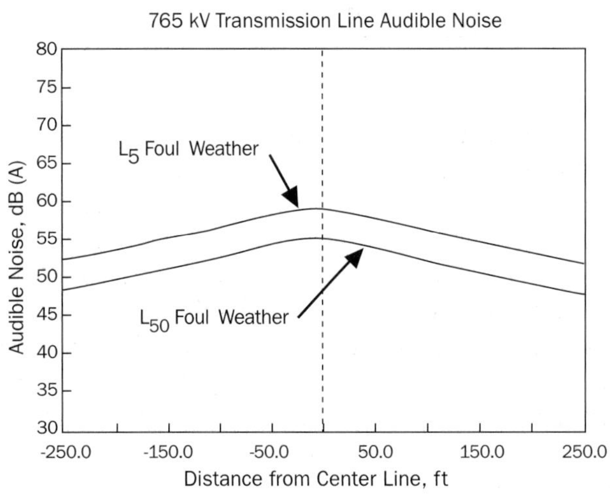

4.2.5 Evaluation of Audible Noise

Many governmental jurisdictions have differing noise abatement ordinances. Some ordinances are based on maximum A-weighted levels, while others use equivalent values, L_{eq}, averaged over a period of time to allow for variations of noise with weather. Some jurisdictions specify day-night limits, L_{dn}, giving more weight to nighttime noise due to its greater annoyance potential.

Without specific local regulations, the information displayed in Figure 4-5 may be applicable[13]. As shown, the threshold of annoyance is around 52 dB(A) under wet conductor conditions. Such conditions are considered to be equivalent to L_{50} foul weather conditions at the edge of the right-of-way. At approximately 58 dB(A), the potential for complaints increases greatly. Although experience has substantiated the general validity of these numbers, some complaints have been received at lower noise levels than those shown.

FIGURE 4-5
ANNOYANCE LEVELS

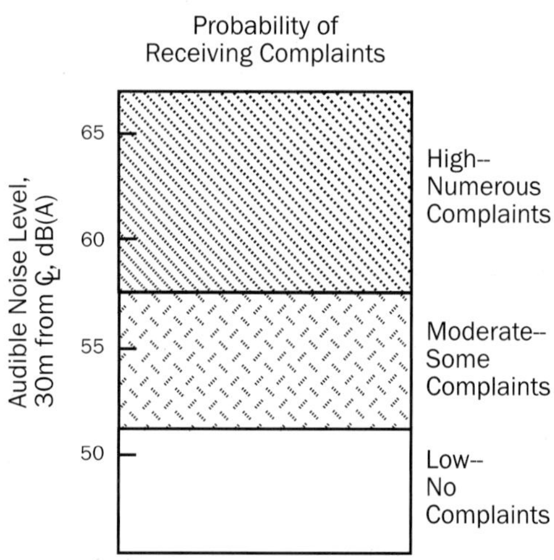

The L_{50} foul weather level, rather than the L_5 level, is used for annoyance evaluation for two reasons. First, the sound of a heavy rain tends to mask the sound of the transmission line. Second, windows are closed and people are dashing for cover during a heavy rain. Under these circumstances, sound from the transmission line is not a major concern.

Transmission line noise can be annoying, however, when the surrounding environment is otherwise quiet. It becomes a factor immediately after a heavy rain, or during light precipitation when people are more likely to be outside or have windows open. Experience has shown that the L_{50} noise level adequately represents the conditions under which annoyance is probable.

4.2.6 Measurement Techniques

Techniques for measuring transmission line audible noise are given in IEEE Standard 656[14]; instrumentation is defined in ANSI Standard S1.4[15]. It is essential to use an ANSI Type 1 precision sound level meter for transmission line work because of the predominance of high frequency sound components. Community noise analyzers are not suitable because they are calibrated for the low end of the audible frequency spectrum where the most common annoying noise sources are concentrated. Traffic and airport noises are examples of sounds that are dominant at these lower frequencies. Only precision sound level meters are sufficiently well calibrated at the high frequency end of the audible spectrum to accurately measure transmission line random noise.

When making audible noise measurements, several precautions should be taken. It is important to make sure the noise being measured is actual noise from the line, and not from rain, birds, or other nearby sources. Careful study of the IEEE Standard and the instruction manual for the instrument will help in avoiding errors caused by misuse of the equipment. When designing a long-term measurement system, it is especially important to ensure that other sounds do not corrupt the data. Proper weather protection of the instrumentation, as well as proper calibration techniques, must be carefully applied.

When making audible noise measurements, it is often desirable to measure profiles with other than "A" weighting. It also can be advantageous to measure the frequency spectrum of the noise using octave band filters. If significant background noise exists, it may be possible to use measurements at the higher octave bands to establish a profile, and then correct it to an approximate "A" weighted value using typical frequency spectra for corona noise.

4.3 ELECTRIC FIELDS

An electric charge produces an effect distant from the charge itself. The attraction of tissue paper to an electrostatically charged comb is one example of such electric phenomena. Similarly, the effect of voltage on a wire extends beyond the dimensions of the wire itself. This effect, called an electric field, makes itself felt near transmission lines in terms of coupled voltages and currents on nearby motor vehicles and other metallic objects. Electric field effects were originally a problem only with EHV transmission lines. Today, however, they must be considered in the design of any new transmission line.

4.3.1 Electric Field Sources

Historically, environmental concern over the electrical effects of power transmission and distribution lines was limited to maintenance of adequate clearances for prevention of dangerous flashovers. This concern is embodied in the National

Electrical Safety Code[16] with its inclusion of "provisions for the safeguarding of persons from hazards arising from the installation, operation, or maintenance of (1) conductors and equipment in electric supply stations, and (2) overhead and underground electric supply and communications lines." While clearance requirements are still of utmost concern, coupling effects of electric fields also have become important, particularly with EHV transmission lines.

Electric and magnetic fields produced by the voltage and current in an electrical conductor are not confined to the conductor itself, but are spread throughout the surrounding space. A power transmission conductor located above the earth may have a potential of 100,000 volts with respect to the earth. This creates a voltage distribution, called a "space potential," throughout the surrounding space. In this case, the voltage potential steadily increases from zero at ground level to 100,000 volts at the conductor. Although space potentials can be computed and mapped with the aid of a space potential probe, the absolute magnitude of a space potential is of little concern. The important variable is incremental change of the space potential over small distances. For the 100,000 volt conductor, the space potential may change only 3000 volts within 1 meter above the earth, but 10,000 volts within 1 centimeter of the conductor.

The negative of the space potential gradient is called the electric field. The electric field is not a mathematical abstract, but is a way of expressing the natural change in space potential with change in distance from the potential source. An overview of power line electric and magnetic fields is given in [3][17].

4.3.2 Prediction Programs

For many calculations at power frequencies, the electric and magnetic fields may be assumed to be static. This "quasistatic approximation" is reasonable since the variation of voltage and current with time at power frequencies is much slower than other phenomena, such as the relaxation time of the earth. This fact allows use of static (dc) field formulas for the calculation of electric and magnetic fields. There are times, however, when it is necessary to apply the time-varying electromagnetic equations[18]. Failure to understand the quasistatic approximation can lead to confusion in understanding certain results, such as elliptical field patterns.

To aid the understanding of the quasistatic approximation, it is useful to consider that one wavelength at 60 Hz is about 3,000 miles. Since all distances relating to fields from power lines are very small compared with a wavelength, near-field conditions apply. This is in contrast to the situation that occurs with radio and television. The wavelength at 100 MHz in the FM broadcast band is 3 meters. Thus, for radio and television, the receiving antenna is many wavelengths from the transmitting antenna. When the receiver is many wavelengths from the transmitter, the receiver is located in the far field of the antenna. The far field is also known as the "radiation field." Some people incorrectly use the term "radiation" for the near field associated with power frequency fields from a power line. If a significant radiation field existed for power frequency electric and magnetic fields, there would be a significant unexplained power loss. Past studies have shown that this loss does not occur. Power line electric and magnetic field effects involve coupled voltages and currents, not radiation.

Computer programs for prediction of electric field profiles are based on the quasistatic approximation. Consequently, the effect of the finitely conducting earth can be represented by image conductors. These image conductors are located at the same depth below the earth's surface as the height of the actual conductors above the surface. At the earth's surface, the effect of the images is to make the field lines perpendicular to the surface. This is important since it is a general principle in electrostatics that electrostatic field lines enter a conductor perpendicular to the surface. Simple electric field programs can then take advantage of this fact to calculate only the vertical field component at zero height.

More accurate programs compute the field profile at a specified height, usually 1 meter. Such programs, however, involve calculation of the field ellipse, more fully described in later sections.

The actual parameter that determines an electric field is conductor charge rather than conductor voltage[17]. If the capacitance matrix of the line and image conductors is known, conductor charge can be calculated from the line voltage. Computer programs calculate the potential coefficient matrix because it is more easily calculated than capacitance. This matrix is then inverted to form the capacitance matrix so the corresponding line charges and resulting electric field can be calculated.

4.3.3 Electric Field Profiles

Power line electric effects are generally related to fields near the earth's surface. These fields, usually called "ground level" fields, are customarily measured and calculated at a height of 1 meter above the earth's surface. Other heights may be considered in special circumstances, such as induction into parallel communications wires. The effect of the electric field is often specified by the value of the field at the edge of the right-of-way.

Power line electric fields are commonly described by a lateral profile of the field that excludes conducting objects that would perturb the field. These conducting objects may be people, animals, plants, buildings, fences or vehicles, etc. Such unperturbed fields are used as a basis of further analysis.

Electric field strength is the change in electric potential in space, and is given in volts per meter (V/m) or kilovolts per meter (kV/m). Figure 4-6 presents calculated electric field profiles for the same 765 kV line design of Figure 4-1.

FIGURE 4-6
TRANSMISSION LINE UNPERTURBED ELECTRIC FIELD LATERAL PROFILE

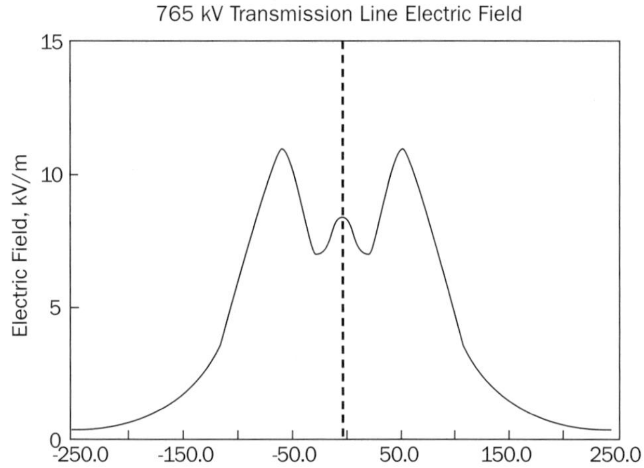

Factors influencing the electric field profiles of overhead power lines are:

- Line voltage
- Midspan conductor height above ground
- Configuration of conductors, including line geometry, conductor spacing, relative phasing of multi-circuit lines, use of bundled conductors, and ground wires
- Lateral distance from the centerline of the transmission line
- Height above ground at the point of field measurement
- Proximity of conducting objects, such as trees, fences, and buildings
- Local terrain
- Right-of-way width

Excluding proximity of conducting objects, all the above variables contribute to the unperturbed field. They are interrelated and must be considered to optimize the line design. It is important to stress how all design variables are interrelated since changes in one can adversely affect another. Many changes to a line design that reduce the ground level electric field increase the electric field at the conductor surface, causing an increase in corona and noise. Following are important variables to be considered.

Line Voltage

For power frequency electromagnetic fields, the electric field is a function of line voltage and the magnetic field is a function of line current. For a given line geometry, the ground level electric field is directly proportional to the line voltage.

Most high voltage transmission lines in the U.S. are three-phase ac, operating at voltages ranging from 115 to 765 kV. The voltage level chosen is the result of the optimization of many variables, including the amount of power to be transmitted and the distance over which it is to be carried. For technical and economic reasons, there is an advantage to using higher voltages. Technically, the advantages are system stability, thermal capacity and voltage control. Economically, there are advantages in conductor and structure size and land usage. Because allowable power flow for surge impedance loading increases as the square of the voltage, one 765 kV line transmits as much power as thirty 138 kV lines, and requires only 1/13 the right-of-way.

Height of Conductors Above Ground

At power frequencies, the earth's conductivity and the relatively low rate of change of the electric field combine to make the earth's surface act as a zero potential plane. For electric field calculation purposes, this plane is represented by imaginary conductors located below ground as a mirror image of the actual line conductors located above ground. The resulting field magnitude at ground level is strongly influenced by the conductor height. Increasing the conductor height decreases the field directly beneath the conductors, but increases it at some distance away from the center line. Increasing conductor height, also increases the structure weight, cost, and the visual intrusion of the line.

Line Geometry

A three-phase ac power transmission line has three conductors, or conductor bundles, with equal magnitudes of voltage and phase angles 120° apart. The resulting electric field is the sum of the field from each of the conductors, and is dependent on the relative distances from the observer to each of the conductors. When the conductors are closer together, there is a greater degree of field cancellation, and the ground level field is reduced.

Optimum conductor spacing is determined by a process that considers line insulation, conductor surface electric field and accompanying corona effects, and motions due to wind and ice. There are minimum practical values of conductor spacing for reliable operation of a line. Advances in knowledge of the factors that limit spacing now allow closer spacings than those formerly used. This is the origin of "compact transmission lines."

The level and shape of electric field profiles vary with conductor configuration. Ground level field profiles are generally lower for delta arranged conductors than horizontal designs. Ground level fields are also generally lower for more compact lines than wider spaced lines.

Lateral and Vertical Distances

The strength of a single conductor electric field diminishes with distance from the source. Taking into account earth conductivity, some three-phase circuit configurations may exhibit a relative minimum value at some distance from the center line, a maximum value at a greater distance, and then diminishing values as the distance increases. Calculated values of electric field strengths based on line characteristics and distance have been shown to correspond closely to measured values.

Natural and Artificial Shielding

Trees provide considerable natural shielding of the electric field from transmission lines. Research has been conducted into ways of providing artificial shielding by using additional conductors, either grounded or with voltages of specially selected magnitude and angle. The costs involved, however, have restricted the application of this method to special situations.

Transmission Line Right-of-Way

Utilities in the U.S. either own or have easements for transmission and distribution right-of-ways. Right-of-way widths are variable, depending on local conditions and practices. They range from zero for overbuilt lines on distribution poles along city streets to hundreds of feet for 500 kV lines. Right-of-way widths must be sufficient to prevent conductors from approaching too closely to nearby objects during extreme wind conditions, and must allow access for maintenance. During past periods, ground level fields were not a concern in setting right-of-way widths. Today, however, certain state regulations have included edge of right-of-way electric field criteria. These criteria may become the dominating factors in setting right-of-way widths.

Specification of a maximum electric field allowed on the right-of-way may require a greater minimum clearance. If conductor height is increased, the edge of right-of-way field may increase, decrease, or remain the same due to the relative minimum value in the lateral profile, as discussed earlier. It is possible, therefore, that a given maximum field limit on the right-of-way could either increase or decrease the required right-of-way width for a given edge of right-of-way criterion.

4.3.4 Capacitive Coupling of Electric Fields

As shown in Figure 4-7, the capacitive coupling of electric fields is the cause of voltages induced into objects such as motor vehicles, people, or sheds. The Thevenin equivalent circuit for electric field induction is composed of a high voltage source in series with a high equivalent impedance. The voltage source is the open circuit induced voltage and the equivalent impedance is the equivalent reactance of the capacitance network relating the transmission line conductors to the nearby object. Because of the small capacitance between line and object, the capacitive reactance is large. Thus, it is preferable to use a Norton equivalent current source and neglect the source equivalent impedance.

FIGURE 4-7
ELECTRIC FIELD CAPACITIVE COUPLING

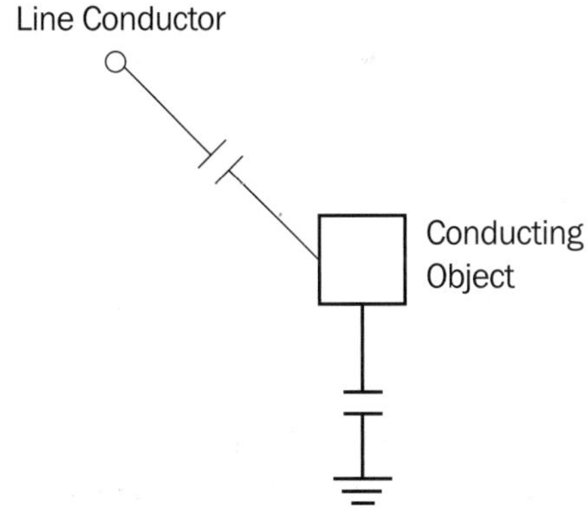

The Norton equivalent current source is given by the electric flux density, or electric displacement vector, D:

$$D = \epsilon E \tag{4-1}$$

where:

E = *the measured or calculated unperturbed electric field*

ϵ = *the permittivity ($\epsilon_0 = 8.854 \times 10^{-12}$ Farads/meter for an object in free space)*

Induced current, I, from an insulated object under the line to a zero resistance ground connection is given by:

$$I = \omega DA \tag{4-2}$$

where:

A = *the equivalent area of the object*
ω = *the angular frequency, $2\pi f$, in radians per second*

The circuit for electric field induction to a motor vehicle is presented in Figure 4-8. Parameters are:

I_{sc} = *short circuit current to ground*
C_v = *capacitance from vehicle to ground*
R_v = *resistance from the vehicle to ground (This may be through tires that deliberately have been made conducting)*
R_p = *resistance of a person contacting the vehicle*

FIGURE 4-8
EQUIVALENT CIRCUIT FOR ELECTRIC FIELD COUPLING

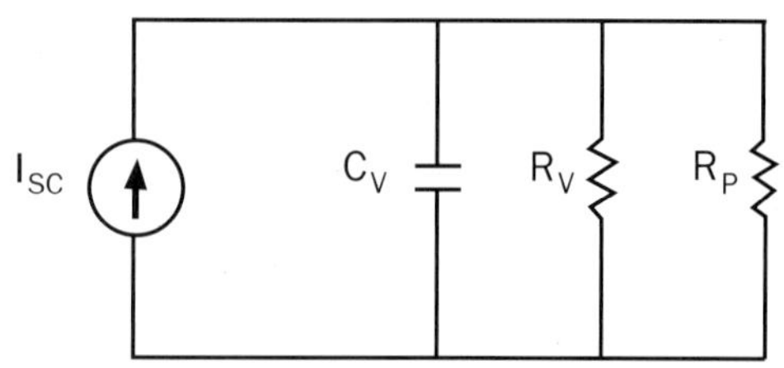

Analysis of Figure 4-8 shows several relevant facts about electric field induction:

- Any conductance to ground reduces the open circuit voltage. The highest possible voltage occurs when there is infinite resistance and only the capacitive reactance is between the vehicle and ground.
- Short circuit current is the most reliably measured experimental variable. Any leakage resistance, even of the order of megohms, reduces the voltage.
- An open circuit voltage of several kilovolts may coexist with a short circuit current on the order of milliamperes.

4.3.5 Evaluation of Electric Fields

Electric field effects engineering for overhead power transmission lines, for other than perceived health effects, encompasses three stages:

Establishment of Criteria

Any electromagnetic environmental study requires careful establishment of the appropriate criteria for assessment. Criteria to be considered must include:

(1) Type of people who may come in contact with an object subject to induction: workers or specially trained personnel; general public; children

(2) Level of concern: safety; annoyance; perception

(3) Probability distribution of the effect with respect to worst case

(4) Location of criteria application: on the right-of-way; at the edge of the right-of-way; off the right-of-way

(5) Special considerations: road crossings; parking lots; farms with long movable irrigation pipes

It is always necessary to meet the National Electrical Safety Code [16] requirement that induced current must be less than 5 mA for the largest vehicle that can be in proximity to the line. This is based on "let-go" considerations to prevent a situation from arising where a person would receive an electric shock and be unable to let go of the contacted object.

Prediction of Effect

The capacitive coupling of electric fields can be analyzed and predicted for a given line configuration. Electric field evaluations are generally based on an unperturbed lateral field profile, 1 meter above ground at midspan with minimum conductor ground clearance. Sometimes field contour plots are developed. After defining the unperturbed field, worst-case induced voltages and currents on appropriately sized objects can be calculated. From this data, a statistical distribution of actual expected voltages and currents can be developed.

Evaluation of Impact

With the criteria and predicted fields established, an evaluation of the significance of the effect must be made. The evaluation may show that changes must be made to achieve an acceptable line design. Changes to be considered would include possible (1) relaxation of criteria, (2) reduction of electric field, and (3) mitigation of induced voltages by grounding and shielding. The first two stages may have to be repeated to achieve an acceptable design.

Transmission line field effects are only one of many variables involved in line design. As such, they must be considered together with all other design variables to arrive at the most favorable design for each particular application. When problems are present in localized areas only, it is possible to provide special local measures, such as underslung shield wires, to reduce the ground level electric field.

The primary concern is the effect on humans through discharges from objects within the electric field. Usually, these objects are vehicles, buildings, or fences that are not effectively grounded. A common situation is a person experiencing a spark discharge between hand and vehicle while touching a vehicle. After good contact is made, a steady current may flow through the body.

The major factors affecting the magnitude of the spark discharge or current are:

- Line voltage
- Size of insulated object
- Location of insulated object with respect to the line
- The degree of grounding on the object and person touching the object
- The probability of exposure

A "worst-case" scenario may be a perfectly ungrounded truck parked directly under an EHV line being touched by a well-grounded person. This is the condition envisioned in the 5 mA requirement in the NESC. Furthermore, if this were in a parking lot where the probability of exposure is high, then safety, annoyance, and perception considerations may all be of concern. The final parameter to be considered is the degree of tolerance that certain groups of people exhibit to these effects[19].

4.3.6 Measurement Techniques

Electric field measurements on a transmission line are customarily made with a free-body meter attached to the end of an insulated stick[20]. When conducting such measurements, it is important to do so in an area where nearby conducting objects do not perturb the electric field. It is equally important to position the instrument so the operator's body does not shield the field and cause an inaccurate reading. Grass and other vegetation also can affect the measurements. In addition, the determination of the location of a virtual ground can be difficult due to local high soil resistivity.

A thorough knowledge of an overhead transmission line's geometry is essential. The exact line voltage and location of each conductor at the time of measurement also must be known. The conductor configuration can be determined from line drawings, but often the conductor height at the measurement location must be determined by on-site measurements. It is also possible that elevation differences along the line may be significant.

4.3.7 Grounding and Shielding

Capacitive coupling of electric fields produced by high voltage lines can induce high open-circuit voltages and low short-circuit currents in objects the size of cars and trucks. The high source impedance results in current source behavior that often makes it possible to mitigate the field effects by use of relatively high resistance, single-point grounds. Typical house construction provides electric field shielding from fields outside the house.

Shielding of electric fields from a power transmission line may involve construction of underslung shield wires in special local situations, such as parking lots. Underground cables are intrinsically shielded from electric fields by the earth.

4.4 MAGNETIC FIELDS

Just as an electric potential produces an effect outside of the wire, so does an electric current. In the case of current, the effect is a magnetic field. Magnetic field effects of transmission lines are as old as the utility industry itself. In the past, the emphasis was on the coordination of induction-related interference with nearby telephone and railway signal lines. While coordination with railroad and telephone systems is still an important consideration, induction into other parallel facilities such as pipelines, irrigation systems, and metallic fences is also of concern today. Calculation of magnetic field profiles is important for assessment of magnetic fields as part of the permitting process for new and uprated lines. Engineering methods have been developed to predict, evaluate and mitigate the effects of magnetic fields.

4.4.1 Magnetic Field Characterization

Both static and time-variant magnetic fields exist at a distance from either a constant or time dependent current source, respectively. These fields are vector quantities with magnitude and direction. The strength of the magnetic field varies according to the geometry of the current elements generating the fields, and can be found by applying the Biot Savart Law. The Biot Savart Law relates the magnetic field at a point in space to the current element source of that field. For an infinitely long, current-carrying straight wire, which approximates a transmission line conductor, the magnitude of the magnetic field is inversely proportionally to the distance from the wire. The direction of the field is determined by the cross-product of the current element and the displacement vector extending from the current element to the point in space where the field is determined. The directional sense of the field can be quickly verified by the right-hand rule. Induction is greater to objects that parallel the power line conductors for a significant distance. Because the magnetic field is a function of line current rather than voltage, magnetic field induction can be a factor for lower voltage lines.

Magnetic fields and above-ground-level electric fields form an elliptical pattern with the major and minor axis magnitudes and orientations being dependent on the geometry of the physical orientation of the transmission lines. These fields lie in a plane perpendicular to a transmission line and are called transverse electromagnetic (TEM) fields. The electric fields at ground level beneath transmission lines have only a vertical component for flat level earth. Magnetic fields at ground level, however, have both vertical and horizontal components for level earth.

The power density of an electromagnetic field is given by the cross product of the electric and magnetic fields. For a TEM wave, if the power density is integrated over an area perpendicular to the line conductors, the power flow in the fields equals the power flow in the conductors. Thus, a power transmission line can be viewed as a waveguide for the power flow in the fields. A consequence of this is that it is impossible to eliminate electric and magnetic fields from power lines. Reducing them at one location increases them at another location if power flow is constant. This also explains why design changes to reduce ground level fields often increase fields, and therefore the electrical stresses near the conductors, possibly resulting in increased corona effects.

4.4.2 Prediction Programs

Computer programs for prediction of magnetic field profiles are based on the quasistatic approximation. However, for magnetic fields, the effect of earth resistivity is such that the earth acts more like a perfect insulator than as a perfect conductor. This effect can be handled in one of three ways:

- By use of Carson's series[21]
- By image conductors where the currents return at a much greater depth than the height of the line conductors above ground
- By image conductors at a complex depth[22]

Near the line, it is acceptable to ignore the currents in the earth entirely. The effect of currents in the earth on the magnetic field profile increases with distance from the line.

The actual parameter that determines the magnetic field intensity is conductor current. Consequently, the matrix inversion operation required to calculate electric field profiles is not required for magnetic fields. However, even at ground level, magnetic fields have a significant horizontal component making it necessary to compute the field ellipse. Computer programs are available to compute the major axis of the field ellipse, called the maximum value of the field, as well as the field resultant. Programs that predict a statistical distribution of the magnetic fields for an assumed amount of current imbalance are also available.

4.4.3 Magnetic Field Profiles

Power line magnetic field induction calculations are generally based on the longitudinal voltage per unit length induced in parallel facilities. The longitudinal voltage has been called longitudinal electromotive force (formerly called longitudinal electric field - LEF). This approach assumes induction from a conductor to parallel facilities. For many considerations, especially biological effects, power line magnetic fields are presented as lateral profiles of unperturbed fields. Most objects are nonmagnetic and do not perturb the field, but the field can be altered by induced circulating currents.

Power line magnetic field profiles are customarily calculated and measured at a height of 1 meter above ground. In special circumstances, such as induction to parallel communications wires, other heights are considered. Magnetic field strength can be given in amperes per meter or more commonly as magnetic flux density in Tesla (SI units) or Gauss (cgs units).

Magnetic fields near ground level are produced by both underground and overhead power lines. The magnetic fields are influenced by the following variables:

- Line current
- Height of conductors above ground
- Configuration of conductors, including line geometry, conductor spacing, and relative phasing of multi-circuit lines
- Lateral distance from the centerline of the transmission line
- Height above ground at the point of field measurement
- Local terrain

- Soil resistivity
- Proximity of current-carrying loops and ferromagnetic objects

A lateral profile of the magnetic field for the same 765 kV transmission line as used for the radio noise profile in Figure 4-1 is shown in Figure 4-9. The magnetic field is measured in mG/kA to allow estimation of the field for different line currents. Because of the different image depth, earth conductivity results in different electric and magnetic field profile shapes for the same line geometry. The effect of the earth can be seen by comparing the electric field profile in Figure 4-6 with the magnetic field profile in Figure 4-9.

FIGURE 4-9
TRANSMISSION LINE MAGNETIC FIELD LATERAL PROFILE

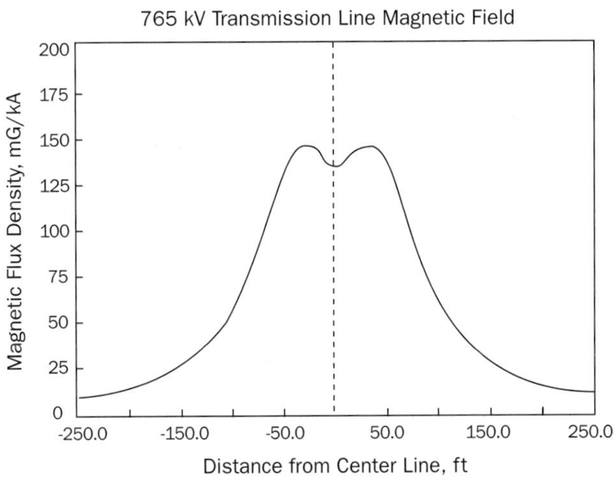

Magnetic fields are current dependent, constantly changing due to changing load currents and system unbalance. Electric fields, however, are voltage dependent, and tend to remain constant over time due to steady line voltage.

4.4.4 Statistical Characterization of Magnetic Fields

Although three-phase transmission lines are typically assumed to be balanced, slight differences in current magnitude and phase angles normally exist. These small differences are not normally monitored, yet can affect the magnetic field, particularly at longer distances from the line.

A Monte Carlo analysis[23] may be used to estimate the error caused by the uncertainty regarding phase conductor current. In this kind of analysis, a base condition of equal current magnitudes and balanced angles is developed. A maximum percentage current magnitude deviation from the base value is specified, as is a maximum number of degrees the angle may vary from the balanced value. A random number generator is then used to select a set of current combinations that fall within the specified range. These current combinations are used to develop a set of magnetic field values for specified line geometry. These magnetic field values reflect possible instantaneous values for the range of line currents developed. A statistical analysis on the ensuing set of magnetic field values at a given location yields the following:

- The magnetic field level that is exceeded 95% of the time (exceedence level L_{95}, essentially a minimum value)
- The magnetic field level that is exceeded 50% of the time (L_{50}, essentially an average value)
- The magnetic field level that is exceeded 5% of the time (L_5, essentially a maximum value)

Such an analysis is presented in Figure 4-10 for a balanced current of 1000 amperes with a magnitude deviation of ± 5% and an angle deviation of ± 3°. The expected accuracy in such a field measurement can be estimated from this spread. Since a large number of possible combinations of current magnitudes and phase angles were used, the results represent the randomness that occurs in nature.

FIGURE 4-10
MAGNETIC FIELD VARIATION DUE TO CURRENT UNBALANCE

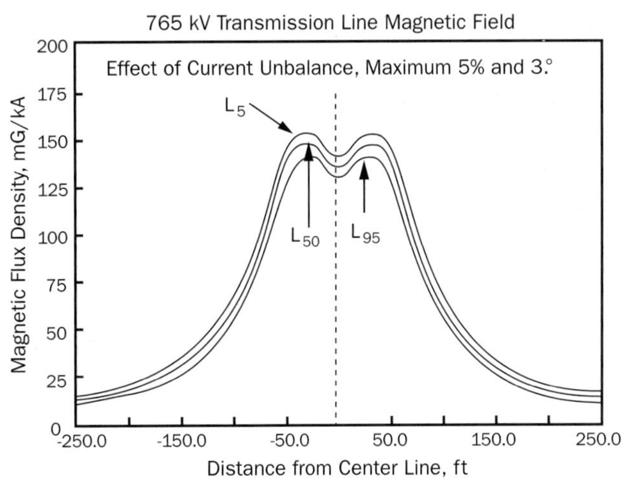

4.4.5 Inductive Coupling of Magnetic Fields

Magnetic field coupling is inductive and is significant for objects that parallel the power line for a significant distance. Examples are pipelines, telephone lines, railroad signal circuits and long fences. The corresponding equivalent circuit for magnetic field induction is given in Figure 4-11.

where:

V_{oc} = *the open circuit induced voltage*
R_s = *the real component of the source impedance*
X_s = *the imaginary components of the source impedance*
R_p = *the resistance of a person contacting the parallel object*

Numerical values of the parameters in Figure 4-11 typically indicate a relatively low-voltage source – on the order of tens of volts behind a low impedance yielding a short-circuit current of several amperes. It is possible that a person contacting the induced voltage on a long parallel object may be the most significant current-limiting impedance in the circuit. This situation is an important consideration in evaluating the implications of electric and magnetic field-induced voltages. Because voltage drops during faults (except for the phenomenon of voltage rise on unfaulted phases), electric field calculations only consider maximum normal operating voltage. In contrast, current increases significantly during a fault, necessitating calculations for both normal load and fault conditions for many applications. Induced voltage may be greatest for single line to ground faults.

FIGURE 4-11
CIRCUIT FOR MAGNETIC FIELD COUPLING

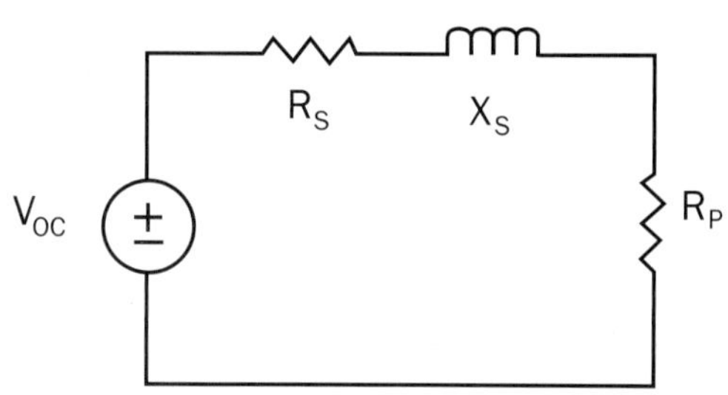

4.4.6 Instrumentation for Magnetic Field Measurement

The basic survey instrument consists of a single sensing coil with associated electronics giving a measurement of the magnetic field along a single axis. Orienting the coil in different directions allows the direction and maximum magnitude to be determined, according to the procedures outlined in IEEE Standard 644[20]. Other instruments contain three mutually perpendicular sensing coils to simultaneously record the magnetic field along all three axes and electronics to combine the three components to obtain the resultant magnitude. When comparing the results of measurements taken with the two types of instruments, it is frequently found that the readings from the three-axis instrument are greater than the single-axis instrument readings, even though both have been calibrated to equal accuracy. The difference is due to the fundamental field ellipse property of 3-phase power line magnetic fields and the way these fields are treated by the instruments. The single-axis instrument measures the major axis of the field ellipse (maximum field), while the three-axis instrument gives the square root of the sum of the squares of the three field vector components (resultant field). It is important to note the type of instrument used for magnetic field measurements.

Obtaining agreement between measured and calculated magnetic field values is frequently difficult. It is important to identify and assess all possible sources of error when making measurements[24][25]. A detailed discussion of measurement instrumentation, along with procedures, is given in IEEE Standard 644[20] and an IEEE committee paper[26]. There are several general characteristics of magnetic field meters that must be considered.

Coil Size

The voltage induced into the instrument coil is a function of the area enclosed by the coil. In a non-uniform field, different size coils average over a different area and could yield different results, but for transmission line fields, coil size is not a significant feature.

Core Material

Magnetic core coils can yield erroneous measurements of non-uniform fields if they have been calibrated in a uniform field because of perturbation of the field by the core. However, both types of coils yield similar results for transmission line magnetic fields.

Coil Current

An accurate measurement is dependent on an insignificant current in the coil. This is accomplished by connection to a voltmeter with a high input impedance.

Frequency Response

Voltage induced in the sensing coil is directly proportional to the frequency of the magnetic field. Regarding frequency, there are three basic types of instruments: linear response instruments that exhibit increasing sensitivity with increasing frequency; flat response types that include an integrator in the circuitry to make the sensitivity independent of frequency; and filter types that use band pass filters to filter out all frequencies except the fundamental power frequency.

Calibration
The calibration of a magnetic field measuring device must be made with the proper size calibration coil. Calibration procedures are described in [20].

Effect of Instrument Precision
The precision of the measuring instrument must always be considered, particularly if magnetic field levels are low. Digital instrument specifications include plus or minus a certain amount in the least significant digit. If the field is sufficiently low that only one or two digits are displayed, instrument precision is an issue.

4.4.7 Measurement Procedures

Besides questions of instrument calibration and concerns related to the design of the instrument, other factors enter into the accuracy of measurements. These factors encompass both the variables that impact magnetic fields and the procedures to be used in measuring these variables.

Physical Location of Line
A thorough knowledge of the overhead transmission or distribution line's geometry is essential. The conductor locations must be accurately known. The orientation and general placement of the line can be determined from line drawings, but often the conductor height at the measurement location must be determined by on-site measurements.

Line Current
The magnetic flux density due to a single infinitely long straight conductor carrying a current, *I*, is given by Ampere's Circuital Law:

$$B = \frac{\mu_0 I}{2\pi r} \tag{4-3}$$

where:

- B = *magnetic flux density, T (Wb/m²). For a single conductor, the direction of B is tangent to a circle of radius r centered on the conductor*
- I = *current in the conductor, amperes*
- r = *radial distance from the conductor to the measurement location, meters*
- μ_0 = *permittivity of free space, $4\pi \times 10^{-7}$ H/m*

From the above, it is seen that the magnetic field intensity is directly proportional to line current and varies with the distance from the conductor axis. Due to the variation of line current as load changes, the ability to compare measured and calculated fields is dependent upon the accurate measurement of the line current at the exact time the magnetic field is measured.

Line Current Unbalance
The effect of current unbalance on the magnetic field is dependent on the distance from the conductors. Where distances are small, approximately equal to the phase conductor spacing or less, the effect of current unbalance on the magnetic field is relatively small. As the distance from the conductor increases to 10 or more times the phase conductor spacing, the magnetic field is strongly influenced or even dominated by the effect of the current unbalance.

Location and Amount of Return Currents
Zero sequence current in its simplest terms is the average of the three phase currents, and is the current that is returning someplace other than in the phase conductors. Some of this current may return in the shield wires, the earth, nearby pipes, or other conducting paths. The location of the return current, however, is significant. It is possible to calculate the fraction of the zero sequence current that returns in the shield wires under certain assumed conditions. The return path or paths for the remaining portion of the zero sequence current are usually unknown.

Even if there is no line current unbalance, it is possible for the induced circulating current in the shield wires to make a 10% or more difference in the magnetic field. This circulating current in the shield wire is not only a function of phase conductor and shield wire geometry, but shield wire characteristics and method of grounding.

Unknown Current Sources
When all precautions have been taken, it is still possible that magnetic field measurements do not match calculations. In such cases, it is possible that there are other currents in the vicinity that were not considered. Possible sources are underground power cables and induced currents on underground pipes. It has been documented that currents in water or gas pipes under streets can produce magnetic fields comparable to those produced by the electric utility facilities.

4.4.8 Grounding and Shielding

The low-source impedance of magnetic field coupling in the circuit shown in Figure 4-11 requires multiple low-resistance grounds to reduce induced voltage. Series electrical breaks may be required in fences and pipelines to limit the maximum induced voltage.

Power frequency magnetic field shielding may involve current loops or thick electrically conducting and/or ferromagnetic material and is generally difficult and costly. Magnetic field shielding is generally impractical for overhead lines and most underground construction, although shielding is provided by the steel pipe in pipe-type underground cable.

4.5 REFERENCES

[1] *IEEE Standard Definitions of Terms Relating to Corona and Field Effects of Overhead Power Lines*, IEEE Standard 539-2005, IEEE, New York.

[2] Loftness, M. O., *AC Power Interference Handbook, 2nd Edition, Revised*, Percival Technology, Tumwater, WA, 2003, ISBN: 0-9653760-3-6. Available from the American Radio Relay League, Newington, CT.

[3] *Transmission Line Reference Book, 345 kV and Above/Second Edition*, Electric Power Research Institute, 1982.

[4] Chartier, V. L., et al, "Investigation of Corona and Field Effects of AC/DC Hybrid Transmission Lines," *IEEE Transactions on Power Apparatus and Systems*, Vol. PAS-100, No. 1, p. 72, January, 1981.

[5] *EHV Transmission Line Reference Book*, Edison Electric Institute, New York, 1968.

[6] Olsen, R. G., Schennum, S. D., Chartier, V. L., "Comparison of Several Methods for Calculating Power Line Electromagnetic Interference Levels and Calibration With Long Term Data," *IEEE Transactions on Power Delivery*, Vol. 7, No. 2, p. 903, April, 1992.

[7] *Limits and Measurement Methods of Electromagnetic Noise From AC Power Systems, 0.15-30 MHz*, Canadian Standards Association, CAN3-C108.3.1-M84, 1984 (R 2005).

[8] "Review of Technical Considerations on Limits to Interference from Power Lines and Stations," IEEE Radio Noise and Corona Subcommittee Report, *IEEE Transactions on Power Apparatus and Systems*, Vol. PAS-99, No. 1, p. 365, January/February, 1980.

[9] *Federal Communications Commission Rules and Regulations*, Part 73, Section 73.182.

[10] "Transmission System Radio Influence," IEEE Committee Report, *IEEE Transactions on Power Apparatus and Systems*, Vol. PAS-84, No. 8, p. 714, August, 1965.

[11] *IEEE Standard Procedures for the Measurement of Radio Noise from Overhead Power Lines and Substations*, ANSI/IEEE Std 430-1986, IEEE, New York.

[12] "A Comparison of Methods for Calculating Audible Noise of High Voltage Transmission Lines," IEEE Committee Report, *IEEE Transactions on Power Apparatus and Systems*, Vol. PAS-101, No. 10, p. 4090, October, 1982.

[13] Perry, D. E., "An Analysis of Transmission Line Audible Noise Levels Based upon Field and Three-Phase Test Line Measurements," *IEEE Transactions on Power Apparatus and Systems*, Vol. PAS-91, No. 3, p. 857, May/June, 1972.

[14] *IEEE Standard for the Measurement of Audible Noise from Overhead Transmission Lines*, IEEE Standard 656-1992, IEEE, New York.

[15] *American National Standard Specification for Sound-Level Meters*, S1.4-1983 (R 2001), American National Standards Institute.

[16] *2002 National Electrical Safety Code*, Accredited Standards Committee C2-2002, The Institute of Electrical and Electronics Engineers, Inc., New York, 2001.

[17] "The Electrostatic and Electromagnetic Effects of AC Transmission Lines," IEEE Tutorial Course Text 79 EH0145-3PWR, IEEE, 1979.

[18] Deno, D. W., "Transmission Line Fields," *IEEE Transactions on Power Apparatus and Systems*, Vol. PAS-95, No. 5, p. 1600, September/October, 1976.

[19] "Electric and Magnetic Field Coupling from High Voltage AC Power Lines — Classification of Short-Term Effects on People," *IEEE Transactions on Power Apparatus and Systems*, Vol. PAS 97, No. 6, IEEE Working Group on Electrostatic and Electromagnetic Effects, November/December, 1979.

[20] *IEEE Standard Procedures for Measurement of Power Frequency Electric and Magnetic Fields from AC Power Lines*, ANSI/IEEE Standard 644-1994, IEEE, New York.

[21] Carson, J. R., "Wave Propagation in Overhead Wires with Ground Return," *Bell System Technical Journal*, Vol. 5, pp. 539-554, 1926.

[22] Dabkowski, J., Taflove, A., "Mutual Design Considerations for Overhead AC Transmission Lines and Gas Pipelines," Palo Alto, CA: Electric Power Research Institute, September 1978, Report EL-904.

[23] Olsen, R. G. , Backus, S.L., Stearns, R.D., "Development and Validation of Software for Predicting ELF Magnetic Fields Near Power Lines, *IEEE Transactions of Power Delivery*, Vol. 10, No. 3, pp. 1525-1534, July 1995.

[24] "Magnetic Fields from Electric Power Lines Theory and Comparison to Measurements," *IEEE Transactions on Power Delivery*, Vol. 3, No. 4, IEEE Magnetic Fields Task Force, p. 2127, October, 1988.

[25] "An Evaluation of Instrumentation Used to Measure AC Power System Magnetic Fields," *IEEE Transactions on Power Delivery*, Vol PWRD-4, No. 1, IEEE Magnetic Fields Task Force Paper, p. 373, January, 1991.

[26] "Measurements of Power Frequency Magnetic Fields Away from Power Lines," *IEEE Transactions on Power Delivery*, Vol. 6, No. 2., IEEE Magnetic Fields Task Force, p. 901, April, 1991.

5 DESIGN OF NEW TRANSMISSION LINES

The design of new overhead transmission lines is a highly technical process involving disciplines in many different areas, such as structural analysis, wind-induced conductor movement, insulator contamination, etc. Although it is almost impossible for a line designer to become an expert in each aspect of line design, it is possible to be aware of the major technical problems that may be encountered. At the very least, the designer will need some knowledge of the wide variety of solutions that may be implemented for each problem.

Overhead transmission lines must meet many different requirements and limits. Clearance to ground and other circuits must be maintained. System considerations regarding maximum MVA loading and reactance per mile must be met. Electrical losses over the life of the line, corona interference and construction cost must also be minimized. Some of these requirements and limits inevitably conflict. Cost-effective placement of structures along a right-of-way may conflict with the desire of adjoining property owners to locate the structures to minimize their visual impact. The use of a large diameter conductor to minimize corona interference may drive up the cost of line and force the redesign of standard structures. Because of this complexity, each line design procedure is unique.

Even though line designs vary, there are several common factors that each design must take into account. Proper insulation coordination of the system, designing for wind-induced conductor motion and economics are essential considerations that must be dealt with in the line design process. For insulation coordination, careful consideration must be given to the type and dielectric strength of insulators chosen for the application, and the impact of different types of flashovers caused by lightning and overvoltages from surges. Wind-induced conductor motion, including aeolian vibration, galloping and subconductor oscillation, can drastically reduce the life expectancy of conductors. As with any design, the entire process must be done in a cost effective manner. In some instances, economic justification may result in emphasis being placed on one aspect while de-emphasizing another aspect.

5.1 NEW TRANSMISSION LINE DESIGN CONSIDERATIONS

Typically, it is system planners who identify the need for new transmission lines. Their decisions are based on continuous system studies of response to contingencies and projected growth of loads and generation. The system planners must first establish the system requirements and then identify those requirements that may become design constraints on the new line. Meticulous planning is required to find alternative ways to rectify or circumvent many of the constraints. The constraints may include a limited number of line designs from which to choose, bypassing physical barriers and meeting stringent environmental and aesthetic concerns.

5.1.1 System Requirements

The initial stage of any new transmission line design involves the determination of the system requirements. Since each new transmission line has specific requirements, it is important that the system planners properly assess the needs during system studies. Presented are some basic considerations that system planners must consider and specify:

- The voltage level of a new line
- Whether the line is to be double or single circuit
- The line's terminal points
- The minimum level of reliability required
- The minimum load capability in terms of thermal stability and voltage drop limits

If only existing standardized transmission line designs are allowed as options, the system planner may have only a limited number of alternatives. In this situation, the design engineer must strategically locate the standard structures along the right-of-way (ROW) and design special structures and foundations for problem areas.

However, increasingly stringent environmental and visual constraints on construction of new lines often limit or prohibit the use of standardized line designs. In this case, the system needs, identified as part of any system planning review, are incorporated into the many design constraints that a new line must meet. In some cases, system requirements may need to be eased to obtain regulatory approval.

Design engineers and system planners must communicate with each other to succeed in adding new overhead transmission lines. Both must be flexible and innovative in preparing for the regulatory approval process. Existing line designs may need to be adjusted to meet changing environmental standards. Circuit reliability may have to be reduced to meet other more pressing constraints. Special non-economic structure placement and altered physical appearance may have to be considered at certain critical locations along the route. Innovations in conductor designs, structure materials and construction techniques may need to be incorporated to meet design constraints. Most importantly, several line designs and routing alternatives may have to be developed and presented early in the regulatory process to gain final approval. The key is to have prepared alternatives for whatever problem may arises.

5.1.2 Environmental Design Constraints

There are three types of environmental constraints under which transmission lines must operate: fundamental, generally applicable and site specific. These constraints, based on the proximity of the line to people and wildlife, are imposed to limit the impact of the line on the surrounding environment as much as possible and to ensure maximum safety.

Fundamental Constraints

Fundamental constraints are those that must be met at every point along the line. These include:

- Assuring public safety by maintaining sufficient clearances to ground and to adjacent structures
- Operating the line on a daily basis with only a rare occurrence of failures related to conductors, hardware, insulators and structures
- Providing adequate load-carrying capacity to ensure reliable system operation under worst-case loadings
- Operating with low levels of electrical failure (flashover) under all foreseeable atmospheric conditions and system switching events

Electromagnetic Environmental Criteria

It is necessary to develop and apply criteria to limit the amount of electrical interference of the proposed line design with nearby populated areas or facilities. For lines in some locations, these criteria may have been previously determined, possibly encoded into law. In other locations, criteria may have to be developed as part of the project.

For a complete evaluation of electromagnetic compatibility of a new transmission line, valid criteria is required to assess the impact of predicted electromagnetic fields and noise at specific locations, typically at the edge of the right-of-way. However, there are also situations where locations on or off the right-of-way must be evaluated. A more complete discussion of this subject is presented in Chapter 4.

Some of the considerations for development of electromagnetic environmental criteria are the following:

- Electric field levels are calculated at a specified height above ground (usually 1 meter) for the line conductors at minimum ground clearance (sag at maximum conductor temperature) and maximum line voltage. Maximum edge of right-of-way electric field levels are customarily specified. Electric field levels on the right-of-way are considered for situations such as road crossings where induced current on vehicles is a concern.

- Magnetic field levels at a specified height above ground (usually 1 meter) for the line conductors at minimum height from ground (sag at maximum conductor temperature). The line current used for the magnetic field calculations varies depending on the jurisdiction. Conductor current may be specified as "typical loading," "maximum predicted loading" or "as equal to the line rating" (e.g. "winter normal" rating). Magnetic field criteria are usually applied at the edge of the right-of-way.

- Radio and television interference are generally specified for a specific weather condition at the edge of the right-of-way to allow reception of certain stations with an appropriate signal-to-noise ratio. Fair or foul weather conditions may be used, depending on the local requirements.

- Audible noise is usually specified at the edge of the right-of-way as random noise in dB(A) for average foul weather conditions.

Site-Specific Constraints
Site-specific constraints occur at specific locations along the ROW. The solution to these constraints may include avoidance of the specific location by re-routing. Some site-specific constraints may include:

- Limitation of the visual impact of overhead lines by choosing structures on the basis of aesthetic appearance. This constraint usually places limits on structure height, location along the ROW, structure materials and finish, and phase spacing.

- Assurance of the safety of wildlife in the vicinity of overhead lines. This limitation may require routing of lines around certain wildlife migratory paths and protected areas.

In some instances, certain design constraints may conflict. For example, attempts to reduce magnetic field levels may involve the use of smaller spacing between phases that may yield greater corona noise. Similarly, reducing the visual impact of the line by reducing the height of its structures may force the use of more structures per mile and reduce the ability to avoid placing structures in certain locations.

5.1.3 Standard Line Design Considerations

In response to conflicting design constraints, engineers have developed some basic design considerations that system planners and line designers should take into account in the line design process. A review of these recommendations/observations, based on assumptions and experience gained in line design, will help avoid many of the known problems that exist in the design of new lines:

- Use structures and foundations designed for ice and wind loading of conductors that have been set by experience, legal requirements and/or codes.

- Conductor fatigue failures can be avoided by the use of low everyday conductor tensions, armor rods and/or dampers.

- Minimum ground clearances are usually determined by using conservative weather conditions in thermal rating calculations. With this approach, clearances may be maintained even under elevated-temperature operation.

- The use of standard structure angle families with standard wind spans, weight spans and blowout limits, regardless of the terrain, will alleviate decision related problems as to which types to use.

- Maintaining standard minimum phase-to-phase and phase-to-tower clearances will prevent flashovers under high winds or ice galloping.

- Maximum conductor operating temperatures are based upon considerations of the conductor's annealing and elevated temperature creep properties.

- Standard guidelines have been written that allow the use of hot-line maintenance techniques.

Clearly, the presentation of a standard design for approval without the consideration of changing requirements and justifiable alternatives yields a low probability of success. System planners and line designers must consider all contingencies and be familiar with the types of situations they may encounter in designing new lines. Failure to obtain the approval of new lines may eventually yield unreliable system operation.

5.2 ELECTRICAL INSULATION COORDINATION

Determination of proper insulation and air gap clearances for a transmission line depends on a careful analysis of (1) the power frequency design requirements, (2) adequate insulation levels for switching surge protection and (3) statistical data on the frequency of lightning strikes. Proper insulation levels are dependent on factors such as insulation contamination, distances between insulators, system conditions and insulator swing. It is important that all these factors be considered to achieve total coordination of the insulation system, a critical element in system reliability and maximum safety.

5.2.1 Power Frequency Design

Two distinct modes of power frequency insulation failure are possible: flashovers across insulators or air gaps related specifically to clearance and flashovers across insulation related specifically to insulator contamination. The first is usually associated with conductors and/or insulators swinging too close to structure surfaces due to wind. This type of flashover usually applies only to I-string configurations, although flashovers to structure surfaces could occur with V-string configurations if clearances are too small. Insulator contamination may be specified as a function of the level of the pollution or salt-deposit density. Recommendations are made accordingly as to the number and type of insulators to use in the design. A discussion of each type of design consideration is presented in the following sub-sections.

Clearance Flashovers

Under high wind conditions, I-string insulators can swing toward the structure in response to wind forces on the adjacent conductor spans. If the insulator swing angle is too great, conductor to structure flashover may occur at normal system voltage. The magnitude of wind-swing angle is a function of wind speed and direction, conductor weight

FIGURE 5-1
TYPICAL POWER FREQUENCY AC FLASHOVER CHARACTERISTICS OF LARGE AIR GAPS

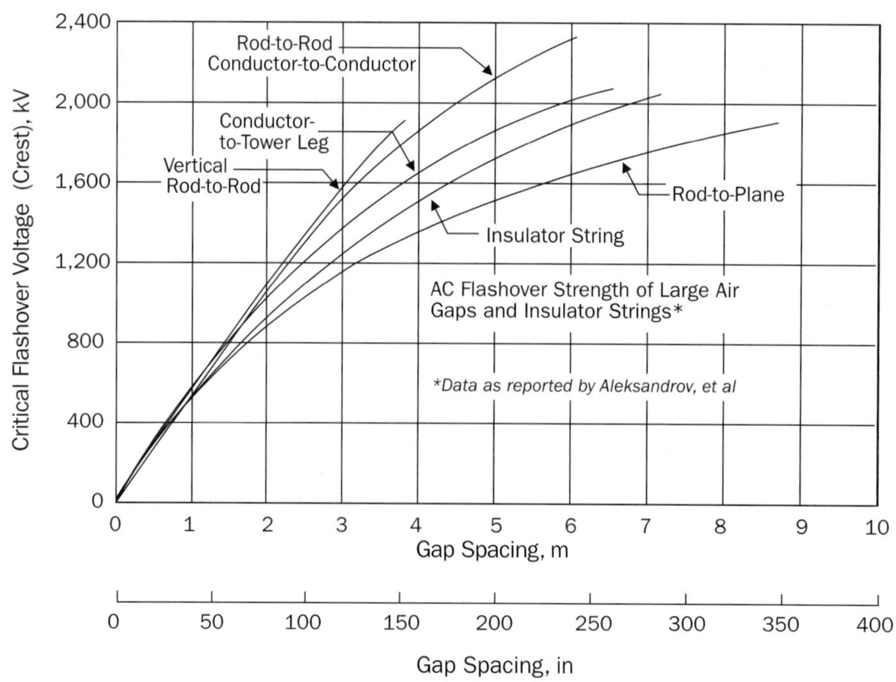

and diameter, and the weight-to-wind-span ratio. The equation for wind swing angle is:

$$\Phi = tan^{-1}\left[\left(\frac{w_W}{w}\right)\left(\frac{S_H}{S_V}\right)\right] \; degrees \qquad (5\text{-}1)$$

where:

w_W = *wind pressure load on the conductor, lbs/ft, as determined from Equations 2-32 or 2-33 of Section 2.3.1*

w = *conductor weight, lbs/ft*

S_H = *wind span, ft*

S_V = *weight span, ft*

The maximum wind speed used for calculations should correspond to the maximum recorded value over a period of at least 10 years, and, more typically, 50 years or more. A 50 year recurrence value may be obtained from 20 to 30 years of peak wind data based upon a Fisher-Tippett extreme value distribution[1] or by other statistical analysis methods. Wind speeds of 70 mph to 120 mph are typically used.

Typical 60 Hz ac flashover characteristics of large air gaps are shown in Figure 5-1[2]. Since the insulating capability of a gap is a function of the relative insulation strength and reflects the effect of temperature, pressure, etc., it is considered prudent to design power frequency clearance requirements based on minimum conditions rather than the statistical approach that is used for switching surge analyses.

Contamination Design

The performance of existing lines, especially those that have performed well and traversed the same area as the proposed line, is one of the best design references to limit insulator contamination flashovers. Alternatively, the International Electrotechnical Commission (IEC) has a guide for the selection of insulators based on four levels of contamination[3]. Minimum recommended leakage distance is specified according to the level of contamination as shown in Table 5-1.

TABLE 5-1
IEC RECOMMENDATIONS FOR LEAKAGE DISTANCE AS A FUNCTION OF CONTAMINATION LEVEL

Level	Characteristic	Leakage mm/kV
I	Light	16
II	Medium	20
III	Heavy	25
IV	Very Heavy	31

For high pollution levels, the IEC standard may suggest that a large number of standard discs be used. However, in practice, such a number would probably not be acceptable. Instead, either high leakage or non-ceramic insulators would be used for the heavy contamination conditions.

Corresponding to the IEC Standard, one IEEE Working Group[4] proposed designs for four levels of contamination and provided specific recommendations for the number of standard insulators. IEEE contamination levels, however, give the resulting leakage distance in terms of the equivalent salt-deposit density (ESDD) as shown in Table 5-2.

TABLE 5-2
IEEE WORKING GROUP INSULATION RECOMMENDATIONS AS A FUNCTION OF CONTAMINATION LEVEL

ESDD mg/cm²	Characteristic	Leakage mm/kV	
		I-string	V-string
< 0.03	Very Light	14.6	12.8
0.03 - 0.06	Light	18.7	15.8
0.06 - 0.1	Moderate	21.6	16.9
0.1 - 0.2	Heavy	25.7	19.3

The EPRI Redbook[5] has a great deal in common with the IEEE Working Group since much of the supporting research was available to both publications, and leading researchers contributed to both. The Redbook also identifies four grades of contamination severity from very light to heavy, with the same levels of ESDD used in the IEEE guide.

5.2.2 Switching Surge Flashover

When loads are switched from one transmission line to another, overvoltages are generated on the line. These switching surge overvoltages typically have peak magnitudes of 1.5 to 3.0 times normal crest line-to-ground voltage, a rise time of about 200 μs, and a decay-time-to-half value of 2000 μs. Historically, transmission line insulation systems were designed to withstand the maximum switching surge overvoltages without flashing over. Modern designs, however, incorporate a statistical design approach that also has great benefit for uprating existing insulation and structure geometry. This statistical approach, which does allow for some flashovers, is based on determining the probability of flashover (PFO) of the line for a given switching configuration.

The PFO of the line is a function of the stresses imposed on the line and the withstand strength of the insulation. It can be calculated by defining the values of these two determinants and mathematically combining them. Line insulation strength is a function of electrical clearance, electrode shape and the number of air gaps, as well as the atmospheric conditions including barometric pressure, temperature and humidity. Line capability is usually represented by a histogram showing the insulation strength versus the frequency of occurrence of switching surges. The stress applied to the line insulation is defined as a distribution of overvoltages resulting from switching operations.

Switching Surge Distributions

Switching surge overvoltages are influenced by system conditions such as line length, X0/X1 ratio (ratio of zero sequence to positive sequence impedance), system strength, line shunt reactors, line surge arrester ratings, circuit breaker closing resistors, etc. The switching operation itself can influence switching surge overvoltages. For example, closing the line with a charge trapped on the line from a previous opening usually results in overvoltages having a higher magnitude than those associated with energizing a line without a charge.

A typical switching surge overvoltage distribution is shown in Figure 5-2. Note that the highest overvoltages have a small likelihood of occurring. Consequently, the statistical approach is a rational one if a small, but finite chance of line flashover is acceptable. In uprating studies, or for compact designs where the overvoltage distribution is limited by surge arrester action, the same magnitude

FIGURE 5-2
TYPICAL SWITCHING SURGE OVERVOLTAGE DISTRIBUTION

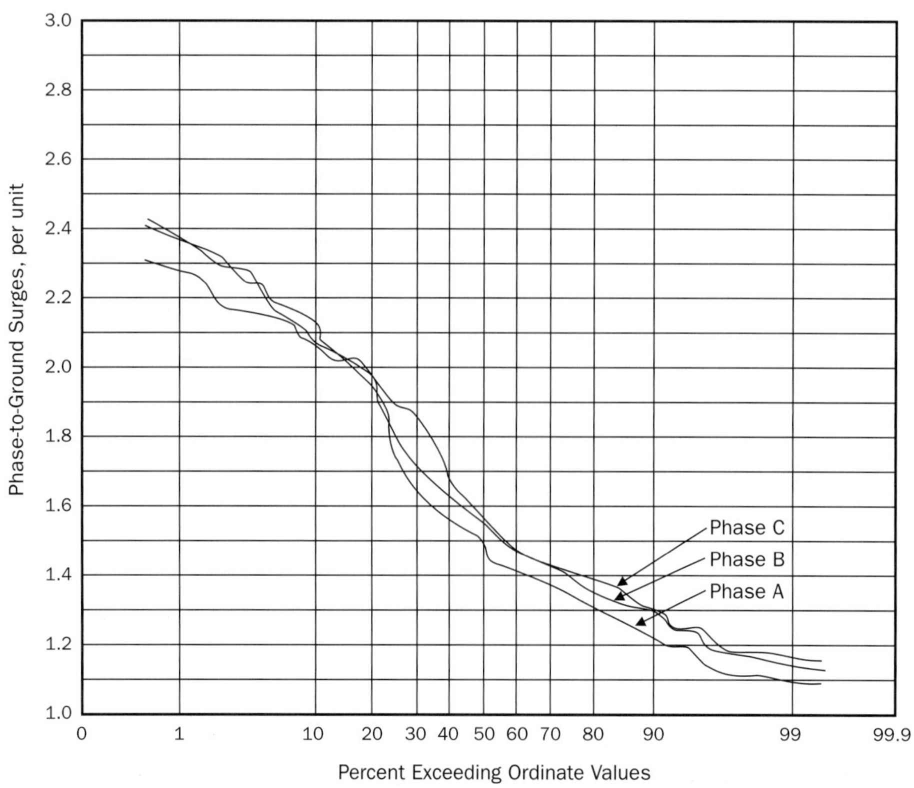

FIGURE 5-3
TYPICAL AIR GAP SWITCHING SURGE CFO'S

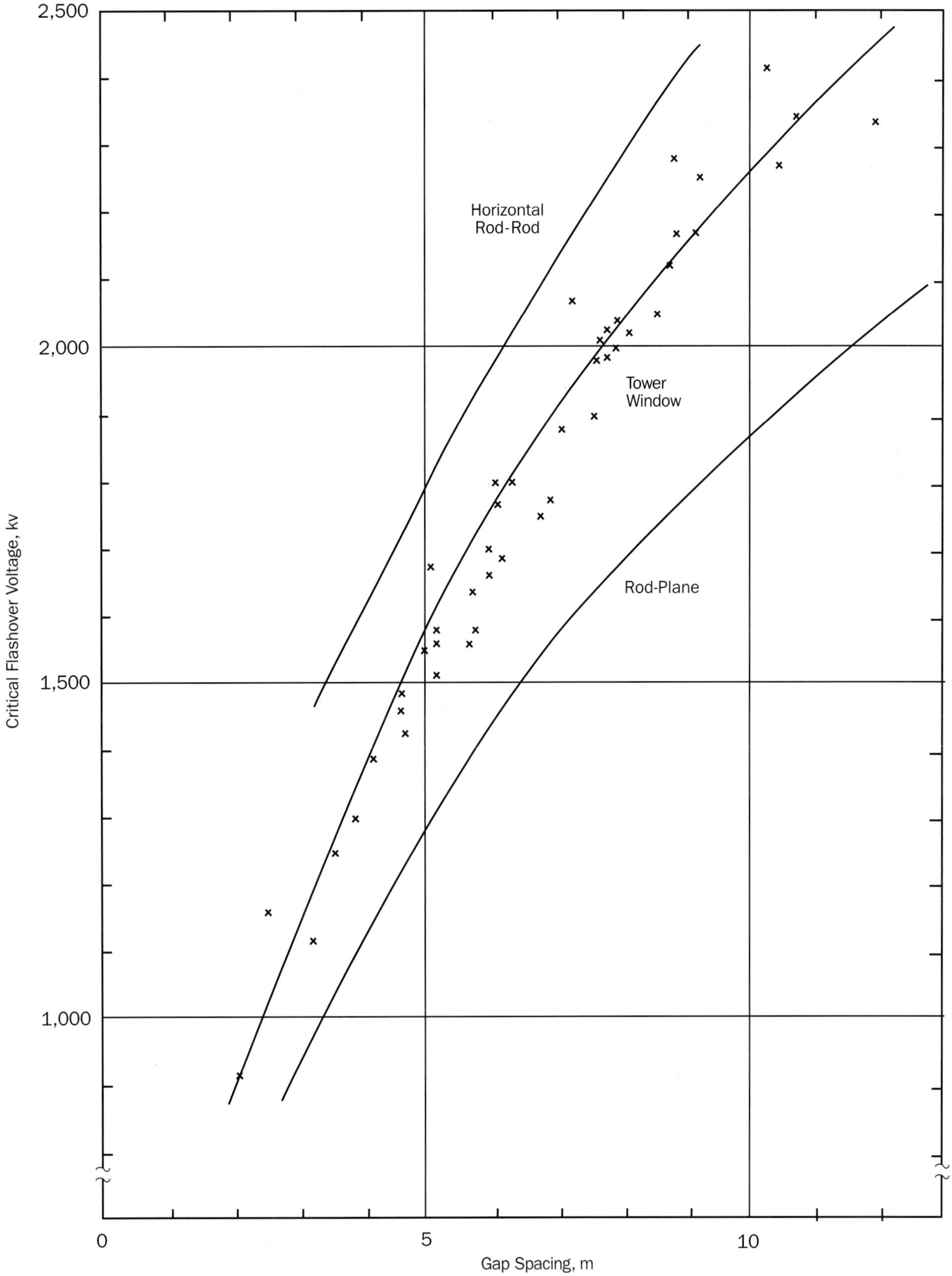

Switching-surge strength of tower window gaps. Square windows, tower width = 1.2 m. Critical time-to-crest. Standard atmospheric conditions (RAD = 1, H = 11 g/m^3)

overvoltage will often appear on each phase. Thus, there is a need to consider the PFO of all three phases simultaneously[6][7].

Flashover Strength

Flashover strength of a given electrode configuration is represented as a Gaussian distribution with a critical flashover level, CFO, the level at which a flashover will occur 50% of the time with a standard deviation of 6%. Insulation strength distributions of air gaps are often determined by full-scale test data, such as that shown in Figure 5-3[5][8][9][10][11][12][13][14]. Subsequent work has indicated that an improved approach may be possible based on a better understanding of the physics of air gap breakdown[15]. In general, only positive polarity surges are considered because tests with negative surges indicate significantly higher flashover voltages.

Wind-Swing Angle Distribution

For free swinging insulators, such as I-strings, it is important to recognize that wind can cause a conductor to swing close to a structure surface, reducing the air gap clearance. The magnitude of wind-swing angle is a function of wind speed and direction, conductor weight and diameter, and the weight-to-wind-span ratio. The calculation of the insulator swing angle resulting from the wind force applied to the conductor/insulator combination is given in Equation 5-1.

$$\Phi = \tan^{-1}\left[\left(\frac{W_W}{W}\right)\left(\frac{S_H}{S_V}\right)\right] \text{ degrees}$$

In keeping with the statistical approach used for switching surge insulation analysis, the line design is not based on the most severe swing occurring at the highest wind, but rather reflects the statistical nature of wind. A distribution of wind speeds affecting conductor motion is used instead. The approach, then, is to statistically reflect the probability of a reduced conductor-to-structure clearance based on the given wind speeds for the area and season. Wind speed statistics can be obtained from an analysis of hourly weather recordings from U.S. or world meteorological groups. Alternatively, some engineers use a constant insulator swing angle for switching surge analysis that is smaller than the extreme wind swing angle used for power frequency insulation analysis to reflect the statistical nature of the wind swing angle distribution.

Relative Insulation Strength

Air gap CFO data, such as that shown in Figure 5-3, is usually given for standard atmospheric conditions: no precipitation, barometric pressure of 29.92 inches Hg, temperature of 77 °F and vapor pressure of 0.6 inches Hg. As the insulation strength changes with air density and absolute humidity, the CFO for a particular set of meteorological conditions may be determined by modifying the CFO at standard atmospheric conditions with correction factors for the relative air density, humidity and rate of rain[2]. The ratio of CFO for a given set of atmospheric conditions to CFO at standard atmospheric conditions is the relative insulation strength (RIS). The statistical influence of weather on switching surge strength of air gaps is then represented by a frequency distribution of RIS computed from hourly surface observations.

PFO Calculation

Given a structure configuration, statistical distributions of switching surge overvoltage, wind speeds and relative insulation strength (RIS), the engineer can calculate the probability of flashover for the line as follows:

- Select a wind speed
- Compute the conductor-to-structure clearance at standard conditions
- Select an RIS value
- Compute the modified insulation withstand capability

The statistical frequency of occurrence of this combination is the product of the statistical frequency of the wind speed and the RIS values used for the specific computation. The probability of flashover (PFO) of the clearance is then determined from the magnitude of the switching surge overvoltage, and multiplied by the frequency of the RIS/wind speed combination and the expected frequency of the overvoltage magnitude used for the PFO calculation.

This computation is performed for each combination of wind speed, RIS and switching surge overvoltage. The overall PFO is the summation of all the PFO-frequency combinations. It may be necessary to determine the overall PFO of the structure as opposed to simply one window. For example, the 3-phase approach for the PFO calculation for a given wind speed and RIS is:

$$PFO_{tower} = \frac{1}{N} \sum_{n=1}^{N} [1 - (1 - pfo_{a,n})(1 - pfo_{b,n})(1 - pfo_{c,n})] \quad (5\text{-}2)$$

where a, b and c are the phase window PFOs and N is the total number of surges in the overvoltage distribution. With k structures exposed to the surge distribution, the overall line performance is then:

$$PFO_{line} = 1 - (1 - PFO_{tower})^k \quad (5\text{-}3)$$

Note that for lines over 100 miles, it is probable that the open, or receiving, end of the line will have a decidedly higher statistical distribution of voltage surge magnitudes than that of the sending end, where the line is being energized. Consequently, it may be reasonable to assume that only half the structures along the line are being exposed to the higher surges. Such an assumption would likely result in a predicted line PFO of about 1/2 the PFO obtained if all the structures were considered. If surge arresters are used to control line overvoltages, then the surge distribution is more uniform along the line.

5.2.3 Lightning

The effects of lightning on transmission line systems can be categorized in three ways: backflashovers, direct strikes to phase conductors and midspan flashovers. The location of the lightning strike determines the classification type. Some lightning strikes may hit shield wires, phase conductors or even hit shield wires and cause flashovers to the phase conductors. Although the magnitude of lightning strikes can vary tremendously, methods have been developed over time to minimize their impact.

FIGURE 5-4
ISOKERAUNIC ACTIVITY - UNITED STATES

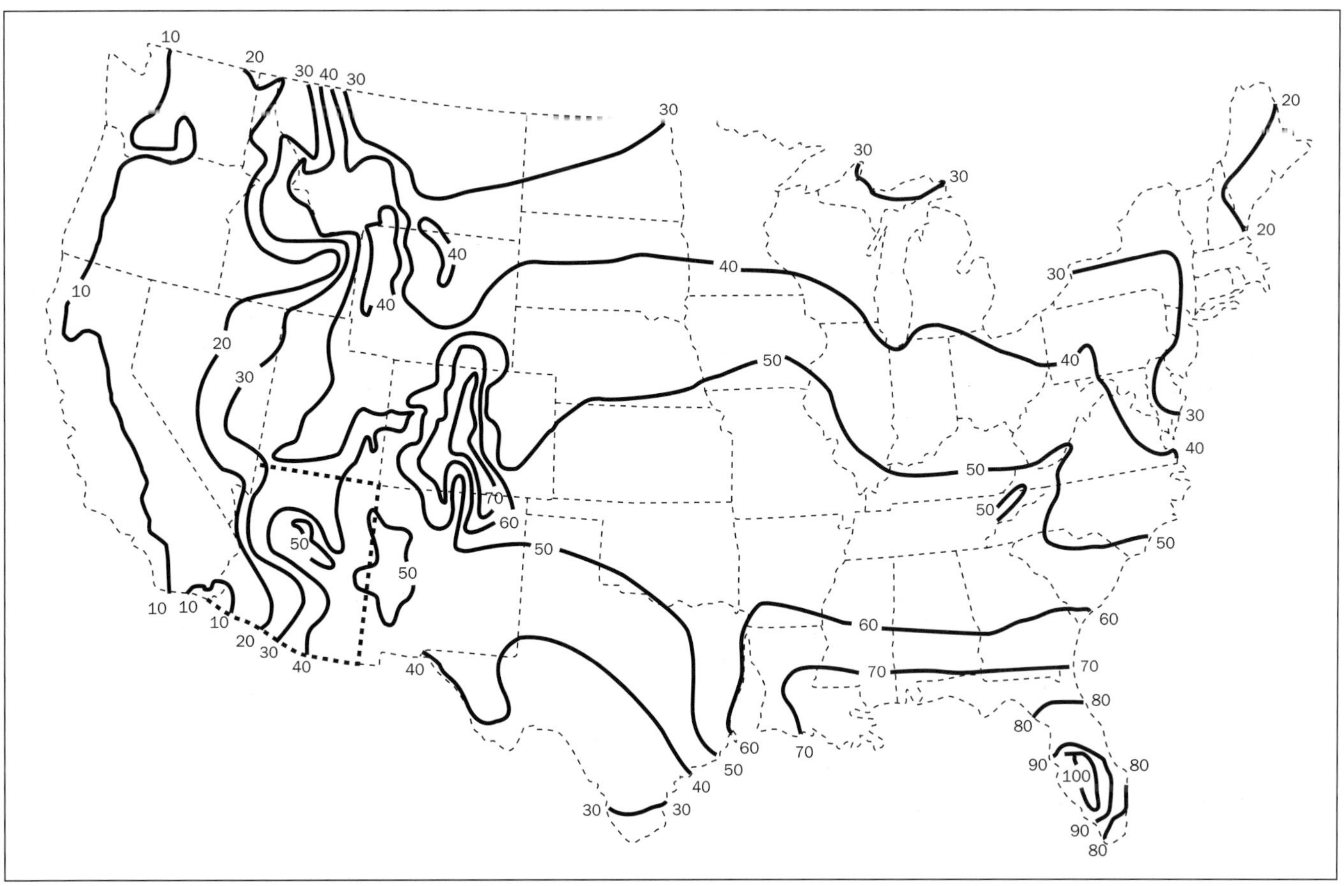

Lightning can be characterized by the keraunic level, stroke magnitude and stroke rise time. The keraunic level is the number of days during which thunder was heard. Estimates of the thunder activity in the United States are shown in the map, prepared by the U.S. Weather Bureau, in Figure 5-4.

The expected ground stroke density is a function of the keraunic level, T,[5]:

$$N = 0.31T \tag{5-4}$$

This formula represents the number of flashes to earth per square mile per year. Stroke magnitude is statistical in nature and can best be described as a log-normal distribution which can be approximated[16]:

$$P_I = \frac{1}{1 + \left[\frac{I}{31}\right]^{2.6}} \tag{5-5}$$

where:

P_I = *the probability that the peak current in a flash will exceed I*

I = *peak lightning stroke current, kA*

The probability that the current in a flash will exceed 31 kA is 50%.Although there is some debate, stroke rise rate, S, is believed by some researchers to be a function of stroke magnitude[5]:

$$S = 3.91(I^{0.55})\ kA/\mu sec \tag{5-6}$$

It should be noted that about 55% of all flashes have more than one stroke, and about 90% will not exceed eight strokes. Usually, the flashes will have occurred in less than thirty cycles.

As indicated, there is still a difference of opinions among engineers about some of the aspects of lightning. Lightning is a powerful natural event that can cause great destruction to homes, people, nature and animals, as well as transmission lines. The three ways lightning can impact a transmission system are presented in the following sub-sections.

Backflashovers

Backflashovers occur when a stroke hits a structure or a shield wire very close to the structure. The stroke travels along the shield wires, as well as down the structure, into the structure footing resistance. As voltage builds at the crossarm (the base of the insulator) due to the current flowing down the structure, the stress across the insulator may be sufficient to flash over the insulator. If this occurs, a portion of the stroke current will travel down the phase conductor and subsequently impact the terminating equipment, usually the circuit breaker. For flashover to occur, the footing resistance and resulting current must be sufficiently high. Therefore, the lower the structure footing resistance, the lower the voltage and the less likely a backflashover will occur.

Structure footing resistance is usually measured at low frequency for evaluation of power frequency grounding of transmission structures. Because of the high crest current and the short-time waveshape of a lightning impulse, the resistance seen by lightning current is often considerably less than the 60 Hz value. Thus, the high frequency resistance value should be used if known. In fact, the impulse resistance for many soil conditions is approximately 25% that of the 60 Hz value[17][18].

Direct Strikes to Phase Conductors

Strikes that miss the shield wires and hit phase conductors are known as shielding failures. On a transmission line, only strikes with a sufficient magnitude result in a flashover across an insulator and a subsequent fault that must be cleared. For currents below that critical level, no line flashover occurs, but the surge can have an impact on station equipment and cause flashovers.

The theory used to predict the maximum stroke current magnitude that can miss the shield wires and hit a phase conductor is known as the electro-geometric model[19]. Using a form of that approach, the maximum stroke current that can penetrate the phase conductor can be computed as follows:

$$I_{max} = \left[\frac{S}{8}\right]^{1.54} \tag{5-7}$$

where:

$$S = \frac{F}{2\cos\alpha}$$

$$F = \sqrt{(Y_s - Y_p)^2 + X_p^2}$$

and

Y_s = *shield wire height, m*

Y_p = *phase conductor height, m*

X_p = *horizontal distance between phase conductor and shield wire, m*

$$\alpha = \tan^{-1}\frac{X_p}{Y_s - Y_p}$$

This model borders on the realm of metaphysics. That is, while there is considerable foundation in physics, the statistics of electrical breakdown in air are not recognized. Instead, the model assumes that the stroke will always seek out the closest path to ground. This is more likely to be true than not, but there are some exceptions. Even if the shortest path to ground is via the path of a shield wire rather than a phase conductor, there is some finite probability that the flash will strike the phase conductor.

Midspan Flashovers

Midspan flashovers have been investigated by a number of researchers over the last 30 years[20][21]. Primarily, the focus of the investigations has been on the mechanism that tends to limit the number of midspan flashovers between shield wires and adjacent phase conductors. The significant mechanism is a sheet of current, or glow discharge[21], that bridges the gap between the shield wire and phase conductor. This glow discharge tends to inhibit flashover between the conductors for approximately 10 μsec. If the rise time of the stroke is long and the reflections from shield wire connections at adjacent structures return quickly enough, no flashover will occur. If conditions are such that a flashover does occur, then part of the stroke travels down the phase conductor and the scenario is similar to that of a shielding failure.

However, during the glow discharge phase, although no flashover occurs, there can still be a considerable current surge that travels along the phase conductor and impacts at the open breaker. As shown in Figure 5-5, the glow discharge has a voltage drop of about 186 kV/ft.

FIGURE 5-5
MIDSPAN FLASHOVER MECHANISM

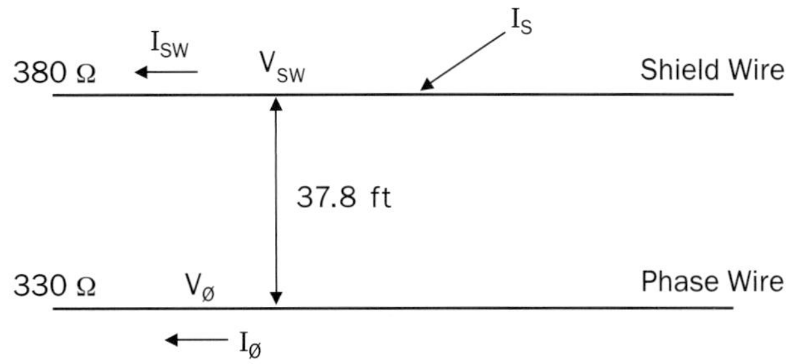

Thus, as a rough approximation, a voltage divider exists with the driving voltage equal to one-half the stroke current, I_s, times the surge impedance of the shield wire (380 Ω, including the corona envelope)[22]. The drop across the glow discharge is constant (186 times a separation distance of 37.8 feet), and the driving voltage minus the glow discharge drop is the voltage, V_Φ, on the phase conductor. The phase conductor current, I_Φ, is then phase voltage divided by the phase conductor surge impedance, 330 Ω including corona. The following equation shows the approximate relationship between the stroke current and the phase current during the glow discharge phase.

$$I_\Phi = \frac{\left(\frac{I_s}{2}\right)380 - 186(37.8)}{330} \tag{5-8}$$

5.3 WIND-INDUCED CONDUCTOR AND INSULATOR MOTIONS

Overhead transmission lines are often subjected to wind-induced conductor motion. Shortened conductor life due to wind-induced conductor motion must be considered in the design stages of any transmission line. In the simplest case, wind can cause the conductor catenary to swing in the direction of the wind. This conductor "blowout" requires conductor clearances be selected to maintain minimum clearance from both the edge of the right-of-way and supporting structures under maximum blowout conditions.

Conductors on overhead transmission lines commonly experience a variety of wind-induced motions. The design of new transmission lines must allow for wind and combined wind/ice-induced conductor and insulator movements.

Aeolian vibration is a major factor in the maximum allowable unloaded conductor tension that determines structure spacing and/or height. Ice galloping is a major determinant of minimum phase spacing, especially for double-circuit and delta-phase arrangements. Subconductor oscillation is an important consideration in bundled subconductor spacing, spacer selection and choice of subspan length. Table 5-3 compares the most common types of oscillatory motion induced by wind.

The most common type of wind-induced motion is aeolian vibration, which occurs at relatively low wind speeds. Fatigue damage to conductors is cumulative, such that broken conductor strands may take many years to occur. Aeolian vibration is normally controlled by (1) installing vibration dampers, (2) stringing conductor at relatively low tension levels to maximize its self-damping or (3) using special types of conductors such as SDC, VR or ACSS.

Ice galloping occurs with both single and bundled conductors and requires both high winds and ice-coated conductors. If uncontrolled, it can yield both phase-to-phase flashovers, due to reduced spacing at mid-span, and hardware damage. In certain instances, extended ice galloping of large amplitude has even been blamed for structural failures. A variety of control methods have been proposed and tested to varying degrees. The most common control method is to allow sufficient phase-to-phase spacing to avoid flashovers when galloping does occur.

Subconductor oscillation only occurs for bundled phase conductors when wind speeds exceed a certain critical velocity. If uncontrolled, it can result in fatigue damage to spacers and suspension hardware. It is usually controlled by keeping bundled conductors at a spacing-to-diameter ratio of about 20 or more and by avoiding uniform spacer spacing.

TABLE 5-3
COMPARISON OF WIND-INDUCED CONDUCTOR MOTIONS

	Aeolian Vibration	**Ice Galloping**	**Subconductor Oscillation**
Types of Overhead Lines Affected	All	All	Limited to lines with bundled conductors
Approx. Frequency Range, Hz	3 to 150	0.08 to 3	0.15 to 80
Approx. Range of Vibration Amplitudes (Peak-to-Peak, Expressed in conductor diameters)	0.01 to 1.0	5 to 300	0.5 to 80
Weather Conditions Favoring Conductor Motion			
Wind Character:	Steady	Steady	Steady
Wind Velocity:	1 to 7 m/sec (2 to 15 mph)	7 to 18 m/sec (15 to 40 mph)	4 to 18 m/sec (10 to 40 mph)
Conductor Surface:	Bare or uniformly iced	Asymmetrical ice deposits	Bare, Dry
Damage			
Approx. Time Required for Severe Damage to Develop:	3 months to 20 years	1 to 48 hours	1 month to 8+ years
Direct Causes of Damage:	Conductor fatigue due to cyclic bending	High dynamic loads on structure and hardware	Conductor clashing accelerated wear in hardware

5.3.1 Conductor Blowout

Transmission lines must be designed not only to provide adequate vertical clearance for electrical and safety considerations, but also to allow for adequate horizontal clearance to tall objects and buildings at the edge of the right-of-way (ROW) under high wind conditions. Conductor displacement as a result of high winds is termed conductor blowout. The maximum displacement of the outermost conductors from the center of the ROW under high wind conditions can be one of the most important variables in determining ROW width. Conductor blowout is primarily a function of conductor weight and the wind force perpendicular to the conductor. However, the calculation of conductor blowout should also include the lateral movement of the suspension insulators. The horizontal wind force acting on a conductor may be estimated by the following equation:

$$F_H = 0.00256 \left(\frac{d}{12}\right) (V_w)^2 \qquad (5\text{-}9)$$

where:

d = *conductor diameter, in*

V_w = *wind speed, mph*

F_H = *horizontal wind force, lbs/ft*

Assuming equilibrium of the moments resulting from the conductor weight and wind force about the attachment point, the angle of conductor blowout, θ, can be determined by:

$$\theta = tan^{-1} \frac{F_H}{w_c} \qquad (5\text{-}10)$$

where:

w_c = *conductor weight per unit length, lbs/ft*

The blowout angle of the conductor catenary and any suspension insulator are assumed to be the same as those indicated in Figure 5-6[23]. This figure demonstrates how the clearance to the edge of ROW must include structure deflection.

FIGURE 5-6
CALCULATION OF CONDUCTOR BLOWOUT

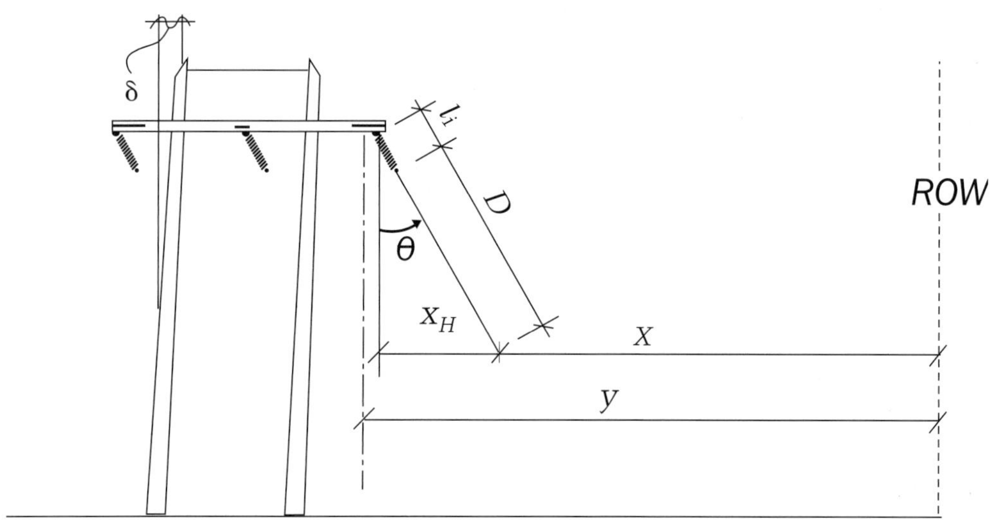

y = total horizontal distance from insulator suspension point to edge of ROW (under still air)

x = required minimum horizontal clearance to edge of ROW under high wind conditions

The total horizontal blowout of the conductor at the midpoint of the span is:

$$X_H = (l_i + D)\ (\sin\theta) + \delta \qquad (5\text{-}11)$$

where:

X_H = *horizontal deflection at midpoint of span, ft*

l_i = *length of insulator string, ft*

D = *midpoint sag of conductor at specified wind and conductor temperature, ft*

δ = *structure deflection, ft*

At lower voltage levels, ROW width may be determined by the need to maintain adequate horizontal clearance. However, at higher voltage levels, the ROW width is typically determined by minimum requirements for environmental effects.

5.3.2 Aeolian Vibration

Aeolian vibration typically occurs in response to low velocity wind (2 mph to 15 mph) blowing steadily perpendicular to the line, as noted in Table 5-3. The resulting conductor vibration, generally less than a conductor diameter in amplitude, is difficult to detect with the bare eye. Vibration amplitudes of less than an inch however, can result in conductor strand fatigue near attachment points for steel, aluminum and copper conductors. To control aeolian vibration, the design everyday tension (EDS) of the conductor is kept as low as possible, and dampers are added when required. While the cost of dampers is typically small, the cost of using low tension levels as a control technique can be expensive in some design situations. Since aeolian vibration leads to conductor failure through the mechanism of strand fatigue, it can be a major factor in line reliability. Lines exhibiting high vibration levels can be very expensive to operate because of the need to inspect and replace conductors and the need to add reinforcing armor rods or other repair hardware. Factors affiliated with aeolian vibration are described in the following sub-sections.

String Mode Vibration of Conductor

An overhead span (catenary) is supported at each end by suspension or dead-end hardware. This hardware is attached to and supported by ceramic or composite insulator strings, which are supported by wood, metal, concrete or steel structures. The aeolian vibration phenomena of this very complex mechanical system is typically represented by ignoring the hardware, insulators and structures, and assuming that the location of the suspension points are fixed. The vibration of the bare conductor may be represented by the vibrating string equation:

$$m\left(\frac{d^2y}{dt^2}\right) - T\left(\frac{d^2y}{dx^2}\right) = 0 \qquad (5\text{-}12)$$

where:

m = *mass per unit length of conductor, lbs/ft*

T = *conductor tension, lbs*

x,y = *horizontal and vertical displacement, respectively, of cable along span, ft*

t = *time, sec*

Using standard mathematical methods, the solutions to this equation consist of traveling waves (transverse disturbances) which propagate along the conductor with a velocity of:

$$V = \sqrt{\frac{T\,(N)}{m\,(kg/m)}} = \sqrt{\frac{T(lbs)\ [32.2\ (ft/sec^2)]}{w\,(lbs/ft)}} \qquad (5\text{-}13)$$

For a typical bare overhead transmission conductor installed to typical tension levels, this mechanical wave velocity is on the order of 400 ft/sec (120 m/sec). For example, if a 795 kcmil – 26/7 ACSR "Drake" conductor is installed to 4725 lbs (15% RBS), the wave velocity is:

$$\sqrt{\frac{(4725)(32.2)}{1.094}} = 373\ ft/sec$$

Combinations of traveling waves on the conductor can lead to standing waves or string modes having certain discrete frequencies. These natural frequencies are related to both the span length, L, and the wave velocity, V:

$$f_n = \left(\frac{n}{2L}\right)V \qquad (5\text{-}14)$$

where:

the modal number, n, is a positive whole number

Using Formula 5-14, when $n = 1$, the conductor vibrates in its lowest frequency mode with no motion (a node) at the ends of the span and maximum motion (an anti-node) at mid-span. The string vibration modes are typically identified by the number of loops per span. This fundamental mode has "one loop per span" and the loop length is equal to the span length.

With a typical wave velocity of 400 ft/sec (120 m/sec), the frequency of the fundamental mode of a 1000-feet (305 m) span is 0.20 Hz. Vibration string modes exist at all multiples of this fundamental mode frequency: 0.4, 0.6, 0.8, 1.0, 1.2 Hz, etc.

The data in Table 5-3 indicate that aeolian vibration is normally observed at frequencies of 3 to 150 Hz, well in excess of these low-order modes. The lack of low frequency vibration is due to the lack of wind vortices forces at these low frequencies, as discussed in the next section.

Oscillatory Wind Forces

The force, which causes aeolian vibration of overhead lines, is created by periodic vortices shedding. This phenomenon has long been recognized as evidenced by Leonardo da Vinci published sketches of vortices shedding behind objects in a flowing fluid. The shedding of vortices from a cylinder has been studied by many investigators. Among the earliest were Strouhal[24] and Benard [24]. Von Karman[24] developed a theoretical explanation. The three primary variables involved are cylinder diameter, the fluid velocity and the kinematic viscosity of the particular fluid. The results of research obtained for various fluids, with different diameter cylinders, can be correlated if velocities are expressed in terms of the Reynolds number, R, a dimensionless variable which is widely used to relate wind phenomena for different fluids, cylinder diameters, and wind velocities. It is defined as:

$$R = \frac{V_W D}{v} \qquad (5\text{-}15)$$

where:

V_W = *wind velocity perpendicular to the conductor, ft/sec*

D = *conductor diameter, ft*

v = *kinematic viscosity of the fluid, ft²/sec*

A dimensionless number that relates velocity, cylinder diameter and vortex shedding is the Strouhal number. The Strouhal equation used to calculate vortices shedding frequency is a function of the wind velocity, V_W, conductor diameter, D, and the Strouhal number. A Strouhal number of 0.185 is suggested by Bate and Callow[24], based on wind tunnel measurements by Chen[25]. Thus the vortex shedding frequency of a fixed conductor is:

$$f_s = 0.185 \left(\frac{V_W}{D}\right) \qquad (5\text{-}16)$$

The Strouhal number varies moderately with the Reynolds numbers between 500 and 20,000.

As is indicated in Table 5-3, aeolian vibration of significant magnitude is usually seen for wind speeds of 2 mph to 15 mph (3 ft/sec to 20 ft/sec). According to the preceding equation, for a 1 inch diameter conductor over this range of wind speeds, periodic vortices shedding forces will occur at frequencies of 7 Hz to 70 Hz. As noted in the previous section, for a 1000 feet span of ACSR "Drake,"

FIGURE 5-7
AEOLIAN VIBRATION BEATS DUE TO MULTIPLE VIBRATION FREQUENCIES

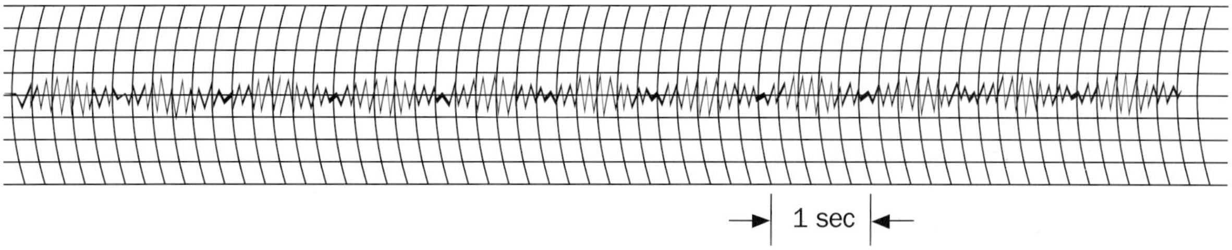

the lower order modes corresponding to less than 10 or 20 loops per span cannot be excited because the Strouhal frequency is already 7 Hz for a 2 mph wind. For a smaller diameter conductor (typical of a shield wire) of 0.35 inch, the vortices shedding frequency is three times as high - 21 Hz to 210 Hz.

A 1000 feet span of 1 inch diameter conductor vibrates at a series of discrete string modes differing by only 0.2 Hz. These modes correspond to wind velocities that differ by only 0.1 ft/sec, and the excitation of adjacent string modes occurs for wind velocities which differ by less than 1%. Thus, it is common to observe aeolian vibration at two or more frequencies. The interaction of multiple frequencies of vibration yields the beat phenomena illustrated in Figure 5-7.

Locking-In Mechanism

The string mode equation indicates that a span of bare overhead conductor can only vibrate at certain discrete frequencies corresponding to an integer number of loops per span. The Strouhal equation indicates that vortices shedding forces can occur for each string mode only at a particular wind velocity. Because natural winds vary rapidly in speed, the energy input to any span mode would be very small indeed if this were strictly true.

If the conductor is allowed to vibrate, the frequency of vortices shedding can be determined by the conductor vibration frequency over a range of wind speeds. This locking-in mechanism allows particular span vibration modes to collect energy from a range of wind speeds and makes the vibration problem of overhead lines much more severe. For example, a 1000 feet span of ACSR "Grosbeak" (0.990 inch OD) exposed to wind speeds ranging from approximately 4 ft/sec to 5 ft/sec could supply energy to the string mode at 4.46 Hz.

Self-Damping of Bare Stranded Conductor

The amplitude of aeolian vibration is essentially determined by the balance of energy input from the wind and countering effects from conductor self-damping, mechanical dampers added to the span, or a combination of the two. Conductor self-damping may be measured in a laboratory span according to the standing wave ratio method described in [26]. Measurements of self-damping are typically of the form presented in Figure 5-8[26]. Power dissipation is usually plotted as a function of frequency and nodal amplitude divided by the conductor diameter.

FIGURE 5-8
SELF-DAMPING OF CONCENTRIC-LAY STRANDED CONDUCTOR

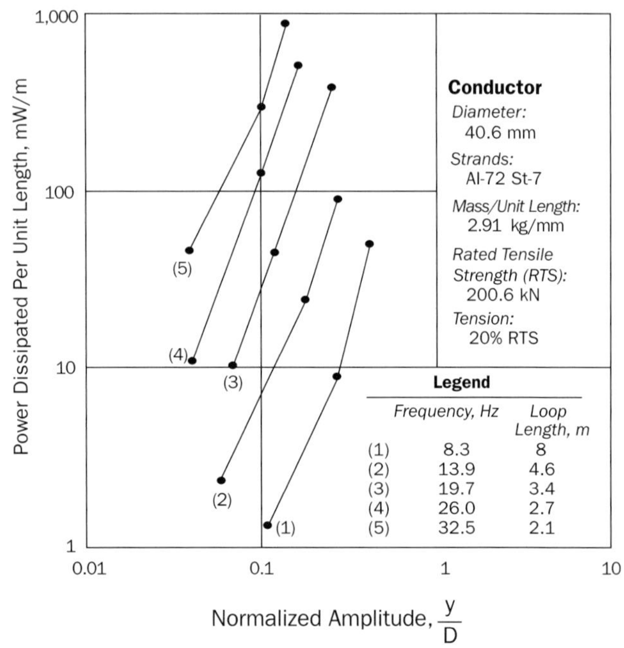

Increasing conductor tension causes a corresponding decrease in the self-dampening property . If the use of higher conductor tensions is economically attractive for overall line design purposes, dampers may be added to compensate for this loss of self-damping ability.

Increasing conductor tension also causes an increase in aeolian vibration-induced bending strain, which can result in fatigue damage to the conductor strands. Figure 5-9[27], developed by Fargo Manufacturing Co., illustrates the relationship between conductor tensions and bending strain at various wind velocities.

Conductor Strand Fatigue

Strand failures due to aeolian vibration fatigue result from millions of cycles of vibration. Fatigue failures usually occur in the vicinity of support points where the combination of static and dynamic strains is greatest. As shown in Figure 5-10[28], fatigue of conductor strands typically requires years to develop, although sufficiently poor line design can yield fatigue failures in a matter of weeks.

FIGURE 5-9
AEOLIAN VIBRATION INCREASE WITH TENSION

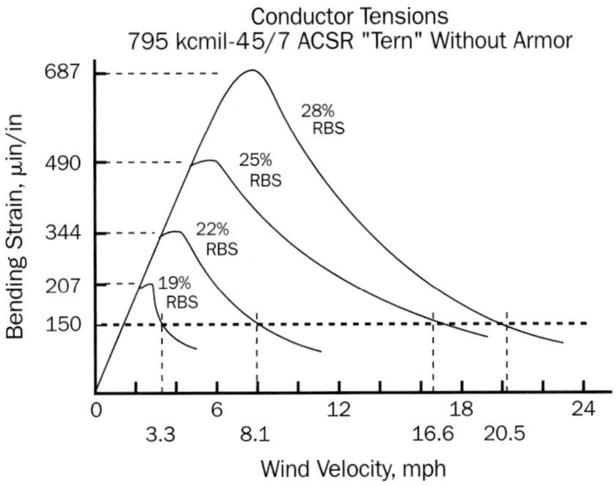

Courtesy Fargo Mfg. Co.

FIGURE 5-10
FATIGUE OF CONDUCTOR vs. TIME

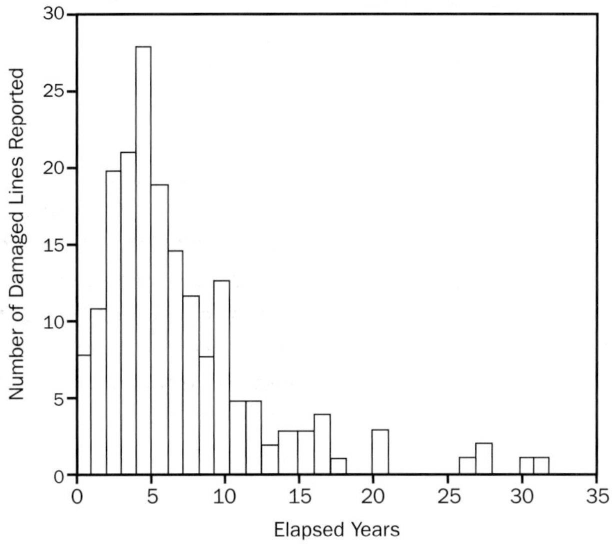

Courtesy Alcoa Conductor Products Company.

Figure 5-11[24] shows a typical plot of fatigue test data for multi-layer ACSR conductors. The stress level which yields no fatigue failures is referred to as the endurance limit of the conductor. Endurance limits for ACSR and all-aluminum conductors are similar, while the endurance limit of 6201 alloy aluminum conductor is somewhat lower.

Field Measurement of Aeolian Vibration

Many different types of vibration recorders have been used to monitor conductor vibration. One of the best known is the Ontario Hydro (OH) recorder. The body of this recorder is mounted below a suspension clamp with an arm extending into the span to a contact point about 3.5 inches from the clamp mouth. As the conductor vibrates, the arm moves; this movement is recorded every 15 minutes by a stylus on 16 mm movie film for a period of 1 second. The stylus marks on the film are then translated into amplitude and cycle data. The Alcoa Scolar, an updated version of the OH recorder, collects data electronically and has an LED display that can be viewed from the ground. These recorders are no longer commercially available, but many are available from private testing groups.

At this time, there are only two recorders that are commercially available. They are the PAVICA, manufactured by ROCTECH, and the VIBREC 400, manufactured by SEFAG AG.

Aeolian Vibration in Bundled Conductors

Aeolian vibration is generally less severe for bundled conductors than for single conductors. End-span damping used with single conductors is not effective with bundled conductors. Vibration control is usually obtained by spacer dampers, rather than Stockbridge or other conventional dampers.

Reducing Vibration with Dampers

The most widely used type of vibration damper for overhead lines is the Stockbridge damper. Mechanical damping is obtained through the flexing of a short length of un-tensioned, end-weighted, stranded cable that is attached to a clamp as shown in Figure 5-12. To operate effectively, the Stockbridge damper must move far enough and fast enough to flex the stranded cable. Proper placement of the damper is necessary. It must be near the node, which always exists at the attachment point, yet not so far into the span that the damper finds itself at a node at which it would be ineffective. The two resonant vibration modes of a Stockbridge damper are illustrated in Figure 5-12. Non-symmetrical Stockbridge-type dampers are available which have more resonant peaks and are more effective over a wide range of vibration frequencies.

Repairing Fatigued Conductors

Conductors with fatigued strands can be repaired in a variety of ways. These include installing preformed armor rods, reducing conductor tension, or installing dampers. The objective of each approach is to reduce the amplitude of vibration to a satisfactory level.

"Safe" Installed Tension Levels

Avoiding strand fatigue failures from aeolian vibration is one of the most important constraints to be considered when calculating sag-tension limits for bare overhead conductors. The NESC[31] suggests unloaded (no ice, no wind) initial tension limits not exceeding 35% of the rated breaking strength (RBS) of the conductor at 60°F, and unloaded final tension limits of 25% RBS at the same temperature. The implication in the code is that dampers may be needed in certain regions at these installed tension levels.

Many attempts have been made to develop "rules of thumb" to limit conductor tensions to safe levels. For example, it is common for utility line designers to use tension limits of the order of 20% initial and 15% final unloaded at 60°F. Reference [24] suggests the following limits:

- 32% RBS initial unloaded at minimum temperature during stringing
- 24% RBS final unloaded, after 10 year creep at 60°F (15°C)

FIGURE 5-11
FATIGUE TESTS OF MULTI-LAYER ACSR[29][30]

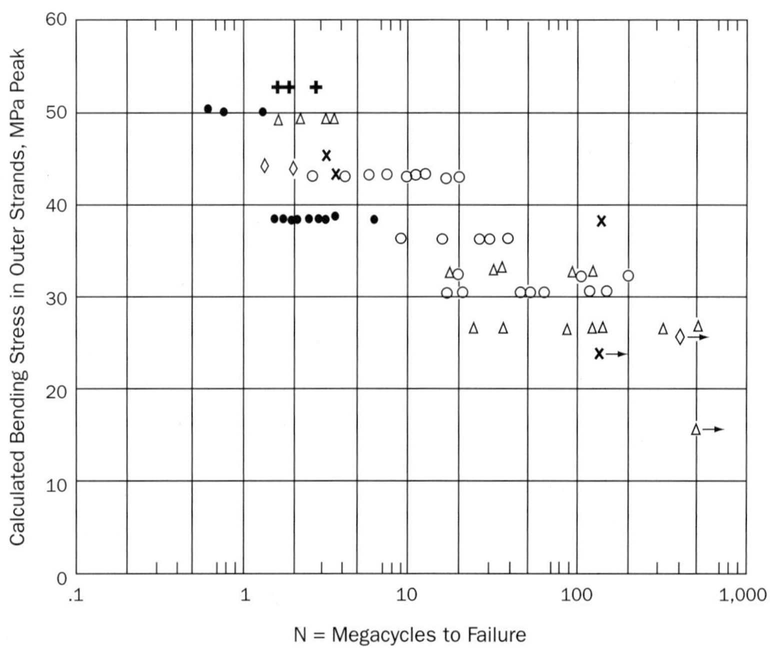

	Conductor	Clamp	Reference
Δ	795 kcmil ACSR 54/7	BM	[28]
O	397kcmil ACSR 30/7	BM	[28]
+	795 kcmil ACSR 45/7	Susp.	[29]
x	954 kcmil ACSR 54/7	Susp.	[28]
•	397.5 kcmil ACSR 26/7	Susp.	[30]
◊	397.5 kcmil ACSR 30/7	Susp.	[28]

Note:

→ Indicates that no failures occurred. 68 tests represented

BM = Aluminum bell-mouthed large radius suspension clamp

Susp. = Commercially available short radius suspension clamp

Courtesy Alcoa Conductor Products Company.

• 27% RBS final unloaded at minimum temperature for the area

• 35% RBS under maximum loaded, worst conditions

With these limits, it is assumed that dampers are added to the line in exposed areas, and it is noted that the 27% RBS limit should be reduced in areas which experience prolonged periods of low temperature. Field data [24] indicates that lines with ACSR conductor, having an everyday unloaded tension of 20% RBS or less, almost never exhibit fatigue failures.

More recent studies have raised serious doubts about both the adequacy of using all-purpose unloaded conductor tension limits in all designs and about the use of %RBS as the design parameter. It is stated in [32] that "There is no direct relationship between conductor rated breaking strength [RBS] and self-damping. Therefore any recommendation based on %RBS is valid only for conductors having the same tensile breaking stress."

As an example illustrating the precaution about the use of %RBS as a design parameter, consider two identical lines, one constructed with AAC and the other with AAAC. If sagged to the same final or initial unloaded tension limit, expressed as %RBS, then the actual tension level in the AAAC conductor is twice that in the AAC. However, the self-damping properties of the two conductors are essentially the same, making the higher-tensioned AAAC more vulnerable to aeolian-induced fatigue damage.

Using fixed tension parameters for all designs does not consider the following important fatigue endurance issues:

• The effect of persistency of wind speed and direction relative to the line section

• Terrain covers of trees, shrubs, buildings and the terrain itself, which have a marked effect on the degree of turbulence

• The dangerous conditions existing for short-span construction with high tensions at very low-temperature operation. However, these conditions may only exist for a month or two during winter

• The different effects on mass and strength of conductor stranding and the amount of steel in ACSR

In light of the above issues, it has been suggested that a tension-over-weight per unit length criterion, H/w, be used instead of %RBS, and that terrain and span length be considered when choosing safe conductor tensions. Recommended maximum values of H/w range from 3000 to 5000 feet. As can be seen in Table 5-4, for ACSR "Drake" in a 1000 feet span, this would correspond to an unloaded initial tension of between 10% and 15% RBS. If dampers are used, then a higher H/w level would be acceptable.

H/w limits for conventional ACSR may not be applicable to special conductors such as ACSS, GTACSR and SDC since these conductors have better self-damping characteristics than standard ACSR. These limits may also not apply to VR™ conductors that have non-circular cross-sections.

Given the complexity of the calculation process for estimating aeolian vibration levels, and given the availability of well documented calculation methods for the prediction of aeolian vibration levels under known terrain, span lengths, conductor type, tension, etc., it is not reasonable to use rules of thumb for setting tension limits when designing new lines in areas where there is no experience regarding aeolian vibration damage. The process of selecting suitable tension limits in conjunction with damper suppliers is the most sensible approach.

FIGURE 5-12
RESONANT MODES OF A STOCKBRIDGE DAMPER

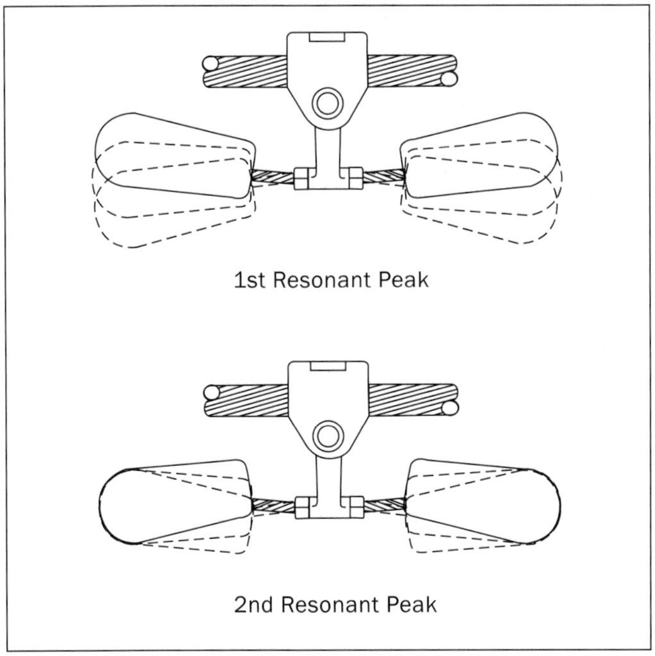

Courtesy Fargo Mfg. Co.

TABLE 5-4
H/W VALUES FOR ACSR "DRAKE" IN A 1000 FEET SPAN

Initial unloaded conductor tension @ 60°F %RBS	Initial unloaded H/w @ 0°F (average temperature for coldest month) ft	Tension @ 0°F with 0.5" ice and 4 lbs/ft² wind load %RBS	Tension @ max ice and wind load lbs	Final sag @ 212°F ft
10	3040	22.6	7,115	49.1
15	4860	31.7	9,980	37.0
20	6825	38.4	12,090	31.5
25	8730	43.5	13,710	28.3

5.3.3 Ice Galloping

Field observations of ice galloping were first reported by Davison of Ontario Hydro in 1930[33]. He described an occurrence of large-scale, ponderous motions of overhead lines which had a non-uniform ice coating.

The mechanism of ice galloping was suggested by Den-Hartog of MIT in 1932[34]. Since then, a considerable amount of time and money has been expended on the problem of analyzing ice galloping. It is probably fair to say that the sheer complexity of the combined non-linear aerodynamic and mechanical equations has defied general analysis. Certain investigators have developed relatively simple mathematical models that serve to guide the development of control methods. Probably the most useful model is that proposed by Den-Hartog, to account for torsional movement.

The first effective means of control was suggested by Davison of Ontario Hydro in 1939[35]. He described the concept of providing sufficient phase-to-phase spacing when new lines are designed so that flashovers are prevented during galloping. Calculation of minimum required phase-to-phase spacing is done in terms of "galloping ellipses" as shown in Figure 5-13[24].

A second effective means of galloping control resulted from the studies performed at MIT under Richardson in the early 1960s[36]. From these studies, a device called the wind damper, or drag damper, was developed and refined over the intervening years. A third effective means of control, the detuning pendulum, was developed at Ontario Hydro in the 1970s and became the object of an extensive field investigation by EPRI in the early 1980s. The effectiveness of this device is supported by a thorough statistical analysis of the field data. Additional methods, which have been employed in recent years, are the use of spoilers or motion resistant conductors, such as VR.™

FIGURE 5-13
ICE GALLOPING ELLIPSE FOR PHASE SPACING

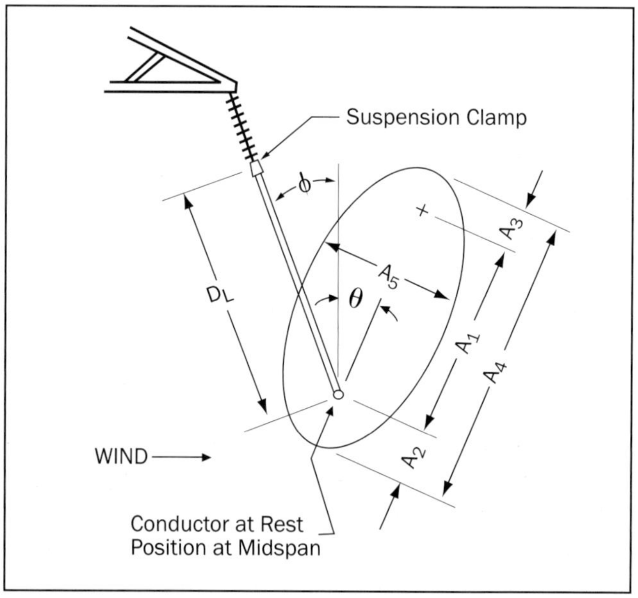

$A_1 = D_L$, the loaded sag

$A_2 = A_1/4$

$A_3 = 0.3$ m (1 foot)

$A_5 = 0.4\,A_4$

$\theta = \phi/2$

$\phi = \tan^{-1}\left(\frac{F_H}{W_C}\right)$

W_C = conductor weight, lbs/ft

F_H = horizontal wind force, lbs/ft

Courtesy of Alcoa Conductor Products Company.

The Causes of Ice Galloping

Ice galloping requires both a buildup of ice on the conductor and wind speeds between 15 mph and 40 mph normal to the conductor. The ice layer on the conductor observed during galloping is typically crescent-shaped and between 0.125 inches and 0.250 inches thick. Even if both conditions are present, ice galloping may not occur.

Den-Hartog suggested a relatively simple explanation that certainly does not explain all forms of ice galloping, but at least gives some insight into the way in which mechanical and aerodynamic forces interact in this complex phenomenon. The iced conductor is assumed to move only vertically; it does not rotate, nor does it move horizontally. As the conductor moves up and down, the lift force acts in the direction of motion, and ice galloping can occur. Note that if there is no lift force, or if the slope of the lift curve with angle of attack is positive, there is no tendency to oscillate. That is, the iced conductor is stable and will not gallop.

Mathematically, Den-Hartog expressed the possibility of ice galloping by the following equation, commonly referred to as the Den-Hartog Criteria for galloping.

$$D(\alpha_o) - \left[\frac{dL}{d\alpha}\right]_{\alpha_o} < 0 \qquad (5\text{-}17)$$

If the lift curve slope $(dL/d\alpha)$ is sufficiently negative for the angle of attack, α_o, then the equation is negative and ice galloping is possible. When the iced conductor is in a region of "positive lift slope," or the negative lift slope is not steep enough, ice galloping does not occur.

A different theory of ice galloping was developed at Ontario Hydro. The basic concept is that galloping results when the iced conductor's torsional span frequency equals a first, second or third multiple of the vertical span frequency. If coincidence exists between the vertical and torsional span frequencies, then a torsional instability could feed energy into the vertical mode.

A number of attempts have been made to look at the iced conductor span as a lumped parameter subject to horizontal, vertical and torsional motions. All such attempts include three-dimensional aerodynamic forces (lift, drag and moment). The results are of limited use because the complex non-linear system subject to non-linear wind forces is modeled by linearized equations. Thus, when any significant galloping motions occur, the linear assumptions break down. The failure of such complex, three-dimensional models is the primary reason for resorting to field test data as a basis for evaluating control devices.

Ice galloping occurs infrequently and is dependent upon geographic factors. The greatest frequency of ice galloping occurs in the north-central part of the U.S. where utilities experience ice galloping annually on one or more lines[24]. According to the same reference, "Utilities in the Atlantic Seaboard States experience galloping rarely or never, except in New York and New Jersey, where galloping may occur every 2 to 10 years."

Ice galloping motions occur at one of the fundamental vertical string modes of the catenary, usually the first, second or third mode. The amplitude of motion associated with the single loop mode is of primary concern because the maximum amplitude of such motion can approach or even exceed the sag of the catenary. As the span length increases, the degree of rotation under ice load increases and the probability of lower order galloping motion increases. This can be seen in Figure 5-14[24] which represents data calculated from the formulas for "galloping ellipses" as proposed by Davison for single loop galloping[33] and Toye for double loop galloping[37]. Thus, short spans of less than 600 feet are more prone to single loop galloping than are long spans of the same conductor.

FIGURE 5-14
EFFECT OF SPAN LENGTH ON ICE GALLOPING MAGNITUDE

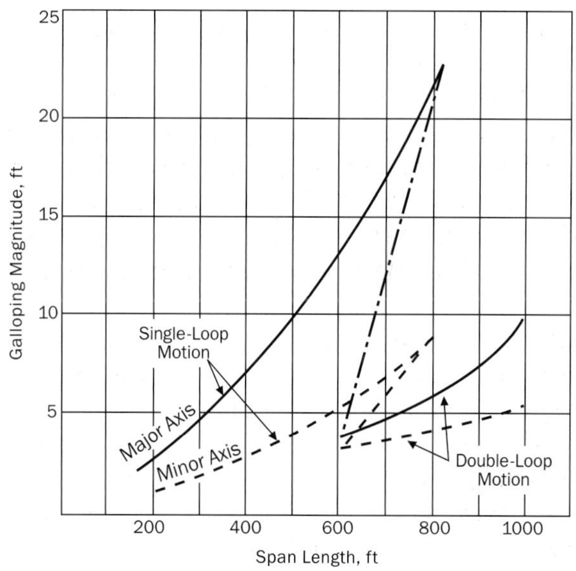

Figure prepared by Power Technologies, Inc. under EPRI project 260. Reprinted with Permission.

Field Observations of Galloping

There are several sources of field data, some formal and semi-quantitative, others informal and qualitative. This data is available for "untreated" lines and for lines treated with various types of control devices. The most comprehensive analysis of field data for untreated spans has been published by Rawlins[38] who studied galloping reports from numerous sources with the purpose of determining parameters that could be used to identify spans prone to galloping. His study concluded that the best parameters

FIGURE 5-15
OBSERVED AMPLITUDES OF GALLOPING

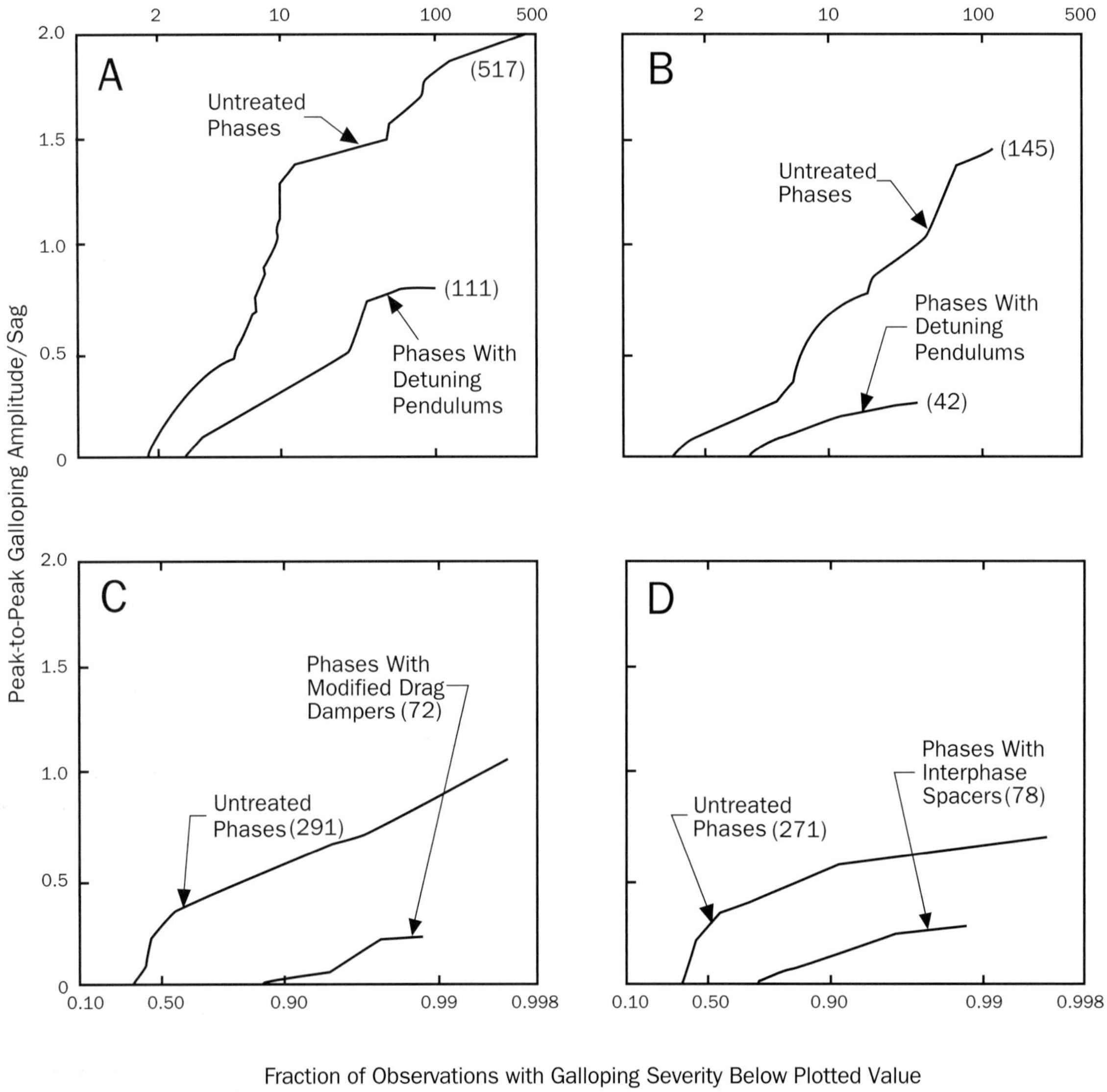

A For detuning pendulums on single conductors from 33 observations.
B For detuning pendulums on bundled conductors from 21 observations.
C For modified drag dampers on single conductors from 8 observations.
D For interphase spacers on single conductors from 8 observations.

Reprinted with permission of the Canadian Electrical Association.

to determine if lines will gallop, and the amplitude of galloping, are T/w (the conductor tension over conductor weight per unit length) and M′ (a catenary's parameter related to span type (suspension or strain) and the unloaded conductor elastic strain).

Controlling or Preventing Ice Galloping

The choice of control measures is certainly diverse, and the best solution for one line may not be best for another. Devices that can be used to reduce the amplitude of galloping, or to prevent flashovers should galloping occur, fall into two categories: mechanical and aerodynamic. Mechanical devices tend to limit the galloping amplitudes by altering the mechanical response of the system to galloping motions. Aerodynamic devices control galloping by changing the aerodynamic forces acting on the iced conductor, thereby preventing the input of wind energy to the galloping instability.

Figure 5-15 illustrates amplitudes of galloping taken from field data observations[38] on treated and untreated conductors. The amplitudes, normalized by dividing by sag, are presented in the form of an extreme values distribution plot. The number of individual spans observed is indicated in brackets for each curve. Curve A plots 33 observations for detuning pendulums for single conductors; Curve B plots 21 observations for detuning pendulums for bundled conductors; Curve C plots 8 observations for modified drag dampers on single conductors; and Curve D plots 8 observations for inter-phase spacers on single conductors.

The following is a brief discussion of the two mechanical and one aerodynamic devices and their effectiveness in controlling ice galloping as indicated in the Figure 5-15. A second aerodynamic device of interest, the air flow spoiler, is also discussed.

(1) Mechanical Devices

In-Span Insulators: As indicated in Curve D, the in-span insulators are effective in reducing galloping. These composite insulators, placed between vertically arranged phase conductors, also help prevent flashovers when galloping does occur. However, this solution may not be appropriate for lines with older structures in questionable mechanical condition. The use of in-span insulators does nothing to reduce the dynamic forces on structures during galloping and may, in fact, serve to increase such forces by coordinating the motion of two phases.

Detuning Pendulums: The detuning pendulum proposed by Havard of Ontario Hydro[39] is the most thoroughly field-tested galloping control device available, and as indicated by Curves A and B, is very effective on both single and bundled conductors. Pendulums are an excellent choice for long spans (600 feet or greater) of either single or bundled conductors, if there is sufficient clearance between the phases. However, the clearance between phases must be adequate to accommodate the increased sag and reduced inter-phase clearance

FIGURE 5-16

GALLOPING DATA FROM SINGLE CONDUCTORS[40]

Normalized amplitude versus percent of observations. Based on 518 data pairs from 28 observations.

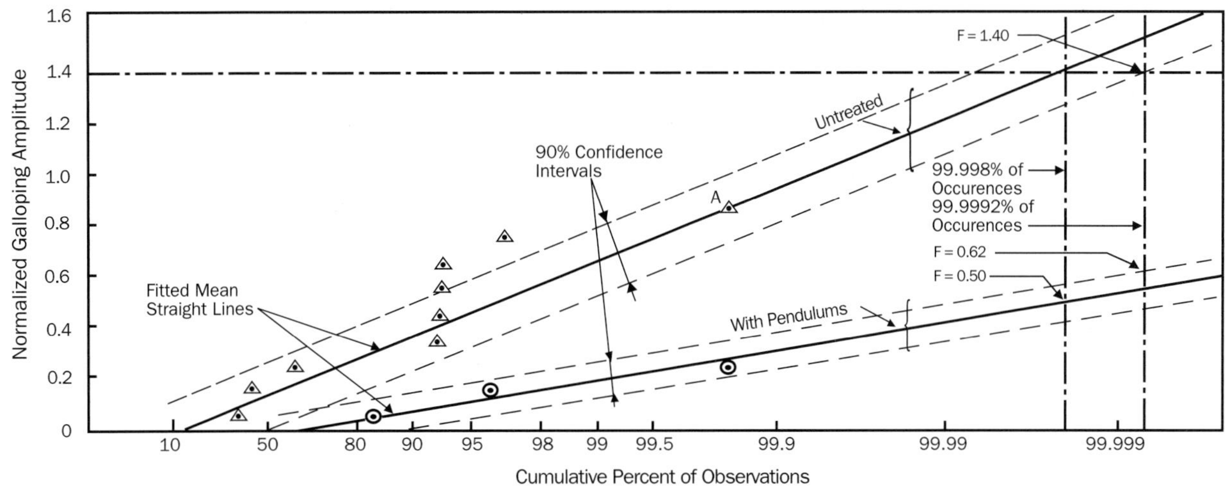

due to the presence of the detuning pendulums. Most field data which support the effectiveness of the detuning pendulums for spans in the range of 300 feet to 1200 feet. The use of pendulums will increase sags and reduce effective interphase spacing.

(2) Aerodynamic Devices

Wind Drag Dampers: The wind drag damper, shown in Figure 5-17[36], is a result of the MIT study sponsored by EEI in the early 1960s. As Curve C indicates, it is an excellent means of controlling galloping, and is normally used in spans greater than 600 feet where there is sufficient phase-to-phase clearance. These dampers are generally installed at the one-third points of the span. The device itself is approximately 10 feet long, is clamped to the conductor at two points, and hangs about 12 to 18 inches beneath the conductor. Field data regarding the drag damper indicates that it is clearly effective in reducing the amplitude of galloping incidents.

Air Flow Spoilers[41]: An air flow spoiler, shown in Figure 5-18, consists of one or more PVC tubes helically wrapped around the conductor per manufacturers recommendations. The PVC tube diameter is approximately equal to the diameter of the conductor on which it is installed. Assuming that 20% of the conductor length per span is "treated" with the spoilers, the maximum structure loads due to wind and ice will be about 10% higher.

The mathematical analysis of this control device is very limited, but the general theory of operation, that it prevents the occurrence of Den-Hartog instabilities, is very plausible. It presents a constantly changing cross-section along the span as shown in Figure 5-18. Field data have been accumulated at a large number of test sites, and although the analysis of this data has not been as thorough as that for the detuning pendulum, there is no doubt that the device is very effective in preventing galloping in many types of lines. The major limitation in the use of air flow spoilers is the problem of PVC degradation due to corona at line voltages of 115 kV and above.

FIGURE 5-17
AERODYNAMIC DRAG DAMPER
(as suggested by A.S. Richardson, Jr.)

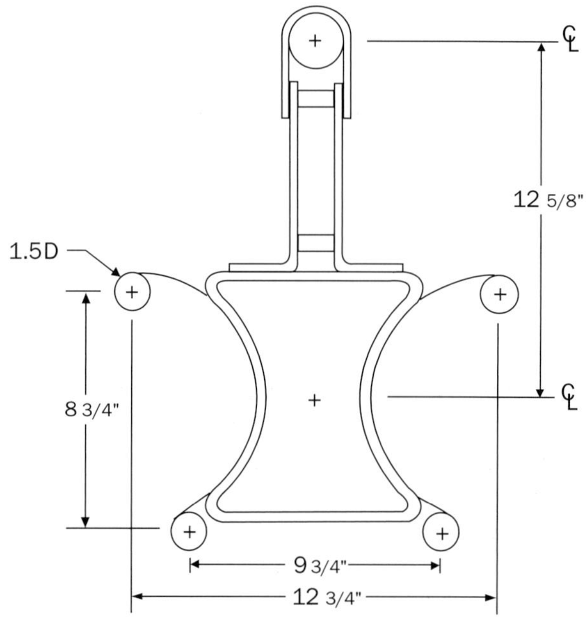

FIGURE 5-18
PREFORMED LINE PRODUCTS SPOILER

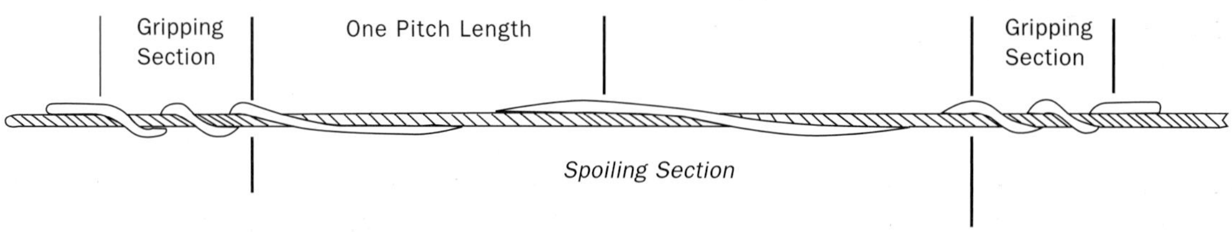

Courtesy of Preformed Line Products, Inc.

FIGURE 5-19
CROSS-SECTIONS OF SPOILER

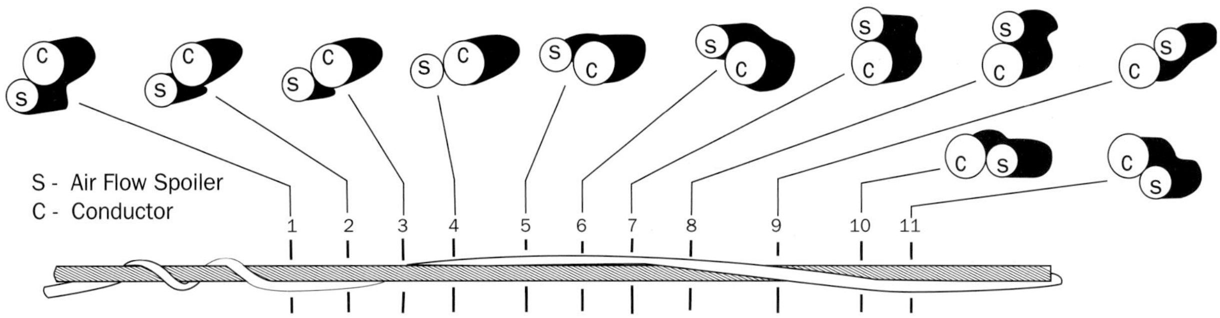

A. Cross-Sections of Artificially Iced Conductor Model With Air Flow Spoiler

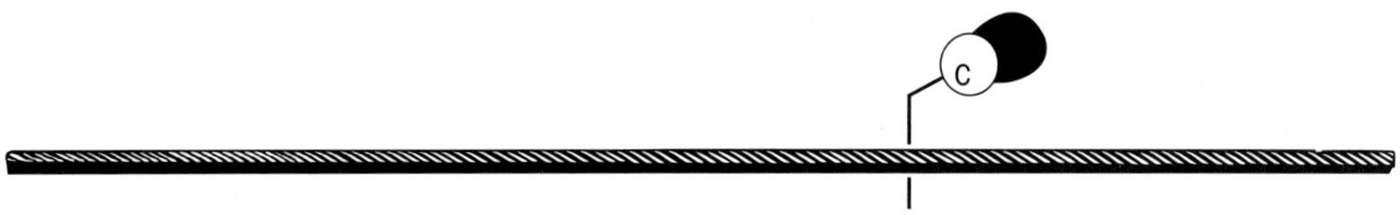

B. Cross-Sections of Artificially Iced Conductor Model Without Air Flow Spoiler

Courtesy of Preformed Line Products, Inc.

Line Design to Avoid Galloping Flashovers

No ice galloping control device is 100% effective under all conceivable weather conditions. Therefore, it is more desirable to make changes in standard line design practices, if possible, to prevent ice galloping rather than trying to control it after the line is built.

From Rawlins' work, it is apparent that the use of higher conductor unloaded tensions and more rigid insulating attachments may reduce the amplitude of potential galloping. Unfortunately, both of these changes will increase static loading on any angle structures, thereby requiring costlier structures.

As an alternative to the use of higher conductor tensions, the use of vibration resistant (VR™) conductor (see Chapter 1) may be considered. The use of VR™ conductor would increase the wind and ice loads about 10% above those of a standard conductor having the same total cross-sectional area. The VR™ conductor is as effective in preventing galloping as a standard conductor with an air flow spoiler, and unlike air flow spoilers, there are no problems with deterioration of PVC or with spoiler droop under high temperature conductor operation.

Operation Procedures to Prevent Galloping

If, under icing conditions, the conductors were kept warm enough to prevent the formation of ice on the conductor, then galloping could not ensue. Alternatively, if galloping occurs, line loadings could be increased to melt the already-formed ice. Both approaches present certain hazards and difficulties. Besides the dubious advisability of increasing the power flow on a line which is likely to trip out, the high wind speeds typical of ice galloping make heating the conductor much above ambient temperature nearly impossible. Operational solutions (other than taking the line out of service until the ice galloping stops) do not appear practical in most situations.

5.3.4 Subconductor Oscillation

Subconductor oscillation[24] only occurs for bundled phase conductors with wind speeds typically above 10 mph, and usually in areas unobstructed by trees and other obstacles. If uncontrolled, these motions can result in fatigue damage to spacers and suspension hardware, but rarely cause flashover of lines. Subconductor oscillation takes a number of different forms. It may be controlled by providing sufficient spacing between the subconductors in a bundle, by tilting the bundle or by avoiding uniform sub-span lengths.

Appearance of Oscillation

Figure 5-20[24] illustrates some of the most common types of subconductor oscillation. The sub-span mode of oscillation is unique and is greatly influenced by the type and spacing of the spacers. The other three types of oscillation shown in the figure are affected very little by the choice or location of the spacers.

FIGURE 5-20
TYPICAL FORMS OF SUBCONDUCTOR OSCILLATION

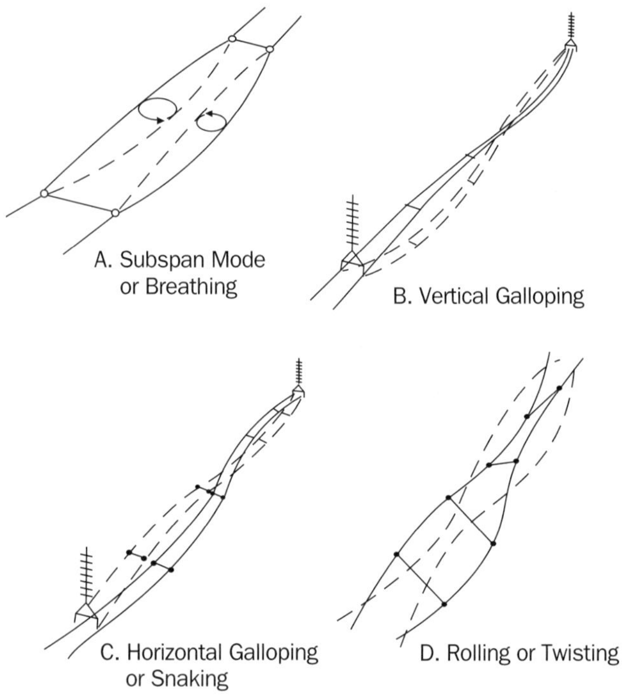

The Causes of Subconductor Oscillation
The causes of subconductor oscillation involve wind forces on the downwind conductor as it moves within the wake of the upwind conductor of the bundle. Figure 5-21 shows the forces acting on a downwind conductor with its static position of rest indicated by R_0. Note that for the hypothetical downwind conductor orbit shown in this figure, the wind force acts in the direction of conductor motion as it orbits, making this orbit unstable. Clearly, if the proposed orbit were moved up or down toward the edges of the wake, the instability would not occur.

Controlling Subconductor Oscillation
The primary means of limiting subconductor oscillation involves keeping the ratio of the conductor spacing-to-conductor diameter to a minimum value of 16 for a 2- or 3-conductor bundle, and 21 for a 4-conductor bundle.

Subconductor bundles can be stabilized if tilted on the order of 10° to 20°. Actual tilt during the occurrence of high winds due to subconductor oscillation is not the same as simple tilt of the bundle at rest. In any event, this method has been shown to be effective in field testing. The most effective form of bundle tilting may well be as drastic as making a 2-conductor bundle vertical rather than horizontal or rotating a 4-conductor bundle 45° to make a star configuration. Subspan oscillation (see Figure 5-20) can also be controlled by the combined use of relatively short subspans[42] and variations in sub-span lengths[43].

5.4 ECONOMIC DESIGN OF OVERHEAD LINES

The goal of economic line design is to design a transmission line which meets or exceeds all constraints for the minimum total cost over the life of the line. The main cost elements are:

- Capital construction cost: includes the purchase and clearing of right-of-way; the purchase and installation of supporting structures, phase and shield conductors, and insulators; and transportation cost. Also included are regulatory and engineering costs. Regulatory cost includes legal, environmental, and publicity issues, and can be enormous in certain developed areas. (These costs can often be controlled or minimized through consideration of esthetic aspects of lines from the very beginning of the design process as discussed in Section 5.5.) Engineering cost includes line design engineering, structure analysis, and preparation of finished drawings.

- Electrical losses cost: primarily due to ohmic (I^2R) losses, but can also include corona losses.

- Maintenance cost: cost of routine repairs and the cost associated with catastrophic failures.

Capital construction, costs are incurred at the time the line is built, but are expensed over the life of the line, just as the costs of electrical losses and maintenance are expensed. Total line costs are usually expressed as the present worth of a series of annual payments. As an example, the capital construction cost of a transmission line is capitalized by the owner and paid as a series of level annual payments, the present worth (PWCC) of which is given by the Equation 5-18:

$$PWCC = \Sigma\left[\left(1+\frac{i}{100}\right)^{-n}\left(C_I\,\frac{F_L}{100}\right)\right] \quad (5\text{-}18)$$

where:

C_I = *initial capital investment in construction*
F_L = *line fixed-charge rate, %*

The present worth *(PW)* of line losses is calculated in a similar manner:

$$PW_{Losses} = \sum\left[\left(1+\frac{i}{100}\right)^{-n}(Losses_n)\right] \qquad (5\text{-}19)$$

where:

$Losses_n$ = *cost of electrical losses in the* n^{th} *year*

i = *interest rate, %*

Capital construction cost, ohmic losses, and similarly calculated maintenance cost are summed to yield the total present worth of costs over the life of the line.

FIGURE 5-21
AERODYNAMIC FORCE FIELD ACTING ON LEEWARD CONDUCTOR

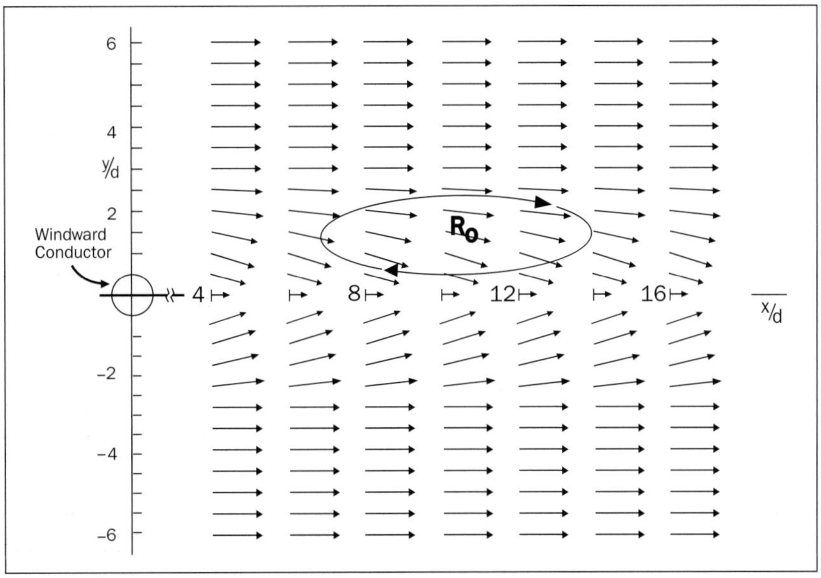

Figure 5 - 21 A. Aerodynamic static force field acting upon leeward conductor. (R_o indicates rest position).

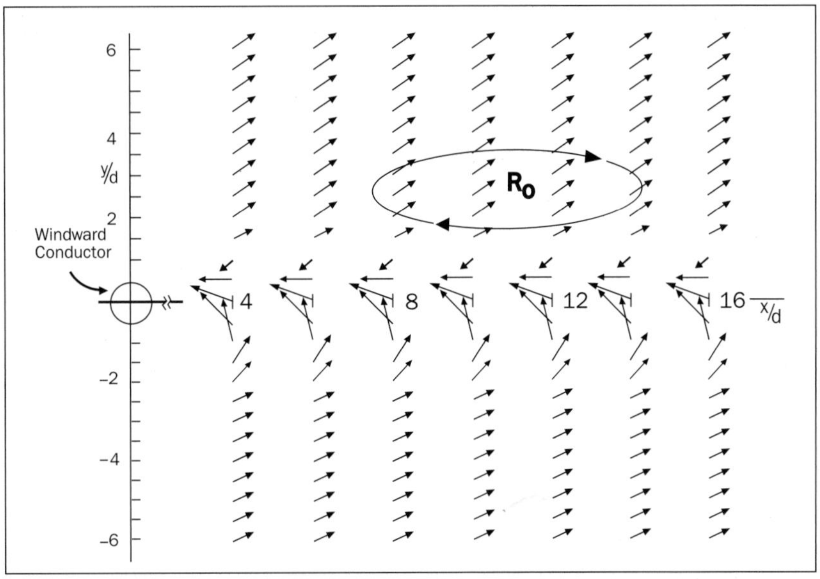

Figure 5-21 B. Variable part of aerodynamic static force field acting upon leeward conductor. (R_o indicates rest position). (Note: scale length of vectors is twice that of Figure 5-21 A).

5.4.1 Present Worth (PW) of Electrical Losses

The flow of electrical current on the phase conductors of a transmission line results in a loss of energy in the form of conductor heating. This ohmic power loss per unit length, P_L, is simply equal to the conductor's resistance per unit length at a specified temperature, T_{cdr}, times the square of the current:

$$P_L = R_{cdr} I^2 \qquad (5\text{-}20)$$

where:

I = *rms current per phase, amperes*

R_{cdr} = *resistance of the conductor at the temperature* T_{cdr}*,* Ω/ft

The power loss per mile of a 795 kcmil-26/7 ACSR "Drake" conductor at 25°C with a resistance of 0.1178 Ω/mile and carrying a constant 60 Hz current of 500 amps is:

$$P_L = (0.1178)\,(500)^2 = 29{,}450 \quad W/mile$$

For a single circuit, balanced 3-phase line, the power loss is three times as large, just over 88 kW/mile. Obviously, the use of a 1590 kcmil conductor would cut these losses in half and a 397.5 kcmil conductor would double them.

The rms powerflow, P, of a 3-phase overhead power line is:

$$P = \sqrt{3}\ I_p\ V_{pp} \qquad (5\text{-}21)$$

where:

P = *rms power, MVA*

I_p = *current per phase, kA*

V_{pp} = *phase-to-phase voltage, kV*

In actual operating transmission lines, the voltage between phases remains essentially constant while the current per phase (and thus the power flow) varies with the season of the year, weather and time of day. In designing new lines, the projected electrical losses for each year are calculated from estimates of annual peak and average power in MVA. "Peak annual load" is the maximum load, which occurs during normal operation, and "average annual load" is the average current, which occurs over the year during normal operation. The ratio of the latter to the former is the annual line load factor, *LoadF*. In calculating annual losses, a loss factor, *LossF*, is used instead of a load factor. The loss factor is related to the load factor as shown in Equation 5-22:

$$LossF = 0.15\ LoadF + 0.85\ (\ LoadF\ )^2 \qquad (5\text{-}22)$$

The annual electrical losses are estimated by multiplying the peak annual losses by the loss factor for each year during the life of the line.

Once the total electrical losses for each year are known, the cost of these losses for each year must be found. As described in [44], there are two costs associated with the energy lost in the line: annual demand charges based on the cost of added generation to supply the losses, and annual energy charges based on the cost of energy to generate the losses.

$$ADC_n = \left(\frac{C_{kW}\,ESC_n}{1000}\right)\left(\frac{F_g}{100}\right) \times \left[1 + \left(\frac{RES}{100}\right) I_L^{\,2} \left(\frac{R}{N_c}\right) N_{ckt}\,N_p\right] \qquad (5\text{-}23)$$

where:

ADC_n = *annual demand charge of year n, \$*

C_{kW} = *installed generation cost, \$/kW*

ESC_n = *escalation cost factor for year n*

F_g = *generation fixed charge rate, %*

I_L = *demand phase current, A/circuit*

R = *single conductor resistance, Ω/mile*

N_c = *number of conductors per phase*

N_{ckt} = *number of circuits*

N_p = *number of phases per circuit*

RES = *required generation reserve, %*

$$AEC_n = 8760\left(\frac{C_{MWh}\, ESC_n}{10^6}\right) \times \left(\frac{L_f}{100}\right) I_L^{\,2} \left(\frac{R}{N_c}\right) N_{ckt}\, N_p \quad (5\text{-}24)$$

where:

AEC_n = *annual energy charge for year n, \$*

C_{MWh} = *cost of generating energy, \$/MWh*

L_f = *loss factor (ratio of average to peak losses)*

5.4.2 Value of Increased Power Flow Capacity

The present value of electrical losses over the life of a power transmission line can be calculated, given an accurate estimate of the annual power flow of the line, as shown in Section 5.4.1. Unfortunately, the present value of an increase in power flow capacity limit is less well defined. Nonetheless, increased power flow capacity, particularly thermal capacity, has real value and makes the use of high temperature, low sag (HTLS) conductors an important consideration in designing and building new transmission lines.

5.4.3 Capital Construction Costs

Capital construction costs include the cost of conductors, supporting structures, insulators, hardware, construction labor and right-of-way. Though the proportion of such costs varies, the costs of conductors, supporting structures and right-of-way predominate in most line designs, each amounting to 20% to 30% of the total cost.

If a standard line design is to be used, then the conductor size and installed tension is fixed, and the primary economic constraints to be investigated consist of comparing routing costs and minimizing the cost of structures through optimal tower spotting. If a new design is to be developed, then a method of estimating supporting structure cost as a function of conductor size and tension must be found. Once the range of optimal conductor sizes is determined, these same supporting structure cost models can be utilized to investigate lower cost combinations of structure height and load capability.

5.4.4 Minimizing Total Line Cost

When minimizing total line costs, it is important to realize which controllable cost parameters have the largest impact on overall line cost. The main capital cost components of any line are the right-of-way, supporting structures, conductor and labor. The cost of right-of-way, conductors and structures is typically 20% to 60% of the total capital cost of a transmission line. A 10% reduction in any of these capital cost components will result in a reduction in overall line cost on the order of 2% to 4%. Reducing the cost of minor items such as conductor hardware, shield wires, dampers, armor rod and counterpoise conductors will produce only small cost reduction benefits.

One of the primary decisions required in economic line design is the selection of phase conductor size. This selection is done by estimating the impact of conductor size on both capital construction cost and electrical losses. Conductor economic analyses normally use the present worth of revenue required (PWRR), which is the sum of the present worth of the annualized construction cost and cost of electrical losses. Figure 5-22 shows a typical plot of PWRR and its component present worth costs for a typical span analysis of a 345 kV line with two 45/7 (Type 7) ACSR conductors per phase.

As seen in Figure 5-22, increasing conductor size increases the capital construction cost, both because of the increased conductor cost and the stronger structures required to support the larger, heavier conductor and associated mechanical loading. The larger conductor cross section, however results in lower resistance and therefore lower losses. If corona losses are considered, they are also reduced as the conductor size increases, assuming other design

FIGURE 5-22
PWRR, PWCC AND PW LOSSES vs.
PHASE CONDUCTOR SIZE

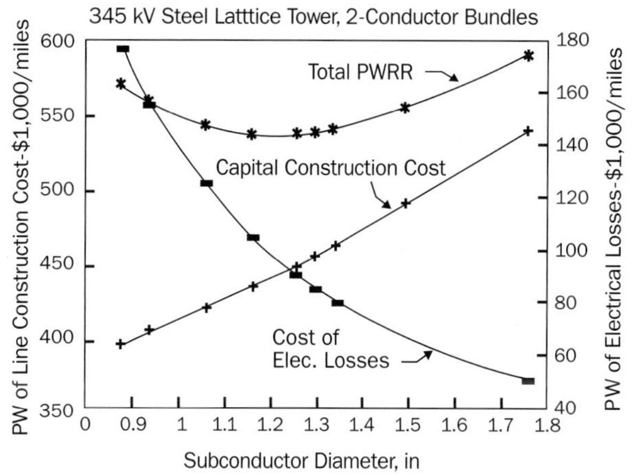

parameters, e.g. phase spacing, remain constant. An overall minimum cost will result for a conductor size where the installed cost forces the PWRR higher for large conductors and the cost of losses forces the PWRR higher for smaller conductors. The "minimum" total cost (PWRR) region is relatively flat. The data indicates that the conductor size has a relatively minor impact on total PWRR. Conductors with diameters of 1 to 1.5 inches yield nearly the same total PWRR, however, the total PW cost of capital investment is much larger for the larger conductor. The smaller conductor is a better financial choice because the cost of construction is noticeably less, even though the total cost for both conductors is nearly the same.

Following this type of analysis, the appropriate size conductor can be selected and the capital cost of the line can be minimized. Changes in structure spacing or conductor sag-tension have no impact on electrical losses and are unlikely to change the conductor size that yields minimum total cost.

FIGURE 5-23
TOTAL PW OF CAPITAL CONSTRUCTION AND
ELECTRICAL LOSSES vs. CONDUCTOR DIAMETER

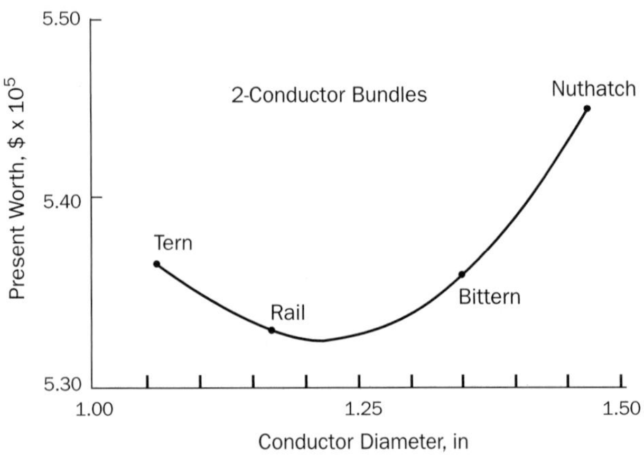

FIGURE 5-24
TOTAL PW OF CAPITAL CONSTRUCTION AND
ELECTRICAL LOSSES vs. STRUCTURE SPACING

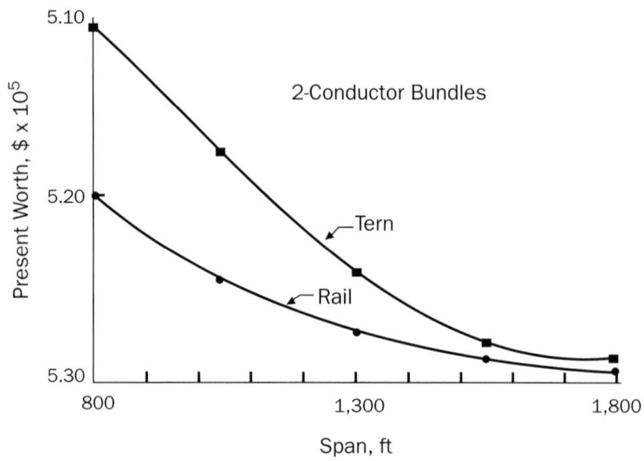

FIGURE 5-25
COMPACT AND STANDARD 345 KV STRUCTURES

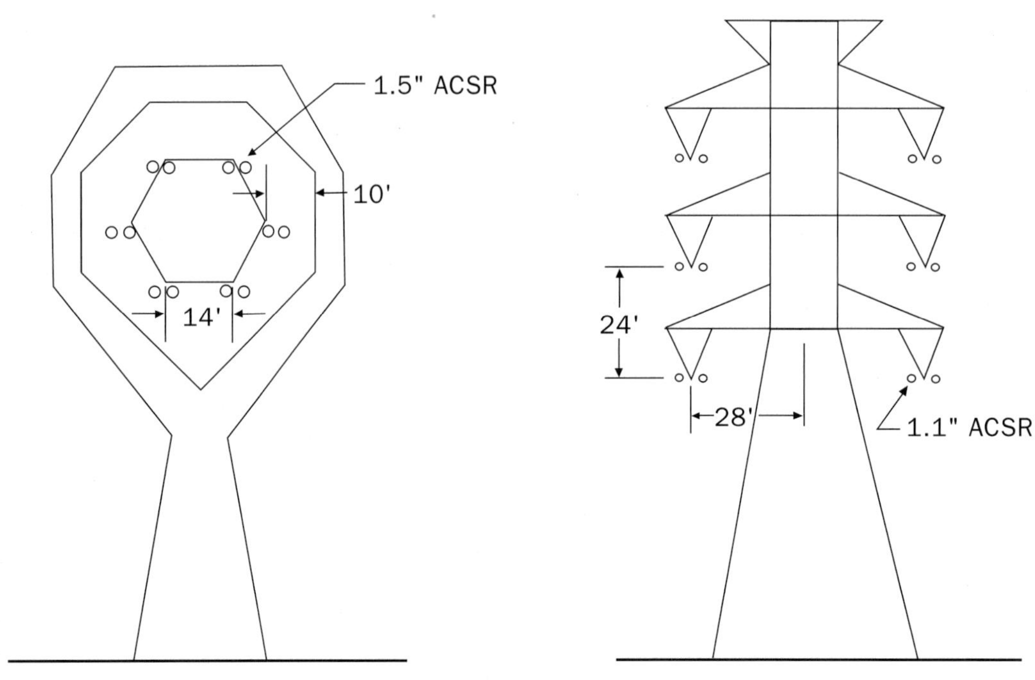

Several design parameters have a major cost impact on the line:

- The maximum phase conductor tension largely determines the cost of dead-end and strain structures.

- Phase conductor sag at maximum temperature largely determines the height of suspension structures.

- Width of ROW is a major cost factor.

- Choice of line route can be a major economic factor.

In the following sections, possible methods of estimating and minimizing line capital costs will be explored.

5.4.5 Supporting Structure Cost

The detailed design of conductor supporting structures and their foundations is a complex process and will be addressed only in general terms. Their cost is clearly both a major component of any new transmission line and a factor in the decisions made regarding the type and size of conductor.

If the design of transmission structures is completely standardized, then the conductor size or type may not be varied, the conductor tension limits are set, and the weight, cost and load capability of all members of the structure family (suspension, light angle, heavy angle and dead-end) are specified. The load capability of the structures is usually expressed in terms of a maximum wind and weight spans and a maximum line angle for each type of standard structure. In this case, economic optimization is limited to finding the least cost solution for spotting the standard structures on a given right-of-way.

If the transmission line structure type (steel pole, steel lattice, wood pole, wood pole H-Frame, concrete pole, etc.) is known, but the structure load capabilities and range of heights are open to modification, then the line designer needs to find a way to estimate structure cost as a function of conductor loads and sag. This estimate can be done by having structural engineers calculate the detailed structure costs for each possible conductor choice. A more practical approach to preliminary structure/foundation costing is the development of an equation that relates structure and foundation cost to conductor loads and structure height. A cost equation may need to be developed for each of the possible structure types and angle family members. While the development of cost equations requires a certain amount of detailed analysis, the subsequent use saves a considerable amount of time.

5.4.6 Conductor Selection with the Typical Span Method

Preliminary calculations regarding the optimum choice of conductor type, size and tension may be performed using a typical span assuming the proposed transmission line consists of a series of identical spans with the same spacing and the same structure height. The structure loading increases with the size of the conductor and structure spacing. Normally the terrain is assumed flat with all spans having the same ground clearance. Radio noise, audible noise and electric and magnetic fields are normally estimated with this same sort of simple typical span analysis and can be calculated at the same time that costs are estimated. The cost of angle structures may be included in these calculations if the number and magnitude of line angles is known, and cost equations for angle structures are known.

Figure 5-23 shows the variation of total PWRR with conductor diameter, and Figure 5-24 shows the variation of total PWRR with the typical structure spacings. As the typical span length increases, the number of structures in the line decreases, but the height of the structures and the maximum loads increase. For larger or stronger conductors, the angle structure loads increase. If the installed tension for a given conductor is increased, the transverse load on angle structures increases, increasing structure cost.

In evaluating the economic considerations of Figure 5-22, it should be noted that in areas where sensitivity to radio/ TV interference is a major concern, the smaller, more economic conductor sizes may not be a suitable design alternative.

As structure spacing increases, structure height also increases. The economic advantages of longer spans may be impractical if the visual impact of the taller towers becomes environmentally undesirable. All alternatives must be weighed individually and as part of a system.

The typical span approach is most useful in jointly evaluating economic and environmental impact of radically different line designs, such as those in Figure 5-25. Both structure designs are intended to support a double-circuit 345 kV line. The larger structure is of conventional lattice-type design, while the smaller structure is a newer compact design. The purpose of analysis, in this case, is to quantify the environmental and economic differences in the two line designs as shown in Table 5-5[45].

This simple typical span model allows comparison of environmental effects versus total present worth of construction cost and losses. The closer phase spacing of the compact design requires a larger conductor to maintain the same corona levels. The economic analysis also indicates that the lower electrical losses of the larger conductor help offset its higher cost. For this design situation, the compact design yields lower electric and magnetic field levels and less visual impact using the same right-of-way as the standard design. This preliminary analysis of innovative designs is one of the best applications of the typical span analysis method.

5.4.7 Economic Structure Spotting

As stated previously, the typical span method of comparing line designs economically is very useful and relatively simple. However, a more realistic estimate of total line cost can be developed though the application of a modern line design computer program, such as PLS-CADD or TL-Pro, in combination with digital survey data. Such software allows the user to evaluate various options as to structure location and load capability while meeting the detailed clearance requirements of each line section[46].

Both manual and numerical tower spotting involves the following parameters for each class of structure: (1) wind span, (2) weight span, (3) H/V ratio, (4) attachment heights — range and interval, (5) structural family, (6) conductor tension limits and uplift, (7) terrain data and (8) ruling span. In the following sub-sections, each of these parameters is defined, and typical values and/or generally-accepted methods of derivation are presented, along with suggestions regarding problems that may arise.

Wind Span

The wind span of any structure equals one half the sum of the distances to adjacent structures. The maximum allowable wind span for a structure must be defined based upon the diameter and weight of the conductor and/or shield wires, the installed tensions and the line angle. It equals

TABLE 5-5
TYPICAL SPAN CALCULATIONS FOR COMPACT AND STANDARD 345 KV LINE DESIGNS

Structure ROW Width, ft Ground Clr, ft Phasing Conductor	**Standard 150 27 Low X "Rail"**	**Standard 150 27 Supr. Bdl. "Rail"**	**Standard 150 37 Low X "Rail"**	**Standard 200 27 Low X "Rail"**	**Standard 200 37 Low X "Rail"**	**Compact 150 27 Low X "Lapwing"**	**Compact 100 27 Low X "Lapwing"**
E_{max} [kV/M]	6.6	7.2	4.0	6.6	4.0	3.5	3.5
E_{row} (kV/M)	0.6	0.4	0.8	0.1	0.1	0.3	0.8
AN_{row} [dB(A)]	47.0	46.0	46.0	46.0	45.0	51.0	52.0
RI_{row} [dB]	42.0	42.0	42.0	39.0	39.0	41.0	46.0
B_{row} [Gauss x10^{-3}]	92	158	76	47	41	14	35
Cost [$1000/mile]	504	504	534	538	567	558	525
PWRR [$1000/mile]	1242	1242	1290	1296	1345	1172	1118

Note:

Low X	=	*low reactance phase arrangement[5]*
Supr. Bdl.	=	*super bundle phase arrangement[5]*
ROW	=	*right-of-way*
E_{max}	=	*maximum ground level electric field within ROW*
E_{row}	=	*ground level electric field at edge of ROW*
AN_{row}	=	*audible noise level for wet conductor at edge of ROW*
RI_{row}	=	*fair weather radio noise level at edge of ROW*
B_{row}	=	*magnetic field at edge of ROW*
Cost	=	*construction cost including conductor, structure, ROW*
PWRR	=	*total present worth of construction and electrical losses over 40 years*

the maximum total wind load that the structure can support, divided by the wind load per foot of span.

The transverse loading forces on a structure results from wind on the conductors, shield wires and the structure itself, plus the horizontal components of the tension in the conductors and shield wires if the structure is placed at a line angle. Both the wind load and the tension load at line angles depends on the size, number and installed tension of the shield wires and phase conductors. If tension loading is decreased, then the wind span is increased.

If the calculated wind spans are consistently less than approximately 80% of the maximum allowable wind span, the selected wind spans may be too high for the terrain, conductor and conductor tension. In this case, replacing the conductor with a higher strength conductor and/or increasing the conductor tension should be considered. If this is not possible, the possibility of reducing the wind spans should be considered. If the calculated wind spans are consistently near the limiting values, the calculation may be repeated with structures having a somewhat higher wind span to see if the total line cost can be reduced.

Weight Span

The weight span of any structure equals the sum of the distances from the structure location to each adjacent point of maximum sag. The maximum allowable weight span for any structure may be defined by the total vertical forces acting on the structure resulting from the weight of the conductor and shield wires (if any), plus tension and environmental loading conditions. It is equal to the maximum total vertical load that the structure can support, divided by the weight load per foot of span.

If the calculated weight spans are consistently less than 80% of the maximum allowable weight spans, the possible savings from reducing the maximum allowable weight span should be considered. This savings is normally less than that obtained from reducing the maximum allowable wind span. If the calculated weight spans are close to the limit, then the maximum allowable weight spans may be increased to evaluate the resulting cost savings.

H/V Ratio

The H/V ratio of any structure equals the maximum ratio of the wind-to-weight span under any of the environmental conditions specified. A maximum allowable H/V ratio, entered for each structure family member, is dependent on the maximum allowable suspension insulator string swing for a high, but not extreme, wind speed. The wind swing angle of the insulator string is equal to the arc tangent of the ratio of horizontal wind force to vertical weight on the insulator string. The horizontal wind force is equal to the product of the wind span and the wind pressure per unit length. The vertical conductor weight force is equal to the weight span times the bare conductor weight per unit length (the weight of the insulator string is normally neglected). Since the wind pressure and conductor weight per

FIGURE 5-26
PLS-CADD DESIGN DETAIL

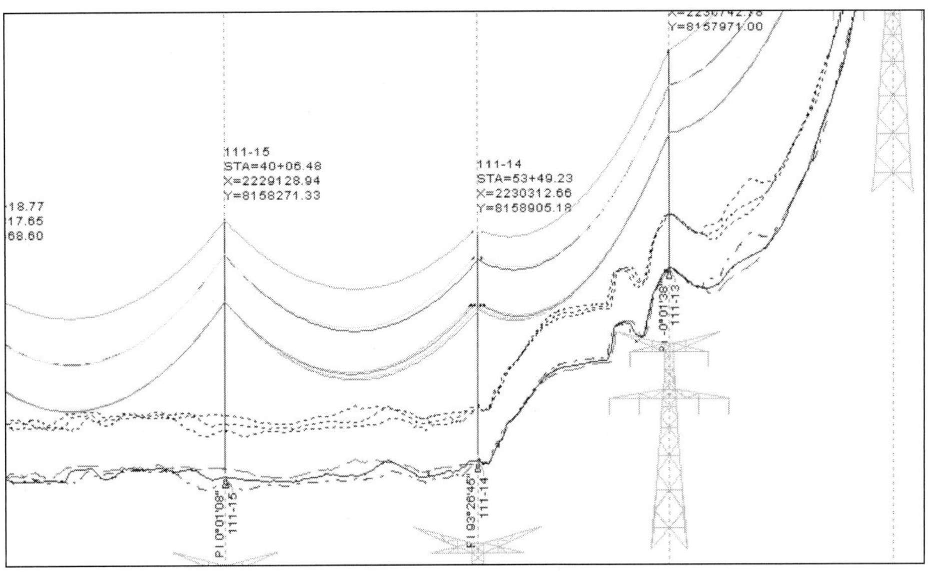

Courtesy of Power Line Systems, Inc.

unit length are the same for all structures, the maximum swing angle is determined by the maximum allowable ratio of wind-to-weight span (H/V ratio). Excluding the use of weights or special brackets, most tangent structures can be expected to have an H/V ratio between 1.3 and 1.7, and small angle suspension structures a ratio of 1.5 to 2.4. The limiting H/V ratio can be increased by the use of weights, by the use of brackets to lower the insulator attachment point, or by increased phase-to-structure spacing.

Attachment Heights

The attachment height of any tangent structure is equal to the height of the bottom crossarm above the structure ground-line center, minus the sum of the insulator string and conductor hardware lengths. Attachment height is commonly referred to as the effective structure height. In flat terrain, the minimum attachment height should be slightly greater than the required terrain clearance plus the conductor sag for a typical span. In hilly or mountainous terrain, the minimum effective structure height should be somewhat less than this, approaching the terrain clearance alone. In general, strain structures should be kept shorter than suspension structures, and the number of available attachment heights minimized.

Structure Family Members

Structures are typically grouped by family member, e.g., 15° suspension or 60° strain structure. Each member of the structure family has a wind span, a weight span, a maximum line angle and a maximum H/V ratio.

Plan and profile drawings based on the route survey are used to locate points along the route where the line must change direction. Changes in line direction as small as 5° or as large as 90° require angle structures. The total number of angle structures over any section of the line's route is determined by the route survey.

A structure capable of supporting the conductors through a 90° angle is much more expensive than a structure capable of supporting the same conductors through a 5° line angle. Therefore, the line designer normally selects a family of angle structures containing members with different maximum allowable line angles. In this way the weaker structures may be used for small angles and the stronger structures for large line angles. The larger the number of such angle structures, the better the fit between line angles and angle structures, but the greater the design and inventory costs. The line designer normally compromises by selecting the minimum family of angle structures consistent with the population of required line angles.

Uplift problems can be solved by use of strain structures, but often may be eliminated by redesigning tangent structures to increase the maximum allowable H/V ratio. This is usually accomplished by increasing the maximum allowable tangent tower heights, or decreasing conductor design tension.

Conductor Tension and Uplift Problems

In flat terrain, as the everyday conductor tension is increased, other changes also occur:

- Conductor sag decreases
- Design structure height decreases
- Average span increases
- Tension load on angle structures increases

On flat terrain, where few angle structures are required, increasing conductor tension has been shown to be an effective way to reduce line cost. On hilly or mountainous terrain, however, increased conductor tension may require suspension structures with excessive H/V ratios and may result in uplift conditions on certain structures.

Conductor sags are often limited by the final tension under minimum conductor temperature conditions when aeolian vibration is at a maximum. If, as a result of high conductor tension and rugged terrain, a tangent structure is subjected to uplift, use of a strain structure may be required.

Ruling Span

The ruling span, *RS*, is defined as the square root of the sum of the cubes of each span divided by the sum of the spans:

$$RS = \sqrt{\frac{S_1^3 + S_2^3 + S_3^3 + \cdots + S_n^3}{S_1 + S_2 + S_3 + \cdots + S_n}} \tag{5-25}$$

where:

S_1, S_2, S_3, *etc.* = *the different span lengths in the ruling span segment of the line.*

Exact determination of the ruling span can only be made after all structures have been spotted. For tower spotting purposes, one or more initial estimated ruling spans must be selected for the calculation of design sag tension data. After spotting structures on the right-of-way, the ruling span should be calculated for each line section between line angles or deadends.

It is important that the actual ruling span be reasonably close to the ruling span values used in spotting the towers. If the calculated ruling span is within 10% of the initial assumed ruling span, then the accuracy is probably sufficient; otherwise, potential clearance problems may exist.

5.4.8 Cost with Respect to Conductor Tension

Tension limitations are specified by the National Electrical Safety Code (NESC)[31] or Rural Utility Service (RUS)[23]. Within these limitations, the design engineer has a choice of a range of everyday tension levels for the phase conductors. The following considerations will influence the final choice:

As the everyday tension level is increased, the potential need for vibration dampers increases, the maximum tension load on angle and dead-end structures increases and the sag at maximum temperature decreases. Therefore, as conductor tension increases, the cost of tangent structures decreases, the cost of angle structures increases, and the possible need for conductor damping increases. As the everyday conductor tension deceases, the potential need for dampers decreases, the maximum tension load on angle structures decreases, and the height of tangent structures increases.

Figure 5-27 illustrates this trade-off for a conventional single-circuit 345 kV line in hilly terrain. The figure shows the impact of capital construction cost to be at minimum for relatively low everyday tension levels. It also shows the optimum tension is lower for stronger conductors. Each point on this plot is the result of a numerical tower spotting run on nearly four miles of terrain using the EPRI Transmission Line Optimization Program (TLOP).

Other possible changes in structure and conductor characteristics should also be investigated as part of the economic optimization of proposed line designs. As an example, consider Figures 5-28A and B that show plots of structure positions corresponding to a weak and a strong structure family, respectively. In Figure 5-28A, the basic tangent structure has a wind span of 800 feet, while in Figure 5-28B, the basic tangent structure wind span has been increased to 1200 feet. Note that the number of structures has decreased as a result of the strengthening, but, the cost of each structure has increased as a results.

FIGURE 5-27
IMPACT OF INSTALLED TENSION ON CAPITAL CONSTRUCTION COST

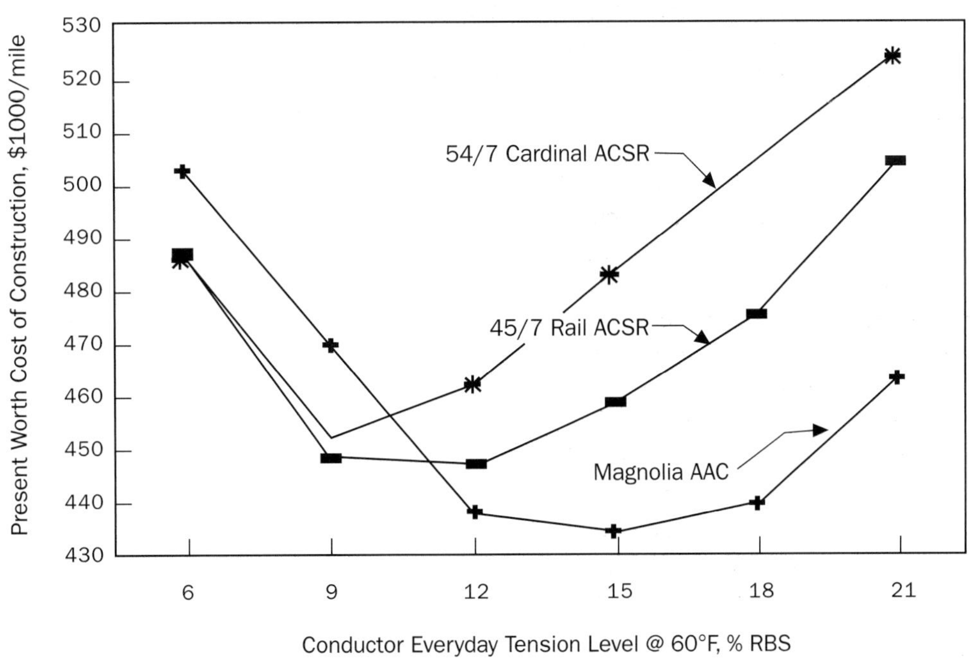

FIGURE 5-28A
OPTIMAL STRUCTURE LOCATION WITH 800 FEET TANGENT WIND SPAN

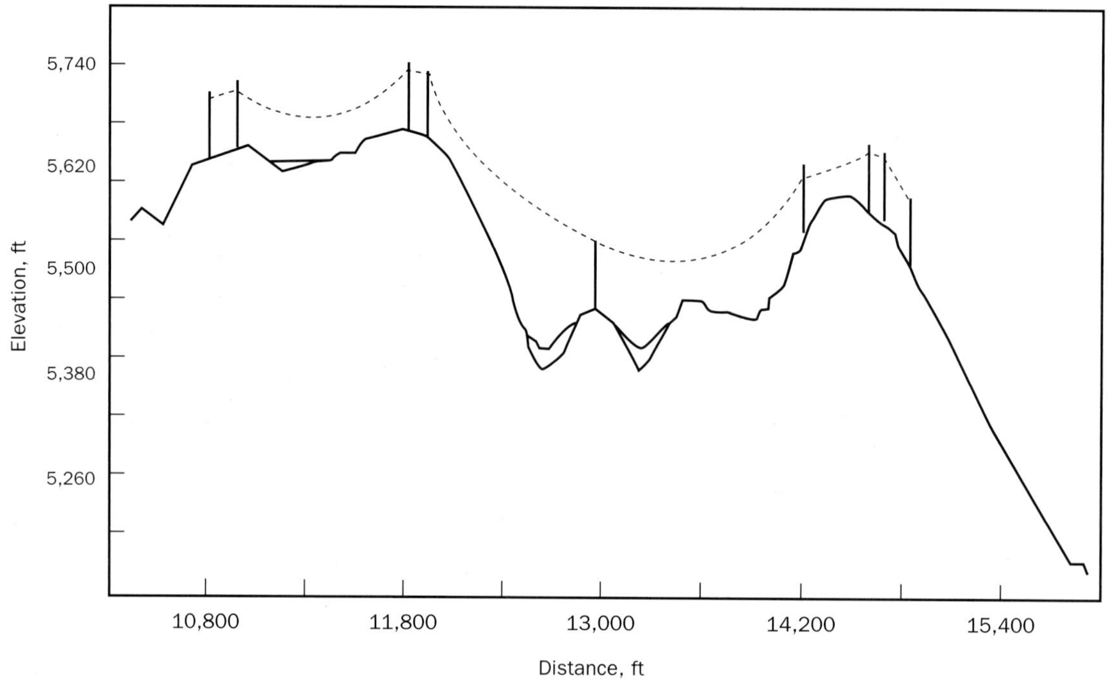

FIGURE 5-28B
OPTIMAL STRUCTURE LOCATION WITH 1200 FEET TANGENT WIND SPAN

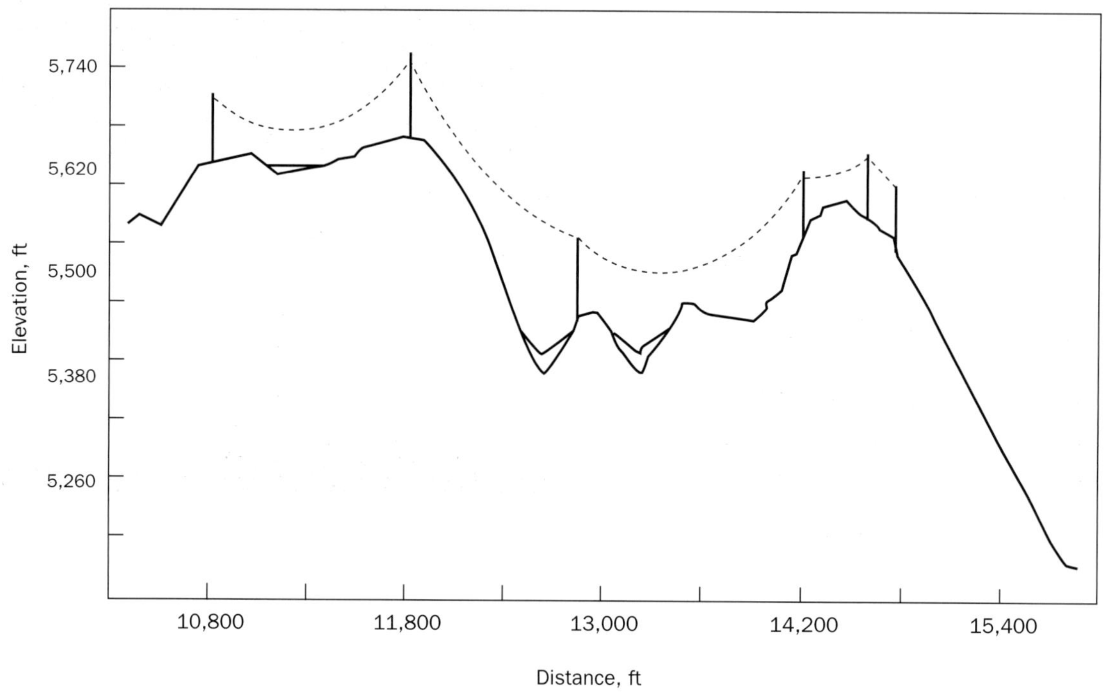

5.5 VISUALLY ACCEPTABLE LINE DESIGNS

This conductor reference book provides a great deal of technical material related to the design and uprating of overhead transmission lines. The goal of the book is to help the transmission line design engineer to design, build and maintain lines that are safe, reliable and economic. So far, however, we have not discussed a matter of great importance to the successful design of new transmission lines - public and regulatory acceptance.

The goal of this section is to describe some of the reasons for public opposition to new lines and to suggest some line design methods that have been successful in dealing with opposition in different areas of the world.

The primary reasons for public opposition to new lines are reasonably well known. People in the vicinity of overhead power lines do not like their appearance, they are concerned about the impact the line may have on their property values, and they find them a source of concern regarding their personal health and safety. As shown in Figure 5-29, audible noise and electrical interference are minor issues since these are normally minimized in the design process.

FIGURE 5-29
REASONS FOR PUBLIC OPPOSITION TO NEW OVERHEAD POWER LINES

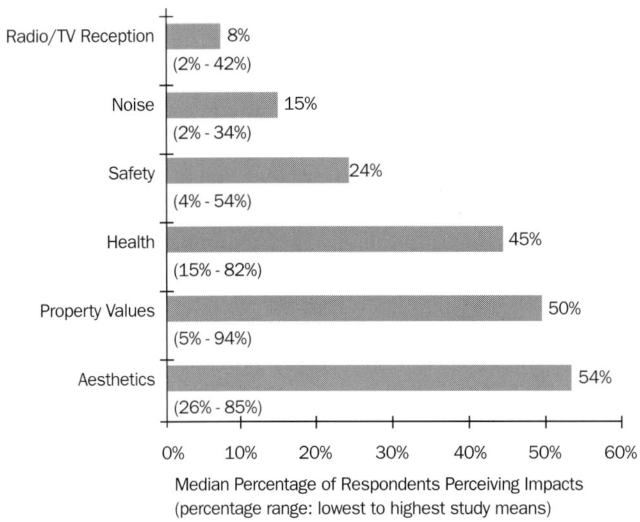

Given these sources of annoyance, it seems obvious that public opposition decreases with distance from the line as shown in the following table.

TABLE 5-6
IMPACT OF DISTANCE UPON PUBLIC OPPOSITION TO OVERHEAD POWER LINES

	Distance from Line	
Public Opinion	**Less than 1 mile**	**More than 1 mile**
Like it	2%	3%
Indifferent to it	33%	72%
Dislike it	65%	25%

While no one has been consistently successful in overcoming public opposition to new power lines near where they live or play, some exceptions have occurred. In some cases, lines are made more acceptable by using playful designs like this well-known example in Orlando, Florida (Figure 5-30).

FIGURE 5-30
STRUCTURE NEAR DISNEY WORLD IN ORLANDO, FLORIDA

Another approach, where there is "no place to hide," has been to make the line structures colorful and cleanly modern as in this 400 kV line that goes through the harbor in Helsinki, Finland, right past the Finnish president's home (Figure 5-31).

In many areas, however, even colorful modern designs may engender opposition. For example, in the state of Vermont, there is a state law to the effect that, in scenic areas, the addition of a new overhead transmission line must not degrade the visual appearance of the area. The ideal response to this legal constraint on new lines would be an invisible overhead line. Since this is impossible, and underground cable is usually considerably more expensive to install, the best response is to make new lines less visible.

In order to make overhead lines less visible, there are a number of obvious design approaches:

- Reduce the height of the structures
- Minimize the number of structures
- Limit conductor and structure visibility
- Reduce phase to phase and phase to ground spacing
- Reduce the right-of-way width
- Treat structures and conductors to reduce visibility

Recently developed materials and component designs allow the design of new lines which are more compact (and thus less visible), while maintaining or improving upon present levels of system reliability and environmental impact. By compacting new lines, certain design constraints are easier to meet. For example, by reducing phase to phase spacing with a vertical or delta phase arrangement, the required pole height is lessened and magnetic field levels at ground level are reduced. Similarly, by using high temperature, low sag conductors, the reduced sag allows shorter structures or greater spacing, and allows occasional high temperature operation which may avoid the need for bundling of phases.

On the other hand, compaction of new lines may make certain design constraints more difficult to meet. For example, reduced phase to phase spacing may yield higher audible noise and radio noise levels, which may lead to the need for larger conductors to compensate. Larger conductors may, in turn, require shorter spans, more visible conductors and higher cost.

FIGURE 5-31
400 KV LINE WITH COLORFUL STEEL POLE STRUCTURES IN HELSINKI, FINLAND

Observations on Esthetics

The esthetic impact of line design is difficult to quantify. Some actions intended to improve the appearance of overhead lines can lead to unacceptable increased cost. It may be useful to review some existing designs that seem to meet the criterion of less visibility and greater public acceptability while not straying too far from designs that have been shown to be reliable, economic and require minimal maintenance over extended time periods (Figures 5-32, 5-33 and 5-34).

FIGURE 5-32
COMPACT SINGLE-POLE STRUCTURE DESIGN WITH POST INSULATORS IN A SUBURBAN AREA

FIGURE 5-33
COMPACT SINGLE-POLE STRUCTURE DESIGN WITH HORIZONTAL V INSULATOR DESIGN ON STEEL POLE FOR LONGER SPANS

FIGURE 5-34
A SIMPLE CLEAN WOOD POLE DESIGN WITH POST INSULATORS IN A SCENIC AREA

It may also be useful to review some designs which, while fully functional, do not meet the criterion of "less visibility" and therefore have less likelihood of public acceptance (Figure 5-35).

Finally, when certain sections of overhead lines need to go underground, the use of graceful, clean transition structures at 69 kV to 230 kV can help make the entire line more acceptable. The use of attractive transition structures is particularly important because they are often placed in highly populated areas (Figure 5-36).

From this brief discussion, it should be obvious that the design of overhead lines that minimize public opposition is not simple, nor is it straightforward. On the other hand, it is common in many regions of the United States and Canada for the costs associated with permitting, environmental investigations and public hearings to easily exceed the costs of materials and labor. Therefore, designing new lines to be clean and attractive and minimizing their visual impact is well worth the effort. It may make the difference between building the line or not.

FIGURE 5-35
TRANSMISSION STRUCTURES WITH LONG SPANS, TALL STRUCTURES, AND BUNDLED PHASE CONDUCTORS

FIGURE 5-36
UNDERGROUND TO OVERHEAD TRANSITION STRUCTURES

5.6 REFERENCES

[1] Fisher, R.A. and Tippet, L. H. C., "Limiting Forms of the Frequency Distribution of the Largest or Smallest Number of a Sample," Proc. Cambridge Phil. Soc., 24, part 2, pp. 180-190, 1928.

[2] Aleksandrov, G. N., Kizvetter, V. Y., Rudakova, V. M., Tushnov, A.N., "The AC Flashover Voltages of Long Air Gaps and Strings of Insulators," *Elektrichestvo*, No. 5, pp. 27-32, 1962.

[3] "Guide for the Selection of Insulators in Respect of Polluted Conditions," International Electrotechnical Commission: IEC Report, Publication 815, 1986.

[4] "Application of Insulators in a Contaminated Environment," IEEE WG Paper, *IEEE Transactions on Power Apparatus and Systems*, Vol. 98, pp. 1676-1695, September/October 1979.

[5] *Transmision Line Reference Book, 345 and Above, Second Edition*, Electric Power Research Institute, EPRI EL-2500, 1982.

[6] Lambert, S. R., "Effectiveness of Zinc Oxide Surge Arresters on Substation Equipment Probabilities of Flashover," IEEE Paper No. 87 SM 561-4, Presented at the IEEE Summer Power Meeting, San Francisco, CA, 1987.

[7] Lambert, S. R., Riberio, J. R. and Wilson D. D., "Protection of Compact Transmission Lines with Metal Oxide Arresters," presented at the CIGRE Leningrad Symposium, June 3-5, 1991.

[8] Kachler, A. J., LaForest, J. J., Zaffanella, L.E., "Switching Surge Flashover of UHV Transmission Line Insulation,"*IEEE Transactions on Power Apparatus and Systems*, Vol. PAS-90, pp. 1604-1611, July/August 1971.

[9] Young, F. S., Schneider, H. M., Gutman, Y. M., Tikhodeyev, N. N., "US-USSR Investigation of 1200-kv Tower Insulation," *IEEE Transactions on Power Apparatus and Systems*, Vol. PAS-99, No. 6, pp. 462-470, March/April 1980.

[10] Paris, L., "Influence of Air Gap Characteristics on Line-to-Ground Switching Surge Strength," *IEEE Transaction on Power Apparatus and Systems*, Vol. PAS-86, pp. 936-947, August 1967.

[11] Menemenlis, C., Harbec, G., "Switching Impulse of EHV Transmission Towers," *IEEE Transactions on Power Apparatus and Systems*, Vol. PAS-94, pp. 255-263, March/April 1975.

[12] Dillard, J. K., Hileman, A. R., " UHV Transmission Tower Insulation Test," *IEEE Transactions on Power Apparatus and Systems*, Vol. PAS-89, pp. 1772-1784, February 1970.

[13] Kalb, J. W., "How The Switching Surge Family Affects Line Insulation," *IEEE Transactions on Power Apparatus and Systems*, Vol. PAS-82, pp. 1024-1031, December 1963.

[14] Atwood, Jr., A. W., Hileman, A. R., Skooglund, J. W., Wittibschlager, J. F., "Switching Surge Test on Simulated and Full-Scale EHV Tower Insulation systems," *IEEE Transactions on Power and Apparatus Systems*, Vol. PAS-84, pp. 293-303, April 1965: Discussion, pp. 517-520, June 1965.

[15] Menemenlis, C., Carrara, G. and Lambeth, P. J., "Application of Insulators to Withstand Switching Surges in Substations. Part I: Switching Impulse Insulation Strength," IEEE Winter Power Meeting, Paper No. 88 WM 077-0, 1988.

[16] Anderson, R.B. and Eriksson, A.J., "Lightning Parameters for Engineering Application," Pretoria, South Africa: CSIR, Report ELEK 170, June 1979.

[17] Bellachi, P.L., Armington, R.E., Snowden, A.E., "Impulse and Sixty-Cycle Characteristics of Driven Grounds-II," *AIEE Transactions on Power Apparatus and Systems*, Vol. 61, pp. 349-363, 1942.

[18] Liew, A., Darveniza, M., "Dynamic Model of Impulse Characteristics of Concentrated Earths," Proceedings IEEE, Vol. 121, No. 2, pp. 123-135, February 1974.

[19] Love, R.R., "Improvements in Lightning Stroke Modeling and Applications to the Design of EHV and UHV Transmission Lines," M.Sc. Thesis, University of Colorado, 1973.

[20] Wagner, C., Hileman, A., "Effect of Predischarge Currents on Line Performance," *IEEE Transactions on Power Apparatus and Systems*, PAS-65, pp. 117-131, 1963.

[21] Los, E.J., "New Studies of the Transient Glow Discharge in Parallel Electrode Gaps," *IEEE Transactions on Power Apparatus and Systems*, Vol. PAS-99, No. 2, pp. 720-728, March/April 1980.

[22] Gary, C., Critescu, D., Dragan, G, "Distortion and Attenuation of Travelling Waves Caused by Transient Corona," CIGRE Studie Committee 33 Technical Brochure, 1989.

[23] RUS Bulletin 1724E-200, "Design Manual for High Voltage Transmission Lines," United Stated Department of Agriculture, Rural Utilities Service, September 2004.

[24] Transmission Line Reference Book, "Wind-Induced Conductor Motion," Electric Power Research Institute, Palo Alto, CA, 1980.

[25] Chen, Y.N., "Fluctuating Lift Forces of the Karman Vortex Streets on Single Circular Cylinders and in Tube Bundles. Part I. The Vortex Street Geometry of the Single Circular Cylinder," Transactions ASME Journal of Engineering for Industry, pp. 603-612, May 1972.

[26] "IEEE Guide on Conductor Self-Damping Measurements," IEEE Standard 563-1978, Published by the Institute of Electrical and Electronics Engineers, Inc., New York, NY.

[27] Torok, J., "Aeolian Vibration of Overhead Conductors," Fargo Manufacturing Company, Inc., Presented April 21,1993.

[28] Data Courtesy Alcoa Laboratories.

[29] Silva, J.M., "An Experimental Evaluation of the Effect of Amplitude Reduction on the Fatigue Life of Overhead Transmission Lines Subjected to Aeolian Vibration, "1976 Annual Conference, Southeastern Electric Exchange, April 1976.

[30] Seppä, T., "Effect of Various Factors on Vibration Fatigue Life of ACSR "IBIS,"" CIGRE Report 22-69 (WG04)-04, 1969.

[31] 2002 National Electrical Safety Code, Accredited Standards Committee C2-2002, The institute of Electrical and Electronic Engineers, Inc., New York, 2001.

[32] "Aeolian Vibration on Overhead Lines,"CIGRE Study Committee No. 22 WG 01, CIGRE Report 22-11, 1970.

[33] Davison, A.E., "Dancing Conductors," AIEE Transactions, Vol. 49, pp. 1444-1449, October 1930.

[34] Den Hartog, J.P., "Transmission Line Vibration Due to Sleet," AIEE Transactions, Vol. 51, pp. 1074-1076, 1932.

[35] Davison, A.E., "Ice-Coated Electrical Conductors," Bulletin Hydro-Electric Power Commission of Ontario, Vol. 26, No. 9, pp. 271-280, September 1939.

[36] Richardson, Jr., A. S., "Design and Performance of an Anti-Galloping Device," IEEE Conference Paper 68 CP 670-PWR, Presented at the 1967 IEEE Summer Power Meeting.

[37] Toye, L. W., "Formulas Determining Conductor Path at Quarter Point," Electrical World, pp. 160, January 29, 1951.

[38] Rawlins, C.B., "Analysis of Conductor Galloping Field Observations - Single Conductors," IEEE Transactions on Power Apparatus and Systems, Vol. PAS-100, No. 8, August 1981.

[39] Havard, D.G., and Pohlman, J.C., "The Economics of Galloping Control," Canadian Electrical Association International Symposium on Overhead Conductor Dynamics, Toronto, Canada, June 8 and 9, 1981.

[40] Havard, D.G., Pon, C.J., and Pohlman, J.C., "Galloping Control Permits Tower Head Compaction and Line Upgrading," IEEE 86 T&D 508-6, 1986.

[41] Havard, D. C., Pon, C. J., Pohlman, J. C., "Galloping Control Permits Tower Head Compaction and Line Uprading," Paper No. 86 T&D 508-6, Presented at the IEEE/PES 1986 Transmission and Distribution Conference, Anaheim, CA.

[42] Whapham, R., "Field Research for Control of Galloping with Air Flow Spoilers," Preformed Line Products Status Report of Field Research Program, May 1, 1983.

[43] Cloutier, L.J., Hardy, C., and Brunelle, J., "The Magdalen Island Test Line: A Research Tool for Studies of Bundle Conductor Dynamics," IEEE Conference Paper C74 370-3, 1974.

[44] Mohajery, M., "Wake-Induced Oscillation of Bundled Conductors - Two, Three and Four-Conductor Bundles," EEI T&D Meeting, San Diego, January 1976.

[45] Beaty, H. W., and Fink, D. G., Standard Handbook for Electrical Engineers, 13th Edition, McGraw-Hill, Inc., 1993.

[46] Douglass, D.A., and Stewart, J.R., New Overhead Line Designs to Meet Electrical Environmental Restrictions," PTI Newsletter, April 1991.

[47] PLS-CADD, Power Line Systems, Inc., Madison, Wisconsin.

6 VOLTAGE UPGRADING OF EXISTING LINES

Increasing transmission system capacity may be accomplished by one of the following approaches:

- *Constructing new transmission line*
- *Increasing the current-carrying capacity or thermal rating of existing lines (See Chapters 3 and 7)*
- *Increasing the operating voltage of existing lines*

The basic premise of a voltage upgrading study is that an existing transmission line may have potential for operation at an increased voltage. This premise is especially true of older lines designed from handbook techniques incorporating significant safety factors. A voltage upgrading study utilizes the best available knowledge developed through EHV, UHV and compact transmission line research to evaluate the parameters of the candidate line. From this data, it is possible to determine the factors that limit the operating voltage of the line and the changes, if any, that need to be made to allow operation at a higher voltage.

Increasing the operating voltage may preclude the need for new construction and provide benefits beyond those obtained from increasing the operating current. When system considerations, such as voltage drop or stability, limit line loading, increasing the thermal rating may not allow any increase in line loading. Increasing the voltage, however, may increase the loading by as much as the square of the ratio of the voltage increase. While increasing current increase power losses, increasing voltage may reduce power losses. Even when a line is thermally limited, voltage upgrading may be economically feasible if it can be done with minimal line modifications. The degree of modification to a transmission line for a voltage upgrading ranges from little or no change to major reconstruction. The choice of current uprating or voltage upgrading, as well as the degree of change necessary to the line design, is determined by the requirements of the specific application.

6.1 FUNDAMENTALS OF VOLTAGE UPGRADING

Many existing transmission lines were designed with insulation and ground clearances much greater than those required today. As a result, there may be adequate margin for the line to operate at a higher voltage with minor modifications. In many respects, a voltage upgrading study parallels a new line design study. The major difference is that an upgrading study examines an existing design to determine its capabilities, rather than defining the requirements and designing a line to meet those requirements. An essential part of a voltage upgrading study is defining the requirements and carefully considering the criteria to be applied for evaluating the design. When upgrading is performed for temporary or emergency use, the reliability criteria may be relaxed. As a result, uprated lines may range from no modifications to complete rebuilding.

The introduction of higher transmission voltage levels has generated new research and increasingly detailed engineering studies. This has resulted in the clearance required for power frequency voltage withstand being reduced with each new voltage level. The ratio of phase-to-phase spacing to withstand clearance (flashover spacing) for typical transmission line operating voltages is in Figure 6-1. Two conclusions may be drawn from the data presented:

(1) If the same design principles used for higher voltages (e.g., 765 kV) were used for the lower transmission voltages (e.g., 115 kV), then existing, lower voltage lines could be upgraded using the smaller phase-to-phase spacings that were allowed in earlier designs. This is the premise behind compact overhead transmission lines.

(2) Margin may exist, especially in lines at the lower transmission voltages, e.g., 69 kV to 230 kV, which could be used to increase the voltage of the line without increasing phase spacing. However, the higher the voltage on an existing line, the less likely that the voltage can be increased without major reconstruction. This margin is the premise behind "voltage upgrading."

Voltage upgrading studies rely heavily on four broad areas of transmission line research: insulation coordination, compact line design, corona and field effects, and structures and foundations. Improved knowledge in these areas in recent years has allowed engineers to reevaluate existing designs to determine if safety margins, originally built into the lines, could be utilized to increase the operating voltage of the line.

FIGURE 6-1
SPACING RATIO vs. VOLTAGE LEVEL

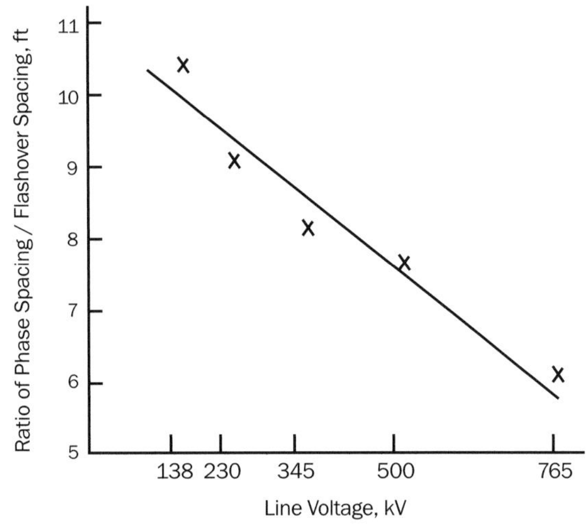

Upgrading Options

The options for increasing the voltage on an existing line range from relatively straightforward and simple to very difficult and costly. Examples of the possible upgrading options include:

- Increasing to the next voltage level without any modification to the line itself (an example is a double-circuit 115 kV line successfully operated at 230 kV[1]).
- Retaining the existing conductor and structures, and adding only a few additional insulators per insulator string.
- Adding a second conductor per phase to make 2-conductor bundles, which will require bracing or reinforcing of the structures.
- Reconstruction of the existing structures to accommodate a new conductor configuration as illustrated in Figure 6-2.

Considerations of Line Components

Re-tensioning of the existing conductors, or addition of a second conductor per phase, often requires reconstruction or replacement of the angle structures, but lesser modifications to the tangent structures. The degree of structure modifications required for voltage upgrading ranges from minor to major, depending upon the upgrading method selected.

Careful attention must be paid to the condition of the conductor, shield wire, insulators, structures, and foundations. Older style insulators, having less leakage or air gap distance than present-day suspension units, may need replacing. A complete evaluation of all structures and foundations is required if any increase in mechanical loading is anticipated. In some cases, increasing the voltage may ease certain design contraints by reducing the operating current, resulting in lower conductor temperature and reduced maximum sag.

6.2 ELEMENTS OF A VOLTAGE UPGRADING STUDY

When considering transmission system modifications, the reasons for the proposed changes must be carefully weighed. For some applications, it may be appropriate to consider current uprating; for others, voltage upgrading may be more advantageous. If increasing the voltage is a consideration, the reasons for the upgrading should be carefully established. Insulation, reliability, noise criteria, and the degree of line modification may be affected by the selected upgrading approach.

6.2.1 Need for Voltage Upgrading

Several factors enter into the loading capability of a given transmission line: conductor thermal capacity, voltage drop and system stability. Conductor thermal capacity is a function of the conductor resistance, the maximum design temperature of the line and weather conditions. Shorter transmission lines tend to be thermally limited, while longer lines tend to be limited first by voltage drop and then by system stability. Voltage drop and stability are functions of line reactance. If voltage drop or stability are the present limits, voltage upgrading or new construction may be the only options. If thermal capacity is the limiting factor, then current uprating, voltage upgrading, or new

FIGURE 6-2
VOLTAGE UPGRADING INVOLVING MAJOR RECONSTRUCTION

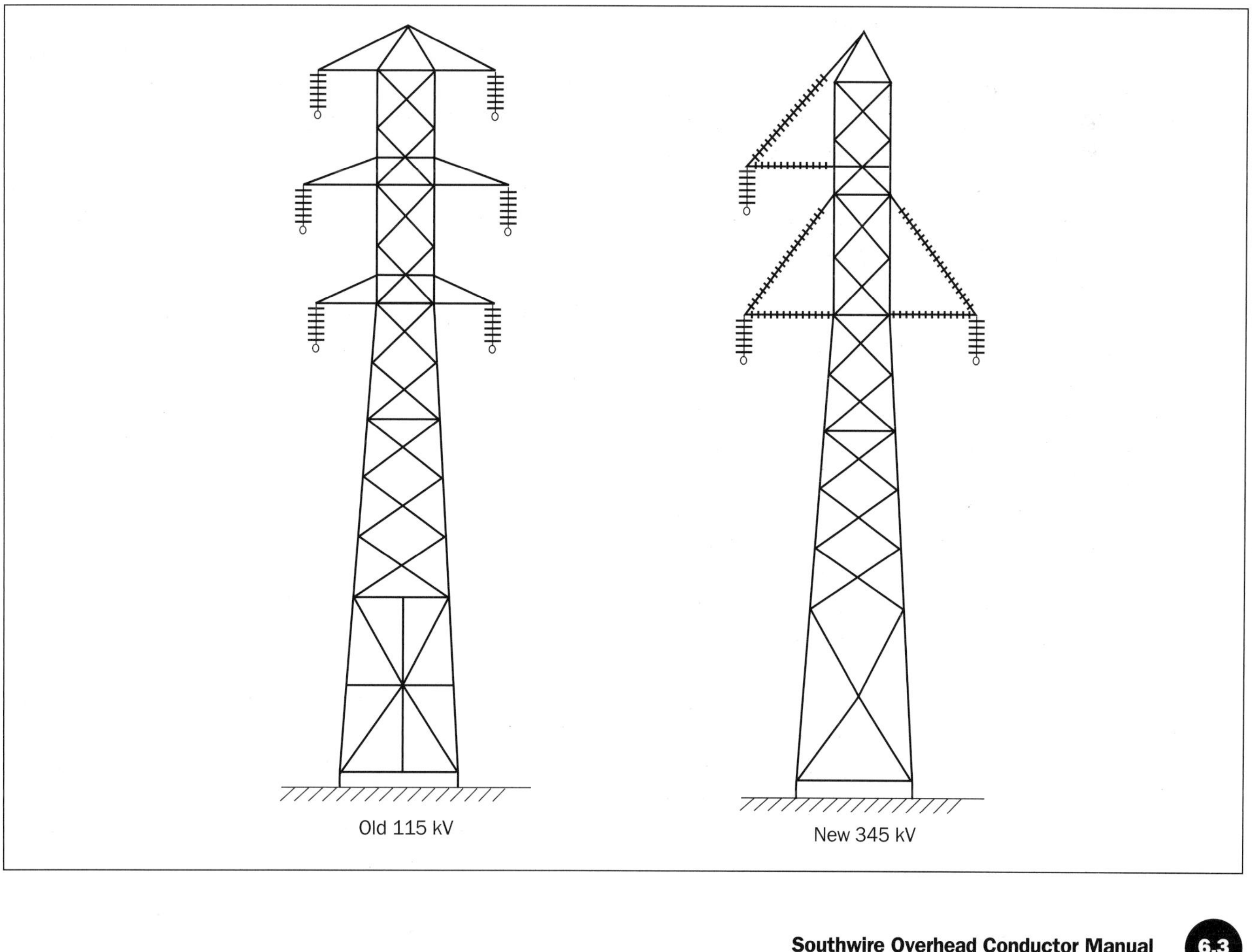

construction may be required. Doubling the line voltage doubles the thermal MVA rating and increases the voltage drop or stability limits on power flow through the line by a factor of up to 4.

6.2.2 Analysis of the Problem

Careful consideration must be given to the reason why the upgrading is being undertaken. Typical reasons are:

(1) An important base-loaded line from a critical generating station requires additional capacity to handle the load from plant expansion.

(2) System problems, such as low voltage or inadequate stability margin, presently occur under normal loading.

(3) Additional transmission capacity is needed for reliability under specified contingency conditions.

(4) Load growth requires upgrading of an entire class of lines in a system[2].

(5) An emergency feed to a station is needed in the event of failure on existing main circuits[3].

Specifics of the line's location, operating history, and both present and anticipated future loading are additional factors to be considered. Allowable noise levels, insulation coordination and other variables are affected by the conditions of the line under study. Criteria for evaluating the final design must be established at the outset.

6.2.3 Feasibility Analysis

A feasibility analysis is normally conducted at the outset. It identifies any major design or operating limitations, as well as those areas requiring more detailed study based on approximate calculations. The feasibility analysis should consider the following:

- Environmental effects related to corona-induced noise and electric and magnetic fields
- Adequacy of insulation for power frequency voltage, switching surges and lightning
- Applicable safety code clearance restrictions
- Structure and foundation condition and integrity
- Conductor and hardware type and condition

If a new voltage level has already been selected, the feasibility study will focus on the anticipated voltage. Some voltage upgrading studies, however, are based on the general need for additional transfer capacity, and the new voltage level is one of the outcomes of the study. If this is the case, results of the feasibility analysis will be used to specify the target voltage. Several alternatives may need to be studied in greater detail. Other factors, such as non-standard voltages, may be identified and will require further detailed consideration.

6.2.4 Detailed Studies

It is not always possible to know at the outset of a voltage upgrading study which area — environment, insulation or structure — will require detailed study. Not all upgradings will require the same depth of analysis in all areas. For example, if the conductor is sufficiently large that corona effects are of no concern, but insulation withstand is marginal, then the bulk of the study may focus on the insulation. Additional system studies may be required, especially if several upgrading alternatives have been identified. System modifications, such as shunt compensation, may affect the final choice of voltage or ultimate circuit rating.

The detailed studies may reach an impasse where it is impossible to meet all the previously established criteria. At this stage, it is appropriate to review the criteria themselves. It may be possible to slightly relax limits on the probability of flashover due to switching surges without greatly impacting the oprerational realiability of the upgraded line. Throughout this process, careful definition of the purpose of the project will guide the decisions.

6.3 INSULATION COORDINATION

A fundamental limitation to voltage upgrading of a transmission line is the withstand capability of the insulation. As discussed in Chapter 5, this includes both power frequency and switching surge design. Voltage stress versus dielectric strength of air, as well as other insulation elements of the line, must be examined.

6.3.1 60 Hz Flashover

A study of power frequency performance considers extreme conditions that temporarily reduce the effectiveness of the insulation system. Two of these conditions are insulator contamination and line operation under extreme wind swing angles. In a voltage upgrading study, it is desirable to increase the line voltage with either no change or minimal change to the existing insulation. In order to successfully accomplish the upgrading, it is necessary to clearly understand the performance of the insulation under the increased stress.

The dielectric strength of the air gap between the conductor and the grounded structural members under extreme wind swing conditions is an essential determinant of 60 Hz insulation performance. For power frequency voltage withstand calculations, the voltage stress is considered constant. The extreme value approach is used to calculate the voltage withstand strength of air gaps. The withstand approach refers to the use of an air gap spacing sufficiently large to prevent flashover under the highest steady-state voltage and the lowest relative insulation strength of the air. Relative insulation strength is a function of altitude and meteorological conditions and varies over time. The withstand approach is in contrast to the probability approach frequently used for switching surge studies. These studies use probability distributions of voltage, relative insulation strength and conductor motion to assess the overall probability of flashover.

The power frequency withstand is based on the minimum clearance between the conductor and tower under extreme wind conditions. Analysis of meteorological data from historical weather records can be used to determine the maximum wind and minimum relative insulation strength of the air. These factors determine the minimum withstand of the air gap and whether it is adequate for the stress of the increased line voltage. If air gap strength under maximum wind swing is inadequate, strut insulators may be added to prevent swinging of the suspension insulators.

Air gap clearances can also be reduced by conductor differential blowout, conductor jumping as a result of the release of accreted ice loads, ice galloping and conductor motion resulting from passage of fault current. These considerations are customarily restricted to compact line designs, but occasionally must be considered in a voltage upgrading study because of the increased electrical stress on the air gap clearances. Details of the analysis procedure for air gap clearances are given in *Transmission Line Reference Book, 115-138 kV Compact Line Design*[4].

6.3.2 Insulator Contamination

Airborne industrial and natural contaminants often reduce the insulation strength of insulators to below that of clean rain. In a voltage upgrading study, the designer has the important advantage of knowing the operating history of the existing line. If a line has a small number of unexplained flashovers before the voltage upgrading, it is likely that the flashover incidents will increase after upgrading. If there is doubt about the state of the existing insulators, one or more strings can be removed from the line and sent to a high voltage laboratory for testing in a contamination chamber.

Insulator contamination performance is related to the leakage, also called creepage, distance of the insulator. The leakage distance is the length, measured along the insulator surface, from the conductor attachment fitting to the structure attachment fitting. A value of 1-inch leakage distance per phase-to-ground kV is frequently used for porcelain insulators in lightly contaminated areas. Even smaller values, based on knowledge of the performance of the existing line and other lines in the area, have been successfully used in uprated lines.

If contamination or insufficient leakage distance prevents operation at a higher voltage level, several options should be considered:

(1) The leakage distance may be increased by increasing the insulator string length by one or more insulator units. When an insulator string is lengthened, it may encroach on required air gap clearances. These clearances are between the conductor and the tower under extreme wind swing conditions for power frequency voltage, and clearance between the conductor and lower tower arms for switching surges. Additional insulator length lowers the conductors and may leave insufficient ground or structure clearance. In an upgrading, there is usually a limitation on insulator string length, as shown in Figure 6-3.

FIGURE 6-3
LIMITATION ON LENGTH OF INSULATOR STRINGS

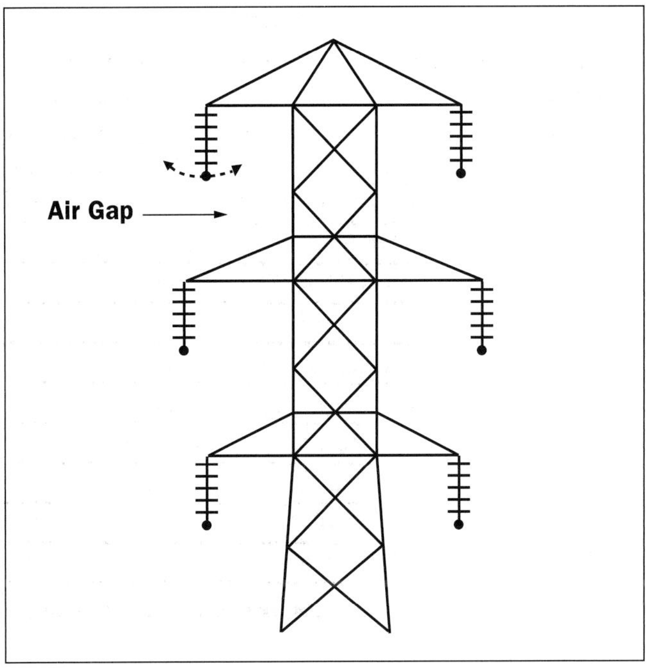

(2) Replace existing insulator units with insulator units having greater leakage distance, such as fog-type insulators. These insulators have the same length between ends, but have wider skirts or deeper grooves to increase the leakage distance.

(3) Replacing porcelain strings with polymers may be an alternative to obtain improved contamination performance with the same insulator length[5].

(4) Insulator washing or application of a hydrophobic coating can be used to improve the contamination performance of the existing insulators.

6.3.3 Switching Surge Flashover

Detailed insulation studies were seldom performed prior to construction of older transmission lines. Switching surge analysis did not become common until the 345 kV level was introduced in the 1950's, and even higher voltage levels in the 1960's. When stressing the line insulation to the maximum with a voltage upgrading, switching surge considerations become important, even at 115 or 230 kV. Therefore, a switching surge study is an important part of a voltage upgrading study.

The frequency of flashover of a transmission line due to switching operations is obtained using a simulation of the transmission line insulation under the effects of various atmospheric conditions. Line insulation strength is statistically dependent on wind swing angle of the insulator strings and atmospheric conditions that affect the insulation strength of air gaps. The insulation stress is represented as a distribution of overvoltages resulting from switching operations, and is a function of the line geometry, length, and source impedance. Probability of line flashover (PFO) is then determined mathematically as a combination of the stress and strength of the transmission line insulation.

When a switching surge study is performed for a new line, a maximum PFO (typically 0.001) is specified, and the line is designed to meet the specification. For a voltage upgrading study, a switching surge analysis is performed on the existing line at the new voltage to determine the probability of flashover. The switching surge is based on statistical distributions of possible overvoltages during line energization. These surge distributions are determined from a computer study, frequently using the Electromagnetic Transients Program (EMTP). If high speed reclosing is applied, the distributions must include the effect of trapped charge remaining on the line after the fault has been cleared. If the desired PFO is exceeded, one or more of the following actions can be taken:

(1) Circuit breakers with closing resistors can be specified to limit the surge distribution. Normally applied to EHV lines, closing resistors have been advantageously applied to upgraded lines as low as 115 kV where clearances and insulator lengths are reduced.

(2) Synchronized closing of circuit breakers can be specified to limit the surge distribution.

(3) Surge arresters may be applied at multiple points along the line to limit the surges to a tolerable level.

(4) The length of the insulator strings may be changed. Longer strings have greater flashover strength along the insulator string. However, longer insulator strings may result in smaller air gap clearances and increased flashovers to the structure.

(5) Reconsideration of the permissible PFO for the uprated voltage level may be required. When all factors are considered, including the consequences of a trip-out and an attempt at re-energization, a higher PFO may be acceptable.

6.3.4 Lightning Performance

The lightning performance of a transmission line is measured by the number of lightning-produced tripouts per year. A lightning stroke can initiate an arc between phases, or more commonly, from one or more phases to a grounded portion of the structure. The arc can sometimes be self-extinguishing, depending on the instantaneous value of the phase voltage or characteristics of the lightning stroke. If the line voltage is sufficient to sustain the arc, the fault must be cleared by circuit breaker operation. For transmission lines, all lightning arcs are considered to be the results of direct strikes.

Direct strike flashovers are classified as either shielding failures or back-flashovers. Shielding failures are due to strokes to a phase conductor bypassing a shield wire, causing a rise in potential of the phase conductor and resulting in line insulation flashover. Back-flashovers are due to strokes to towers or shield wires causing a rise in potential of the local ground and subsequent flashover to the phase conductors.

Lightning tripout rate is a function of line insulation, tower geometry, shield wires, span length, and tower footing resistance. A voltage upgrading study has the advantage over a new line design in that a number of years of operating experience are available. This experience can be used to calibrate the results of an analysis of the existing line design with a computer program to predict lightning performance. A comparison of lightning program output for the line before and after the upgrading with the operating record of the existing line will give a better indication of the expected lightning performance than merely a calculation of the performance after upgrading.

Increased operating voltage across the insulators and air gap space will cause the tripout rate to increase somewhat. The increase is normally very slight because the power frequency component is a small part of the total lightning impulse voltage. Also, for half the ac voltage wave cycle, the increased voltage has a polarity that opposes the lightning voltage and reduces the insulator stress. An increase in insulator string length for power frequency or switching surge voltages will also improve lightning performance. These effects combine to make the effect of voltage upgrading on lightning performance a secondary consideration.

Perhaps the most important single parameter that affects lightning performance is footing resistance. Additional ground rods and/or counterpoise can offset the increase in lightning tripouts that would otherwise result from voltage upgrading. The installation of underbuilt shield wires is also beneficial in spot locations, such as crossings of exposed rocky ridges, where frequent lightning tripouts are expected.

6.4 CORONA LIMITATIONS

As discussed in Chapter 4, corona effects present a fundamental limitation to the voltage applied to any line design. While all the analyses presented in that chapter may be required for the permit process, certain corona-related effects are especially relevant to transmission line voltage upgrading. Two particularly important effects are radio noise and audible noise. Under certain conditions, the electrical loss due to corona also may be an important consideration.

6.4.1 Radio and Audible Noise

Corona presents a fundamental limitation to the voltage that may be applied to any conductor type and configuration since corona-induced radio noise and audible noise will become excessive at some voltage level. If the desired increase in voltage is above that at which either type noise is a limiting factor, some modification to the line design will be necessary.

An evaluation of radio and audible noise performance of the existing line before a voltage upgrading will help to indicate the degree of the noise problem to be expected after the voltage upgrading. Combined with noise sources on the existing line, new sources resulting from the increased electrical stress due to the upgrading will obviously increase noise levels. It is to be expected, at the very least, that it will be necessary to locate these sources and reduce them to within acceptable levels.

The evaluation of conductor corona radio, television and audible noises related to voltage upgrading involves the same three stages applied to new line construction: selection of appropriate criteria, prediction of the effect, and evaluation of the effect with respect to the criteria. For a radio noise evaluation, appropriate radio station signal strength, applicable signal-to-noise ratio, weather conditions, and measurement location are considerations involved in the selection of criteria. Once the criteria are established, conductor noise predictions are then made with the aid of a radio noise computer program. The predictions are then compared to the criteria to evaluate the radio noise performance of the line at the new voltage level. If radio noise performance is inadequate, consideration should be given to reconductoring or adding a second conductor to make a 2-conductor bundle per phase. If the decision is to add a second conductor to each phase, it will almost certainly be necessary to reinforce the structures. Regardless of whether the decision is to reconductor or to add a second conductor per phase, the modification must be evaluated by using the noise prediction program.

Radio-noise computer models usually deal with conductor corona only, assuming that noise from hardware components and insulators is insignificant[6]. This assumption is usually acceptable for new line construction, but may not be valid for a voltage upgrading. It is possible the new voltage may only be 115 or 230 kV, but the electrical stress at the hardware surface may be equivalent to that normally produced at 345 or 500 kV levels. These higher stress levels usually result from closer conductor spacings in the uprated line than are typical for 115 or 230 kV conventional designs. The closer spacing, together with higher voltage, produces a larger electric field that creates greater electrical stress. Under these conditions, the existing hardware may produce excessive corona noise and must be replaced with corona-free hardware of the type used for EHV lines.

When hardware corona is a concern, a sample of the hardware can be removed from the existing line and tested in a high voltage laboratory. Traditional corona tests, however, may not give a valid indication of the corona performance of the hardware at the higher electrical stress level. These tests are based simply on a radio noise voltage measurement with the hardware energized at a set percentage above operating voltage, not taking into account the increased electrical stress that may result from closer conductor spacing. To evaluate the existing hardware for use in the upgraded line, therefore, a more accurate testing procedure based on measurement of the approximated electrical field at the surface of the hardware must be used.

When evaluation of corona noise performance has been properly made, experience indicates that over 90% of radio interference complaints and nearly all television interference complaints result from spark sources. Increasing line voltage increases electrical stress on all transmission line components and may result in new spark noise sources. It is desirable, therefore, to check the existing line for possible spark noise sources and to identify locations where sparking may occur after upgrading[7]. This is an especially important consideration for wood structures where sparking has been found at a number of unique locations. An example of such locations is at staples making poor contact with down leads.

Although radio noise is the main concern for most existing transmission lines below 345 kV with typical phase-to-phase spacing and conductor size, audible noise is generally a factor for conventional lines above 345 kV. When considering a voltage upgrading, the conductor surface electric field and resulting corona are often pushed near the levels where audible noise becomes a limiting factor. It is important, therefore, to evaluate audible noise performance whenever considering a voltage upgrade. Even if audible noise is not anticipated to be a limiting factor, an evaluation probably will be necessary to pass the permit procedure.

6.4.2 Corona Loss

Electrical loss due to corona discharge, commonly called corona loss, is usually limited to insignificant levels because of the limitations placed on corona-induced radio and audible noise. There may be occasional applications, however, that may arise in a remote location where noise is not a major factor due to sparse population. In such cases, the conductor surface electric field may be taken to higher values than are customary, and corona loss may become a factor. Whether the desired voltage upgrading for a given line is feasible may depend on the ambient conditions. Since moisture increases corona activity, a drier climate will yield a lower annual energy loss for the same line than a wetter climate. Data that can be used to estimate corona loss and its effects is found in *Transmission Line Reference Book, 345 kV and Above, Second Edition*[6].

6.5 ELECTRIC AND MAGNETIC FIELDS

An increase in line voltage causes a proportional increase in the electric field surrounding the line. Since some governmental entities place specific limits on electric fields, the increase in the electric field resulting from a voltage upgrading must be seriously considered. These limits usually apply to the level measured at the edge of the right-of-way, but are sometimes given as maximums anywhere on the right-of-way.

Fortunately, ground-level electric fields are not often a limiting factor in a voltage upgrading. In cases where upgrading results in closer conductor spacing than customary for that voltage class, the electric field near ground level is reduced due to a mutual partial cancellation of the individual electric fields of the conductor. It is often the case therefore that a voltage-upgraded line exhibits a lower ground-level field than that of new constructions at the same voltage level.

Electric fields are a concern related to transmission lines because of the electrical currents these fields may induce into nearby objects. The National Electrical Safety Code[8] stipulates that no more than 5 mA can be induced in the largest vehicle located near the line. A voltage upgrading study should include, therefore, the calculation of the open-circuit induced voltage and short-circuit induced current on such a vehicle under worst-case conditions. The effects of spark discharge and steady currents on a person touching the vehicle can then be determined.

Magnetic fields seldom become a limiting factor in voltage upgradings. Since the strength of a magnetic field is a function of line current rather than voltage, the effect of the voltage upgrading on the magnetic field depends on the line loading following the upgrading. If an upgraded line were to be operated at the same power flow level as before the voltage upgrading, less current would be required, resulting in a reduction in the magnetic field. If the line were to be operated at the same thermal limit, or current flow, as before the upgrading, the magnetic field would remain the same even though power flow would be increased. It should be kept in mind, however, that increases in current-carrying capacity, or the utilization of certain conductor configurations, will result in an increase in the magnetic field.

6.6 GROUND CLEARANCE

It is imperative that all relevant codes be checked in any voltage upgrading to ensure that ground clearances are adequate for the proposed voltage. Some transmission lines in the United States, built in accordance with earlier editions of the National Electrical Safety Code, have very conservative ground clearances. It is quite possible these lines could carry higher voltages and still be within the clearances outlined in recent code editions. Table 6-1 shows the gradual reduction in overall clearance requirements for terrain without vehicle access made possible through each code revision. The 1984 (Seventh Edition) code clearance requirements had the greatest influence on upgrading since it specifically permitted reduced clearance where switching surges are controlled. The table shows that for a 500-ft span, a line designed for 115 kV could operate within code requirements at 345 kV with surge control. Surge control would likewise permit a 230 kV line to operate at 345 kV and, provided that a 345 kV line slightly exceeded its original code requirements, it could be uprated to 500 kV. The foregoing applies, of course, only to line-to-ground clearance requirements.

Code editions prior to the Sixth Edition contained a provision for minimum spacing between conductors of the same circuit. This requirement was eliminated in the Seventh Edition with the statement that "No requirement is specified for clearance between conductors of the same circuit when rated above 50 kilovolts." This change is important because it allows higher voltages for the same conductor spacing. After the Sixth Edition, phase spacing is limited only by the factors discussed in this chapter.

If the clearances are inadequate for the increased voltage due to clearance or code reasons, sags must be reduced or the structures raised. Perhaps re-tensioning the line is all that is necessary to provide adequate ground clearance. If re-tensioning is considered, however, attention must be paid to the possibility of increased conductor vibration and the increased loadings on angle and dead-end structures. In rare cases, when only a few spans are limiting, necessary ground clearances have been attained by grading the underlying terrain.

6.7 RECONDUCTORING

Corona performance of an existing conductor may not allow the line to operate at a higher voltage. Since corona-induced radio or audible noise may be excessive, or in special cases, corona loss may be unacceptable. Increasing the voltage on a conductor causes greater electrical stresses on the conductor, thereby possibly causing corona or noise. These problems along with others relative to voltage upgrading can be alleviated by reconductoring the existing line.

Selection of the new conductor may result in improved thermal properties, allowing for greater current transfer. An increase in thermal capacity allows for an increase in the total power flow of the conductor. Reconductoring a line can range from (1) reconductoring the line with a larger standard conductor, to (2) reconductoring the line with a special conductor that is capable of higher unloaded tension, to (3) conductor bundling.

The primary considerations specific to voltage upgrading are the conductor surface electric field and corona performance with the new conductor. However, when selecting a conductor for reconductoring, considerations should be given to the following:

(1) A conductor with lower electrical resistance will allow operation at lower temperatures. Lower operating temperatures yield decreased loss of conductor strength, lower electrical losses and reduced sags.

(2) A conductor with lower thermal elongation will result in less sag, thereby allowing an increase in the maximum allowable conductor temperature of the line.

TABLE 6-1
NATIONAL ELECTRICAL SAFETY CODE CLEARANCE REQUIREMENTS
(ROADS, STREETS, ALLEYS, ETC. 500-FT SPANS ASSUMED)

	Clearance Requirements per Voltage Class, ft			
Code Edition	**115 kV**	**230 kV**	**345 kV**	**500 kV**
4th	25.18	28.95	31.72	35.44
5th	25.93	28.70	31.47	35.19
6th	25.80	28.01	30.22	33.21
7th	24.80	26.01	29.22	32.21
7th*	—	23.39	25.69	29.71

** With switching surge control to 2.2 p.u.*

(3) A conductor capable of operating at higher temperatures has a higher conductor annealing temperature permitting greater maximum allowable conductor temperature while retaining the same conductor loss of strength limitations.

(4) A conductor with improved self-damping and motion-resistant properties allows the conductor to be strung to a higher unloaded tension, thereby reducing the sag at the maximum allowable temperature.

Each approach of reconductoring offers distinct advantages. However, the benefits and consequences of each method must be assessed before making a decision regarding which approach to take. Three methods of reconductoring are the use of:

Larger-Diameter Standard Conductors
Reconductoring a line with a sufficiently larger-diameter conductor reduces the possibilities of corona. The larger conductor has reduced electrical stress and lower electrical resistance allowing for a lower operating temperature. Usually, the selected conductor is of the same type. In some cases, however, some all-aluminum or copper lines are reconductored with larger ACSR conductors. ACSR composite conductors offer the advantage of reduced resistance, lower thermal elongation and creep characteristics, and improved loss-of-strength properties.

Using a larger conductor may require structural modification. Although in most cases, it does not. If structural modification is required, the cost is primarily dependent on the diameter of the chosen conductor. Chapter 7 discusses these factors and criteria in further detail.

Special Conductors
Special conductors are designed for the purpose of yielding improved performance under specific conditions. Self-damping conductors, for example, are designed to specifically reduce conductor fatigue caused by aeolian vibrations. Self-damping conductors can be strung at much higher tension levels than standard conductors yielding reduced sag. There are many other advantages to self-damping conductors, and the many other special conductors used in line design. These special conductors, along with their advantages and disadvantages, are also discussed in Chapter 7.

Horizontal and Vertical Bundling
Conductor bundling is achieved by combining a new conductor with the existing conductor of a line to achieve greater voltage (or thermal) capacity. Most new transmission lines are constructed with 2-conductor bundles with the two conductors per phase placed in a horizontal (flat) configuration. In a horizontal configuration, both subconductors are placed at the same height. However, when adding a second conductor per phase for a voltage upgrading, the following should be considered:

(1) When a second conductor is added per phase, significantly more space is required for the phase conductors than for the original single conductor. In lines having a horizontal phase configuration, the use of a vertical bundle will avoid reducing the phase-to-phase and phase-to-structure spacing. In lines having a vertical phase configuration, the use of a horizontal bundle will avoid reducing the phase-to-phase spacing but will still reduce the phase-to-structure spacing.

(2) Whether horizontal or vertical, the sag of the new bundle subconductor will increase whereas the existing subconductor is largely in its final sag state. This difficulty can be dealt with by pre-stressing the new conductor. Generally, small sag differences are less noticeable in vertical than in horizontal bundles.

(3) Horizontal bundles are limited to a minimum spacing of approximately 12 times the subconductor diameter or greater to avoid wake-induced subconductor vibrations in wind. These vibrations result when one downwind conductor oscillates in the wake of the upwind conductor. With a vertical bundle, neither conductor is downwind of the other, and wake-induced vibration is eliminated.

It may be appropriate to chance occasional wake-induced bundle movement for a horizontal bundle by using a subconductor spacing less than the commonly applied value of 18 inches. Bundle spacings of 9 inches and 12 inches have been used for vertical bundles where one conductor is not in the wake of the other conductor. Because a voltage upgrading study is a balancing act between a number of often conflicting variables, spacings smaller than 18 inches may be beneficial in achieving the optimum overall design. Radio noise performance is affected only slightly by the reduction in bundle spacing, especially if vertical bundles are used. Regardless of bundling type, structure loads are changed and their implications must be evaluated in the process.

The addition of a second identical conductor per phase basically doubles the vertical and transverse loads on the structures, and may require reinforcement of the structures to handle the increased loading. With wood pole or lattice H-frame structures, the addition of an X-brace and bracing of the crossarm may be sufficient. Structure guying or other reinforcement may also be needed.

6.8 SIX-PHASE CONVERSION OF EXISTING DOUBLE-CIRCUIT LINES

An option for voltage upgrading of double-circuit 3-phase lines is the conversion to single-circuit 6-phase lines with an increased phase-to-ground voltage. This option is particularly applicable when corona is the limiting factor. Because a double-circuit 3-phase line has six conductors in place, all the conductors are already present for a 6-phase line. A 6-phase line has 60 electrical degrees between adjacent phase conductors and a lower conductor surface electric field than the 3-phase alternatives. Conversion to 6-phase allows a voltage increase without exceeding the allowed conductor surface electric field.

For any power transmission line, the power delivered can be written in terms of phase-to-ground voltage as:

$$P = NVI\,(PF) \tag{6-1}$$

where:

- P = *total power, MVA*
- N = *number of phases (six phases for two 3-phase circuits)*
- V = *phase-to-ground voltage, kV*
- I = *phase current, kA*
- PF = *power factor*

Use of phase-to-ground voltage is convenient when other than normal 3-phase circuits with 120° phase angles are considered. For a double-circuit 3-phase line, anything that increases the phase-to-ground voltage will increase the maximum power capacity of the line. A 6-phase transmission line involves six equal magnitude voltages spaced 60° apart. Conversion of an existing double-circuit 3-phase transmission line to single-circuit 6-phase at the same phase-to-ground voltage reduces the electric field at the conductor surface and, consequently, the corona and noise.

Research of existing double-circuit lines indicates that some are adequately designed to allow operation at a higher voltage without modifications. Some lines have such small conductors that a voltage increase is not practical without reconductoring. However, a number of lines are capable of being uprated if converted to 6-phase to alleviate corona and noise[9]. A demonstration installation of this type conversion applied to the voltage upgrading of a double-circuit 115 kV transmission line was constructed near Binghamton, New York[10].

Any 6-phase conversion requires substation modifications and the application of advanced relay protection methods. These modifications are well within the boundaries of present practice[11]. The concepts of phase conversion transformers have been understood at least since the 1930s, and several different types are available for this application[12].

An engineering study for a 6-phase voltage upgrading of an existing double-circuit transmission line follows the same procedure as a voltage upgrading study for a 3-phase voltage increase. This includes electrical environmental effects, insulation coordination, system studies, and economic analysis[13]. Substations and protection must be modified for any voltage conversion, so the additional complexity of a 6-phase system is one of detail, not one involving a major departure from a 3-phase system.

6.9 GENERAL CONCERNS

Economic analysis of voltage upgrading alternatives provides a rational basis for choice. Other factors, however, related to safety and reliability may determine the overall best choice. In addition to different voltage upgrading schemes for a particular project, other considerations may include new construction or thermal uprating. The choices based on technical merit may dictate either increasing the voltage on an existing circuit or constructing an entirely new circuit at the increased voltage. The economic analysis may reduce to the comparison of construction costs of upgrading versus the construction costs of the new line. If the selection of a conductor is involved, whether for the upgrading or new construction alternative, the present worth of the cost of electrical losses plays an important role in the choice. In a few cases, voltage upgrading may be considered, or justified by, the savings due to lesser electrical losses.

Major capital and operational cost elements associated with upgrading the voltage on a transmission line include the following:

- Replacement or addition of insulators and hardware
- Replacement and reinforcement of existing structures

- Purchase, stringing, sagging and clipping of new conductor
- Increased maintenance or increased cost of more difficult maintenance

In the rare case where a particular line can operate without modification at the next higher voltage class, there are no capital costs associated with the project. However, there may be increased maintenance costs.

Compared to the existing voltage, increasing the line voltage results in a potentially significant reduction in the cost of electrical losses. Other economic benefits relate to increased transfer capability and the increased flexibility of operation to optimize the overall system.

Many transmission line upgrading projects are not justifiable on a purely economic basis. Considerations of safety or system reliability are as likely to prompt the decision to voltage upgrade a line as is the opportunity to reduce electrical losses or to operate the transmission and generation grid at a marginally lower cost per kw-hr. On tie lines between systems or on radial feeds from low-cost generators where major savings from the purchase of lower cost power or increased income from the sale of power is at stake, economic justification of upgrading is more likely.

If the capacity of certain lines is increased, then the probability that the system operator will be forced to take emergency actions, such as load shedding or initiating quick startup generation, is reduced. This increase in reliability of power supply to interruptible customers and reduced use of relatively expensive generation has a real cost savings potential.

The decision of whether to proceed or not with a voltage upgrading is made based on one or more factors: technical, economic, permitting or system operation. The degree of difficulty of obtaining approval and a right-of-way for the project may be the dominating concern. Before proceeding with a voltage upgrading, all factors and criteria must be considered.

6.10 REFERENCES

[1] Nabet, G., "Reduced Insulation for Overhead 230 kV Transmission Line,"Southeastern Electric Exchange Transmission Section, April 18, 1975.

[2] Broschat, M., and Clayton, R., "Compaction Techniques Applied to Subtransmission Line Uprating 41.6 kV to 115 kV," *IEEE Transactions on Power Apparatus and Systems*, Vol. PAS-100, No. 4, p. 1959, April, 1981.

[3] Koller, J. P., "Uprate 34-kV Line to 138-kV Emergency Tie," *Transmission and Distribution*.

[4] *Transmission Line Reference Book, 115-138 kV Compact Line Design*, Electric Power Research Institute, 1978.

[5] *IEEE Guide for Application of Composite Insulators*, ANSI/IEEE Std 987-2001.

[6] *Transmission Line Reference Book 345 kV and Above/Second Edition*, Electric Power Research Institute, Palo Alto, CA, 1982.

[7] Loftness, M. O., *Power Line Interference, A Practical Handbook*, National Rural Electric Cooperative Association, 1992.

[8] *National Electrical Safety Code®*, C2-2002, 2002 Edition, The Institute of Electrical and Electronic Engineers, Inc.

[9] Kallaur, E., and Stewart, J. R., "Uprating Without Reconductoring - The Potential of Six-Phase," Canadian Communications & Energy Conference, Conference Record, IEEE Publication 82CH1825-9, October 13-15, 1982.

[10] Dorazio, T. F., Stewart, J. R., Wilson, D. D., "Six-Phase Power Transmission: First Utility Application," CIGRE Paper 600-01, CIGRE Symposium Compacting Overhead Transmission Lines, Leningrad, June, 1991.

[11] Rebbapragada, R. V., Brown, M. T., Dorazio, T. F., Stewart, J. R., "Design Modification and Layout of Utility Substations for Six Phase Transmission," *IEEE Transactions on Power Delivery*, Vol. 8, No. 1, p. 24, January, 1993.

[12] Stewart, J. R., Oppel, L. J., Thomann, G. C., Dorazio, T. F., Rebbapragada, R. V., "Transformer Winding Selection Associated With Reconfiguration of Existing Double Circuit Line to Six-Phase Operation," *IEEE Transactions on Power Delivery*, Vol. 7, No. 2, p. 979, April, 1992.

[13] Stewart, J. R., Oppel, L. J., Thomann, G. C., Dorazio, T. F., Brown, M. T., "Insulation Coordination, Environmental, and System Analysis of Existing Double Circuit Line Reconfigured to Six-Phase Operation," *IEEE Transactions on Power Delivery*, Vol. 7, No. 3., p. 1628, July, 1992.

7 THERMAL UPRATINGS OF EXISTING LINES

The need for additional power transmission capacity has traditionally been met by constructing new lines and substations. In many areas of the United States, as more and larger lines have been built, public opposition has increased to the point where it is difficult to obtain permission for new facilities. This opposition has been created by such issues as land use, esthetics and electrical environmental effects. An additional deterrent to the construction of new lines is the uncertainty in projecting future load growth on particular circuits which is a natural result of the concept of "open access" for generators.

When increased power transmission is required, both problems may be circumvented by the thermal uprating of existing lines. Within the power industry, the phrase "thermal rating" is used to denote the maximum allowable current that a line can accommodate and remain within the maximum allowable conductor temperature limit. By definition, thermal uprating encompasses all possible methods to increase the maximum allowable electrical current without damaging the line or violating safety clearance requirements. These methods may range from (1) simply reassessing very conservative estimates of variables employed in the calculation of maximum allowable conductor temperature to (2) real-time thermal rating and monitoring of the line to (3) reconductoring, resagging, and/or modifying structures in the existing line.

The need for an increase in the thermal capacity of a particular transmission circuit is usually identified during system planning studies. Such studies attempt to answer both short- and long-term system planning questions, ranging from how to survive the next winter or summer peak, to how to best handle the power transmission from a proposed major generation addition. Often, these system effects are complicated and involve several utilities. Short- and long-range load flow, fault assessment, and stability studies are all required to predict future transmission needs and identify circuit uprating candidates. Such studies are normally carried out by system planning personnel who then pass the results to line-design engineers.

7.1 INTRODUCTION TO LINE UPRATING

The selection of the best method for uprating to meet new system requirements involves several factors. Consideration must be given not only to the cause of the present thermal limitations, but to physical limitations, institutional constraints and economic conditions, as well.

7.1.1 Present Limitations

The design engineer attempting to select the best uprating method needs to know the cause of the present system operation limitations. Possibilities include inadequate thermal capacity during normal operating conditions, insufficient safety margins for low probability contingencies and inability to meet seasonal load demand. Additional factors to be considered are the accuracy of the projected load growth, whether the line can be taken out of service for a time, can the line be worked "hot," etc.

7.1.2 Major Economic Factors

Selection of the most economic thermal uprating method requires consideration of the capital cost and the costs of electrical losses and maintenance. One alternative may require greater immediate capital cost than another, but may be more economically viable due to lesser electrical losses during the life of the uprated line. A detailed analysis must incorporate estimated changes in energy costs, interest rates, and other factors to evaluate the sensitivity of the optimum solution to changes in economic parameters. Generally, the projection of significantly higher energy costs will favor the use of larger transmission conductors.

Often financial, rather than economic, factors dominate. A project may be economically advantageous but be prohibitive for financial reasons. An example is the oft-repeated claim that it would be economically justified to replace all distribution transformers with more efficient ones to reduce the electrical losses. While an economic case could be made, the financial aspects relating to such a large expenditure are prohibitive. There are situations, therefore, when the ideal and optimum solutions do not coincide.

7.1.3 Physical and Institutional Constraints

Physical and institutional considerations strongly influence the optimum solution to the thermal uprating of transmission lines. If the indicated approach results in increased structural requirements due to the use of larger conductors or an additional conductor per phase, the capability of the line's existing structures and foundations must be re-evaluated. Reconductoring frequently requires rebuilding of angle structures, while tangent structures often require only minimal changes.

Consideration must also be given to the present institutional constraints. It may be impossible to gain regulatory approval, for instance, if significant structural changes are required. In this situation, reconductoring may be limited to those conductors that can be supported using the existing structures. It is possible, however, that the increased reliability of the proposed uprated approach will outweigh the objection to structural modifications.

7.2 DETERMINATION OF THERMAL RATINGS FOR OVERHEAD LINES

Limiting conductor temperature is necessary in the design and operation of overhead lines because of two design considerations — the avoidance of conductor and hardware failures under maximum ice or wind loads and the maintenance of minimum ground clearances. The maximum allowable conductor temperature is traditionally specified to limit loss of strength to no more than 10% during the life of the line, and to limit the component of sag attributed to thermal expansion.

The traditional (static) thermal rating of a transmission line is normally calculated under assumed weather conditions of low wind speed and high air temperature. Actual weather conditions may result in greater heat transfer than the assumed conditions. Thus, the conductor temperature may be less than the maximum allowable value used in calculating the thermal rating, even when current on the line is equal to the thermal rating.

For most transmission lines, even those that are in need of thermal uprating, everyday load currents do not normally approach the line's thermal rating. Thus, if the conservative weather conditions assumed in calculating the static thermal rating of the line actually occur periodically, and the current on the line is equal to its thermal rating, the conductor temperature would be significantly below the maximum allowable temperature most of the time. Thermal overload is normally a concern only under certain contingency conditions where current may increase to a value well above normal loading.

The maximum allowable power flow in a power transmission circuit, measured in MVA, is a function of the maximum allowable phase current A_p (kA), the phase-to-ground voltage V_{pg} (kV) or the phase-to-phase voltage, V_{pp} (kV):

$$MVA = 3\, V_{pg}\, A_p = \sqrt{3}\, V_{pp}\, A_p \tag{7-1}$$

The maximum allowable phase current is directly dependent on the maximum allowable conductor temperature. The determination of the maximum allowable conductor temperature is extremely important. To err on the liberal side in making this determination may cause conductor loss of strength, physical damage to the hardware, increase of conductor sag (decrease clearance) beyond acceptable limits, or excessive line loss. To err on the conservative side can cause an unwarranted limitation of power transmission, resulting in great financial loss to the utility.

7.2.1 *Thermal Rating Parameters*

The temperature of an overhead conductor is primarily a function of current flow; ambient conditions (wind speed and direction, air temperature, and solar radiation); and the physical properties of the conductor (electrical resistivity, emissivity, absorptivity and specific heat).

- Ohmic (I^2R) line loss, the main source of heat input to the conductor, is a function of electrical current and conductor resistance. Heat input from solar radiation is a much smaller source, even during daylight hours, and is impacted by the conductor's absorptivity.

- Heat is lost from the conductor primarily through convection and radiation. The rate of convected heat loss is influenced by conductor diameter, air temperature, wind speed and wind direction. The rate of radiated heat loss is not dependent on wind speed or direction, but is dependent on conductor emissivity and diameter.

- The temperature of the conductor is an indirect measurement of the thermal energy (heat) stored within the conductor. As thermal energy is added, much of it is stored within the conductor, raising the conductor temperature. This thermal storage capacity is a function of the specific heat of the conductor material. These heat losses are shown in Figure 7-1.

FIGURE 7-1
PICTORIAL DIAGRAM OF CONDUCTOR HEAT BALANCE [1]

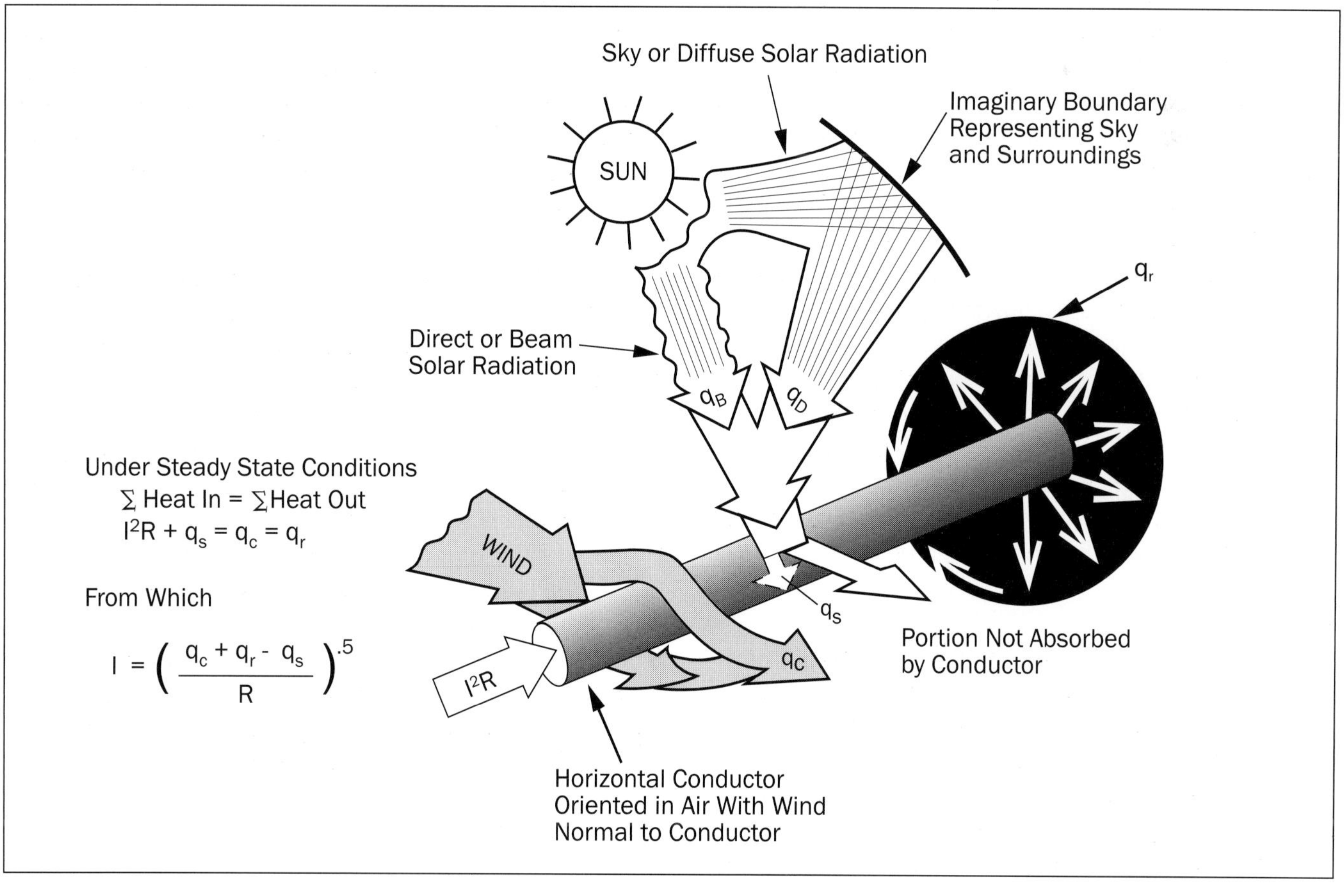

7.2.2 Thermal Rating Terms and Definitions

Conductor temperature and its consequences — increased sag and loss of strength — can be measured, as can current flow. The thermal rating of a conductor, however, is not a measurable quantity. It is defined as the electrical current which would produce the maximum allowable conductor temperature under the assumed or measured conditions. Since several definitions are commonly used, the following definitions of thermal rating terms are presented for clarity.

Static Thermal Ratings

The word static means constant or unchanging. As applied to thermal rating calculations, a static (or sometimes "book") thermal rating implies the rating is calculated using fixed, normally conservative, values for weather variables. Separate static ratings may be calculated for summer and winter with a higher air temperature for summer and a lower air temperature for winter. Power utility thermal line ratings — normal (continuous) and emergency — are static unless indicated otherwise.

Ambient-Adjusted Thermal Ratings

If thermal ratings are recalculated for a daily or hourly time interval, they are referred to as "ambient" adjusted ratings. Typically, only variations in air temperature and, for transient ratings, load current are considered. The use of air temperature and load current for these shorter periods of time results in a thermal rating that is generally higher than an annual or seasonal static thermal rating.

Real-Time Thermal Ratings

Real-time thermal rating calculations are based on continuous direct measurements of weather parameters in the vicinity of the line, conductor temperature and tension. Real-time thermal ratings may be calculated with a time interval as short as a few minutes. Because of the short recalculation interval, the conductor temperature can be determined either by measurement or calculation during times of severe overload.

Real-time ratings require both measuring devices and communication links. Instruments to measure the line tension, ground clearance (or conductor sag) and weather conditions in the vicinity of the line are commercially available. The use of tension and sag monitoring packages (which also measure weather conditions) has the advantage of measuring the conductor's physical behavior directly and is the best method for determining electrical clearance. Weather monitors are typically less expensive and more easily ported from one location to another, but the electrical clearance must be calculated, a process that may be less accurate than direct measurement.

Maximum Allowable Loss of Strength

Bare overhead stranded conductors normally consist of aluminum or copper strands which can lose tensile strength at temperatures above 94°C. Transmission owners typically limit this loss of total conductor strength to no more than 5% to 10% over the life of the line. Thermal line ratings, particularly normal continuous ratings, should be determined to avoid excessive loss of tensile strength in aluminum or copper strands.

Minimum Allowable Clearance

The minimum clearance is the shortest allowable distance from the conductor to objects below the line. This clearance requirement varies with the line voltage and the type of object. The minimum clearance to ground or other circuit lines normally occurs at the maximum allowable conductor temperature. However, certain conductors such as ACSS do not have these loss of strength issues and can be run at higher temperatures.

7.2.3 Exact Heat Balance Equation

In the most general case, the balance between the heat added to the conductor by ohmic losses and solar heating, the heat lost through convection and radiation, and the heat retained through increased storage within the conductor may be expressed by the following equation:

$$mC_p\left(\frac{dT_c}{dt}\right) = I^2 R_{T_c} + q_s - q_c - q_r \qquad (7\text{-}2)$$

where:

T_c = *conductor temperature, °C*

t = *time, sec*

I = *conductor current, amperes*

R_{T_c} = *resistance of conductor, Ω/ft*

q_s = *solar heat gain, W/ft*

q_r = *radiated heat loss rate, W/ft*

q_c = *convection heat loss rate, W/ft*

mC_p = *heat capacity of the conductor, W-sec/ft-°C*

If the conductor temperature is constant, the derivative of conductor temperature with respect to time is zero. Under equilibrium conditions, therefore, the heat balance equation becomes:

$$I^2 R_{T_c} = q_c + q_r - q_s \qquad (7\text{-}3)$$

As explained in Chapter 3, the radiation and convection cooling terms can be expressed as functions of the conductor temperature and certain conductor physical properties and weather parameters. This steady-state equation, therefore, can be used to calculate the conductor temperature given a known or assumed conductor current. Conversely, the current can be calculated if the conductor temperature is known or assumed.

7.2.4 Approximate Thermal Rating Calculations

In discussing thermal line uprating, it is useful to derive a simplified linear form of the steady-state heat balance equation. Such an approximate derivation can be made if the following simplifications are made:

(1) Based upon the paper by Black and Rehberg[2], the terms concerning convection and radiation can be approximated by linear functions of the conductor temperature rise above air temperature, $T_c – T_a$.

(2) The conductor temperature rise is due to both Ohmic losses and solar heating. Test data indicates the amount of solar temperature rise, ΔT_s, can vary from 1°C to 10°C [2]. For simplification, an average value of 5°C rise in air temperature will be assumed for ΔT_s.

(3) The conductor resistance increase resulting from the conductor temperature increase is assumed to be 4% for every 10°C change in conductor temperature.

The use of these simplifications yields the following form of the steady-state equation:

$$I^2 = \frac{K_c(T_c - T_a - \Delta T_s) + K_r(T_c - T_a - \Delta T_s)}{R_{Tc}} \tag{7-4}$$

$$= K(T_c - T_a - \Delta T_s)$$

where:

$$K = \frac{K_c + K_r}{R_{Tc}}$$

K_c, K_r and K are conductor and weather dependent constants

T_c = *conductor temperature, °C*

T_a = *air temperature,°C*

ΔT_s = *conductor temperature rise due to solar heating, °C*

This equation obviously does not allow the calculation of thermal ratings given T_c and T_a, nor the calculation of conductor temperature given I and T_a. However, given a steady-state thermal rating calculated by a more precise method, the approximate equation provides a simple method of determining the effect of changes in the current on conductor temperature, and the effect of changes in maximum allowable temperature on thermal rating.

$$I_{new} = I_{old}\sqrt{\frac{R_{75}}{R_{100}}}\sqrt{\frac{T_{new} - T_a - \Delta T}{T_{old} - T_a - \Delta T_s}} \tag{7-5}$$

As an example of the application of this equation, consider a sag-limited transmission line designed for a maximum allowable temperature of 75°C, with a R_{75} of 1 ohm/ft. The present thermal rating of the line is 1000 amperes at an air temperature of 40°C and a 2 ft/sec crosswind with sun.

If the structures were raised to the point where the maximum allowable temperature of the line is 100°C, the approximate equation allows an estimate of the new thermal rating in terms of the old rating:

Then per Equation 7-5:

$$I_{new} = I_{old}\sqrt{\frac{1}{1.1}}\sqrt{\frac{100 - 40 - 5}{75 - 40 - 5}}$$

$$= 1.29\,(I_{old})$$

It is concluded, therefore, that the 25°C increase in maximum allowable conductor temperature, from 75°C to 100°C, yields a relatively large increase of about 30% in line thermal rating.

This approximate equation also allows some insight into the conductor temperature rise resulting from a given current. For example, in the original line design, 1000 amperes produced a conductor temperature rise of 35°C, including the 5°C solar heat rise. If the normal load on the line is 300 amperes, the conductor temperature is:

$$T_c = T_a + \Delta T_s + \left(\frac{I_{normal}}{I_{emergency}}\right)^2 (T_{new} - T_a - \Delta T_s) \tag{7-6}$$

Then:

$$T_c = 40 + 50 + \left(\frac{300}{1000}\right)^2 (75 - 40 - 5) = 48°C$$

Thus the temperature rise due to normal line current is only about 3°C (48°C - 45°C), while the assumed typical solar heat rise is 5°C, almost twice the temperature rise due to normal line current.

7.2.5 Influence of Primary Variables on Thermal Rating Calculations

To show the relative importance of the various steady-state heating and cooling flows, several thermal ratings and conductor temperatures for a 795 kcmil - 26/7 ACSR "Drake" conductor are presented in Table 7-1. The data reflects typical summer and winter ambient parameters to show the effects of wind speed, wind angle, solar radiation and ambient temperatures on (1) the thermal rating, given various maximum allowable conductor temperatures, and (2) the conductor temperature given various current loads.

TABLE 7-1
INFLUENCE OF PRIMARY VARIABLES ON THERMAL RATINGS
FOR 795 KCMIL - 26/7 ACSR "DRAKE"

OD = 1.108 in
R_{25} = 0.1166 Ω/mile @ 25°C
R_{75} = 0.1391 Ω/mile @ 75°C
Emissivity = 0.5

Absorptivity = 0.5
Line Elevation = 100 ft above Sea Level
Span Direction = East/West, 43° North Latitude
Sun at 12 Noon, Clear Atmosphere

Wind Speed ft/sec (m/sec)	Wind Angle Degrees*	Air Temp. °C	Solar Heating Y/N	Max. Cond. Temp. °C	Special Conditions	Thermal Rating amps	% Change
2(0.6)	90	40	Y	75	None	730	—
0	NA	40	Y	75	None	480	−34
4(1.2)	90	40	Y	75	None	880	+21
2(0.6)	10	40	Y	75	None	510	−30
2(0.6)	90	10	Y	75	None	1050	+44
2(0.6)	90	40	N	75	None	840	+15
2(0.6)	90	40	Y	50	None	180	−75
2(0.6)	90	40	Y	150	None	1330	+82
2(0.6)	90	40	Y	100	None	990	+36
2(0.6)	10	40	Y	100	None	750	+3
2(0.6)	90	10	Y	100	None	1230	+68
2(0.6)	45	40	Y	75	None	670	−8
4(1.2)	90	10	Y	75	None	1240	+70

* 90° means the wind is perpendicular to a vertical plane passing through the conductor axis.

Some useful general observations that can be drawn from this data follow:

- **The thermal rating of bare overhead conductors is very sensitive to wind speed changes under low wind speed conditions.** With a wind angle of 90°C, an ambient temperature of 40°C and a conductor temperature of 75°C with solar heating, the ampacity for "Drake" with a 2 ft/sec crosswind is 52% higher than for still air (730 vs. 480 amps).

- **Changing the angle of the wind with respect to the conductor has a large impact on its thermal rating.** For a 2 ft/sec wind with solar heating, the thermal rating for "Drake" at 100°C decreases from 990 to 750 amps (24%) as the angle of the wind changes from perpendicular to 10° from parallel.

- **With all other variables remaining constant, the thermal rating increases linearly with decreases in air temperature.** The thermal rating of "Drake" at 100°C, with 2 ft/sec crosswind and sun, increases from 990 amps at 40°C ambient to 1230 amps at 10°C ambient, approximately 8 amps per 1°C decrease in air temperature. However, it is important to note that this is only valid if none of the other variables change, which is seldom the case. For instance, there is considerable test data that shows wind speed is related to ambient temperature. Typically, as ambient temperature decreases, so does wind speed. This can offset, or negate, any increase in thermal rating. Therefore, this approximation should be used with caution.

- **The maximum allowable conductor temperature has a major impact on the thermal rating.** For this conductor, given a 2 ft/sec crosswind with sun, the thermal rating increases from 180 to 1330 amps as the maximum allowable temperature is increased from 50°C to 150°C.

It is apparent from these observations that the choice of weather parameters and maximum conductor temperature are very important in determining thermal ratings. Even small changes in these factors can yield very significant variations in the calculated results.

7.2.6 Influence of Secondary Variables on Thermal Rating Calculations

Certain factors included in thermal rating calculations have a relatively small impact. These parameters include emissivity, absorptivity, small changes in conductor resistance due to slight variations in conductor diameter (cross-sectional area), line elevation, and air clarity. The influences of some of the variables are presented in Table 7-2. As shown, large variations in these parameters yield small variations in thermal rating calculations.

TABLE 7-2

INFLUENCE OF SECONDARY VARIABLES ON THERMAL RATINGS FOR 795 KCMIL - 26/7 ACSR "DRAKE"

OD = 1.108 in
R_{25} = 0.1166 Ω/mile @ 25°C
R_{75} = 0.1391 Ω/mile @ 75°C
Line Elevation = 100 ft above Sea Level
Span Direction = East/West, 43° North Latitude

Sun at 12 Noon
2 ft/sec Crosswind
40°C Air Temperature
75°C Conductor Temperature

Emissivity (ϵ)	Absorptivity (α)	Solar Heating, Qs W/ft	Special Conditions	Thermal Rating amps	% Change
0.5	0.5	4.4	None	730	—
0.5	0.5	0.0	None	835	+14
0.7	0.9	7.9	None	680	–7
0.7	0.9	0.0	None	870	+19
0.5	0.5	4.5	30° N Lat	730	0
0.5	0.5	4.8	OD = 1.219 in	750	+3
0.5	0.5	3.5	Dirty Air	755	+3

Although not yielding direct comparisons, rough interpolations of the data below will lead to the conclusion that these parameters are indeed minor determinants of thermal ratings. It should be noted that emissivity and absorptivity [3][4] approach values of 1.0, especially in areas with severe air pollution. Though the impact on thermal rating is small, conservative values of $\epsilon = 0.7$ and $\alpha = 0.9$ are often used for aged conductors.

7.3 UPRATING EXISTING LINES WITHOUT RECONDUCTORING

Given the difficulty in building new lines and the need to provide adequate transmission capacity, there has been a great deal of activity since 1980 to increase line capacity without spending large sums of money. One widely utilized approach involves the reassessment of line rating assumptions, which has led to the use of higher maximum conductor temperatures and less conservative weather conditions for static ratings. Other approaches include reducing clearance safety buffers, selective tower modifications to raise attachment points and/or reducing high temperature sag in critical spans. These approaches were successful because of the conservative design and rating of lines built prior to the deregulation of utility construction in 1980.

Having applied these "low cost" uprating methods widely (and in some cases too aggressively), transmission line operators now find themselves facing more expensive and disruptive methods of uprating, such as re-conductoring and re-building structures. Fortunately, beginning with the introduction of ACSS in the 1970's, conductors with low thermal elongation and higher operating temperatures, up to 250°C, have become available.

Various uprating methods are discussed in the following sub-sections. They should be considered by starting with the simplest and cheapest approaches and progressing to the more expensive and complex methods. Even if the existing line is to be reconductored, there are numerous alternatives ranging from the use of larger conventional conductors with lower resistance to conductors that permit higher maximum allowable operating temperatures. The selected method(s) of uprating must provide a sufficient increase in line loading capacity to satisfy system planning requirements without leading to a reduction in the physical reliability of the line.

7.3.1 Reassessment of Weather Parameters for Static Ratings

Static thermal ratings for transmission and distribution lines are based in part upon conservative estimates of air temperature, wind speed, and solar heating. Thermal uprating can sometimes be accomplished after a careful review of actual weather conditions in the vicinity of the candidate lines. Typical sources of weather data can be found at airports. However, wind speeds taken in such unobstructed locations may vary from winds along entire transmission line right-of-ways because many lines are often partially shielded by trees or terrain. Because transmission line ratings apply to an entire line that can extend over many miles, application of weather data from a single weather station must be adjusted for variations over the line length. In short, the acceptance of less conservative weather assumptions for thermal rating calculations requires a thorough understanding of how weather station data relates to weather along the line.

Consider a transmission line with a 795 kcmil - 26/7 ACSR "Drake" conductor having a summer static thermal rating of 990 amperes based on a maximum allowable temperature of 100°C, an air temperature of 40°C, a wind speed of 2 ft/sec (0.6 m/sec) blowing perpendicular to the conductor axis, with noontime solar heating. Using hourly weather data, supplied by the U.S. Weather Bureau, during two consecutive summers at a nearby airport, conductor temperatures were calculated with the IEEE method[iv] for an east-west span of ACSR "Drake" carrying a constant current of 990 amperes. The resulting conductor temperature distribution is shown in Figure 7-2.

FIGURE 7-2
THERMAL RATING DISTRIBUTION WITH A CONSTANT CONDUCTOR TEMPERATURE OF 100°C

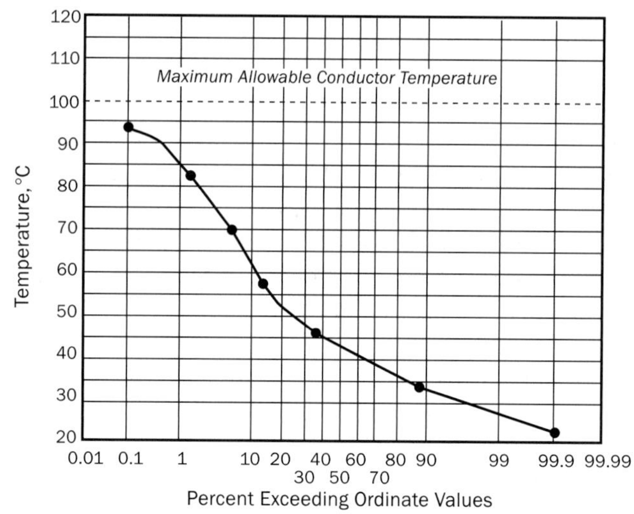

As shown, even though the current in the conductor is equal to the summer static thermal rating, the calculated conductor temperatures are typically well below the maximum allowable temperature of 100°C. In other words, the weather conditions assumed in the static rating calculation are indeed conservative for a conductor at this location. The conductor temperature is less than 43°C 50% of the time.

Alternatively, one may calculate hourly thermal ratings using the same IEEE method and weather data, assuming a continuous conductor temperature of 100°C. The distribution of hourly thermal ratings for the same time period is shown in Figure 7-3.

FIGURE 7-3
CONDUCTOR TEMPERATURE DISTRIBUTION FOR A CONSTANT CURRENT OF 990 AND 1130 AMPERES

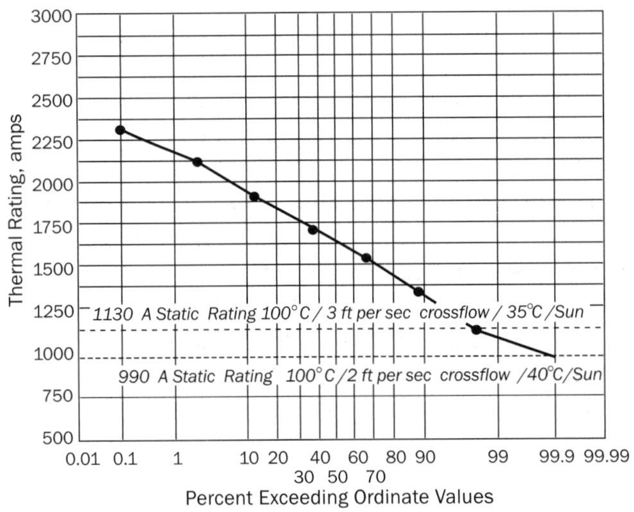

The hourly thermal ratings using actual weather data are higher than the summer static thermal rating of 990 amperes 99.9% of the time and exceed 1650 amperes 50% of the time. One might be tempted at this point to use a less conservative wind speed and air temperature. If a wind speed of 3 ft/sec and an air temperature of 35°C are assumed, the static thermal rating of the conductor increases from 990 to 1130 amperes. The hourly thermal ratings based on weather data now exceed this higher thermal rating only 95% of the time rather than 99.9%.

Plots such as those in Figures 7-2 and 7-3, based on readily-available airport weather data, have generated much enthusiasm for considering the possibility of using less conservative weather conditions to achieve thermal upratings of existing lines. There are, however, a number of cautions that must be noted before less conservative weather conditions are specified. These precautions, based on the preceding analysis, include: (1) collecting weather data at an airport where there are no obstacles to reduce wind speeds; (2) using weather data taken at a single location in the calculation of thermal ratings for a line, whereas the line may be miles in length and is subject to varying weather conditions; (3) using a single conductor orientation when calculating the thermal rating for lines that change direction as they move around obstacles and property boundaries.

Clearly, the calculated temperature and thermal rating distributions of Figures 7-2 and 7-3 would be much more realistic if they were based upon weather data from multiple locations along the line. In that case, the conductor thermal rating would be calculated at each location and the line rating would be the minimum of the ratings. If this is not done, then the following observations must apply to the calculation of line thermal ratings based on single point weather data.

Wind Sheltering by Trees and Terrain

Candidate lines for thermal uprating are often lower voltage lines where trees and terrain can have a major impact on wind speed. While a quantitative estimate of wind speed reduction due to obstructions is not presently available, clearly such effects are real. Reference [5] provides some rough estimates of the sheltering effects of trees when the span is surrounded by a solid stand of trees. Reflecting the height of conductors above ground as a function of voltage, the estimates of thermal rating reduction from unobstructed wind speeds are 30%, 20% and 10% for 115 kV, 230 kV and 500 kV lines, respectively. Other observations regarding wind sheltering are:

- For deciduous trees, the wind speed reductions in winter would be less than in summer.
- Sheltering by trees along the sides of the right-of-way would primarily affect winds that flow across the span, perpendicular to the conductor.
- Sheltering of conductor lengths as short as tens of feet can result in local annealing which may have a minimal affect on sag clearances.

Long, Partially Sheltered, Multi-Directional Lines

A long line may be viewed as a group of dead-end line sections, each having a particular directional orientation. The thermal ratings of long lines are generally less than short lines because:

- Line sections change direction to avoid obstructions and to conform to property boundaries. Because the wind cannot be blowing perpendicular to the conductor in all line sections, the thermal rating for the line can never be higher than the lowest rating for a single-span having non-perpendicular flow.

- The multi-span line will likely have sections that will be sheltered by terrain or trees, resulting in lower wind speeds.

- Long lines can experience different weather conditions in the various spans due to fast moving weather fronts[6]. All things being equal, the longer the line, the lower the thermal rating should be.

As an example of the variation in line rating between line sections of long transmission lines, consider the line ratings for four sections of a 230 kV line in the Southwestern United States as shown in Figure 7-4. The conductor is 954 kcmil - 45/7 ACSR "Rail" with a maximum allowable conductor temperature of 95°C. The rating of each section is calculated based on load cell (conductor tension) measurements. Note that as the day progresses the location of the minimum line rating moves between the sections.

It seems from the preceding observations that the reevaluation of static thermal ratings based upon single location weather data without accounting for wind sheltering, weather variation along the line and line orientation changes, is likely to be overly optimistic. The use of multiple weather stations or the equivalent use of conductor temperature or tension monitors along the line is likely to yield a much more realistic estimate as a basis for possible re-evaluation of line thermal ratings.

FIGURE 7-4
TENSION-BASED 75°C NORMAL RATINGS
9/30/97 - 10/1/97

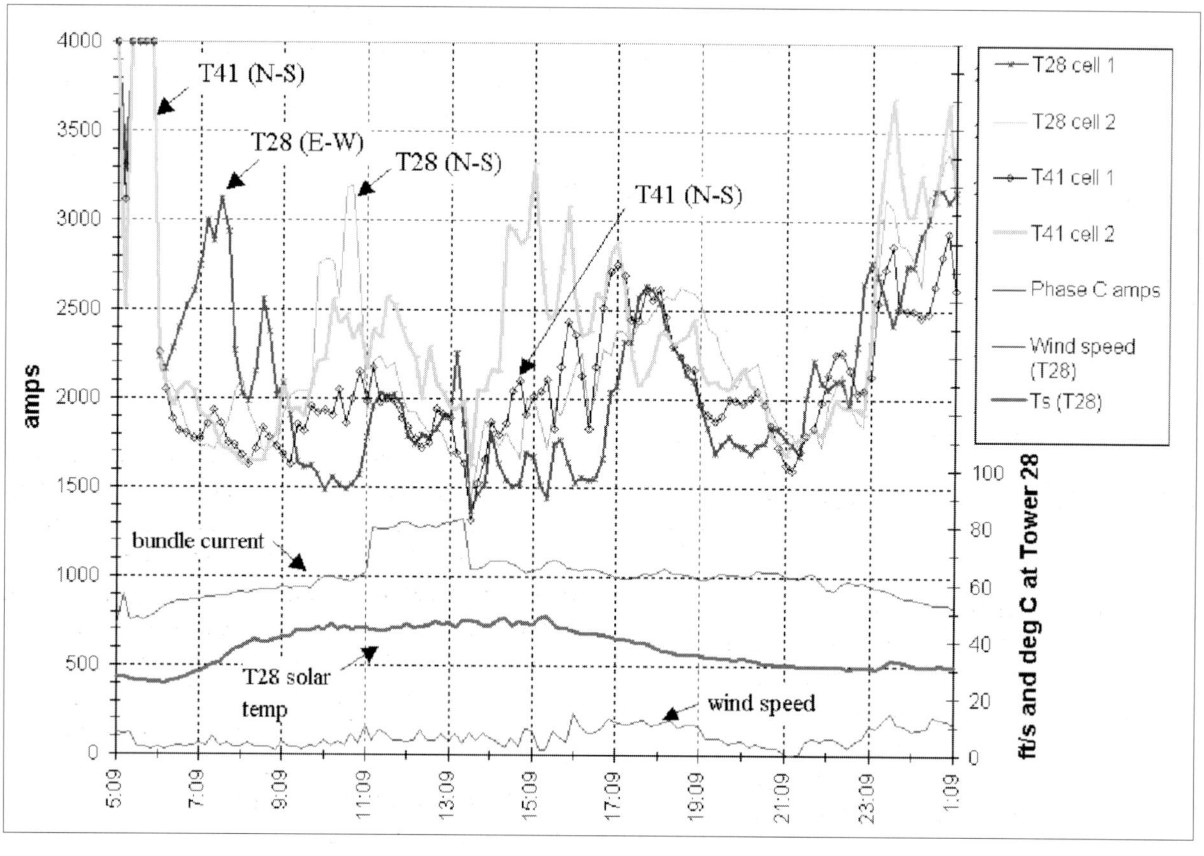

7.3.2 Real-Time Thermal Rating of Overhead Lines

Transmission lines have traditionally been rated by using conservative static weather variables. The preceding section discussed the possible re-evaluation of these weather variables based upon a review of historical weather data along the line. Real-time thermal rating (RTTR) methods use actual, direct measurements of the ambient variables that impact conductor temperature, or direct measurement of conductor temperature or tension to calculate thermal ratings that vary with time. By monitoring these data, the thermal rating of the line is almost always higher than is indicated by a conservative static thermal rating. The use of RTTR methods promise an increase in the effective thermal rating of the line without large capital expenditures and outage times.

The RTTR for overhead lines can be accomplished by several methods. These methods range from monitoring weather conditions at various line locations, to monitoring the conductor temperature at multiple locations with temperature sensing devices mounted on the conductor, to monitoring the conductor tension in multiple-line sections using a tension monitor in series with dead-end insulators. The second and third approaches require the monitoring of weather conditions as well. There are distinct advantages and disadvantages associated with each real-time rating approach.

Multi-Weather Station Method

Besides the continuous monitoring of the electrical load, this method employs multiple weather stations placed at various locations along the line. These stations are used to measure the values of ambient variables that, along with line current, are used in the calculation of conductor temperature. Experimental studies have verified that conductor temperature can be calculated with good accuracy for any location if the weather parameters and electrical load are known for that particular location[7][8]. Several variations of this method have been developed during recent years and can be explored by additional readings[9][10].

Conductor Temperature-Sensing Devices

Given values for conductor current, I_{MEAS}, temperature, T_{MEAS}, and air temperature, T_a, the maximum allowable current, I_{MAX}, that would yield the maximum allowable temperature, T_{MAX}, may be calculated. In approximate terms, the continuous steady-state conductor rating may be calculated using a modification of Equation 7-5:

$$I_{MAX} = I_{MEAS} \sqrt{\frac{R_{T_{MEAS}}}{R_{T_{MAX}}}} \sqrt{\frac{T_{MAX} - T_a - \Delta T_s}{T_{MEAS} - T_a - \Delta T_s}} \tag{7-7}$$

To take advantage of this method of determining the maximum allowable current, devices that can be attached directly to overhead conductors without taking the line out of service have been developed to monitor actual conductor temperature. These monitors contain one or two thermocouples to measure the conductor temperature, a current transformer to measure the electrical current, and a transmitter to relay the measurements to a ground station up to five miles away. Coupled with ambient temperature sensors, these devices are deployed to critical spans that are likely to attain maximum temperatures. The most critical values measured for these spans are then used to determine real-time thermal ratings for the line. This method eliminates the errors resulting from the use of estimated variable values by static and dynamic thermal rating methods.

Conductor Tension Monitors

Minimum ground, or underbuild, clearance of overhead lines is maintained by controlling conductor sag. Viewed as parabolic, conductor sag in its simplest form is a function of span length, S, the horizontal component of conductor tension, H, and conductor weight, w, as expressed by the following equation:

$$Sag = \frac{wS^2}{8H} \tag{7-8}$$

To yield acceptably accurate sag results, this simplistic equation must be expanded to include conductor temperature, stress-strain relationship, thermal elongation properties, and ice and wind loading of the conductor. The inclusion of the estimates, or even measurements, of these variables in sag calculations introduces many sources of error.

Fortunately, a device has been developed that can be placed in series at the low-voltage end of a dead-end insulator string to monitor the actual tension of a line. This device consists of a strain-gauge, adapted from monitors used on construction crane loads, and a data transmitter. It may be attached to the insulator string without taking the line out of service. The ruling span concept allows use of such tension devices in a single span to determine the sag in all spans of a dead-end section. This is a viable concept because the free movement of the insulator strings in the axial direction of the line equalizes the tension in all spans of the dead-end section.

It is also possible to use the measured conductor tension to calculate the average conductor temperature in the dead-end section. It should be noted, however, that this is indeed an average temperature, not the maximum conductor temperature. Hot spots due to localized sheltering at non-monitored tension points will not be revealed.

Assessing Risk with Real-Time Rating Methods

The purpose of any real-time rating method is to permit an increase in the electrical loading of the line. Regardless of the method used, increased electrical loading will have certain consequences that must be considered:

(1) An increase in load increases the operating temperatures of the conductor, splices and clamps. This is true even though the maximum allowable temperature may remain the same.

(2) The average conductor temperatures may be increased if the maximum allowable temperature is increased.

7.3.3 Increasing the Maximum Allowable Conductor Temperature

Although voltage drop, line losses (I^2R) and hardware types may be considerations, the maximum allowable current, or thermal rating, is predominantly a function of conductor temperature. The necessity for limiting the conductor temperature is primarily due to its adverse effects on conductor strength and sag.

Conductor loss of strength is controlled by the time-temperature-related metallurgical characteristic of the conductor called the annealing response. This characteristic determines the amount of conductor strength lost after being exposed to a given temperature for a given period of time. For overhead conductors, the historical maximum allowable continuous operating temperature has been taken as that temperature at which lifetime exposure (usually 30 years) will result in a maximum 10% loss of strength.

As the effects of time and temperature on the annealing response are cumulative, a higher temperature for a short time can be equated to a lower temperature for a longer time. It may be possible to uprate an existing line by defining a new static thermal rating using a higher conductor temperature and also by assigning this rating a time limit over the

TABLE 7-3
SAG-TENSION VALUES FOR A **15%** FINAL TENSION AT 60°F FOR 795 KCMIL - 26/7 ACSR "DRAKE"

Area = 0.7264 in^2
Diameter = 1.108 in
Rated Strength = 31,500 lbs

Span = 1000 ft

NESC Heavy Loading District

Creep is not a factor

Temp. °F	Ice in	Wind lbs/ft^2	NESC Safety Factor lbs/ft	Resultant Weight lbs/ft	Final		Initial	
					Sag ft	Tension lbs	Sag ft	Tension lbs
0	0.50	4.00	0.30	2.509	30.07	10479	30.07	10479
32	0.50	0.00	0.00	2.094	30.56	8607	29.94	8785
–20	0.00	0.00	0.00	1.094	24.09	5694	22.77	6023
0	0.00	0.00	0.00	1.094	25.38	5406	23.90	5739
30	0.00	0.00	0.00	1.094	27.26	5036	25.59	5363
60	0.00	0.00	0.00	1.094	29.07	4725*	27.25	5039
90	0.00	0.00	0.00	1.094	30.82	4460	28.87	4758
120	0.00	0.00	0.00	1.094	32.50	4232	30.45	4513
167	0.00	0.00	0.00	1.094	34.36	4005	32.85	4187
212	0.00	0.00	0.00	1.094	35.62	3865	35.05	3928

* Limiting Design Condition

life of the line. Thus, a maximum allowable conductor temperature of 75°C may be used for normal continuous ratings, but a higher conductor temperature of 100°C could be used for an emergency rating with the understanding that the line load will not exceed the normal continuous rating for more than four hours in each occurrence nor more than 24 hours per year. Methods of calculating loss of strength through annealing are discussed in some detail in Chapter 3.

Reconductoring also may be considered as a method for increasing the maximum allowable conductor temperature. As explained in Chapter 1, aluminum conductor, steel supported (ACSS) can tolerate significantly higher temperatures without loss of strength and with little or no increase in sag.

7.3.4 Modifications to Sag-Limited Lines

Uprating transmission lines with stringent sag limitations requires a meticulous review of existing line clearances and careful inspection of conductor and associated hardware. Slight increases in the thermal rating of these lines may result in increased sag that would cause violation of marginal safety clearances. An increased thermal rating may be achieved, however, by compensating for the increased sag through increasing the height of the structures or by increasing the tension of the conductor.

Increasing Structure Height

Relatively small increases in structure height can offset additional sag created by an increase in thermal rating. For example, consider a 795 kcmil - 26/7 ACSR "Drake" line with assumed weather conditions of 2 ft/sec wind and an ambient temperature of 40°C. Increasing the maximum allowable emergency temperature from 100°C to 110°C results in an increase in the line's emergency static thermal rating of 8%. For a 1000 feet span of this conductor, strung to an unloaded final tension of 20% of its rated breaking strength at 16°C, the final sag at a maximum allowable temperature of 100°C is 29.5 feet. Assuming the emergency conditions occur infrequently, increasing the maximum allowable temperature to 110°C increases the maximum sag to 30.0 feet. This 6 inch increase in sag allows for an 8% increase in the thermal rating of the line.

Increasing Conductor Tension

The NESC recommends three different maximum tension limits for line design. Stringing conductors to these limits

TABLE 7-4
SAG-TENSION VALUES FOR A **17.5%** FINAL TENSION AT 60°F FOR 795 KCMIL - 26/7 ACSR "DRAKE"

Area = 0.7264 in^2 | **Span = 1000 ft**
Diameter = 1.108 in
Rated Strength = 31,500 lbs | *NESC Heavy Loading District*

Creep is not a factor

Temp. °F	Ice in	Wind lbs/ft²	NESC Safety Factor lbs/ft	Resultant Weight lbs/ft	Final Sag ft	Final Tension lbs	Initial Sag ft	Initial Tension lbs
0	0.50	4.00	0.30	2.509	26.60	11834	26.60	11834
32	0.50	0.00	0.00	2.094	26.95	9749	26.21	10026
−20	0.00	0.00	0.00	1.094	19.54	7011	18.07	7579
0	0.00	0.00	0.00	1.094	20.91	6556	19.19	7141
30	0.00	0.00	0.00	1.094	22.93	5981	20.90	6559
60	0.00	0.00	0.00	1.094	24.89	5512*	22.62	6061
90	0.00	0.00	0.00	1.094	26.79	5125	24.34	5635
120	0.00	0.00	0.00	1.094	28.62	4800	26.04	5270
167	0.00	0.00	0.00	1.094	30.80	4463	28.63	4798
212	0.00	0.00	0.00	1.094	32.12	4281	31.01	4432

* Limiting Design Condition

helps to minimize conductor failure due to fatigue and over-stressing under various load conditions. The limits, specified in terms of the conductor's rated breaking strength, are as follows:

(1) An initial unloaded (no ice or wind load) maximum tension of 35% of the rated breaking strength at 16°C.

(2) A final unloaded maximum tension of 25% of the rated breaking strength at 16°C after a certain time, usually 10 years.

(3) An initial loaded (with ice and wind) maximum tension of 60% of the rated breaking strength at a temperature dependent on geographic location.

These limits are essential in establishing maximum conductor sags to ensure minimum ground clearance at the conductor's maximum allowable temperature. It is possible, however, that older lines may have been installed at tensions less than that provided by the current NESC. If so, retensioning of the line will result in less sag. Lesser actual sags allow engineers to increase the maximum allowable temperature of the line while still maintaining adequate ground clearances. This concept is illustrated by Tables 7-3 and 7-4 that show typical sag-tension data for 795 kcmil - 26/7 ACSR "Drake" in a NESC heavy loading area. The data in Table 7-3 represents a final tension equal to 15% of the conductor's rated breaking strength of 31,500 lbs, while that in Table 7-4 represents a final tension of 17.5% of rated breaking strength. Note that use of the higher tension yields the same final sag at 212°F (100°C) as originally occurred at 50°C. Thus, by increasing the final tension of the existing conductor from 15% to 17.5% of rated breaking strength, the maximum allowable temperature can be increased from 50°C to 100°C while maintaining the same ground clearances at the higher conductor temperature.

When employing this approach, several other considerations must be addressed:

FIGURE 7-5
THERMAL RATING AS A FUNCTION OF CONDUCTOR AREA AND MAXIMUM TEMPERATURE

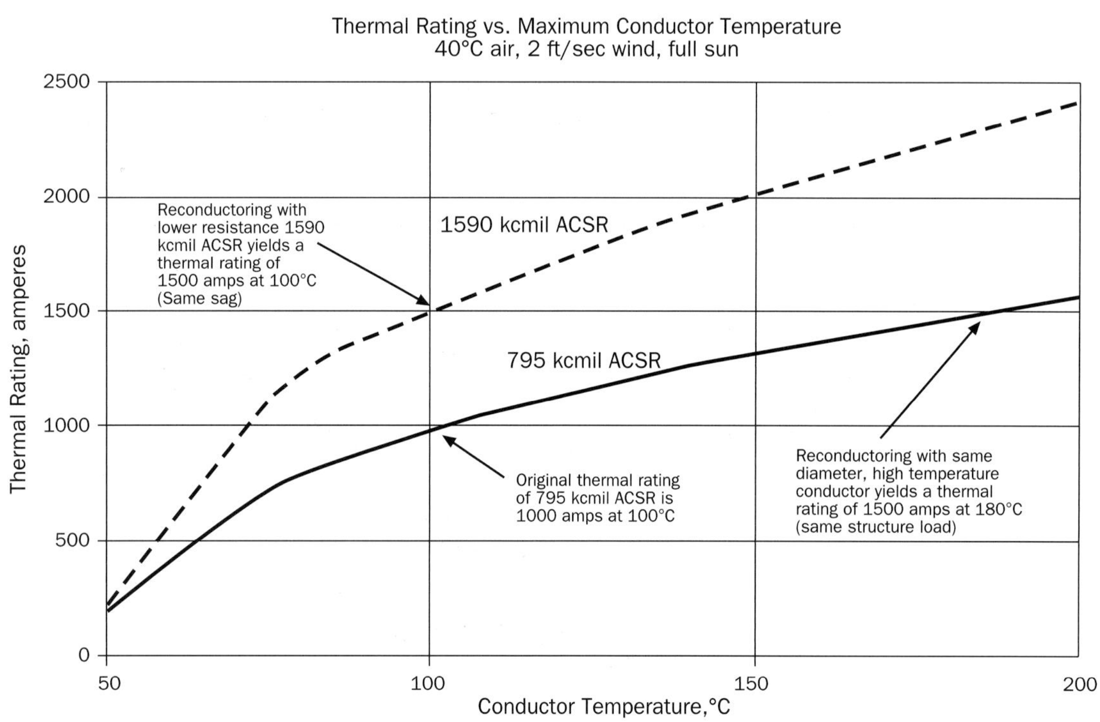

(1) The higher maximum conductor tension under ice and wind load requires a review of mechanical load capacity at all angle and dead-end structures.

(2) The increased tension could yield higher levels of aeolian vibration that may require additional damping.

(3) In the retensioning calculations, the initial stress-strain curve used for a new conductor cannot be used because it does not account for the creep history of the conductor.

7.4 RECONDUCTORING EXISTING TRANSMISSION LINES

Existing lines may be uprated by any of the methods discussed in the preceding sections of this chapter, all of which involve using the original conductor. In certain applications, it may make sense to replace the original conductor. This section reviews the various reconductoring choices and provides some insight into line uprating applications.

The low cost, high conductivity, and low density of aluminum make it the preferred material for overhead conductors. Therefore, any "lower resistance" replacement conductor must have an increased crossectional aluminum area which may increase the wind, ice, and tension loads on the existing structures.

Figure 7-5 shows how the thermal rating of an existing line may be increased by about 50% by using a replacement conductor with twice the original conductor's aluminum area. Note that this doubles the original strain structure loads and increases transverse wind and ice conductor loads on suspension structures by about 40%. Such large load increases would typically require structure reinforcement or replacement.

A number of "High-Temperature, Low-Sag" (HTLS) conductors, capable of continuous operation at temperatures above 100°C with stable tensile strength and creep elongation properties, are currently commercially available and others are under development. Practical temperature limits of up to 250°C have been specified for some conductors. As is also shown in Figure 7-5, use of an HTLS conductor (having the same diameter as the original) at 180°C also increases the line rating by 50%, but without any significant change in structure loads.

The use of a lower resistance conductor has two advantages - losses are reduced and operating temperatures remain modest. The primary advantage of using HTLS conductors is that structure modifications are minimized.

FIGURE 7-6
TYPICAL TRANSMISSION LINE SAG AS A FUNCTION OF TIME, LOAD AND TEMPERATURE

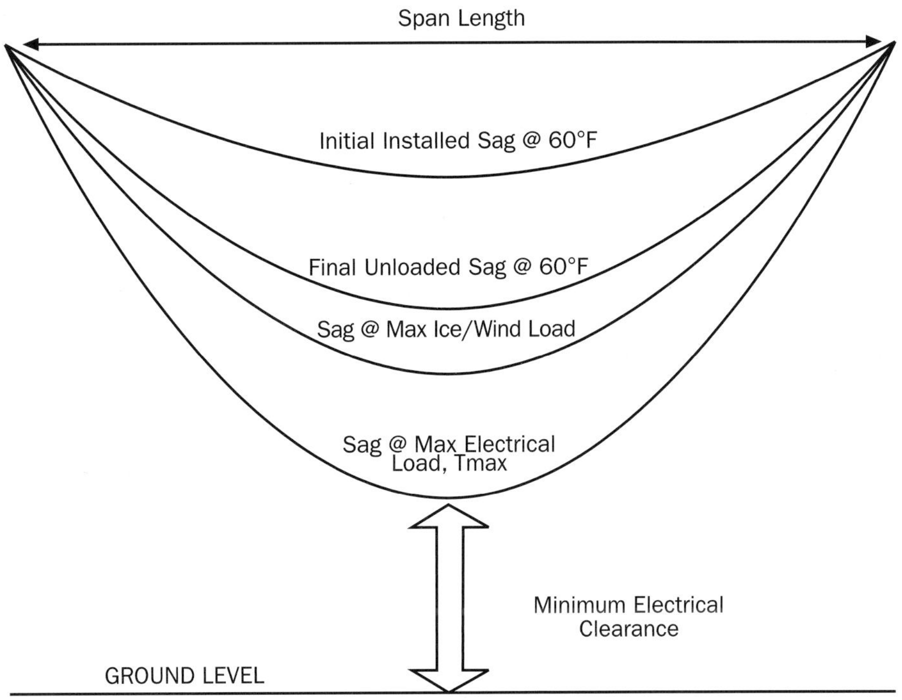

In either case, the replacement of existing conductor should improve the mechanical reliability of the line since conductor, connectors, and hardware are all new.

7.4.1 Sag Constraints for Reconductoring

As shown in Figure 7-6 the original conductor's "Initial installed sag" increased to a final unloaded "everyday" sag condition (typically at 60°F with no ice and wind) as a result of both occasional wind and ice loading events and the "normal" creep elongation of tensioned aluminum strands over time. The final unloaded sag temporarily increases further (but reversibly) due to ice and wind loading or high electrical current loads. For most transmission lines, maximum final sag occurs under high electrical loads rather than high mechanical loads.

Any replacement conductor must be installed such that, over time, its final sag under maximum electrical or mechanical load does not exceed the original conductor's design maximum final sag. Otherwise, existing structures will have to be reinforced, raised, or replaced.

7.4.2 Reconductoring with Conventional Conductors

To increase the static thermal rating of a line, the existing conductor may be replaced with one capable of higher thermal capacity. This process is normally referred to as reconductoring the line. Reconductoring may range from replacing the existing conductor with a larger conductor of the same type, to replacing the existing conductor with a different type of conventional conductor (ACSR in place of AAC or copper), or to replacing the existing conductor with a special conductor having desirable properties (TW conductors allow reduced resistance for the same overall diameter). The selection of the type conductor to be used for reconductoring is dependent on the original line design.

Larger-Diameter Standard Conductor

Uprating by reconductoring with conventional transmission conductor is most often accomplished by replacing the existing conductor with a larger one of the same type having lower resistance. However, in those lines with adequate structural strength, some all-aluminum or copper lines are reconductored with larger steel-cored ACSR conductors. These conductors offer the advantages of reduced resistance, lower thermal elongation and creep characteristics, and improved loss-of-strength properties.

Small increases in conductor diameter can often be tolerated without major structural upgrades. If structural modification is necessary, the cost of such modification is primarily dependent on the diameter of the new conductor, as seen in Figure 7-7. For many existing lower voltage lines, a new conductor diameter 10% larger than the existing conductor could be installed without significant structure modification costs. For these same lines, the use of a conductor with a diameter increase of more than 10% may result in prohibitively high structure modification costs.

FIGURE 7-7
COST OF UPRATING vs. NEW CONDUCTOR DIAMETER

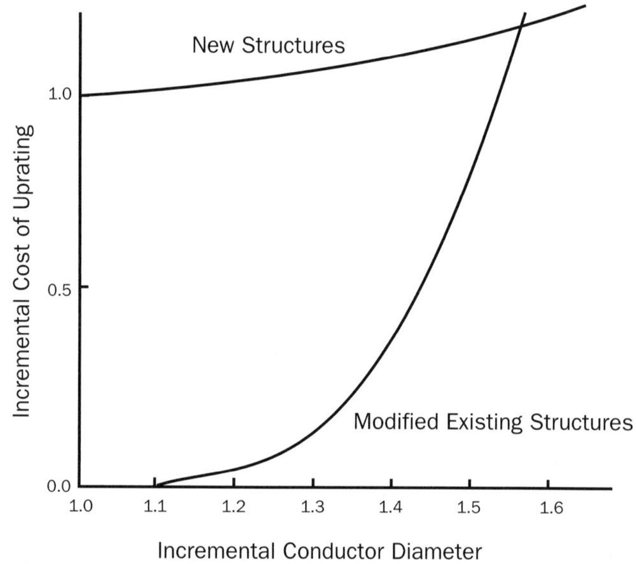

Bundling New and Old Conductors

The old conductor of an existing line can be combined with a new conductor to form a bundle. This technique, known as conductor bundling, can be used in both thermal uprating and voltage line upgrades, and may be accomplished in several ways[11]:

(1) String the new conductor below the existing conductor so that minimum ground clearance is maintained.

(2) String the new conductor next to the existing conductor so the final sags of the horizontal bundle match.

(3) Remove sections of conductor from the existing line and bundle them with other existing sections. Install new conductor bundles in the sections from which the existing conductor was removed.

The advantages and disadvantages of bundling new and old conductors can be illustrated with the following example:

Consider a 115 kV line that is to be thermally uprated. The present conductor is a single, 795 kcmil - 26/7 ACSR "Drake." The choice for reconductoring is to make a 2-conductor bundle by the addition of another "Drake" conductor, or to replace the existing "Drake" conductor with a single, 1590 kcmil - 45/7 ACSR "Lapwing." The ramifications are:

(1) The cost of bundling new and old "Drake" conductors probably will be less than reconductoring with "Lapwing," because the old "Drake" conductor has low scrap value and the cost of the new "Lapwing" is twice that of a new "Drake." The higher installation costs for the "Lapwing" balances the costs for additional hardware for a second "Drake."

(2) The static thermal rating of the 2-conductor "Drake" bundle is significantly higher than the single "Lapwing." The respective ampacities are 1810 vs. 1350 amps for 75°C conductor temperature with 25°C ambient and 2 ft/sec crosswind.

(3) The structure wind and ice loading is less for the single ACSR "Lapwing." Vertical loading is 5.04 lbs/ft for the "Drake" bundle vs. 3.48 lbs/ft for the single "Lapwing." Also, the transverse wind load of the "Drake" bundle is approximately 50% higher.

(4) The electric field at the conductor surface is about 10% lower for the "Drake" bundle and the reactance is about 20% lower.

The increase in thermal capacity gained by bundling has the disadvantage of increasing structural loading. Increasing thermal capacity by bundling, therefore, is desirable only if the required structural reinforcement is economical. Conductor bundling is usually a primary consideration only when the goal is both the thermal uprating and voltage upgrading of the line.

Self-Damping Conductor

A self-damping conductor (SDC)[12] has a steel core like an ACSR, but the innermost layers of aluminum, and sometimes outer layers, are made up of trapezoidal shaped strands. They are sized so a gap exists between the steel core and the innermost layer of aluminum. The mass and tension of the aluminum and steel components yield different natural vibration frequencies which results in reduced aeolian vibration when the two components impact.

There are several benefits to reconductoring with a self-damping conductor. A SDC can be tensioned to much higher levels than a standard conductor, resulting in reduced sag and eliminating the need for vibration dampers. The trapezoidal strands used in a SDC result in a 5% reduction in conductor diameter when compared to a standard conductor of the same ac resistance per mile. The reduced diameter results in reduced structural wind and ice loading.

Compact (TW) Conductor

A trapezoidal wire (TW) conductor is characterized by the use of trapezoidal shaped aluminum wires in AAC, ACSR and ACSS conductors. As compared to a standard conductor, the compactness made possible by using the trapezoidal-shaped wire results in a conductor that has a smaller diameter and equal resistance (area-equivalent conductor), or a conductor of the same diameter with lower resistance (diameter-equivalent conductor). The smaller diameter equates to less ice and wind loading; the lower resistance to increased thermal capacity. The increase in structural loading due to the increased density of TW conductor is minimal.

Vibration-Resistant Conductor

A vibration-resistant (VR™)[13] conductor consists of two conductors, of any type, twisted around each other forming a rotating "figure 8." The length of twist, lay length, is approximately 9 feet.

Reconductoring with a VR™ conductor can reduce the magnitude and frequency of occurrence of large amplitude ice-galloping motion and small amplitude aeolian vibration that cause fatigue-inducing strains near clamps. This reduction in vibration tendency and severity allows the stringing of the conductor to higher tension. VR™ conductors are subject to wind and ice loadings that are about 11% higher than standard conductor of the same ac resistance per mile.

7.4.3 Reconductoring with High Temperature Conductors

Two categories of high temperature conductor are available: High-Temperature, Low-Sag (HTLS) conductors (including ACSS, ACCC, ACCR, TACIR and GTACSR) and conductors that use aluminum alloys capable of operation at high temperature (such as TACSR) without significant loss of strength. More information on each of these can be found in Chapter 1. HTLS conductors are capable of continuous operation at temperatures above 100°C with stable tensile strength and creep elongation properties. Reconductoring with a HTLS conductor can increase ratings by 50%, or more, without requiring significant change in structure loads.

With the exception of ACSS, these conductors have seen limited usage in the United States to date. Due to the limited data available on many of these conductors, this section will focus on reconductoring with ACSS.

ACSS

Aluminum Conductor Steel Supported (ACSS), as described in ASTM B 856, consists of fully annealed strands of aluminum (1350-O) stranded around a stranded steel core. Steel core wire coatings include galvanized zinc, aluminum coated (AZ), aluminum clad (AW) and Zinc-5% Aluminum-Mischmetal Alloy "Galfan" (MA). Available strength ratings for the core wires include standard, high, and ultra high (Southwire HS285).

ACSS is typically available in three constructions: round wire, area equivalent Trap Wire, and diameter equivalent Trap Wire. In appearance, ACSS conductors are essentially identical to standard ACSR conductors. However, the thermal elongation, creep rate, and maximum operating temperature are quite different.

ACSS has a number of advantages for reconductoring compared to conventional conductors. The key advantages of ACSS include:

- Higher operating temperatures, up to 250°C, with no loss in strength
- Lower thermal elongation
- Better self-damping ability
- No creep elongation over time

Higher Operating Temperature

Conventional conductors, which depend entirely on aluminum (1350-H19) or aluminum alloy (6201-T81) for their strength rating, can be operated at temperatures up to 94°C without significant loss of tensile strength. Aluminum conductors with a steel-reinforcing core can be operated at temperatures of up to 150°C for limited periods. Because the aluminum (1350-O) in ACSS is already fully annealed, and the ACSS is "supported" rather than "reinforced" by its steel core, it can operate continuously at temperatures up to 250°C without significant loss of strength.

The Table 7-5 shows a comparison of the continuous operating ampacity (thermal) ratings (with 2 ft/sec cross-wind, 40°C air temperature, and full sun) for an ACSR and equivalent ACSS conductors. Note that the ampacity of an ACSS conductor operating at 250°C is nearly twice that of an ACSR of the same cross sectional area operating at 100°C.

Lower Thermal Elongation

Aluminum strands elongate thermally at approximately twice the rate of steel. In the case of ACSS, the tension level in the aluminum strands is very small, thus transferring the majority of the load to the steel core. Therefore, the conductor elongates thermally as though it were steel. Thus, the rate of increase in sag due to increased temperature is less for ACSS than ACSR.

As an example of the lower thermal elongation of ACSS, consider the data in Table 7-6. The sag increase from 15°C

TABLE 7-5
CONTINUOUS AMPACITY OF EQUIVALENT ACSR AND ACSS CONDUCTORS
AS A FUNCTION OF MAXIMUM ALLOWABLE CONDUCTOR TEMPERATURE

Conductor Temp. °C	Ampacities			
	"Drake" ACSR[1]	"Drake" ACSS	"Drake" ACSS/TW[2]	"Suwannee" ACSS/TW[3]
75	730	740	725	815
100	990	1000	980	1110
150	—	1350	1315	1490
200	—	1600	1555	1760
250	—	1805	1755	1990

[1] ACSR is usually limited to temperatures of 75°C for continuous and 100°C for emergency to avoid annealing

[2] Area equivalent to ACSS "Drake"

[3] Diameter equivalent to ACSS "Drake"

to 150°C with ACSS is the same as the sag increase from 15°C to 100°C with ordinary ACSR. Therefore, reconductoring an ACSR clearance-limited line with ACSS increases the thermal rating of the line by about 34% without the need to raise or reinforce structures.

Self-Damping
The tension of conductors in overhead lines is normally determined to limit aeolian vibration-induced fatigue. It is normal to limit final unloaded tension to no more than 20% of the rated breaking strength in order to limit vibration levels. Because it has better self-damping than ordinary ACSR, ACSS may be installed to higher tensions.

Low Creep Elongation
When reconductoring with ACSR, allowance must be made for creep elongation over time, and, with the exception of high steel content ACSR conductors, consideration must be given for the possibility of accelerated creep at high operating temperatures. ACSS does not experience long term creep. Thus, for ACSS:

- There is little to no difference between initial and final sag
- The initial sag is less than that of standard ACSR
- The change in sag due to temperature is less than that of standard ACSR

7.5 THERMAL UPRATING CASE STUDY

Consider the case of a double circuit, 115 kV transmission line with single, 795 kcmil - 26/7 ACSR "Drake" conductors that has been in operation approximately 40 years. The terrain traversed by the line is flat and open. Design and operating information provided for the line:

- The spans on this line range from 800 to 1100 feet and the ruling span is approximately 1000 feet.
- The line was originally designed with a final unloaded tension of 25% of the conductor's Rated Breaking Strength (RBS) at 0°F (–18°C).

FIGURE 7-8
SAG-TENSION OF EXISTING ACSR "DRAKE"

```
                         Southwire Company
                  Using ALCOA Sag and Tension Data

                        Chapter 7 Case Study
                    Existing "Drake" Conductor


Conductor: 795.0 Kcmil  26/7 Stranding ACSR "Drake"

Area = 0.7264 Sq. in    Diameter = 1.108 in    Wt = 1.094 lb/f   RTS = 31500 lb
Data from Chart No. 1-537
Aluminum Compression was calculated with Built in Aluminum Stress = 2500
Using Exact Catenary Equations

Span = 1000.0 feet              NESC Heavy Load Zone
Creep is NOT a Factor           Rolled Rod
    Design Points                         Final             Initial
 Temp    Ice    Wind    K   Weight  Sag  Tension        Sag  Tension
   F      in     psf  lb/f  lb/f     ft       lb         ft       lb
   0.    .50    4.00  .30  2.509  23.91    13142      23.02    13650
  32.   1.00     .00  .00  3.715  29.08    16026      29.08    16026
  32.    .50     .00  .00  2.094  24.12    10877      22.21    11807
 -20.    .00     .00  .00  1.094  16.04     8532      13.52    10120
   0.    .00     .00  .00  1.094  17.39     7875*     14.40     9506
 120.    .00     .00  .00  1.094  25.43     5392      20.73     6607
 167.    .00     .00  .00  1.094  28.34     4841      23.43     5850 -> 730A @ 75°C
 176.    .00     .00  .00  1.094  28.88     4752      23.94     5725
 212.    .00     .00  .00  1.094  30.04     4569      25.97     5279
 257.    .00     .00  .00  1.094  31.42     4370      28.45     4822 -> 1180A @ 125°C
 302.    .00     .00  .00  1.094  32.78     4190      30.83     4452
 306.    .00     .00  .00  1.094  32.90     4175      31.04     4422 -> 1350A @ 152°C
 *  Design Condition
```

TABLE 7-6
THERMAL ELONGATION COMPARISON OF ACSS vs. ACSR

Conductor Temp.	Sag of ACSR "Drake"		Sag of ACSS "Drake"		Ampacity
°C	ft	m	ft	m	Amperes
15	31.0	9.4	31.0	9.4	—
100	37.6	11.5	35.3	10.8	1110
150	—	—	37.8	11.5	1490

- The line originally met NESC clearances for a maximum conductor temperature of 120°F (49°C). The ruling span final sag was 25.4 feet as shown in Figure 7-8. Although sag was traditionally calculated at 120°F, the thermal rating of the line was typically calculated at 167°F (75°C). To accommodate the additional sag at 167°F, a 6 feet sag buffer was added to the 25.4 feet of sag.

FIGURE 7-9
INITIAL THERMAL RATING OF ACSR "DRAKE"
USING SWRATE

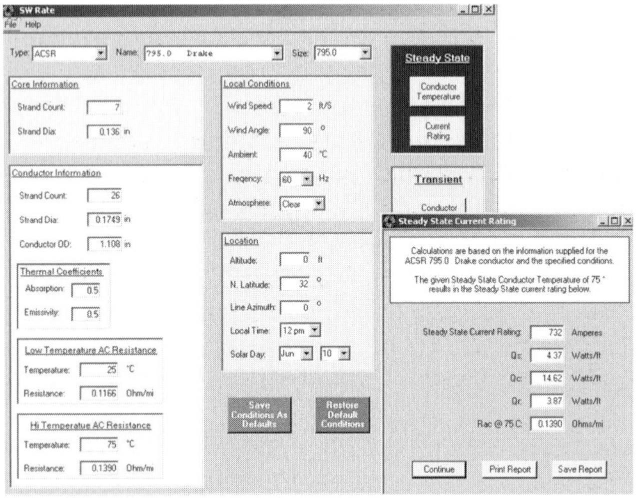

- The initial thermal rating of the line, calculated at 167°F (75°C), was 730 amps, as shown in Figure 7-9, using 2 ft/sec perpendicular wind, full sun, and 104°F (40°C) air temperature. The sag at this operating temperature is 28.3 feet.

- Ten years ago, the maximum allowable conductor temperature of the line was increased from 120°F (49°C) to 203°F (95°C), which increased the high temperature sag to 29.8 feet, thus using up most of the 6 feet buffer. The new thermal rating calculated at 203°F (95°C) is 945 amps using 2 feet/sec perpendicular wind, full sun, and 104°F (40°C) air temperature.

- The conductor sag under an extreme ice load of 1.0 inch of ice at 32°F (0°C) and no wind is 29.1 feet at a tower tension of 16,000 lbs.

- A recent survey of the line indicates that the everyday final sags are reasonably close to those calculated at the time of construction.

- The original line had vibration dampers installed, but a few broken strands under the suspension clamps have been discovered over its 40-year life.

- The new continuous rating requested for the line is 1350 amperes, and there would be additional economic value in having a rating above 1500 amperes.

- Rebuilding the line or increasing the tower height is not considered an option since this would extend outage times and require extensive public hearings.

- Sag and tension data for the existing conductor is shown in Figure 7-8.

The available options appear to be:

(1) Increase the operating temperature of the original conductors.

(2) Use less conservative weather conditions in calculating the thermal rating.

(3) Increase the maximum temperature and use less conservative weather conditions.

(4) Re-tension the line to decrease the high temperature sag.

(5) Reconductor the existing line with a TW conventional conductor.

(6) Reconductor with ACSS conductor.

(7) Reconductor with ACSS/TW conductor.

(8) Reconductor with Southwire HS285 ACSS/TW conductor.

The following constraints have been identified for any uprating design:

- The maximum conductor sag cannot exceed 31.4 feet under either high temperature or extreme ice load conditions.

- The maximum tension of the conductor cannot exceed 16,000 lbs by more than 10% (17,600 lbs maximum).

- The outside diameter of the conductor cannot exceed 1.108 inches by more than 5% (1.163 inches maximum).

- The replacement conductor must avoid worsening the vibration fatigue problems.

OPTION 1

Operate the line at 257°F (125°C) instead of 203°F (95°C).
Increasing the operating temperature of the line to 257°F (125°C) would increase the sag of the existing ACSR "Drake" to the maximum allowed, 31.4 feet as shown in Figure 7-8. At this operating temperature, the conductor would be rated for an ampacity of 1180 amps. This is far short of the 1350 amps required, so this option is not acceptable.

OPTION 2

Use less conservative weather assumptions.
The present weather conditions used to calculate line ratings could be considered to be excessively conservative, so the possibility of using a wind speed of 3 ft/sec and a reduced air temperature of 95°F (35°C) is to be explored. The final sag would be unchanged (the maximum allowable conductor temperature is the same: 203°F (95°C) at 29.8 feet) and the structure loads would remain unchanged. However, the line thermal rating would only be 1090 amps, as shown in Figure 7-10, far short of the 1350 amps required by planning.

FIGURE 7-10
ACSR "DRAKE"AMPACITY USING
LESS CONSERVATIVE WEATHER CONDITIONS

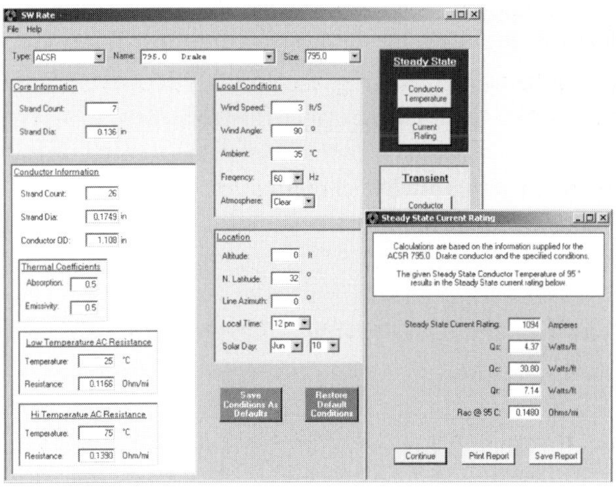

OPTION 3

Operate the line at 257°F (125°C) and use less conservative weather conditions.
This option, using the less conservative weather conditions from Option 2, would yield a current rating of 1330 amps for the ACSR "Drake" conductor. The sag at this temperature would be 31.4 feet which is acceptable. If 1330 amps, rather than the 1350 amps desired, is acceptable, there are other issues that should be considered:

- Operation at this temperature would result in some loss of conductor tensile strength over time. Loss of strength can be calculated based on the methods presented in Chapter 3.

- Under full load, the temperature of the line could increase to 374°F (190°C) in still air. This increases the temperature risk from 30°C to 65°C. If this occurs, the sag will increase to 34.9 feet, which significantly exceeds the limit of 31.4 feet.

- There will be no beneficial economic impact as compared to other options that generate additional revenue by allowing the line to operate at 1500 amps or above.

The acceptability of this option should take into consideration the probability and expected frequency of occurrence of high current loadings. Based on this and the temperature risk, a determination can be made as to whether excessive strength loss or a violation of the electrical clearance may occur. One way to minimize this risk would be to install real-time weather and/or sag monitoring equipment in critical spans.

If the risk is acceptable, this option could be acceptable. It requires little capital investment and can be implemented immediately, but it is not the optimum solution and is therefore not desirable.

OPTION 4

Re-tension the line.
Re-tensioning the existing conductor to a higher unloaded everyday tension would decrease the conductor's self-damping and worsen the fatigue failure problem. Therefore, in this case where broken strands have been found with dampers already in use, any increase in tension is not an acceptable approach. In a different line uprating study, where dampers had never been used and broken strands never discovered, the existing conductor could be re-tensioned to the NESC limit of 25% final at 60°F and installing dampers as needed.

OPTION 5

Reconductor with diameter equivalent ACSR/TW and operate at 257°F (125°C).
Using a trapezoidal ACSR having the same diameter as the original ACSR "Drake," an ACSR/TW "Suwannee," the conductor resistance can be reduced by about 20% and, by going to 257°F (125°C), the line rating can be increased to 1300 amps. The increased weight of the conductor increases the tension on the towers to 17,500 lbs under the extreme ice condition, which is within the 17,600 lbs maximum allowable limit. The same loss of strength issues are present as outlined in Option 3.

While close, this option will not achieve the desired rating of 1350 amps. Even if the 1300 amps were to be considered acceptable, subsequent options will show that higher ratings can be achieved for approximately the same cost. Therefore, this option is not desirable.

OPTION 6

Reconductor with ACSS/MA/HS "Drake."

Replacing the existing ACSR "Drake" conductor with ACSS/ MA/HS "Drake" allows a maximum increase in rating to 1600 amps at 392°F (200°C). The sag at both heavy ice loading and maximum current loading is approximately 31.2 feet, as shown in Figure 7-11. The maximum tension is only 15,000 lbs and the diameter is the same as the original. At 1600 amps, in still air, this conductor will reach a temperature of 491°F (255°C), which exceeds the 250°C maximum conductor temperature rating, resulting in sag of 34.1 feet. At 1500 amps, in still air, the maximum sag is still exceeded. So, even though there is no loss of strength concern for ACSS, there is still some temperature risk that must be considered.

At the required rating of 1350 amps, the sag (28.5 feet) is approximately the same as that of the existing "Drake" at 75°C (28.3 feet). Only when the line is operated at or above 1500 amps does the operating temperature cause the sag to approach the maximum allowable limit. At 1350 amps, in still air, this conductor will reach a temperature of 387°F (197°C) resulting in sag of 31.0 feet. Thus, there is no temperature risk associated with operating this conductor at the required ampacity.

FIGURE 7-11
SAG-TENSION FOR "DRAKE/ACSS/MA/HS"

```
                    Southwire Company
            Using ALCOA Sag and Tension Data

                   Chapter 7 Case Study
                         Option 6


Conductor: 795 Kcmil  26/7 Stranding "Drake/ACSS/HS"

Area = 0.7264 Sq. in    Diameter = 1.108 in    Wt = 1.093 lb/f   RTS = 28000 lb
Data from Chart No. 3-945
Using Exact Catenary Equations

Span = 1000.0 feet    NESC Heavy Load Zone
Creep is NOT a Factor
    Design Points                    Final          Initial
 Temp   Ice   Wind   K   Weight  Sag  Tension     Sag  Tension
  F      in    psf  lb/f  lb/f    ft     lb        ft     lb
   0.   .50   4.00  .30  2.508  26.41   11901     24.30   12927
  32.  1.00    .00  .00  3.714  31.17   14955     31.17   14955
  32.   .50    .00  .00  2.093  26.86    9767     22.39   11708
   0.   .00    .00  .00  1.093  19.55    7000*    13.63   10032
 167.   .00    .00  .00  1.093  24.41    5611     18.66    7330
 212.   .00    .00  .00  1.093  25.77    5316     20.77    6590
 257.   .00    .00  .00  1.093  27.13    5050     23.13    5918
 302.   .00    .00  .00  1.093  28.50    4810     25.65    5341 -> 1350A @ 150°C
 354.   .00    .00  .00  1.093  30.06    4561     28.61    4790 -> 1500A @ 179°C
 387.   .00    .00  .00  1.093  31.04    4418     29.59    4634 -> 1350A Still Air
 392.   .00    .00  .00  1.093  31.19    4397     29.73    4611 -> 1600A @ 200°C
 446.   .00    .00  .00  1.093  32.78    4186     31.31    4381 -> 1500A Still Air
 491.   .00    .00  .00  1.093  34.08    4028     32.61    4207 -> 1600A Still Air
 *  Design Condition
```

This option meets all the sag and tension requirements of the basic 1350 amp load. If the temperature risk for operation above 1500 amps is acceptable, this option would also have an economic benefit. However, there are other conductor designs which eliminate this risk without significantly increasing the cost.

OPTION 7

Reconductor with diameter equivalent ACSS/MA/HS/TW "Suwannee."

Replacing the existing conductor with ACSS/MA/HS/TW "Suwannee" allows a maximum increase in rating to 1740 amps at 385°F (196°C). The sag at this operating temperature is 31.4 feet, leaving no sag buffer. The maximum tension is 16,730 lbs under extreme ice loading. In still air, this ampacity will generate a temperature of approximately 482°F (250°C), causing sag to increase to 34.3 feet.

However, this conductor can be operated at lower temperatures and sags as shown in Figure 7-12. In fact, the conductor can deliver the 1500 amp desired rating with sag (29.1 feet) that is approximately the same as the sag (28.3 feet) of the existing "Drake" operating at 167°F (75°C). This would allow a sag buffer of approximately 2 feet. In still air, this ampacity would generate a temperature of 392°F (200°C) causing sag to increase to 31.6 feet. This virtually eliminates the temperature risk associated with operation at 1500 amps.

This option meets all the tension and sag requirements while virtually eliminating the risk of excessive sag due to still air conditions at a current of 1500 amps. However, there are other conductor designs that would allow for even higher ampacities with less thermal risk.

FIGURE 7-12
SAG-TENSION FOR "SUWANNEE/ACSS/MA/HS/TW"

```
                    Southwire Company
            Using ALCOA Sag and Tension Data

                   Chapter 7 Case Study
                         Option 7


Conductor: 959.6 Kcmil Type 16 HS Steel "Suwannee/ACSS/MA/HS/TW"

Area = 0.8762 Sq. in    Diameter = 1.108 in    Wt = 1.318 lb/f   RTS = 33100 lb
Data from Chart No. 3-945
Using Exact Catenary Equations

Span = 1000.0 feet    NESC Heavy Load Zone
Creep is NOT a Factor
    Design Points                    Final          Initial
 Temp   Ice   Wind   K   Weight  Sag  Tension     Sag  Tension
  F      in    psf  lb/f  lb/f    ft     lb        ft     lb
   0.   .50   4.00  .30  2.722  24.97   13658     23.28   14650
  32.  1.00    .00  .00  3.939  29.54   16726     29.54   16726
  32.   .50    .00  .00  2.318  25.64   11329     21.75   13344
   0.   .00    .00  .00  1.318  19.94    8275*    14.13   11669
 167.   .00    .00  .00  1.318  24.84    6650     19.51    8458
 212.   .00    .00  .00  1.318  26.20    6305     21.69    7611
 257.   .00    .00  .00  1.318  27.57    5995     24.09    6856
 266.   .00    .00  .00  1.318  27.84    5936     24.58    6718 -> 1350A @ 130°C
 307.   .00    .00  .00  1.318  29.08    5685     26.87    6148 -> 1500A @ 153°C
 338.   .00    .00  .00  1.318  30.01    5510     28.62    5776 -> 1600A @ 170°C
 342.   .00    .00  .00  1.318  30.13    5488     28.87    5727 -> 1350A Still Air
 385.   .00    .00  .00  1.318  31.41    5267     30.28    5460 -> 1740A @ 196°C
 392.   .00    .00  .00  1.318  31.61    5232     30.49    5424 -> 1500A Still Air
 428.   .00    .00  .00  1.318  32.67    5065     31.54    5245 -> 1600A Still Air
 484.   .00    .00  .00  1.318  34.29    4828     33.15    4991 -> 1740A Still Air
 *  Design Condition
```

OPTION 8

Reconductor with ACSS/MA/TW/HS285 "Cardinal."

Replacing the existing conductor with ACSS/MA/TW/HS285 "Cardinal" allows an increase in rating to 1740 amps at 396°F (202°C). The sag at this operating temperature is 30.0 feet, as shown in Figure 7-13, allowing a sag buffer of over 1 feet. The maximum tension is 16,240 lbs under extreme ice loading, slightly less than that in Option 7. In still air, this ampacity will generate a temperature of 492°F (256°C), which exceeds the 250°C maximum conductor temperature rating, causing sag to increase to 32.8 feet. Therefore, there is less temperature risk at a load of 1740 amps for this option than for Option 7.

This conductor can deliver up to 1600 amps with sag that is over 1 feet less than the existing "Drake" operating at 167°F (75°C). This would allow a sag buffer of approximately 3 feet. In still air, this ampacity would generate a temperature of 437°F (225°C) causing sag to increase to 31.2 feet. This virtually eliminates the temperature risk associated with operation at 1600 amps, an increase of 100 amps over Option 7.

This option meets all the tension and sag requirements while virtually eliminating the risk of excessive sag due to still air conditions up to 1600 amps and has less sag at all operating loads being considered.

Note: The current in this option is limited to 1740 amps to allow for a comparison to Option 7. The maximum current for this conductor would be 1850 amps at a temperature of 437°F (225°C) resulting in sag of 31.2 feet.

FIGURE 7-13
SAG-TENSION FOR "CARDINAL/ACSS/MA/TW/HS285"

```
                         Southwire Company
                  Using ALCOA Sag and Tension Data

                       Chapter 7 Case Study
                             Option 8

Conductor: 954.0 Kcmil Type 13 Galfan Coated HS285 Steel "Cardinal/ACSS/MA/TW/HS285"

Area = 0.8464 Sq. in    Diameter = 1.084 in    Wt = 1.227 lb/f   RTS = 32300 lb
Data from Chart No. 3-954
Using Exact Catenary Equations

Span = 1000.0 feet    NESC Heavy Load Zone
Creep is NOT a Factor
    Design Points                    Final            Initial
 Temp   Ice   Wind    K   Weight  Sag  Tension      Sag  Tension
  F      in    psf  lb/f  lb/f     ft      lb        ft      lb
   0.   .50   4.00  .30  2.618  24.64   13313     22.67   14469
  32.  1.00    .00  .00  3.819  29.49   16240     29.49   16240
  32.   .50    .00  .00  2.212  25.14   11026     20.79   13325
   0.   .00    .00  .00  1.227  19.02    8075*    12.74   12042
 167.   .00    .00  .00  1.227  23.45    6555     16.54    9282
 212.   .00    .00  .00  1.227  24.72    6219     18.25    8416
 271.   .00    .00  .00  1.227  26.41    5824     21.07    7291 -> 1350A @ 133°C
 315.   .00    .00  .00  1.227  27.68    5558     23.50    6542 -> 1500A @ 157°C
 347.   .00    .00  .00  1.227  28.60    5380     25.35    6067 -> 1600A @ 175°C
 348.   .00    .00  .00  1.227  28.63    5375     25.41    6053 -> 1350A Still Air
 396.   .00    .00  .00  1.227  30.01    5129     28.24    5448 -> 1740A @ 202°C
 399.   .00    .00  .00  1.227  30.10    5114     28.50    5400 -> 1500A Still Air
 437.   .00    .00  .00  1.227  31.19    4937     29.58    5204 -> 1600A Still Air
 492.   .00    .00  .00  1.227  32.75    4704     31.12    4948 -> 1740A Still Air
 *  Design Condition
```

TABLE 7-7
SUMMARY OF ACCEPTABLE OPTIONS AND ASSOCIATED RISK

	1350 Amps		1500 Amps		1600 Amps		1740 Amps	
	Final Sag[1] ft	Still Air Risk[2] ft	Final Sag[1] ft	Still Air Risk[2] ft	Final Sag[1] ft	Still Air Risk[2] ft	Final Sag[1] ft	Still Air Risk[2] ft
Option 6 "Drake" ACSS/MA/HS	28.5	None	30.1	1.4	31.2	2.7	—	—
Option 7 "Suwannee" ACSS/MA/HS/TW	27.8	None	29.1	0.2	30.0	1.3	31.4	2.9
Option 8 "Cardinal" ACSS/MA/TW/HS285	26.4	None	27.7	None	28.6	None	30.0	1.4

1 Final Sag is based on the conductor temperature generated by the indicated current with a 2 ft/sec wind speed.

2 Still Air Risk is the amount of sag in excess of the 31.4 ft maximum allowable sag limit which results from the increased conductor temperature due to the indicated ampacity in still air (0 ft/sec wind speed).

7.6 REFERENCES

[1] Murray, W.D., "Nomographic Computation of the Ampacity Rating of Ariel Conductors," *IEEE Transactions on Power Apparatus and Systems*, Vol. PAS-89, No. 3, pp. 387-398, March 1970.

[2] Black, W.Z. and Rehberg, R.L., "Simplified Model for Steady-State and Real-Time Ampacity of Overhead Conductors," *IEEE Transactions on Power Apparatus and Systems*, Vol. 104, pp. 29-42, October, 1985.

[3] Beers, G.M., Gilligan, S.R., Lis, H.W., Schamberger, J.M., "Transmission Conductor Ratings," *IEEE Transactions on Power Apparatus and Systems*, No. 68, pp. 767-775, October, 1963.

[4] House, H.E. et al, "Emissivity of Weathered Conductors After Service in Rural and Industrial Environments," *AIEE Transactions*, Vol. PAS-81, pp. 891-896, February, 1963.

[5] *IEEE Standard for Calculation of Bare Overhead Conductor Temperature and Ampacity Under Steady-State Conditions*, ANSI/IEEE Standard 738, 2006.

[6] Pearson, J.H.A., "The Ontario Hydro On-Line Method for Transmission Ampacity Evaluation," Panel Discussion on Dynamic Thermal Line Ratings held at IEEE/PES Summer Meeting, 1982.

[7] Black, W.Z., Byrd, W.R., Bush, R.A., Champion, T.C., III, "Experimental Verification of a Real Time Program for the Determination of Temperature and Sag of Overhead Lines," 1982 IEEE/PES Winter Meeting, Paper 83 WM 144-3, January, 1982.

[8] Chisholm W.A. and Barrett, J.S., "Ampacity Studies on 49°C-Rated Transmission Line," *IEEE Transactions on Power Delivery*, Vol. 4, No. 2, pp. 1476-1485, April, 1989.

[9] Radan, M., "Ontario Hydro Experience with On-Line Transmission Ampacity," IEEE/PES T&D Conference and Exposition, New Orleans, April, 1989.

[10] Seppa, T.O., Cromer, E., Whitlach, W.F., Jr., "Summer Thermal Capabilities of Transmission Lines in Northern California based on a Comprehensive Study of Wind Conditions," 1992 IEEE Summer Meeting, paper No. 92 SM 567-8, PWRD, 1992.

[11] Bonny, B.R., "Matching New Conductor with Old in the Same Bundle," *T&D Magazine*, pp. 46-52, November, 1986.

[12] Kirkpatrick, L.A., McCulloch, A.R. and Pue-Gilchrist, A.C., "Ten Years of Progress with Self-Damping Conductor," 1979 IEEE/PES Summer Meeting, Paper No. F79 736-0, 1979.

[13] Douglass, D.A. and Roche, J.B., "T2 Wind Motion Resistant Conductor," *IEEE Transactions on Power Apparatus and Systems*, Vol. PAS-104, No. 10, pp. 2879-2887, October, 1985.

APPENDIX

TEMPERATURE CONVERSION CHART – CENTIGRADE TO FAHRENHEIT

Formula: °F = 9/5 °C + 32

Temp., °C	0	1	2	3	4	5	6	7	8	9
					Temp., °F					
−40	−40.0	−41.8	−43.6	−45.4	−47.2	−49.0	−50.8	−52.6	−54.4	−56.2
−30	−22.0	−23.8	−25.6	−27.4	−29.2	−31.0	−32.8	−34.6	−36.4	−38.2
−20	−4.0	−5.8	−7.6	−9.4	−11.2	−13.0	−14.8	−16.6	−18.4	−20.2
−10	+14.0	+12.2	+10.4	+8.6	+6.8	+5.0	+3.2	+1.4	−0.4	−2.2
−0	+32.0	+30.2	+28.4	+26.6	+24.8	+23.0	+21.2	+19.4	+17.6	+15.8
+0	32.0	33.8	35.6	37.4	39.2	41.0	42.8	44.6	46.4	48.2
10	50.0	51.8	53.6	55.4	57.2	59.0	60.8	62.6	64.4	66.2
20	68.0	69.8	71.6	73.4	75.2	77.0	78.8	80.6	82.4	84.2
30	86.0	87.8	89.6	91.4	93.2	95.0	96.8	98.6	100.4	102.2
40	104.0	105.8	107.6	109.4	111.2	113.0	114.8	116.6	118.4	120.2
50	122.0	123.8	125.6	127.4	129.2	131.0	132.8	134.6	136.4	138.2
60	140.0	141.8	143.6	145.4	147.2	149.0	150.8	152.6	154.4	156.2
70	158.0	159.8	161.6	163.4	165.2	167.0	168.8	170.6	172.4	174.2
80	176.0	177.8	179.6	181.4	183.2	185.0	186.8	188.6	190.4	192.2
90	194.0	195.8	197.6	199.4	201.2	203.0	204.8	206.6	208.4	210.2
100	212.0	213.8	215.6	217.4	219.2	221.0	222.8	224.6	226.4	228.2
110	230.0	231.8	233.6	235.4	237.2	239.0	240.8	242.6	244.4	246.2
120	248.0	249.8	251.6	253.4	255.2	257.0	258.8	260.6	270.4	264.2
130	266.0	267.8	269.6	271.4	273.2	275.0	276.8	278.6	280.4	282.2
140	284.0	285.8	287.6	289.4	291.2	293.0	294.8	296.6	298.4	300.2
150	302.0	303.8	305.6	307.4	309.2	311.0	312.8	314.6	316.4	318.2
160	320.0	321.8	323.6	325.4	327.0	329.0	330.8	332.6	334.4	336.2
170	338.0	339.8	341.6	343.4	345.2	347.0	348.8	350.6	352.4	354.2
180	356.0	357.8	359.6	361.4	363.2	365.0	366.8	368.6	370.4	372.2
190	374.0	375.8	377.6	379.4	381.2	383.0	384.8	386.6	388.4	390.2
200	392.0	393.8	395.6	397.4	399.2	401.0	402.8	404.6	406.4	408.2
210	410.0	411.8	413.6	415.4	417.2	419.0	420.8	422.6	424.4	426.2
220	428.0	429.8	431.6	433.4	435.2	437.0	438.8	440.6	442.4	444.2
230	446.0	447.8	449.6	451.4	453.2	455.0	456.8	458.6	460.4	462.2
240	464.0	465.8	467.6	469.4	471.2	473.0	474.8	476.6	478.4	480.2
250	482.0	483.8	485.6	487.4	489.2	491.0	492.8	494.6	496.4	498.2
260	500.0	501.8	503.6	505.4	507.2	509.0	510.8	512.6	514.4	516.2
270	518.0	519.8	521.6	523.4	525.2	527.0	528.8	530.6	532.4	534.2
280	536.0	537.8	539.6	541.4	543.2	545.0	546.8	548.6	550.4	552.2
290	554.0	555.8	557.6	559.4	561.2	563.0	564.8	566.6	568.4	570.2
300	572.0	573.8	575.6	577.4	579.2	581.0	582.8	584.6	586.4	588.2
310	590.0	591.8	593.6	595.4	597.2	599.0	600.8	602.6	604.4	606.2
320	608.0	609.8	611.6	613.4	615.2	617.0	618.8	620.6	622.4	624.2
330	626.0	627.8	629.6	631.4	633.2	635.0	636.8	638.6	640.4	642.2
340	644.0	645.8	647.6	649.4	651.2	653.0	654.8	656.6	658.4	660.2
350	662.0	663.8	665.6	667.4	669.2	671.0	672.8	674.6	676.4	678.2
360	680.0	681.8	683.6	685.4	687.2	689.0	690.8	692.6	694.4	696.2
370	698.0	699.8	701.6	703.4	705.2	707.0	708.8	710.6	712.4	714.2
380	716.0	717.8	719.6	721.4	723.2	725.0	726.8	728.6	730.4	732.2
390	734.0	735.8	737.6	739.4	741.2	743.0	744.8	746.6	748.4	750.2
400	752.0	753.8	755.6	757.4	759.2	761.0	762.8	764.6	766.4	768.2

For Interpolation:

°C	0.1	0.2	0.3	0.4	0.5	0.6	0.7	0.8	0.9	1.0
°F	0.18	0.36	0.54	0.72	0.90	1.08	1.26	1.44	1.62	1.80

TEMPERATURE CONVERSION CHART – FAHRENHEIT TO CENTIGRADE

Formula: °C = 5/9 (°F – 32)

Temp., °F	0	1	2	3	4	5	6	7	8	9
					Temp., °C					
–40	–40.0	–40.6	–41.1	–41.7	–42.2	–42.8	–43.3	–43.9	–44.4	–45.0
–30	–34.4	–35.0	–35.6	–36.1	–36.7	–37.2	–37.8	–38.3	–38.9	–39.4
–20	–28.9	–29.4	–30.0	–30.6	–31.1	–31.7	–32.2	–32.8	–33.3	–33.9
–10	–23.3	–23.9	–24.4	–25.0	–25.6	–26.1	–26.7	–27.2	–27.8	–28.3
–0	–17.8	–18.3	–18.9	–19.4	–20.0	–20.6	–21.1	–21.7	–22.2	–22.8
+0	–17.8	–17.2	–16.7	–16.1	–15.6	–15.0	–14.4	–13.9	–13.3	–12.8
10	–12.2	–11.7	–11.1	–10.6	–10.0	–9.4	–8.9	–8.3	–7.8	–7.2
20	–6.7	–6.1	–5.6	–5.0	–4.4	–3.9	–3.3	–2.8	–2.2	–1.7
30	–1.1	–0.6	+0	0.6	1.1	1.7	2.2	2.8	3.3	3.9
40	4.4	5.0	5.6	6.1	6.7	7.2	7.8	8.3	8.9	9.4
50	10.0	10.6	11.1	11.7	12.2	12.8	13.3	13.9	14.4	15.0
60	15.6	16.1	16.7	17.2	17.8	18.3	18.9	19.4	20.0	20.6
70	21.1	21.7	22.2	22.9	23.3	23.9	24.4	25.0	25.6	26.1
80	26.7	27.2	27.8	28.3	28.9	29.4	30.0	30.6	31.1	31.7
90	32.2	32.8	33.3	33.9	34.4	35.0	35.6	36.1	36.7	37.2
100	37.8	38.3	38.9	39.4	40.0	40.6	41.1	41.7	42.2	42.8
110	43.3	43.9	44.4	45.0	45.6	46.1	46.7	47.2	47.8	48.3
120	48.9	49.4	50.0	50.6	51.1	51.7	52.2	52.8	53.3	53.9
130	54.4	55.0	55.5	56.1	56.6	57.2	57.8	58.3	58.9	59.4
140	60.0	60.5	61.1	61.7	62.2	62.8	63.3	63.9	64.4	65.0
150	65.5	66.1	66.6	67.2	67.8	68.3	68.9	69.4	70.0	70.5
160	71.1	71.6	72.2	72.8	73.3	73.9	74.4	75.0	75.5	76.1
170	76.6	77.1	77.8	78.3	78.9	79.4	80.0	80.5	81.1	81.6
180	82.2	82.8	83.3	83.9	84.4	85.0	85.5	86.1	86.6	87.1
190	87.8	88.3	88.9	89.4	90.0	90.5	91.1	91.6	92.2	92.8
200	93.3	93.9	94.4	95.0	95.5	96.1	96.6	97.1	97.8	98.3
210	98.9	99.4	100.0	100.5	101.1	101.6	102.2	102.8	103.3	103.9
220	104.4	105.0	105.5	106.1	106.6	107.2	107.8	108.3	108.9	109.4
230	110.0	110.5	111.1	111.6	112.2	112.8	113.3	113.9	114.4	115.0
240	115.5	116.1	116.6	117.2	117.8	118.3	118.9	119.4	120.0	120.5
250	121.1	121.6	122.2	122.8	123.3	123.9	124.4	125.0	125.5	126.1
260	126.6	127.2	127.8	128.3	128.9	129.4	130.0	130.5	131.1	131.6
270	132.2	132.8	133.3	133.9	134.4	135.0	135.5	136.1	136.6	137.2
280	137.8	138.3	138.9	139.4	140.0	140.5	141.1	141.6	142.2	142.8
290	143.3	143.9	144.4	145.0	145.5	146.1	146.6	147.2	147.8	148.3
300	148.9	149.4	150.0	150.5	151.1	151.6	152.2	152.8	153.3	153.9
310	154.4	155.0	155.6	156.1	156.7	157.2	157.8	158.3	158.9	159.4
320	160.0	160.6	161.1	161.7	162.2	162.8	163.3	163.9	164.4	165.0
330	165.6	166.1	166.7	167.2	167.8	168.3	168.9	169.4	170.0	170.6
340	171.1	171.7	172.2	172.8	173.3	173.9	174.4	175.0	175.6	176.1
350	176.7	177.2	177.8	178.3	178.9	179.4	180.0	180.6	181.1	181.7
360	182.2	182.8	183.3	183.9	184.4	185.0	185.6	186.1	186.7	187.2
370	187.8	188.3	188.9	189.4	190.0	190.6	191.1	191.7	192.2	192.8
380	193.3	193.9	194.4	195.0	195.6	196.1	196.7	197.2	197.8	198.3
390	198.9	199.4	200.0	200.6	201.1	201.7	202.2	202.8	203.3	203.9
400	204.4	205.0	205.6	206.1	206.7	207.2	207.8	208.3	208.9	209.4

For Interpolation:

°F	0.1	0.2	0.3	0.4	0.5	0.6	0.7	0.8	0.9	1.0
°C	0.06	0.11	0.17	0.22	0.28	0.33	0.39	0.44	0.50	0.56

CONVERSION FACTORS
English and Metric Weights and Measures

Multiply	By	To Obtain
LENGTH – English		
mils	0.001	inches
mils	0.0254	millimeters
inches	1000.	mils
inches	25.40	millimeters
inches	2.54	centimeters
feet	30.48	centimeters
feet	0.3048	meters
feet (thousands of)	0.3048	kilometers
yards	0.9144	meters
miles	1.6093	kilometers
LENGTH – Metric		
millimeters	39.37	mils
millimeters	0.03937	inches
centimeters	0.3937	inches
centimeters	0.032808	feet
meters	39.37	inches
meters	3.2808	feet
meters	1.0936	yards
kilometers	3280.83	feet
kilometers	0.62137	miles
AREA – English		
square mils	1.2732	circular mils
square mils	0.000001	square inches
circular mils	0.7854	square mils
circular mils	0.0000007854	square inches
circular mils	0.0005067	square millimeters
square inches	1,000,000	square mils
square inches	1,273,240	circular mils
square inches	645.16	square millimeters
square inches	6.4516	square centimeters
square feet	0.09290	square meters
square yards	0.8361	square meters
AREA – Metric		
square millimeters	1973.52	circular mils
square millimeters	0.00155	square inches
square centimeters	0.155	square inches
square meters	10.7638	square feet
square meters	1.19599	square yards
VOLUME – English		
cubic inches	16.38716	cubic centimeters
cubic feet	0.028317	cubic meters
quarts (liquid U.S.)	0.9463	liters
gallons (U.S.)	3.7854	liters
VOLUME – Metric		
cubic centimeters	0.06102	cubic inches
cubic meters	35.3145	cubic feet
liters	1.05668	quarts (liquid U.S.)
liters	0.26417	gallons (U.S.)

CONVERSION FACTORS
English and Metric Weights and Measures

Multiply	By	To Obtain
WEIGHT – English		
ounces	28.3495	grams
pounds (Av.)	453.59	grams
pounds (Av.)	0.45359	kilograms
tons (short)	907.9	kilograms
tons (long)	1016.05	kilograms
WEIGHT – Metric		
grams	0.03527	ounces
grams	0.002205	pounds
kilograms	35.274	ounces
kilograms	2.2046	pounds
kilograms	0.001102	tons (short)
kilograms	0.0009842	tons (long)
MISCELLANEOUS - English		
pounds per 1000 feet	1.48816	kilograms per kilometer
pounds per mile	0.28185	kilograms per kilometer
pounds per square inch	0.0007031	kilograms per sq. millimeter
pounds per square inch	0.07031	kilograms per sq. centimeter
pounds per cubic inch	27.68	grams per cubic centimeter
feet per second	18.288	meters per minute
feet per second	1.09728	kilometers per hour
miles per hour	1.60935	kilometers per hour
ohms per 1000 feet	3.28083	ohms per kilometer
ohms per mile	0.62137	ohms per kilometer
decibels per 1000 feet	3.28083	decibels per kilometer
decibels per mile	0.62137	decibels per kilometer
decibels	8.686	nepers
MISCELLANEOUS - Metric		
kilograms per kilometer	0.67197	pounds per 1000 feet
kilograms per kilometer	3.54795	pounds per mile
kilograms per square millimeter	1422.34	pounds per square inch
kilograms per square centimeter	14.2234	pounds per square inch
grams per cubic centimeter	0.03613	pounds per cubic inch
meters per minute	0.05468	feet per second
kilometers per hour	0.91134	feet per second
kilometers per hour	0.62137	miles per hour
ohms per kilometer	0.3048	ohms per 1000 feet
ohms per kilometer	1.6093	ohms per mile
decibels per kilometer	0.3048	decibels per 1000 feet
decibels per kilometer	1.6093	decibels per mile
nepers	0.1153	decibels
TEMPERATURE		
°fahrenheit	5/9 (°F – 32)	°centigrade
°centigrade	9/5 °C + 32	°fahrenheit

DC Circuit Characteristics

Ohm's Law:

$E = IR \qquad I = \frac{E}{R} \qquad R = \frac{E}{I}$

E = voltage impressed on circuit (volts)
I = current flowing in circuit (amperes)
R = circuit resistance (ohms)

In direct current circuits, electrical power is equal to the product of the voltage and current:

$$P = EI = I^2R = \frac{E^2}{R}$$

P = power (watts)
E = voltage (volts)
I = current (amperes)
R = resistance (ohms)

AC Circuit Characteristics

The instantaneous values of an alternating current or voltage vary from zero to a maximum value each half cycle. In the practical formulae which follow, the "effective value" of current and voltage is used, defined as follows:

Effective value = 0.707 × maximum instantaneous value

Impedance:
Impedance is the total opposition to the flow of alternating current. It is a function of resistance, capacitive reactance and inductive reactance. The following formulae relate these circuit properties:

$$X_L = 2\pi fL \qquad X_c = \frac{1}{2\pi fC} \qquad Z = \sqrt{R^2 + (X_L - X_c)^2}$$

X_L = inductive reactance (ohms)
X_c = capacitive reactance (ohms)
Z = impedance (ohms)
f = frequency (cycles per second)
C = capacitance (farads)
L = inductance (henrys)
R = resistance (ohms)
π = 3.14

Ohm's Law for AC Circuits:

$E = I \times Z \qquad I = \frac{E}{Z} \qquad Z = \frac{E}{I}$

Power Factor

Power factor of a circuit or system is the ratio of actual power (watts) to apparent power (volt-amperes), and is equal to the cosine of the phase of the circuit:

$$PF = \frac{\text{actual power}}{\text{apparent power}} = \frac{\text{watts}}{\text{volts} \times \text{amperes}} = \frac{KW}{KVA} = \frac{R}{Z}$$

KW = kilowatts
KVA = kilovolt-amperes = volt-amperes × 1,000
PF = power factor (expressed as decimal)

Single-Phase Circuits

$KVA = \frac{EI}{1{,}000} = \frac{KW}{PF} \qquad KW = KVA \times PF$

$I = \frac{P}{E \times PF} \qquad E = \frac{P}{I \times PF} \qquad PF = \frac{P}{E \times I}$

P = E × I × PF
P = power (watts)

Three-Phase Circuits, Balanced Star or WYE

$I_N = 0 \qquad I = I_p \qquad E = \sqrt{3}\,E_p = 1.73\,E_p$

$$E_p = \frac{E}{\sqrt{3}} = \frac{E}{1.73} = 0.577E$$

I_N = current in neutral (amperes)
I = line current per phase (amperes)
I_p = current in each phase winding (amperes)
E = voltage, phase-to-phase (volts)
E_p = voltage, phase-to-neutral (volts)

Three-Phase Circuits, Balanced Delta

$I = 1.732 \times I_p \qquad I_p = \frac{I}{\sqrt{3}} = 0.577 \times I$

$E = E_p$

Power: Balanced 3-wire, 3-phase Circuit, Delta or Wye

For unity power factor (PF = 1.0):

$P = 1.732 \times E \times I$

$I = \frac{P}{\sqrt{3}E} = \frac{0.577 \times P}{E} \qquad E = \frac{P}{\sqrt{3} \times I} = \frac{0.577 \times P}{I}$

P = total power (watts)

For any load:

$P = 1.732 \times E \times I \times PF \qquad VA = 1.732 \times E \times I$

$$E = \frac{P}{PF \times 1.73 \times I} = \frac{0.577 \times P}{PF \times I}$$

$$I = \frac{P}{PF \times 1.73 \times E} = \frac{0.577 \times P}{PF \times E}$$

$$PF = \frac{P}{1.73 \times I \times E} = \frac{0.577 \times P}{I \times E}$$

VA = apparent power (volt-amperes)
P = actual power (watts)
E = line voltage (volts) phase-to-phase
I = line current (amperes)

Power Loss: any AC or DC Circuit

$P = I^2R \qquad I = \sqrt{\frac{P}{R}} \qquad R = \frac{P}{I^2}$

P = power heat loss in circuit (watts)
I = effective current in conductor (amperes)
R = conductor resistance (ohms)

ALTERNATING CURRENT FORMULAE

To Find	Value Known	SYSTEM			
		Direct Current	Single-Phase	*Two-Phase-Four Wire	Three-Phase
Amperes I	Hp	$I = \frac{746\ Hp}{E \times eff}$	$I = \frac{746\ Hp}{E \times eff \times pf}$	$I = \frac{746\ Hp}{2 \times E \times eff \times pf}$	$I = \frac{746\ Hp}{1.73 \times E \times eff \times pf}$
Amperes I	Kw	$I = \frac{1000\ Kw}{E}$	$I = \frac{1000\ Kw}{E \times pf}$	$I = \frac{1000\ Kw}{2 \times E \times pf}$	$I = \frac{1000\ Kw}{1.73 \times E \times pf}$
Amperes I	Kva		$I = \frac{1000\ Kva}{E}$	$I = \frac{1000\ Kva}{2\ E}$	$I = \frac{1000\ Kva}{1.73\ E}$
Kilowatts Input, Kw		$Kw = \frac{I \times E}{1000}$	$Kw = \frac{I \times E \times pf}{1000}$	$Kw = \frac{I \times E \times 2 \times pf}{1000}$	$Kw = \frac{I \times E \times 1.73 \times pf}{1000}$
Kilovolts-Amperes, Kva			$Kva = \frac{I \times E}{1000}$	$Kva = \frac{I \times E \times 2}{1000}$	$Kva = \frac{I \times E \times 1.73}{1000}$
Horse-power Output		$Hp = \frac{I \times E \times eff}{746}$	$Hp = \frac{I \times E \times eff \times pf}{746}$	$Hp = \frac{I \times E \times 2 \times eff \times pf}{746}$	$Hp = \frac{I \times E \times 1.73 \times eff \times pf}{746}$

I == Line current in amperes
E == Line-to-line volts
Kw == Input in kilowatts
Kva == Input in kilovolt-amperes
Hp == Output in horsepower

eff == Efficiency in decimals
pf == Power factor in decimals

* In two-phase, three-wire, balanced circuits, the amperes in the common conductor are 1.41 times that in either of the other two.

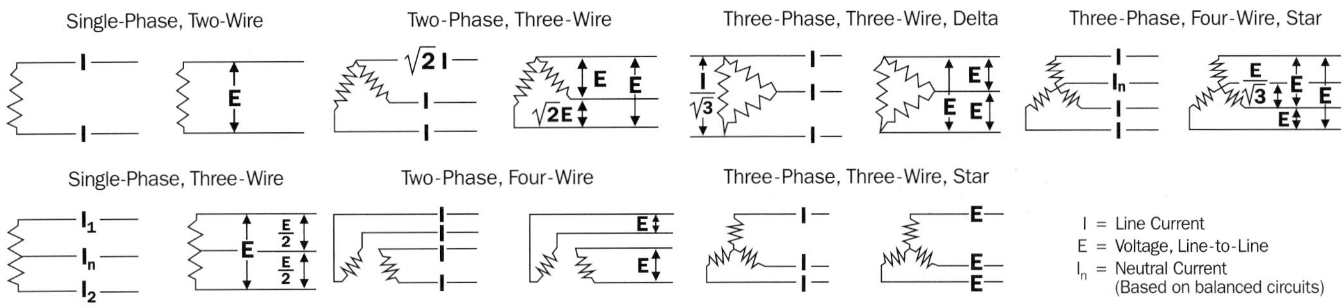

I = Line Current
E = Voltage, Line-to-Line
I_n = Neutral Current (Based on balanced circuits)

FORMULAE FOR DETERMINING AMPERES IN AC CIRCUITS

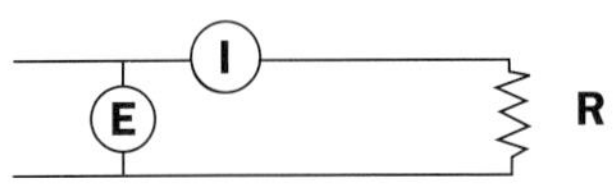

$$I = \frac{E}{R}$$

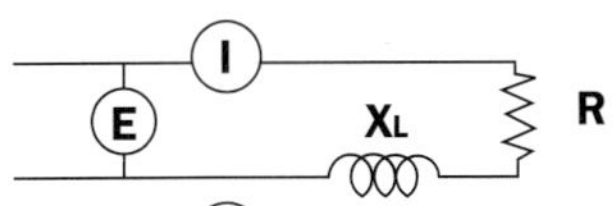

$$I = \frac{E}{\sqrt{R^2 + X_L^2}} = \frac{E}{Z}$$

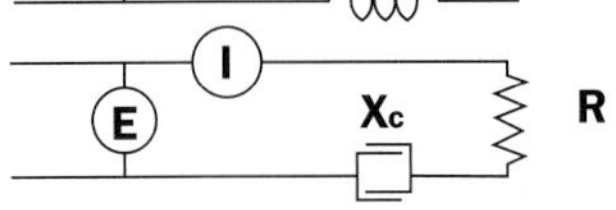

$$I = \frac{E}{\sqrt{R^2 + X_c^2}} = \frac{E}{Z}$$

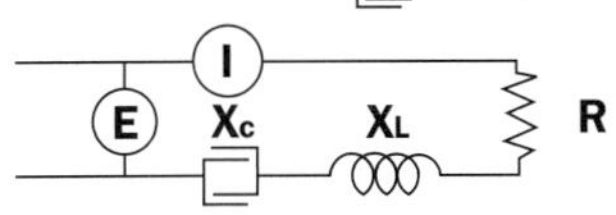

$$I = \frac{E}{\sqrt{R^2 + (X_L - X_c)^2}} = \frac{E}{Z}$$

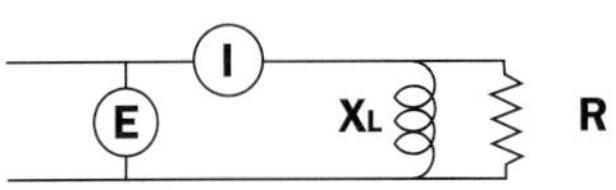

$$I = \frac{E}{\frac{R X_L}{\sqrt{R^2 + X_L^2}}} = \frac{E}{Z}$$

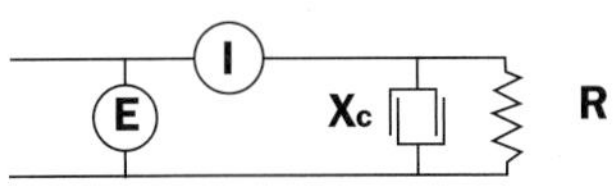

$$I = \frac{E}{\frac{R X_c}{\sqrt{R^2 + X_c^2}}} = \frac{E}{Z}$$

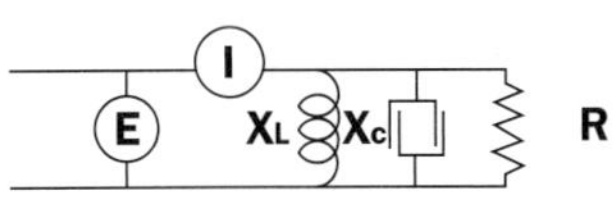

$$I = \frac{E}{\frac{R X_L X_c}{\sqrt{X_L^2 X_c^2 + R^2 (X_L - X_c)^2}}} = \frac{E}{Z}$$

R = Resistance in ohms
Z = Impedance in ohms
I = Current in amperes
E = Pressure in volts
X_L = Inductive reactance in ohms = $2\pi f L$
X_c = Capacitive reactance in ohms = $\frac{1}{2\pi f C}$
f = Frequency in cycles per second
L = Inductance in Henrys
C = Capacitance in Farads

GREEK ALPHABET

Name	Large	Small
Alpha	Α	α
Beta	Β	β
Gamma	Γ	γ
Delta	Δ	δ
Epsilon	Ε	ϵ
Zeta	Ζ	ζ
Eta	Η	η
Theta	Θ	θ
Iota	Ι	ι
Kappa	Κ	κ
Lambda	Λ	λ
Mu	Μ	μ
Nu	Ν	ν
Xi	Ξ	ξ
Omicron	Ο	ο
Pi	Π	π
Rho	Ρ	ρ
Sigma	Σ	σ
Tau	Τ	τ
Upsilon	Υ	υ
Phi	Φ	ϕ
Chi	Χ	χ
Psi	Ψ	ψ
Omega	Ω	ω

BASIC METRIC UNITS MAGNITUDE (POWERS OF 10)

Prefix	Symbol	Multiplication Factors
tera	T	$1\,000\,000\,000\,000 = 10^{12}$
giga	G	$1\,000\,000\,000 = 10^{9}$
mega	M	$1\,000\,000 = 10^{6}$
kilo	k	$1\,000 = 10^{3}$
hecto	h	$100 = 10^{2}$
deka	da	$10 = 10^{1}$
		$1 = 10^{0}$
deci	d	$0.1 = 10^{-1}$
centi	c	$0.01 = 10^{-2}$
milli	m	$0.001 = 10^{-3}$
micro	u	$0.000\,001 = 10^{-6}$
nano	n	$0.000\,000\,001 = 10^{-9}$
pico	p	$0.000\,000\,000\,001 = 10^{-12}$
femto	f	$0.000\,000\,000\,000\,001 = 10^{-15}$
atto	a	$0.000\,000\,000\,000\,000\,001 = 10^{-18}$

NOTES

NOTES

NOTES

NOTES

NOTES

NOTES